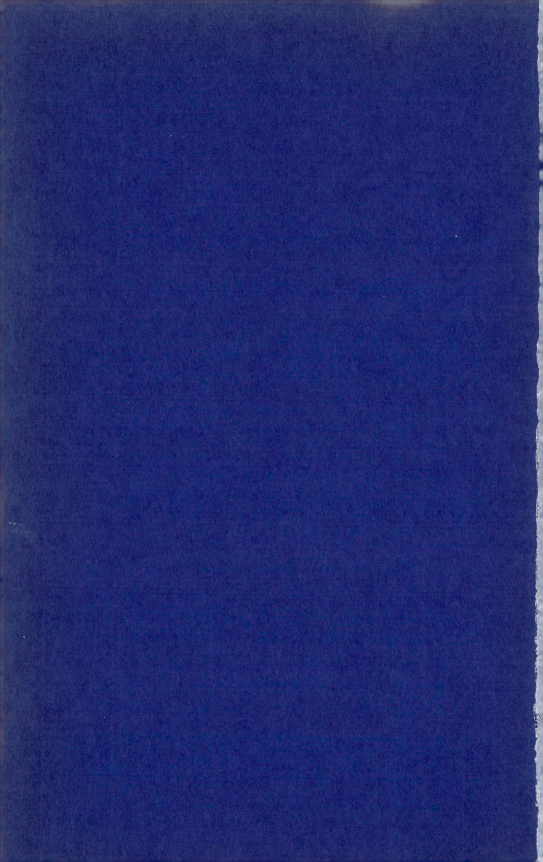

Between Pacific Tides

Between Pacific Tides

Fifth Edition

Edward F. Ricketts, Jack Calvin, *and*
Joel W. Hedgpeth
Revised by David W. Phillips

Stanford University Press, Stanford, California 1985

Between Pacific Tides was first published in 1939. A revised
edition prepared by the original authors, with a Foreword
by John Steinbeck, was published in 1948. Subsequent
editions, with revisions by Joel W. Hedgpeth, appeared in
1952, 1962, and 1968. For the present edition, once again
completely redesigned and reset, David W. Phillips has
prepared new chapters, incorporated further extensive
revisions and expansions, and introduced many new
photographs. These changes are elaborated in the Preface.

Stanford University Press   Stanford, California

Printed in the United States of America
First published 1939   Fifth edition 1985

CIP data appear at the end of the book

Contents

REFERENCE MATTER

Prefaces

Preface to the Original Edition, 1939

The enormous wealth of life that occurs between the upper and the lower limits of the tide is a phenomenon of intense interest to the biologist and to the layman alike. Here strange plants and bizarre, brilliantly colored animals grow in such abundance that the most casual visitor to the seashore cannot fail to notice some of them. Almost invariably his curiosity is aroused: Is that gorgeous flowerlike thing in the tide pool a plant or an animal? What is it called? What does it eat? How does it defend itself and reproduce its kind? Will it hurt me if I touch it?

And while the visitor is puzzling over his first sea anemone, a score of crabs may scurry away at his footfall or may rear up and offer battle in defense of life and liberty. When he turns to watch the crabs he may see a bed of urchins, their bristling spines half concealed by bits of seaweed and shell. He may stoop to pick up a snail, only to have the creature roll from the rock at the approach of his hand, tumble into a pool, and scramble away at a very unsnaillike pace. He hears scraping sounds and clicks and bubblings, perhaps sharp cracks like tiny pistol shots. Jets of water shoot up. Everywhere there is color, life, movement.

In short, our visitor to a rocky shore at low tide has entered possibly the most prolific life zone in the world—a belt so thickly populated that often not only is every square inch of the area utilized by some plant or animal but the competition for attachment sites is so keen that animals settle upon each other—plants grow upon animals, and animals upon plants.

To supply such a person with as much as possible of the information that he wants is the chief aim of this handbook. The arrangement, therefore, is the one which we believe can be most readily grasped by the person who has had little or no biological training. The treatment is ecological and

inductive; that is, the animals are treated according to their most characteristic habitat, and in the order of their commonness, conspicuousness, and
interest. . . .

Edward F. Ricketts
Jack Calvin

Foreword by John Steinbeck, 1948

Periodically in the history of human observation the world of external
reality has been rediscovered, reclassified, and redescribed. It is difficult
for us to understand the reality of Democritus, of Aristotle, of Pliny, for
they did not see what we see and yet we know them to have been careful
observers. We must concede then that their universe was different from
ours or that they warped it and to a certain extent created their own real-
ities. And if they did, there is no reason to suppose that we do not. Possibly
our warp is less, owing to our use of precise measuring devices. But, in
immeasurables, we probably create our own world.

The process of rediscovery might be as follows: a young, inquisitive,
and original man might one morning find a fissure in the traditional tech-
nique of thinking. Through this fissure he might look out and find a new
external world about him. In his excitement a few disciples would cluster
about him and look again at the world they knew and find it fresh. From
this nucleus there would develop a frantic new seeing and a cult of new
seers who, finding some traditional knowledge incorrect, would throw out
the whole structure and start afresh. Then, the human mind being what it
is, evaluation, taxonomy, arrangement, pattern making would succeed the
first excited seeing. Gradually the structure would become complete, and
men would go to this structure rather than to the external world until eventually something like but not identical with the earlier picture would have
been built. From such architectures or patterns of knowledge, disciplines,
ethics, even manners exude. The building would be complete again and no
one would look beyond it—until one day a young, inquisitive, and original
man might find a fissure in the pattern and look through it and find a new
world. This seems to have happened again and again in the slow history of
human thought and knowledge.

There is in our community an elderly painter of seascapes who knows
the sea so well that he no longer goes to look at it while he paints. He dis-
likes intensely the work of a young painter who sets his easel on the beach
and paints things his elder does not remember having seen.

Modern science, or the method of Roger Bacon, has attempted by
measuring and rechecking to admit as little warp as possible, but still some
warp must be there. And in many fields young, inquisitive men are seeing
new worlds. And from their seeing will emerge not only new patterns but
new ethics, disciplines, and manners. The upheaval of the present world
may stimulate restive minds to new speculations and evaluations. The new

eyes will see, will break off new facets of reality. The excitements of the chase are already felt in the fields of biochemistry, medicine, and biology. The world is being broken down to be built up again, and eventually the sense of the new worlds will come out of the laboratory and penetrate into the smallest living techniques and habits of the whole people.

This book of Ricketts and Calvin is designed more to stir curiosity than to answer questions. It says in effect: look at the animals, this is what we seem to know about them but the knowledge is not final, and any clear eye and sharp intelligence may see something we have never seen. These things, it says, you will see, but you may see much more. This is a book for laymen, for beginners, and, as such, its main purpose is to stimulate curiosity, not to answer finally questions which are only temporarily answerable.

In the laboratories, fissures are appearing in the structure of our knowledge and many young men are peering excitedly through at a new world. There are answers to the world questions which every man must ask, in the little animals of tidepools, in their relations one to another, in their color phases, their reproducing methods. Finally, one can live in a prefabricated world, smugly and without question, or one can indulge perhaps the greatest human excitement: that of observation to speculation to hypothesis. This is a creative process, probably the highest and most satisfactory we know. If only in the process one could keep the brake of humor in operation, it would be even more satisfactory. One has always to keep in mind his own contribution to the world of reality. Aristotle built a world and we are building one. His was a true world, and ours is. And the two need not meet and quarrel. His world worked for him and for his people and ours works for us. A Greek thinker built a world that operated, and, given that man and that society, it would still work. We build a motor and it runs. It will always run if the principle involved is followed correctly, but it is not now impossible to imagine a world wherein the principle of the internal-combustion engine will become inoperative because it is no longer important.

This book then says: "There are good things to see in the tidepools and there are exciting and interesting thoughts to be generated from the seeing. Every new eye applied to the peep hole which looks out at the world may fish in some new beauty and some new pattern, and the world of the human mind must be enriched by such fishing."

Preface: About This Book and Ed Ricketts, 1952

When a cannery fire spread to the Pacific Biological Laboratories in 1936, Ed Ricketts bounded out just ahead of the flames with the only two items he had had time to save—a pair of pants and the bulky typescript of

Between Pacific Tides. Later he regretted the pants, remembering more important things that he might have grabbed as he ran, but it was characteristic of him that he shrugged off the loss of the biological supply house that was his means of livelihood and cheerfully went on with the job that really mattered. *Between Pacific Tides* was almost ready for the printer.

Obviously many people have helped to shape this book, supplying knowledge, assistance, criticism, encouragement. But the qualities that make the book truly unique in its field came from the lively intellectual curiosity, the warm humanness and, quite possibly, the touch of genius, that constituted the mind of Ed Ricketts. The junior author, Jack Calvin, was the literary collaborator and photographer for the original edition. He emerges now from his happy obscurity in southeastern Alaska long enough to help to write this preface, and to explain that, being no zoölogist, he could not, if he would, do the work of preparing this new edition for the press. This task has fallen to one who has undertaken it, not as a zoölogist, but as a friend.

The revised edition of 1948 was admittedly an interim printing, since a large amount of wartime research was not yet accessible. The editor of this edition has tried to bring things up to date and to make as many corrections and necessary changes as possible. Some of the photographs and drawings have been replaced, and some new illustrations added. The appendix has been overhauled. The text remains essentially unchanged, though new paragraphs have been added here and there, and some old ones have been rewritten. As with previous editions, there have been many helping hands, and for their work acknowledgment will be found on page 404 [of the Third Edition].

Edward F. Ricketts was a <u>fine example</u> of the idea that naturalists are born, not made. A less-promising environment, especially for a seashore biologist, could scarcely be imagined than the streets of west-side Chicago, where Ed was born on May 14, 1897. His parents were staunch pillars of the Episcopal Church near Garfield Park and his father had studied for holy orders, but Ed's interests lay elsewhere. He spent many hours in the city parks whetting his taste for biology, and many hours roaming the city at night studying that subspecies of Homo sapiens, the city man.

Ed was never much of a conformist, and his education in zoölogy was informal and haphazard: a year at Illinois State Normal and two or three years at Chicago, where the zoölogy courses were the only ones that mattered. He did not complete his Bachelor of Arts degree but, what is more important, he came under the influence of W. C. Allee. He became one of a group of "Ishmaelites," as Dr. Allee remembers them, "who tended sometimes to be disturbing, but were always stimulating."

Ed came to the Pacific coast in 1923 as a partner in a biological supply business—a business admirably adapted to his talents and temperament, for it offered a variety of work, working hours determined by the tide, or the mood, periods of tranquillity alternated with periods of intense activity. Best of all, it enabled him to work constantly with his beloved animals of the seashore, to travel up and down the Pacific coast, seeing more of its tide pools and beaches than any other single person. Standing in a tide pool, watching some small animal, or dropping it tenderly into a collecting jar (see Plate IV [Fig. 67 in the Fifth Edition]), telling his companions something of the creature's habits or accomplishments, he glowed with happiness.

And there were nearly always companions, for Ed possessed to a rare degree the ability not only to make friends but to draw them along with him, whether it was physically, as on collecting trips, or spiritually, as on his flights into philosophy, mysticism, or science. But he was as ready to follow as to lead if a companion had an idea that promised to develop into a brain-teasing discussion, a pleasant trip, or an amusing party.

In the things that were important to him—music, literature, automobiles, women, and, in the days when he wore one, his beard—Ed insisted gently and very firmly on having the finest quality available. He would drive the hundred and forty miles from Pacific Grove to San Francisco because he knew a posh barber there who was an artist at beard trimming. Many a time he made that drive, with a car full of friends, to hear the singing of midnight Mass at the Russian Orthodox Church, although buying gasoline for the trip might take some doing. The things that were unimportant to him might as well not have existed, except as the law or other circumstance beyond his control forced him to give them a bare minimum of attention. His own clothing was a notable example. He usually dressed like a tramp, sometimes in nothing more than a tattered shirt, a pair of pants held together inadequately with safety pins, and shoes that weren't held together at all. In towns away from the Monterey peninsula the contrast between Ed's fine car and informal clothes sometimes startled cops into making inquiries.

In the friends who accompanied him on his collecting trips, Ed stirred something of his own interest in the luxuriant seashore life, and soon they began urging him to put some of his thumbnail sketches of the tide-pool dwellers into writing. Just a little pocket-size handbook for the casually curious. . . .

At first he was diffident. Writing was for the Carmel long hairs, not for him. But the idea took root, or perhaps it had been growing there for some time, and his "no" became "well, maybe," if he had help and somebody to make him apply the seat of his pants to the seat of his typewriter chair. But once started he needed no prodding. His enthusiasm grew, and so did the book, which quickly ceased to bear any resemblance to a pocket-size handbook. There was just too much material in scattered papers and in Ed's own fund of tide-pool lore. The urge to communicate to others the things that he knew, or believed, or surmised, kept him going during the years that it took to write the book and see it through the frustrating delays of its first publication.

By the time *Between Pacific Tides* was published, Ed was less interested in the moderately profitable business of selling bits and pieces of beautifully preserved seashore life to school laboratories (he had long since bought out his partner) than in the unprofitable pastime of writing about it. At the time of his death the supply business had a very poor Dun & Bradstreet rating, but its owner was working on a book on the Queen Charlotte Islands.

Ed died on May 11, 1948, after his car was struck by a train almost in front of his laboratory. As he was being extricated from the wreckage he told the policeman that the accident was not the engineer's fault. At the hospital he complimented the nurse on duty on the new way she was doing her hair. No one with his health and vitality (he looked ten years younger than his almost fifty-one years) gives up life easily, but few have done so more gracefully.

This is not the place for literary criticism, but one book that Ed did not write will have to be mentioned, because it was written about him. How much of John Steinbeck's *Cannery Row* is true? Was Ed really like Doc? To avoid an irritating yes and no answer as far as possible, we might say, Yes, *Cannery Row* is a true story, but it is not the whole truth. The laboratory, the phonograph records, the beer milkshake, and the establishment across the street—these details are true. And Ed was really like Doc, but Doc is only a one-dimensional portrait of Ed. Who would know from *Cannery Row* that Ed was the devoted father of three children, and that he was a hard-working biologist who managed to get a great deal done between his records and that solemn procession—Ed would have called it a lovely procession—of wine jugs? We mean no criticism of John in suggesting that the Doc of *Cannery Row* is half Ricketts the man and half Steinbeck the author. Doc remains a delightful character, and people who did not know Ed can achieve a considerable acquaintance with him by reading the lusty story of Doc of *Cannery Row*.

However, if Ed Ricketts has achieved a trace of immortality, we believe that it is because of his ability to plant in the minds of others not facts, since many can do that, but the essential truths beyond facts: the shadowy half-truths, the profoundly disturbing questions that thinking men must face and try to answer. This ability was the mature fruit of the quality that Dr. Allee saw when he noted that Ed tended sometimes to be disturbing, but was always stimulating.

Jack Calvin
Joel W. Hedgpeth

Preface to the Fourth Edition, 1968

In 1853 there appeared in England a book on the marvels of seashore life, illustrated directly on the lithographic stone by the author. This book, *A Naturalist's Rambles on the Devonshire Coast*, by Philip Henry Gosse, was the first of its kind in any language, and all others have been imitators and followers of this prototype. Yet only a few of the later books have been based on original observation done at the seashore. *Between Pacific Tides*, first published in 1939, is one of those few. It is also one of the few to survive an original edition and to remain in demand for still another generation of readers and tide-pool watchers. There are many reasons for this sustained popularity; but perhaps the chief one is that people want to know about seashore animals, and no one else has presented this information in terms of the way of life—the ecology, if you will—of the seashore in such a readable manner. Where Gosse was a particularizer, dealing with limited localities, E. F. Ricketts was a generalizer—with an eye for particulars, to be sure, yet interested in all that went on along a coast of more than a thousand miles.

The author of *Between Pacific Tides* and P. H. Gosse would have much in common if they could meet on some Elysian shore; but that is unlikely, since, according to their respective religious orientations, they could not both be destined for the place Gosse fervently believed in and into which he firmly anticipated admission. But they did have in common a love for the creatures of the seashore and an interest in telling people what they had learned from their own observation. Gosse, despite his narrowly doctrinaire concept of special creation and his resistance to the ideas of a certain Charles Darwin (whom he nevertheless graciously supplied with information on request), stimulated both popular and professional interest in the seashore life of Britain. His books were for some time the primary source of information, and even now, more than a century later, they are still cherished on the shelves of professional biologists. This book by Ed Ricketts has had very much the same place in the development of serious marine biology on the Pacific coast of North America. At the same time it has continued to introduce the interested bystander to a strange and complex way of life. As John Steinbeck wrote in his foreword to an earlier edition, "This is a book for laymen, for beginners, and as such, its main purpose is to stimulate curiosity, not to answer finally questions which are only temporarily answerable."

Many books about seashore life appeared in the 86 years between *A Naturalist's Rambles* and *Between Pacific Tides*, and in the last 30 years there have been as many more. (Quite a few of the more recent books, of course, have been about shores that were previously inadequately known.) Books of this sort cannot be written without some firm background of scientific knowledge of the marine life in the region concerned, and one of the unique features of *Between Pacific Tides* was (and is) an appendix that summarized the state of the literature. This appendix is in part an expression of the personality of Ed Ricketts: he evidently liked to do this sort of thing, although arid pedantry was one trait he did not have. But the preparation of *Between Pacific Tides* was a long, arduous process, which involved ransacking the scattered literature on marine invertebrates and corresponding with specialists, as well as observing in the field. It took a lot of hard work to find out all these things, and obviously it seemed best to share this work with others who would use the book. It must be remembered that Ed Ricketts was not, in one sense, a professional biologist, and he was not concerned with formal degrees or with conforming to what society might expect of him. For example, he did not bother to graduate from the University of Chicago after studying there for two or three years. Had he done so, he would probably have turned into an obscure instructor somewhere, and *Between Pacific Tides* would never have appeared. Instead, and in the ideal sense, Ed was a professional naturalist; and once started, he discovered that he wanted to communicate.

Ed was other things as well, as John Steinbeck has so eloquently said in the Preface to *Log of the Sea of Cortez*. It would seem that those of us who have some personal memory of Ed, who shared some private part of the world with him and knew Cannery Row before it was converted to less utilitarian purposes than canning fish, are fewer now. Indeed we are, not because most of us have disappeared, but because there are so many more people at Monterey, and everywhere on the Pacific coast. Perhaps it is a good thing that people no longer make pilgrimages to the old place on Cannery Row—or even a good thing that literary critics have ceased to worry about John Steinbeck's soul and which part of his mind might really have been Ed's, or about who really wrote some of those philosophical passages in *Sea of Cortez*. In his own way Ed was a born teacher, with the instincts of Aristotle, and John Steinbeck was his most devoted student. At least this is the feeling I now get when rereading Steinbeck's memoir of Ed.

From all this, the reader may wonder whether Jack Calvin, the junior author, really exists, and what he had to do with the making of this book. He exists indeed, and lives among the fjords of Alaska, where he operates a printing business. Although not a zoologist, he still has a lively interest in tide-pool life from his days at Pacific Grove. He took most of the photographs for the original edition, and helped with the wearisome business of putting things together in words. Many of his photographs were fine in their day, but photography has improved in recent years, and we have replaced quite a few of them for this edition; those familiar with earlier editions, however, will recognize some old friends. We have been able to revise the text completely for the first time, so that it is difficult to decide, here and there, who is now responsible for particular words. I am responsible for revisions of factual statements and for the annotated bibliographies in their present state. Nevertheless, the book is still essentially as Ed Ricketts intended it to be.

In the days when *Between Pacific Tides* was an idea in the heads of Ed Ricketts and the friends who encouraged him to write it, there were only three institutions on the Pacific coast that officially paid much attention to things of the seashore: the University of Washington, with its summer station at Friday Harbor; Hopkins Marine Station at Pacific Grove, a short

Edward F. Ricketts   S. F. Light

walk from Cannery Row; and Scripps Institution of Oceanography at La Jolla, where not much of anything happened in those days. It is perhaps symptomatic of how things were that the only adequate book for professional and amateur alike was Johnson and Snook's *Seashore Animals of the Pacific Coast*, written by a professor at San Diego State College and a high school biology teacher from Stockton. This is still a fine book, although time has not dealt kindly with parts of it; but it was also symptomatic of those days that the publishers remaindered it.

In those days more was being done for the cause of what we now so glibly call marine biology by a somewhat old-fashioned professor at Berkeley, S. F. Light, who conducted field trips to such places as Moss Beach attired in his gray business suit, complete with vest and starched collar; a pair of rubber boots was his only concession to the environment. Dr. Light was interested in all sorts of invertebrates, and he welcomed students who would study any sort of seashore animal. He was, as one of my friends once remarked, an "inspired pedagogue," and, through his students, he has left his mark on virtually every institution of learning on the Pacific coast. It was he who persuaded the Stanford University Press, which was at first disconcerted by the annotated systematic index, to retain it as a valuable part of this book. He and his students produced the essential professional guide to central California, familiarly known as "Light's Manual."

It is pleasant to think of some implausible Elysian seashore, with Philip Henry Gosse, in an "immense wide-awake, loose black coat and trousers, and fisherman's boots, with a collecting basket in one hand, a staff or prod in the other," leading Ed and Sol Felty Light in search of zoophytes and corals in some still unexplored tide pool—Ed in one of his favorite wool shirts, a battered hat, and drooping rubber boots, and Light impeccable under his fedora, with boots neatly supported over his business suit. Surprisingly enough, it would be Gosse who would strip off his clothes to reach some almost inaccessible bright spot in a deep tide pool, as he did to find that *Balanophyllia* was still a living entity on the shores of Britain. Ed would probably not remove his clothes, but would wade in anyhow; and Light, interested but aloof, would stand on the rocks.

At least that is the way I prefer to think of them: Gosse, whose writings are still alive and fresh despite his extremely narrow theological Weltanschauung; Ed, the friend of many visits at Pacific Grove during those days I drove my mother down to see her old friend who lived where there is now a supermarket; and Light, my professor who perhaps never quite approved of me. I am not sure that Light completely accepted anyone, despite Ted Bullock's faithful reminiscence, since I do remember his startled glance when, as a vestryman, he ushered me to a seat in the church at Berkeley of which, following their chairman's lead, a number of zoologists were staunch members. There was indeed something more than words can express to that brief encounter on a Sunday morning—perhaps because it was never alluded to afterwards—the feeling that perhaps I had caught my professor out at something he was not sure his students should know about him, as well as amazement that I should turn up. He almost forgot to give me a program.

All of this seems so long ago. Dr. Light never lived to see the marine station of his dreams, but there are now marine stations up and down this coast, almost as close together as Father Junípero Serra wanted to have his missions. There are at least 10 such establishments open all year around, and almost as many more open only during the summer—counting only those associated with colleges and universities. The study of marine life is no longer something for the chosen few; it has become fashionable among both professors and students. Inevitably, all sorts of new and interesting things are being observed and recorded. Some of them now seem so obvious that it is a bit embarrassing to many of us that we didn't see them on our first trip to the beach. But, in retrospect, that is asking too much. There is still much to learn, and no one can be expected to learn it all by himself.

One of the disconcerting things to me, as the editor of this work, is the frequency with which it has been quoted in the scholarly literature as a source of something that subsequent research has demonstrated to be erroneous. It would be easy to say that many of these errors have been left in this edition in order to stimulate still further research, but it would be fairer to say that I just don't know all the mistakes that still lurk in these pages. I am especially aware that many statements of range are inaccurate from the viewpoint of the growing fraternity of divers; but since the book remains essentially a guide for those who do not dive but who approach the sea-shore from the land in boots or old tennis shoes, I have left most of the ranges as they were in earlier editions. Perhaps someone will produce the logical companion volume, *Below Pacific Tides*, for those who explore the shallow sea, as Rupert Riedl has done for the Mediterranean.

Well, that is for someone else to do, some time from now. There is still much to be learned about the shore, and my most pleasant memory of Ed Ricketts is a morning when he was sorting some pickled beasts into cans for shipment, in the basement of his place on Cannery Row. Ed had just taken someone from the East on his first field trip to the Pacific tide pools, and the visitor had almost immediately found some creature that Ed had never seen before. Ed was pleased that his guest had found this animal. He had no resentment, just pleasure in remembering someone who had made a discovery that Ed could share with him. Remember this when you use this book. It is still mostly Ed Ricketts, and he would have been pleased, as will I who have inherited it, to know that something new and more inter-esting has been learned about what is going on between Pacific tides.

Joel W. Hedgpeth

Preface to the Fifth Edition

The Fifth Edition of *Between Pacific Tides* is still Ed Ricketts's book. Its fundamental purpose and philosophy are those of the original, and it re-mains a book for all who find the shore a place of excitement, wonder, and beauty, and who want to know more. As before, the text introduces many of the animals, and a few of the plants, common on Pacific seashores, de-scribing them, their habits, and their habitats. As before, a detailed Anno-tated Systematic Index and General Bibliography summarize the state of the literature and offer an abundance of sources for further inquiry. Like any good introduction, *Between Pacific Tides* invites exploration of many kinds, and the reference lists, with their wealth of information, are one of its invitations to the curious mind.

Although the basic purpose and structure of Ricketts's book remain unchanged, this edition represents a comprehensive revision, in the sense that all of the text, the illustrations, and the references have been reviewed, and a multitude of changes made. Almost all of the text sections have been revised to some extent, perhaps simply to substitute a more current scien-tific name, to record an updated geographic range, or to correct an inac-curacy, but often to add new information of interest, as well. There has also been some shifting of information from one section to another. For instance, the animals described collectively in the Fourth Edition under the heading "Newcomers to the Pacific Coast" have now been dispersed to the appropriate habitat sections.

Not much was deleted outright from the Fourth Edition's treatment of animals and habitats. A few sections of text that seemed to have outlived their usefulness were trimmed, and I confess to having deleted or to toning down a few Rickettsian statements that, although of long standing, seemed by now too anthropomorphic, teleological, or metaphysical; readers who want their Ricketts pure can still find it in the earlier editions. Certainly less was deleted than was added, and despite efforts to keep the size of this edition "within bounds," the book has grown by about 20 percent. Our specialized knowledge of marine invertebrates has increased dramatically over the past two decades, and that amount of growth in this book is a conservative indication of the growth of the field.

Many new photographs and drawings are included. Some of these replace illustrations in the previous editions. Others are additions that expand or modernize the text: photographs of species not previously illustrated, scanning electron micrographs of gastropod radulae, and the like.

The Annotated Systematic Index and General Bibliography remain structurally as they have been, for the most part, but they too have been updated and expanded. The years since the publication of the Fourth Edition have been a time of great activity in the critical application of taxonomy, as well as in marine biology generally. In matters of nomenclature and taxonomy, the view of specialists was usually adopted, and many genus and species names have had to be changed, although in a few cases I chose not to accept a new name until the passage of time allows a few more opinions to be voiced. Some of the name changes have been the result of new work on phylogenetic groupings (limpets listed in the Fourth Edition as belonging to the genus *Acmaea* are now distributed among the genera *Collisella*, *Notoacmea*, and *Acmaea*), and some changes address previous taxonomic oversights (the snail *Littorina planaxis* of previous editions has had its name changed, to *L. keenae*, in order to comply with the rules of the International Code of Zoological Nomenclature). In some taxonomic groups, as many as a third or more of the names have changed, in others only a very few.

The intense research activity pursued since the preceding edition appeared has spawned an immense literature of other sorts, as well, and the number of entries has grown despite the deletion of many older references. Over 2,000 references are now listed, a majority of these from 1968 or later. For some taxonomic groups, the listings are fairly complete; for others, they represent only a small sampling of what is known. Still, the references and their annotations contain much additional information, and it is hoped that readers will at least browse through the listings as one might survey and sample a fine buffet.

The increasing role of marine invertebrates as experimental subjects will be apparent from the reference lists. Certain species have become standard laboratory subjects because of their unique features or ready availability: sea urchins and *Urechis* for research on the processes of development, squid and giant barnacles for the physiology of excitable membranes and muscle contraction, and opisthobranchs for nervous system integration and the neural correlates of behavior. Ecologists interested in the principles of community organization have also found intertidal plants and animals nearly ideal for some kinds of experimentation and study in the field. The intertidal communities on the Pacific coast are diverse and subject to a wide variety of physical and biological influences. Furthermore, they are readily accessible and easily manipulated to isolate important variables, features that have sparked a boom in field studies aimed at elucidating basic mechanisms of ecological organization and function. Many research papers from this ecological literature are included, primarily in the General Bibliography.

In addition to the expanded text and reference lists, one entirely new chapter has been added to the Fifth Edition, and one has been deleted. Neither change affects the basic structure of the book, which since the First Edition has consisted of two basic parts—the text sections on animals and habitats, and the Annotated Systematic Index and Bibliography—with new chapters being inserted between the two parts in successive editions to cover topics of currency and special interest. Because the years since the Fourth Edition have been characterized by a concentration on the principles of ecological organization, on experimentation, and on factors that influence distributions and abundances, especially biological interactions, I have written for this edition a chapter intended to reflect the scientific disposition of the time. The chapter is arranged topically and addresses, in a general way, the various factors that are seen to influence shore organisms and their interactions with the environment, both physical and biological. One might argue that this chapter on "principles," or organizing factors, should have been placed at the beginning of the book, perhaps as an expanded introduction, instead of at the end, as a summary of sorts. The chapter is placed where it is for two reasons: the first is the book's historical structure and my reluctance to tamper with it; the second is my belief that the richness of natural history detail, presented in Chapters 1–10, lays the foundation upon which these principles are built. Readers can, of course, begin the book wherever they wish.

Again, one chapter has been dropped. In the interests of cohesiveness and keeping an already expanded volume from bursting at the seams, the chapter considering oceanographic data (Chapter 15 of the Fourth Edition) has been deleted from the Fifth. This was done with considerable regret, relieved chiefly by the knowledge that those who want to know can still find the information in the Fourth Edition and in other sources.

With each edition of this work (there have actually been six although

Joel W. Hedgpeth

this is formally the fifth), it has become more and more difficult to decide who is responsible for particular passages or phrases. Certainly, however, Joel W. Hedgpeth's many contributions to the original work by Ricketts and Calvin must be acknowledged. Hedgpeth's labors on the Third, the Third Revised, and the Fourth editions have provided the sturdy foundation on which the present edition rests. *Between Pacific Tides* grew in both size and stature during the several revisions he undertook, and the present edition is replete with examples of his eloquence, literary breadth, and scientific acumen. Beyond his many additions to the text sections and reference lists, much of the present chapter on Intertidal Zonation (Chapter 11) is entirely his, although I have added some discussion of tidal mechanics and taken the liberty of rearranging some parts.

The absence of Hedgpeth's active involvement in the Fifth Edition breaks a link with the past, and it is an important link because *Between Pacific Tides* is truly a historical document as well as a modern, scientific account. Some feeling for the origins and development of *Between Pacific Tides* can be gathered from John Steinbeck's Foreword to the Revised Edition (1948), from Hedgpeth and Calvin's Preface to the Third (1952), and from Hedgpeth's Preface to the Fourth (1968), all reprinted here. Unfortunately, I did not know Ed Ricketts. Nor am I a historian, so it would be presumptuous of me to comment further. I will, however, refer readers to two other works, Hedgpeth's *The Outer Shores*, Parts 1 and 2 (Eureka, Calif.: Mad River Press, 1978, 1979), and his essay "Philosophy on Cannery Row" in R. Astro and T. Hayashi, eds., *Steinbeck: The Man and His Work* (Corvallis, Ore.: Oregon State University Press, 1971). Both are fine reading and provide a rich sense of Ricketts, his life, and his philosophy.

Although several hands have now stirred the stew, any errors that remain are, of course, my responsibility. Many specialists have been consulted in an effort to reduce the number carried over from previous editions and to minimize the number added, but some errors no doubt remain. One can hope that these will stimulate investigation. Perhaps they will also stimulate the flow of reprints and comments; both would be appreciated.

The preparation of this Fifth Edition is now complete, but of course the revision itself is a continuous process—new information is added daily, new patterns emerge, and perspectives change. The revision is also a collective process to which many people contribute, and I am grateful for having had the opportunity to give back some of what I have gained on the shore and in the library, and to present formally one slice of the collective effort. I thank all who have offered their encouragement and their science, and I hope that this edition will stimulate others to probe the pools and contribute their discoveries and enthusiasm. Much remains to be learned between Pacific tides, and that is exciting.

Over the years, many people have helped with this book—identifying material, correcting names, suggesting references, and supplying other information. From the very beginning, *Between Pacific Tides* could never have been written without the assistance of people at Hopkins Marine Station; these included Rolf L. Bolin, Walter K. Fisher, Harold Heath, and Tage Skogsberg, all no longer with us. Fortunately, the tradition has continued, and it is a pleasure to acknowledge the great help of Donald P. Abbott in preparing this edition.

Many colleagues have contributed in a variety of ways to this Fifth Edition, providing references, critical acumen, advice, and encouragement. Those to whom thanks are due include Isabella A. Abbott, Gerald J. Bakus, J. L. Barnard, Charles H. Baxter, David W. Behrens, Patricia R. Bergquist, Charles Birkeland, Robert Black, E. L. Bousfield, Ralph O. Brinkhurst, Gary J. Brusca, Richard Brusca, Theodore H. Bullock, C. Burdon-Jones, Debby Carlton, James T. Carlton, Victor Chow, Eugene V. Coan, Valerie Connor, Peter Connors, Howell V. Daly, Anthony D'Attilio, Paul K. Dayton, Christopher M. Dewees, John T. Doyen, Daphne F. Dunn, J. Wyatt Durham, Thomas A. Ebert, William G. Evans, Howard M. Feder, Antonio J. Ferreira, William S. Fisher, Jonathan B. Geller, Daniel W. Gotshall, Richard W. Grigg, Gary D. Grossman, Eugene C. Haderlie, Cadet Hand, Josephine

F. L. Hart, Joel W. Hedgpeth, Paul Illg, Rebecca Jensen, A. Myra Keen, David R. Lindberg, Donald L. Maurer, Bill Maxwell, Gary R. McDonald, Patsy A. McLaughlin, James H. McLean, Claudia E. Mills, Delane A. Munson, Barbara W. Myers, A. Todd Newberry, William A. Newman, John S. Pearse, Vicki B. Pearse, Diane Perry, Margaret S. Race, Gordon A. Robilliard, Pamela Roe, Greg Ruiz, Paul C. Schroeder, Kenneth P. Sebens, Janice B. Sibley, Norm A. Sloan, Dorothy F. Soule, John D. Soule, Jon D. Standing, A. Lynn Suer, Ellen Tani, Teresa Turner, Harvey D. Van Veldhuizen, Steven K. Webster, Lani A. West, Daniel Wickham, Mary K. Wicksten, William G. Wright, and perhaps others I have regrettably overlooked.

Of course, much of the information, advice, and assistance provided by others for former editions is still part of the book; but it might have been misleading to have again thanked those people here, since the final version of the information is my responsibility. All who have contributed, however, have my sincere gratitude. In addition, I thank all who have sent me offprints of their papers; whenever appropriate, these have been added to the bibliographies. I am sure that I would be much farther behind than I am were it not for this courtesy.

Except as indicated below, the illustrations in this book are those used in the original edition: the photographs by the late Jack Calvin, and the line drawings by the late Ritchie Lovejoy.

The late Nick Carter of San Francisco provided photographs for the following: Figs. 12, 15, 19, 20, 24, 37, 39 (left), 68, 71, 99, 102, 104, 135 (top), 177, 178, 180, 188 (above), 211, 213, 221, 243, 251, 252, 258, 294, 295, 299, 325, 326.

We have retained from previous editions the pictures taken by Woody Williams for Figs. 16, 98, 230, 253 (above), and 284.

Other photographs were contributed by the following persons: Fig. 170 by C. D. Snow of the Oregon Fish Commission; Fig. 204 by Joel W. Hedgpeth; Fig. 210 by Boyd Walker; Fig. 338 by John LeBaron; Fig. 273 (left) by Wheeler North; Fig. 303 by G. P. Wells; Fig. 312 by C. B. Howland (contributed by Josephine F. L. Hart). Willard Bascom provided the photograph of E. F. Ricketts.

Drawings and charts were done by the following people for various figures. Darl Bowers: 172–74. Sam Hinton: 292. Mrs. Carl Janish: 191, 329 (left), 330–37. Joel W. Hedgpeth: 5, 8, 10, 11, 41, 175, 179, 192, 212, 218, 290, 305, 308, 319 (left). R. J. Menzies: 191, 215, 219. Frank A. Pitelka and R. E. Paulson: 316, 318. Emily Reid: 216. Gunnar Thorson: 69 (above). Victoria White Hand: 311.

Other illustrations have been borrowed from various papers in the "open" literature (usually with some modifications): the drawings of *Velella*, originally by George O. Mackie and W. Garstang; diagrams from papers by Joseph Connell and Peter Glynn; and illustrations from the work of J. L. Barnard, L. A. Zenkevich, and O. B. Mokievsky.

Many people have contributed new photographs to this edition. Most were originally in color, and black-and-white reproduction has unfortunately not done them justice. In all cases the photographers' original color contributions were splendid, and we accept responsibility for how they appear in this book. The following people provided numerous photographs for this edition.

David W. Behrens: Figs. 97, 103 (below), 108, 109, 112, 150, 250, 269, 280, 289, 310.

Daniel W. Gotshall: Figs. 77, 78 (above), 81, 94, 135 (center), 158, 159, 168, 254, 283, 348, 351.

Gary R. McDonald: Figs. 47, 48, 49, 50, 52 (above), 83, 84, 85, 101, 103 (above), 107, 110, 111, 113, 114, 115, 120, 126, 135 (left), 136, 139 (left), 148 (left), 156 (below), 157, 162, 183, 197, 201, 240, 246, 253 (below), 319 (left).

Robert H. Morris: Figs. 14, 22, 27, 29, 34, 39 (above), 40, 42, 54, 55, 57, 59, 62, 63, 72, 74, 76, 79 (right), 82, 88, 89, 90, 91, 93, 95, 96, 105 (left), 106, 116, 117, 118, 122, 127, 128, 130, 131, 132, 134, 138, 143, 147, 155, 160 (right), 161, 165, 167, 185 (center), 189, 194, 202, 206, 207, 208, 214, 220, 222, 236, 237, 238, 239, 248, 255, 262, 266, 270, 274, 288, 293.

David W. Phillips: Figs. 1, 7, 13, 36, 79 (above), 171, 176, 196, 203, 265, 272, 317, 321, 339, 340, 341, 343, 344, 345, 352, 353, 355, 359, 360.

Other new photographs were also contributed. Robert Ames: 51. Eldon E. Ball: 245. Robert D. Beeman: 268. Debby Carlton: 349. James T. Carlton: 190, 346, 361, 364. Melbourne R. Carriker: 277. F.-S. Chia: 73. Victor Chow: 342, 354, 358. Eugene V. Coan: 152. Bertram C. Draper: 69 (right), 70. Robert Evans: 145, 146. J. J. Gonor: 31. Carole S. Hickman: 9, 25, 26, 148 (right), 185 (below). Holger Knudsen: 350. James H. McLean: 149. Claudia E. Mills: 271. Carlos Robles: 347. Peter A. Roy: 188 (right). Steven Obrebski contributed the photograph of Joel W. Hedgpeth; the print was courtesy of Photo Mural Service, Seattle.

We also wish to thank the following journals and publishers for permission to reprint figures: Fig. 31 from the *Fishery Bulletin*; Fig. 51 from *The Biological Bulletin*; Figs. 69 (right), 70 from *The Veliger*; Fig. 73 from *Systematic Zoology*; Fig. 170 from *The Journal of the Fisheries Research Board of Canada*; Fig. 174 from *Pacific Science* by permission of the University of Hawaii Press; Fig. 277 by Plenum Publishing Corporation; Fig. 303 by The Zoological Society of London; Fig. 323 from *Pacific Seashores* (1977), by Thomas Carefoot, by permission of the University of Washington Press; Fig. 347 from *Oecologia* by permission of Springer-Verlag; Fig. 356 from *Ecology*. We thank the University of California Press for permission to use various illustrations from *Seashore Life of the San Francisco Bay Area and Northern California* (1962), by Joel W. Hedgpeth, and from *Light's Manual: Intertidal Invertebrates of the Central California Coast* (3rd ed., 1975), edited by Ralph I. Smith and James T. Carlton; specifically, in this edition they are Figs. 21, 33, 38, 41, 56, 123, 124, 125, 172, 173. Stanford University Press has allowed the inclusion of drawings of marine algae by Mrs. Janish originally done for Gilbert M. Smith's *Marine Algae of the Monterey Peninsula* (1944) and by Susan Manchester (Fig. 329, right), originally published in *Marine Algae of California* (1976), by Isabella Abbott and George J. Hollenberg. The pictures by Robert Morris were previously published by Stanford University Press in *Intertidal Invertebrates of California* (1980), by R. H. Morris, D. P. Abbott, and E. C. Haderlie, and are reprinted here with permission.

Finally, I wish to thank my wife, Carol A. Erickson, for many things—for love and encouragement, for believing in me and in the value of doing this book, for financial support as family funds were spent on endless rolls of typewriter ribbon, stamps, and film, and for practical assistance and skills cheerfully donated on many inopportune occasions. It was she who spent long hours in the darkroom producing the best possible black-and-white print from a color slide or negative, her artful dodging applied to the photographs and not to my requests for assistance.

D.W.P.

June 1985

Between Pacific Tides

Introduction

The intertidal zone of the Pacific coast of North America supports an extraordinarily rich assortment of plants and animals. Within this narrow strip of shore, colorful and intriguing representatives of nearly every group of invertebrate animals can be found. Probably in no other natural setting can so many different kinds of organisms be observed so readily. One factor contributing to this diversity is seasonal upwelling (February to July), which brings to the surface nutrient-rich cold water from the ocean depths. These nutrients stimulate the growth of coastal algae, which in turn provide food for many animals. Another factor favoring the wealth of seashore life on our coast is the great variety of habitats available. There are wave-swept rocks, surf-pounded sandy beaches, protected boulder fields, and quiet-water flats of soft mud (Fig. 1). Each of these habitats bears a characteristic assemblage of organisms.

Despite the richness and complexity of Pacific coast intertidal life, three coordinate and interlocking factors can be identified as the primary determinants of how these shore invertebrates are distributed. They are the degree of wave shock, the type of bottom (whether rock, sand, mud, or some combination of these), and the tidal exposure. Of course, many other factors (several are considered in Chapter 12) also can, and do, greatly influence the local distribution of shore invertebrates. But these three are of fundamental importance, and each will be considered in turn.

On the Pacific coast the degree of wave shock is of particular importance. According to the physics of wave motion, the size of the unbroken water area (and to a limited extent the depth), the wind velocity, and the wind direction determine the size of waves. Assuming that wind velocities are the same in the Atlantic and the Pacific, the great unobstructed expanses of the Pacific make for larger wave possibilities. When in addition there is a bluff, unprotected coast with few islands or high submarine ridges for thousands of miles, as from Cape Flattery to Point Conception, and this stretch experiences prevailing northwesterly to westerly winds,

Fig. 1. The diversity of Pacific coast shores and habitats: *upper left*, a rocky section of semi-exposed outer coast; *upper right*, an exposed sandy beach; *lower left*, a protected boulder field; *lower right*, a mud flat.

wave shock is probably more powerful than in any other part of the Northern Hemisphere.

Various factors nevertheless modify the force of the waves in particular regions and cause correlative changes in the animal assemblages. The extreme modification comes in closed bays, sounds, and estuaries, where there is almost no surf at all. Given all gradations in wave shock from the pounding surf of 6-meter groundswells to the quiet waters of Puget Sound, where the waves barely lap the shores, one could enumerate an indefinite number of gradient stages. For the purposes of this book, however, we use only three, although even in Puget Sound it is quite possible to recognize sheltered and exposed positions and to note corresponding faunal distinctions. Ours is a classification of convenience. In spite of this arbitrariness, however, visitors to the shore will find that there is often a reasonably sharp dividing line, as where the surf beats against a jutting point that protects a relatively quiet bay; and that there is a fairly high degree of correlation between these divisions and the animal communities. There is often overlapping of the animals, of course, but if test counts show that 75 percent of the total observed number of a certain seastar are found on surf-swept rocky points, then we feel no hesitation in classifying it as an animal that belongs predominantly in that environment, noting also that individuals stray to more protected shores and to wharf piling.

These qualifying statements apply with equal force to the other two factors—type of bottom and tidal exposure—on which we base our classification. It must be understood that infinite variations exist, that few regions belong purely to one or the other of our divisions, and that any animal, even the most characteristic "horizon marker," may occasionally be found in totally unexpected associations.

The intergradation in types of bottom is too obvious to stress beyond remarking that in the cases of the innumerable variations between sand, muddy sand, sandy mud, and mud we have begged the question somewhat by using only two headings—sand flats and mud flats—leaving it to the judgment of the observer to decide where one merges into the other.

The third important aspect of habitat, tidal exposure, categorizes animals according to the relative lengths of their exposure to air and water (bathymetrical zoning)—in other words, the level at which the animals occur. A glance at any beach will lend more weight than would much discussion to an understanding of the extreme variations that exist between the uppermost region at and above the line of high spring tides, where the animals are wetted only a few times in each month by waves and spray, and the line of low spring tides, where the animals are uncovered only a few hours in each month. Many animals adhere closely to one particular level, and all have their preferences; but again the overlapping of one zone with another should be taken for granted. Also, local conditions may affect the animals' level, as along the coast of Baja California, where the high summer temperatures force the animals down to a lower level than they would normally assume. Even there, however, the animals' relative positions remain the same. There is merely a compressing of the life zones into something less than the actual intertidal zone. With these provisos, then, our system of zonation is equally applicable to Bahía de San Quintín, where the extreme range of tides is less than 2.4 meters (8 ft), and to Juneau, where it is more than 7.0 meters (23 ft).

These three aspects of habitat—wave shock, type of bottom, and tidal level—which so tremendously influence the types of animals that occur, suggest the scheme of topic organization adopted for this book, a scheme that may at first glance seem cumbersome but that is in fact quite convenient, and certainly more practical than a single taxonomic sequence of the animals themselves. The most unscientific shore visitor will know whether he is observing a surf-swept or a protected shore, and whether the substratum is predominantly rock or sand or mud, although he probably would not know the classification by which zoologists arrange the animals he is interested in. He will further know, or can easily find out, whether the tide at a given time is high or low; and when it is low he can easily visualize the four zones or levels of the bared area. The observer possessing this information should be able, with the aid of the illustrations and descriptions provided, to identify most of the animals that he is likely to find and to acquire considerable information about each.

Users of this book will accordingly find it profitable to fix in mind the following classifications of shore habitats:

I. Protected Outer Coast. Under this division we treat the animals of the semisheltered coast and open bays, where the force of the surf is somewhat dissipated. These shores, which are rich in animals, are generally concave, and are characteristically protected by a headland or a close-lying island (Fig. 2). Bahía de Todos Santos, Laguna, Santa Barbara, much of Monterey Bay, Half Moon Bay, Bodega Bay, and Point Arena are good examples. However, stretches of shore that appear at first glance to be entirely open to the sea may actually belong in this category. An offshore reef of submerged rocks or a long, gradually sloping strand is sufficient to break the full force of the waves and will greatly modify the character of the animal assemblages. Even distant headlands, outlying bars, or offshore islands, if they are in the direction of the prevailing winds, provide a degree of protection that makes the animals correspond to our protected-outer-coast division. An offshore kelp bed will serve the same purpose, smoothing out the water to such an extent that many years ago small coasting vessels often used to seek shelter inside the beds and take advantage of their protection whenever possible in landing goods through the surf. The 1889 edition of the *Coast Pilot* mentions many places where the kelp beds were frequently so utilized, and in recent years fishermen and an occasional yachtsman have kept the tradition alive. Today the beds are favorite places for venturesome divers and small-boat sailors.

In many places the protection is not provided by headlands or offshore rocks directly, but by the refraction of waves as they reach headlands or rocks (Fig. 3). Because the waves tend to approach parallel to the shore,

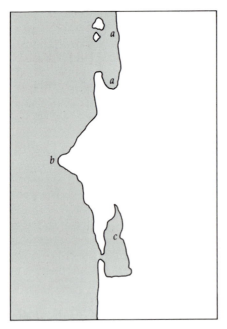

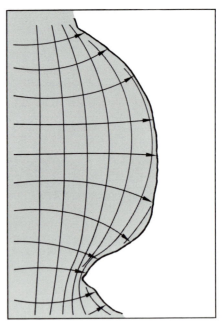

Fig. 2. Types of coastline: *a–a*, a typical stretch of protected outer coast; *b*, a characteristic headland on the open coast; *c*, an enclosed bay and estuary.

Fig. 3. Wave refraction on an irregular shore. A beach need not be in the lee of an offshore feature to be protected from heavy wave action.

they tend to bend around a headland, meeting it headlong from all sides, so to speak. Where two headlands lie close together, the energy of the waves along the shore between them is reduced. But this energy can be concentrated on the sides, and sometimes the apparently protected rear, of headlands and offshore rocks. Such places can be dangerous, and no one should venture on an unfamiliar shore without first observing for several minutes the patterns of waves against the shore. Many dangerous places along the coast are now marked by park services, and the warnings should be taken seriously.

Here on the Pacific we have only two types of shore in this division, rocky shores and sandy beaches.

II. Open Coast. Entirely unprotected, surf-swept shores, though by no means as rich in animals as partially protected shores, support a distinctive and characteristic assemblage of animals that either require surf or have learned to tolerate it. This type of shore is generally convex (Fig. 2*b*), varying from bold headlands to gently bulging stretches, and fairly deep water usually lies close offshore. Pismo Beach, the Point Sur and Point Lobos outer rocks, and the outer reefs of Cypress Point and Point Pinos are all good examples from the central California coast. And most of the coast of northern California and Oregon falls in this category. Obviously there are no muddy shores in this division; so again we have only rocky shores and sandy beaches.

III. Bays and Estuaries. Animals of the sloughs, enclosed bays, sounds, and estuaries, where the rise and fall of the tides are not complicated by surf, enjoy the ultimate in wave protection and are commonly different species from the animals inhabiting the open coast and the protected outer coast. Where the same species occur, they frequently differ in habit and in habitat. The shores embraced in this division are sharply concave; that is, they have great protected area with a relatively small and often indirect opening to the sea (Fig. 2c). Bahía de San Quintín in Baja California is one example; moving northward we find other such areas in San Diego, Newport, Morro, San Francisco, Tomales, and Coos bays, in Puget Sound, and in all the inside waters of British Columbia and southeastern Alaska. This time we have a greater variety of types of shore, including rocky shores (for example, the San Juan Islands and north), sand flats (quiet waterways along the entire coastline), eelgrass (growing on many types of completely protected shore along the entire coast), and mud flats (found in nearly all protected waterways).

IV. Wharf Piling. In addition to many animals found elsewhere, wooden pilings in these areas support numerous species, such as the infamous *Teredo*, which is seldom or never found in any other environment. The nature of piling fauna justifies its division (for convenience, and probably in point of fact) into exposed piling and protected piling.

Each of the four habitat types is also subdivided into zones (Fig. 4), according to the extent of tidal exposure.

Zone 1. Uppermost horizon: from the highest reach of spray and storm waves to about the mean of all high tides, which is +5 feet (1.5 meters) at Monterey. In this infrequently wetted zone, only organisms like the land-aspiring periwinkles and pill bugs occur. This is the splash, spray, supralittoral, or *Littorina* zone of various authors.

Zone 2. High intertidal: from mean high water to about the mean flood of the higher of the two daily lows, a bit below mean sea level. This is the home of barnacles and other animals accustomed to tolerating more air than water. In some zonation schemes, our Zone 2 is the upper part of the balanoid or midlittoral zone. On the Pacific coast this is the zone above the mussel beds.

Zone 3. Middle intertidal: from about mean higher low water to mean lower low water—the zero of the tide tables. Typically covered and uncovered twice each day, this belt extends from +2.5 feet (0.8 meter) to 0 on the California coast, and corresponds to the lower part of the balanoid or midlittoral zone of many schemes of classification. Most of these schemes recognize one general midlittoral zone, but in practice we can recognize the two major subdivisions that we refer to as Zones 2 and 3. The animals found in Zone 3 have accustomed themselves to the rhythm of the tides, and many actually require it.

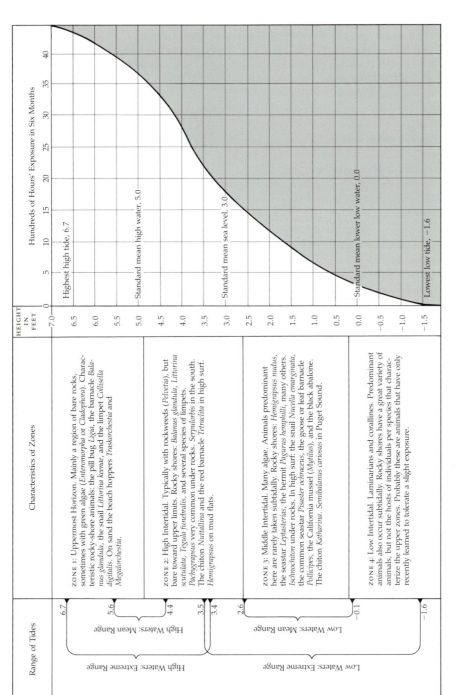

Fig. 4. The vertical zoning of animals on the Pacific coast.

Zone 4. Low intertidal: normally uncovered by minus tides only, extending from 0 to −1.8 feet (−0.6 meter) or so at Pacific Grove, and corresponding to the upper parts of the *Laminaria* zone or infralittoral fringe of some schemes of classification. This zone can be examined during only a few hours in each month and supports animals that work up as far as possible from deep water. Most of them remain in this zone, forgoing the advantages of the less-crowded conditions higher up, because they are unable to tolerate more than the minimum exposure incident to minus tides.

These zones, however, are not immutably fixed according to tide levels, but tend to spread wider and higher toward the region of heavier wave action (Figs. 5, 339). Often this can be seen on exposed coasts, along a vertical face set at right angles to the main direction of the waves. The conspicuous mussel zone will be wider toward the sea. What is not so obvious to the casual observer is that in very sheltered regions the whole zonal pattern may be lowered such that the uppermost zone is actually below the highest water mark of the year, as the figure suggests. The generalized zonal pattern described here is essentially a summary of observations of a semi-protected coastal situation; most specifically, of the zonation observed on the central California coast (see Chapter 11).

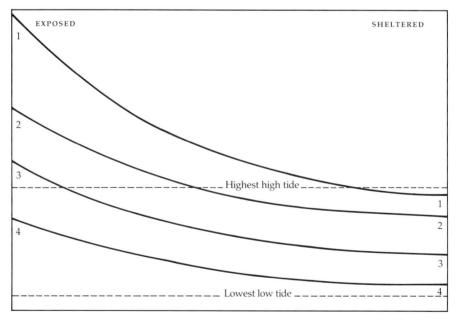

Fig. 5. The zones are displaced upward as one proceeds toward the more exposed part of the coast; such a displacement is often seen on rocky headlands. This schematic diagram does not indicate actual height.

The Table of Contents may be consulted as a summary of the Divisions and Zones within which the living creatures here described are classified.

This is a book for the observer or student who is limited to the shore and without equipment. The scope had to be limited in some way, and this choice accommodates the largest field. Though we have attempted to give a fair showing to the entire coast from Mexico to Alaska, as described by the maps in Figure 6, it may well be that the central California area has received undue attention. Expediency has dictated that this should be so, despite summer trips to Puget Sound, British Columbia, and southeastern Alaska, and frequent winter excursions south. The writers have themselves captured and observed practically all the animals listed—except as otherwise stated—but some we may not have identified correctly.

Absolute beginners will do well to devote their primary attention to large, common, and spectacular animals, which may be easily identified merely by reference to the illustrations. The more sophisticated will find that the book says a good deal also about the nondescript, the unusual, and the secretive. But the beginner, as well as the experienced, will often find that a creature he has observed is not mentioned in this book. There must be several thousand relatively common species of invertebrates along this coast. One hopeful ecologist, inspired by the Eltonian dictum that it would be best to learn as much as possible about a simple community rather than a little about a more complicated one, set himself to studying the apparently simple and limited group of creatures found among and around the little tufts of the red seaweed *Endocladia* at the higher tide level. He soon found that he had more than 90 species—five of them undescribed, and several previously unreported from the Pacific coast!

The Annotated Systematic Index lists every species covered in the book according to its zoological classification; here the reader will be directed to books or papers that provide more information. We have cited, from the widely scattered literature, only the most recent general taxonomic papers covering the region; smaller papers likely to be overlooked because of publication in obscure or foreign journals; and important references to the natural history, embryology, life history, and occasionally even the physiology of common Pacific types or of comparable forms occurring elsewhere under supposedly similar circumstances. Papers concerned exclusively with pelagic, dredged, parasitic, or very minute species have rarely been cited unless they also carry information on related littoral forms. To the General Bibliography we have added papers of a general ecological nature and books on seashore life.

One more word of explanation is necessary before we consider the animals themselves. It would have been desirable in some respects to depend on scientific names less frequently than we have been forced to do in this account. Certainly it would make pleasanter reading for the layman if a common name were provided in each case, and if he did not have to read (or to skip over) long tongue-twisting names compounded of Latin and Greek words and the latinized names of scientists—or even of boats that have carried scientific expeditions. Unfortunately, few of the marine animals have popular names that are of any diagnostic value. Popular names seldom have more than local significance, and often the same name is applied to different animals in different regions. Just as often, a particular animal goes by a different name in each region where it occurs. We have accordingly given, for what they are worth, any popular names already in use, along with their scientific equivalents; as for the rest, persons who believe that magnetos, carburetors, cams, and valves are all "gadgets" are free to designate them, indifferently, as bugs, worms, critters, and beasties.

We are, alas, no longer in the halcyon days of carefree collecting and unspoiled abundance of life on our seashores. We must hope that those interested enough to obtain a copy of this book and use it for a guide will also be interested enough to watch intertidal life between the tides, rather than bringing it away to die in buckets or suffocate in poorly managed aquaria. We also hope that teachers will reconsider the utility of bringing brigades of students to some choice spot and turning them loose to trample paths among the seaweeds and the urchins. There is probably no hope for those teachers and counselors who stand placidly by while robust, undirected adolescents throw sea urchins at each other or stomp the gumboot chiton to death, but all should be aware that the life of the seashore was never adapted to withstand the pressure of hordes of people. Places that once abounded with urchins are barren of them now; the abalone, once a dominant animal of the intertidal region, is now common only on inaccessible offshore rocks or islands; and the rock scallop *Hinnites*, once common in the lower intertidal, is now a rare animal.

This has happened before. Edmund Gosse wrote, in his moving story of life with his father, P. H. Gosse: "All this is long over and done with. The ring of living beauty drawn about our shores was a very thin and fragile one. It had existed all those centuries only in consequence of the indifference, the blissful ignorance, of man. These rock basins, fringed by corallines, filled with still water almost as pellucid as the upper air itself, thronged with beautiful, sensitive forms of life—they exist no longer, they are all profaned, and emptied, and vulgarized. An army of 'collectors' has passed over them and ravaged every corner of them. The fairy paradise has been violated; the exquisite product of centuries of natural selection has been crushed under the rough paw of well-meaning, idle-minded curiosity." (*Father and Son*, Chap. VI.)

At least we seem to realize the danger more, nowadays. Perhaps Edmund Gosse sacrificed a bit of actuality to literary force, perhaps not. In any event, on the Pacific coast we are establishing more and more reserves,

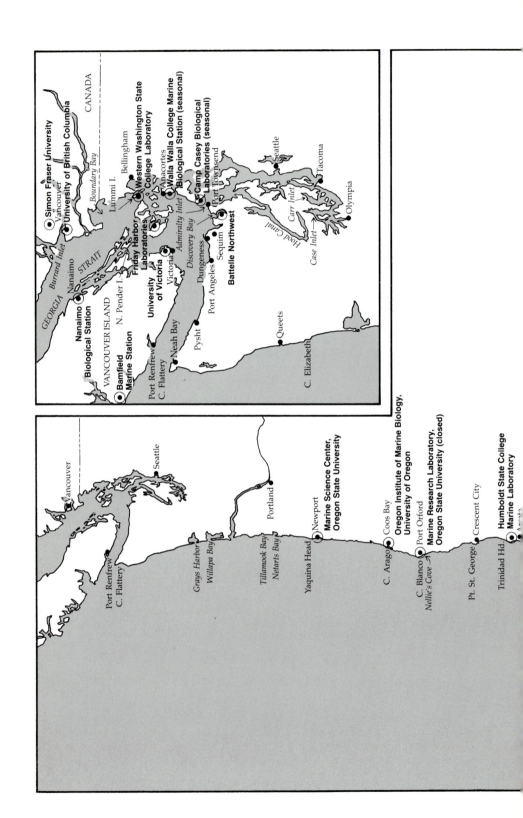

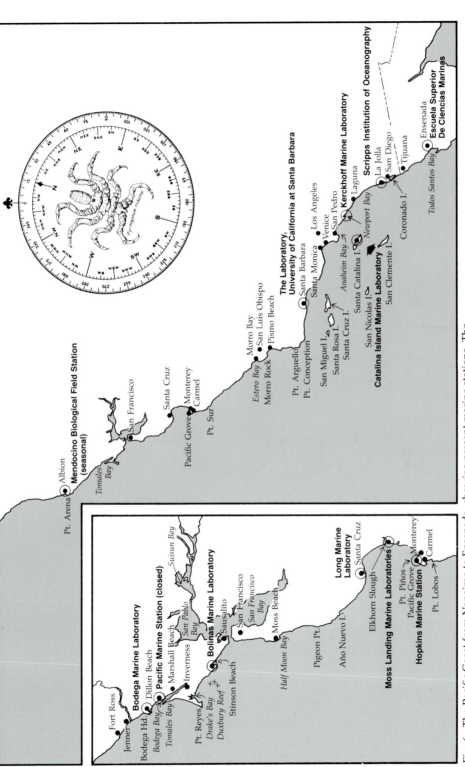

Fig. 6. The Pacific Coast from Nanaimo to Ensenada, showing present marine stations. The insets show collecting localities in Puget Sound and central California.

state parks, and national seashores where collecting is forbidden; and there are still inaccessible areas to serve as sources for reproduction. But many of our seashore animals do not move around very much, quite a few of them may lead extraordinarily long lives, and the process of recovery is slow; not everyone can be privileged to visit an unspoiled area. To be sure, it is necessary to take specimens for study and identification, but it is not necessary to take buckets of them; nor is it necessary for each of a group of 50 schoolchildren to collect "one of each" as part of his field trip. It is especially unnecessary to remove the larger, conspicuous animals, which give our shore its unique character. Not only should you replace the rocks you overturn, you should leave the animals where they are unless you have a good reason for taking a very few of them.

PART I

Protected Outer Coast

From the standpoint of the field observer, the rocky tide flats of the protected outer coast are the most important of all seashore regions. Rocky-shore animals and plants are abundant, easy to find, and frequently spectacular in their bright colors and unexpected shapes. So keen is the competition for living space here that not only is every square centimeter of shore surface likely to be utilized, but the holdfasts and stipes of kelp are also occupied; and such forms as sponges, tube worms, and barnacles often take up positions on the shells of larger animals.

Rocky-intertidal animals are characterized by interesting physiological, morphological, and behavioral features that offer methods of attachment, means of surviving wave shock or of coping with an alternate exposure to air and water, and techniques of offense and defense—all intensely specialized to meet the crowded environment. Obviously, there is no need for the elaborate devices developed by sand-flat and mud-flat animals to avoid suffocation or to retain orientation within the substratum. But the rocky shore presents its own special problems. Rocky-shore animals, unbuffered by water-retaining sediments, are more exposed to environmental changes at low tide, and crowded conditions accentuate the effects of competition and predation. Patterns of zonation often reflect these influences.

One example of the specialization developed in this environment is that by which a captured crab or brittle star deliberately breaks off an appendage in order to free itself. Losses occasioned by this process, called autotomy, are usually replaced by subsequent growth, or regeneration. This is a trait of great survival value to animals that may be imprisoned by loose rocks overturned by wave action.

The sandy beaches of the protected outer coast, often small pockets less than a few hundred meters long between rocky outcrops, are far more barren than regions of comparable exposure and intertidal height on rocky shores. Only a few species of large animals are characteristically found on such beaches, and considerable digging is usually required to bring them into view.

1 🐟 Outer-Coast Rocky Shores

Rocky-shore collecting is best managed by turning over small rocks by hand and lifting big ones with a bar. Moreover, it is highly important that the shore naturalist, in the interests of conservation, carefully replace all rocks right side up in their depressions; otherwise many of the delicate bottom animals are exposed to fatal drying, sunlight, or wave action. The animals on the bottom of rocks are adapted specifically to this particular habitat and could no more survive on top than could forms from the lower shore survive if moved to the uppermost zone, or rocky-shore forms transferred to a sandy beach. Whoever doubts the necessity for this care should examine a familiar intertidal area directly after it has been combed by a biology class that has failed to observe the precaution of replacing overturned rocks, and should visit it again a few days later. At first the rocks will simply look strange and scarred. In a few days whole colonies of tunicates, solitary corals, and tube worms will be found dead, and a noticeable line of demarcation will set off the desolate area from its natural surroundings. It takes months and sometimes a year or more for such a spot to recover.

Zone 1. Uppermost Horizon

This is a bare rock area in the main (Fig. 7), but sometimes there are sparse growths of green algae. It is inhabited by hardy, semiterrestrial animals. At Pacific Grove the zone lies between +2.1 or 2.4 meters (7 or 8 ft) and +1.5 meters (5 ft).*

§1. The small dingy snails that litter the highest rocks are not easy to see until one can distinguish their dry, dirty-gray shells from the rocky background of the same tone. Their habitat, higher on the occasionally wet-

---

*Most measurements of a scientific nature are now reported using the metric system, and measurements throughout this edition have been converted to metric. However, in two common situations—tide tables and sport-fishing regulations—measurements are still printed mostly in feet and inches.

Fig. 7. Rocky shores of the protected outer coast: *above*, flat benches and vertical walls; *below*, boulder fields. Together, these areas provide probably the richest diversity of habitats and organisms for the shore visitor.

ted rocks than that of almost any other animal, gives them their Latin name of *Littorina*, "shore-dwellers"—a more diagnostic term than the popular "periwinkle." Marine animals though they are, the littorines keep as far as possible away from the sea, staying barely close enough to wet their gills occasionally. In regions of moderate surf they may be a meter or more above the high-tide line, scattered over the rock face by the thousands or clustered along crevices. Specimens placed in an aquarium show their distaste for seawater by immediately crawling up the sides until they reach the air.

It seems probable that some species of *Littorina* are well along in the process of changing from sea dwellers to land dwellers, and they have, in consequence, an extraordinary resistance to nonmarine conditions. Individuals of *Littorina keenae* (Fig. 8), formerly known as *L. planaxis* and the commonest species on these high rocks, have been kept dry experimentally for 2 months without being damaged. They can stand immersion in fresh water, which kills all true marine animals, for several days. We once subjected a specimen to a shorter but even more severe test: we fed it to a sea anemone (*Metridium*). There was no intention, however, of testing *Littorina*'s endurance; we merely wished to feed the anemone something to keep it alive in captivity, and assumed that its powerful digestive juices would circumvent the difficulty of the shell. The anemone swallowed the snail promptly, and our expectation was that in due time the empty shell would be disgorged. But it was an intact and healthy *Littorina* that emerged, like Jonah, after a residence of from 12 to 20 hours in the anemone's stomach. When first discovered, the disgorged snail was lying on the bottom of the dish, and since it must have been in some doubt as to just where it was, its shell was still tightly closed. When it was picked up for examination, however, it showed signs of life; and after being returned to the dish, it crawled away at its liveliest pace. It had apparently suffered no harm whatever, but its shell was beautifully cleaned and polished.

Such resistance to distinctly unfavorable conditions is made possible by a horny door, the operculum, with which *Littorina*, like many other marine snails, closes its shell. This door serves the double purpose of retaining moisture in the gills and keeping out fresh water and desiccating wind; that it can also exclude the digestive enzymes of an anemone is a considerable tribute to its efficiency. Closing the shell does present a problem, however. With the foot withdrawn and the operculum in use, the snail must find some other way to attach to the rocks. To this end, the snail secretes a thin sheet of mucous glue around the aperture of the shell, providing at once attachment and a second barrier against unfavorable conditions.

Like other marine snails, *Littorina* possesses a wicked food-getting instrument, the radula—a hard ribbon armed with rows of filelike teeth (Fig. 9). The radula of this particular animal is unoffendingly used to scrape detritus and microscopic plants from almost bare rocks—an operation that necessitates continual renewal by growth—but the radulae of several other snails are used to drill holes through the hard shells of oysters, mussels, and other animals. Though unspectacular in the short term, the grazing activity of littorines on coarse or loosely bound sandstone may in time abrade the surface, perhaps as much as 5 cm in a century.

During excursions on the rocks, individuals tend to follow the trails of mucus laid down by other littorines. The snails in some way are able to perceive from the trail the direction of travel of the trail maker, which they

Fig. 8. *Littorina keenae* (§1), a dirty-gray periwinkle inhabiting the highest intertidal rocks in California; this snail, shown about 2½ times natural size, is often attached by a mucous film.

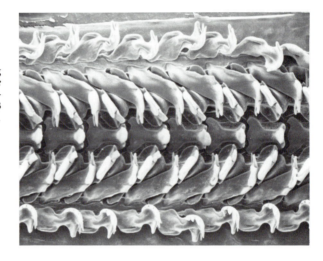

Fig. 9. The food-scraping radula of *Littorina keenae* (§1); this scanning electron micrograph is 173 times natural size.

consistently turn to follow. This behavior undoubtedly has as one of its benefits the bringing together of potential sexual partners, although individuals of the same sex will also follow each other's trails.

The sexes of the littorines are separate. In central California, at least, some specimens may be found copulating at almost any time of the year; but reproduction in mass seems to be confined to the spring and summer months, when it is difficult to find, among thousands, a dozen specimens that are not attached in pairs. Eggs are laid in an elongate, gelatinous mass by animals recently wetted or immersed by the rising tide. After several hours in water, the matrix of the mass disintegrates, liberating several thousand eggs, each individually packaged in its own tiny, disk-shaped capsule.

The extreme-high-water form, *Littorina keenae*, ranges from Charleston, Oregon, to southern Baja California (but not from Puget Sound, as sometimes, and erroneously, reported), but it is most abundant on the California coast. North of Cape Arago, Oregon, it is replaced by *L. sitkana*

Fig. 11. *Littorina scutulata* (§10),
a somewhat smaller, more conical
snail often with checks or dots
on the shell, is common at
slightly lower levels all along the
coast and in a variety of habitats;
about 2½ times maximum size.

Fig. 10. *Littorina sitkana* (§195),
which replaces *L. keenae* as the
common periwinkle on high intertidal rocks from Oregon
north; about 2½ times natural
size.

(§195; Fig. 10). The slightly smaller, taller-spired *L. scutulata* (Fig. 11) is found all along the coast, usually at slightly lower levels. It appears to be more tolerant of reduced salinities than the other two, and is commonly found on rocks and piling in bays; it is, for example, the littorine of such places as the sheltered parts of Tomales Bay, or Coyote Point in San Francisco Bay.

§2. In some places specimens of a large isopod, the rock louse *Ligia occidentalis* (Fig. 12), up to 25 mm long, can be found scurrying around on the rocks at distances so far removed from the water that one is led to believe it is another animal in the very act of changing from a marine to a terrestrial habitat. *Ligia* (including a more northern, open-coast species, §162) is very careful to avoid wetting its feet, and eventually will drown if kept under water. When the tide is out, it may venture down into the intertidal area, but when there is any surf it will be found just above the spray line. The animal cannot stray too far up the shore, however, for it needs some means to wet its gills and to replace the moisture lost from its body by evaporation. It apparently accomplishes both by periodically dipping its tail into a tide pool.

Ligia occidentalis is recognized by the presence of two long forked spines projecting behind—which are used, apparently, as a rudder to help the animal steer its rapid and circuitous course over the rocks. Specimens rarely occur singly; it seems there will be hundreds or none. Some may be seen moving about in broad daylight, but more become active late in the afternoon and most are active at night. Associated with this cycle is a diurnal rhythm of color change, presumably related to protective coloration; during the day the animals are dark, matching their rocky background, but at night they become paler. Specimens have been taken at Pacific Grove in March, May, and June that had the incubatory pouch on the underside of the thorax turgid with developing eggs. The range is from Sonoma County

Fig. 12. *Ligia occidentalis*, the
rock louse (§2), a fast-moving
scavenger or scraper of algal
films on high intertidal rocks.

to Central America. It is common on bay shores and occurs in the Sacramento River in almost fresh water.

Still another child of the ocean that has taken to living on land is the
beach hopper, or "sand flea," *Traskorchestia traskiana* (formerly *Orchestia*).
Along the steep cliffs west of Santa Cruz we have found it more than 6 meters above tidewater and above the usual spray line, living practically in the
irrigated fields. It is found at Laguna in a brackish-water slough that runs
through marshy fields and rarely receives an influx of tidewater. Like other
hoppers, it is a scavenger, and may often be found about decaying seaweed
that has been thrown high up on the shore. The animal's body color is dull
green or gray-brown, with the legs slightly blue. Except for its small size (1
cm or a little more in length) and its short antennae, this form is similar in
appearance to the large, handsome hoppers of the more open beaches. The
hoppers are members of the family Talitridae; this family includes at least
one strictly terrestrial species, which has spread from its native New Zealand and become a resident of greenhouses in many parts of the world.

On islands to the far north, foxes are important predators on *Traskorchestia* and other amphipods. Indeed, vertebrate visitors to the shore
should be kept in mind as potentially important predators on the more-or-
less permanent marine residents that provide the focus of our account. This
general notion warrants some emphasis here, for although such predation
may often occur, it is easily overlooked. The mammalian predators that frequent the shore in some locations—raccoons, mice, foxes, and others—
tend to be nocturnal, doing their predatory deeds when we are away from
the shore, and foraging birds tend to stay far enough away from human
visitors to make detailed observations difficult. However, one has only to
follow for a short while a small flock of black oystercatchers to be impressed
with the pile of limpet shells (§6) that accumulates in the wake of these
birds as they forage among the rocks (Fig. 13). Likewise, the stomach of a

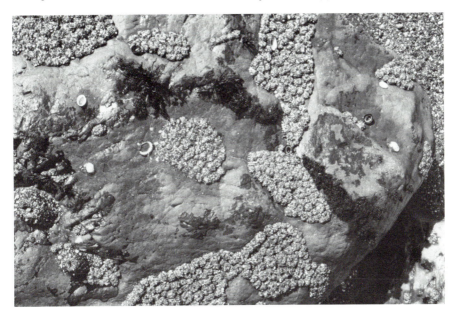

Fig. 13. Empty limpet shells (*Notoacmea scutum*, §25, and *N. fenestrata*, §117) after two black oystercatchers moved through the area. At least six shells are visible on the flat top of this rock. Also present are clumps of the anemone *Anthopleura elegantissima* (§24) and the alga *Gigartina*.

fox containing the remains of several hundred beach hoppers is convincing testimony to an effect easily overlooked, but nevertheless important.

§3. A conspicuous rocky-shore animal that keeps a considerable distance above mean sea level is the large flatworm *Alloioplana californica* (Fig. 14), which is up to 4 cm long. Thick-bodied and firm, almost round in outline, these animals are rather beautiful with their markings of blue-green, black, and white. The markings trace in large part the outline of the highly branched digestive tract, and presumably vary in color according to the food recently consumed by the animal. These flatworms are relatively abundant, near Monterey at least, but are not often seen, since their habitat is restricted, when the tide is out, to the undersides of large boulders on gravel that is damp but seldom wet. The casual observer should feel properly thrilled at finding the several specimens that usually turn up, and will easily remember them from their size and beauty.

Like most of the flatworms, *Alloioplana* is hermaphroditic, each animal developing first male and then female gonads. The egg masses, found almost throughout the year, form round encrusting bodies up to 6 mm in diameter under boulders and in rock crevices. *Alloioplana* is known to feed on minute snails, since portions of snails' radulae, or rasping "tongues," have been found in its digestive tract. This flatworm has been found

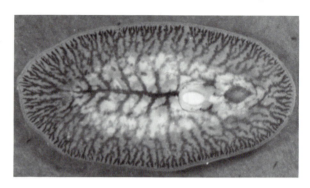

Fig. 14. The large flatworm *Alloioplana californica* (§3). The intricate dark markings are branches of the digestive tract, which leads from a more or less centrally located mouth (white in the photograph) on the ventral surface; the dark, diamond-shaped structure posterior to the mouth is part of the reproductive system.

throughout California, and we have taken an identical or similar form in Baja California.

§4. Small acorn barnacles, with small limpets interspersed among them, occur abundantly near the extreme upper limit of the intertidal zone. Except at extreme high tide, both have dried, dead-looking shells of the dingy-gray color so common to high-tide rocks. The barnacles, usually *Balanus glandula* (Fig. 15), are actually crustaceans, more nearly related to the crabs and shrimps than to the shelled mollusks with which the tyro associates them. The sharp projecting shells, usually 12 mm or less in diameter, are sometimes crowded together like cells in a honeycomb on the sloping or vertical rock faces. They are necessarily exposed a great deal to sun, rain, or wind; but at such times the operculum, a set of hinged plates, is kept tightly closed to protect the delicate internal anatomy. At extreme high tide, sometimes for only a few hours in a week, each animal throws open its operculum and rhythmically sweeps the water with its brightly colored appendages, respiring and searching for the minute organisms that constitute its food. A dry cluster of barnacles will "come to" very quickly if immersed in a jar of fresh seawater, and they live well in aquaria.

Balanus glandula, one of the most abundant solitary animals on the coast, is the Pacific's high-intertidal representative of a group of animals distributed throughout all oceans and in all depths. Its habitat requirements are unspecialized; hence it is able to tolerate a broad range of conditions and ranges all the way from the Aleutian Islands to Bahía de San Quintín, Baja California. It thrives not only in the constantly aerated ocean waters along the protected outer coast (and even on violently surf-swept points), but in the quiet waters of Puget Sound. On the quiet beaches of Tomales Bay individual barnacles perch on pebbles the size of walnuts; in such circumstances, even solid rock is obviously unnecessary.

The reproductive habits of barnacles are noteworthy. The eggs and embryos are brooded within the shell of the parent, to be discharged as free-swimming larvae called nauplii. A nauplius (Fig. 16) is a one-eyed, one-shelled, microscopic animal with three pairs of legs. After undergoing five intermediate transformations it becomes a cypris (Fig. 16)—an animal

Fig. 15. *Balanus glandula* (§4), the acorn barnacle, in a characteristic grouping; one of the most abundant animals on the protected outer coast, this crustacean also thrives in quiet waters and along the surf-swept open coast. A limpet (*Collisella digitalis*) is also visible.

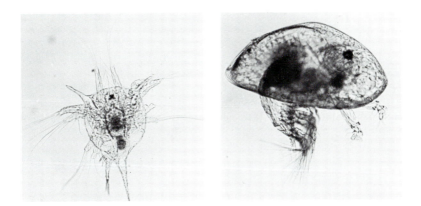

Fig. 16. Larval stages of a barnacle (§4): *left*, nauplius; *right*, cypris. Both pictures are greatly magnified.

with three eyes, two shells, six pairs of legs, and an inclination to give up the roving habits of its youth and settle down. The cypris attaches itself, by specially modified antennae, to a rock or other object, whereupon it secretes a cement and begins to build the limy protective shell of an adult barnacle. Now it is a blind animal, fastened by its head and feeding with its curled, feathery legs. It has been debated whether the animal is able to take food during its free-swimming stages; but now it is known that the nauplius feeds, whereas the cypris does not.

Like their relatives the crabs, barnacles molt at regular intervals, shedding the thin, skinlike covering of the body and appendages rather than the familiar jagged shell. In the quiet waters of Puget Sound and British Columbia, these almost transparent casts may be seen floating on the water in incalculable numbers. Specimens of *Balanus glandula* grow to about 7 mm in diameter and reach sexual maturity by the end of the first year. Adults, which may live for 8 to 10 years, produce two to six broods during the winter and spring, the size of the brood, up to 30,000 larvae, increasing with the size of the parent.

Recently settled barnacles initially occupy a broad vertical expanse of shore; however, their many predators (*Nucella*, §§165, 200; *Pisaster*, §157; *Leptasterias*, §68) and competitors (other barnacles, §§168, 194; *Mytilus*, §158) soon remove nearly all of these from the lower shore in many locations. The breeding population of barnacles is thus often limited to a narrow band on the upper shore beyond the reach of their antagonists.

It should be mentioned in passing that we may occasionally have mistaken *Solidobalanus hesperius* (formerly *Balanus*) for *B. glandula*, although no specimen from our random shore collections has been so identified. We have taken undoubted specimens of *S. hesperius* only from boat bottoms and from deep water. In any case, it makes little difference; the two are so similar that only specialists are thoroughly competent to tell them apart.

§5. Ranging southward from the neighborhood of San Francisco is a much smaller gray or brown barnacle, *Chthamalus fissus*, whose shell has less than half the diameter of that of *Balanus*. Everything else being equal, this tiny barnacle will be found gradually but evenly replacing its larger relative as one moves southward, until in northern Baja California it becomes the predominant form. In some sheltered localities, or on gradually sloping beaches, it is the only barnacle to be found; but where there is a bit of surf the larger *Balanus* steps in again. There is one wave-pounded cliff on a jutting headland about 60 km below the Mexican border that has *Balanus* only, although the smaller form is found within a few kilometers on either side.

An Italian investigator, experimenting with another species of *Chthamalus*, found that it prefers atmospheric respiration, using the air dissolved in spray, and that consequently it thrives best on wave-battered reefs. While water remains on the rock, the cirri, or feeding legs, are frequently protruded into the air. This investigator kept 100 individuals on his laboratory table for 3 years, immersing them for 1 to 2 days about every 3 months. They were immersed a total of only 59 days out of the 1,036, and yet only 10 to 12 of them died each year. Other lots lived as long as 4 months continually immersed in fresh water, which kills nearly all marine animals. Some lived 2 months completely immersed in vaseline! Under either natural or artificial conditions, periods of drying are as necessary to them as periods of immersion.

§6. The limpets *Acmaea*, *Collisella*, and *Notoacmea* are gastropods ("stomach-foot"), and are related to the common marine snails, however little the limpets' lopsidedly conical shell resembles theirs. The height of a limpet's shell sometimes is associated with its position in the intertidal zone. Within a species individuals inhabiting relatively low positions on the shore have low shells, whereas those inhabiting higher (and hence drier) habitats have taller shells. This association does not extend well to comparisons between species, however; our tallest limpet for instance, *Acmaea mitra*, frequents the lowest intertidal.

Limpets occur in great abundance, plastered so tightly against the rocks that specimens can be removed undamaged only by slipping a knife blade under them unawares. If the animal receives any warning of an impending attack, it draws its shell into such firm contact with the rock that a determined attempt to remove it by force will sometimes break the shell, hard as it is. So powerful is the foot's attachment, once it has taken hold, that an estimated 32-kilogram (or 70-pound) pull is required to remove a limpet with a basal area of about 6.5 cm² (1 square inch).* The sexes of limpets are typically separate, and fertilization in our temperate Pacific species occurs in the sea, since the eggs and sperm are shed freely into the water.

Some limpets stay pretty much at home during the day, but other species move about in daylight as long as they are covered with water. Much can be learned, however, from a night trip to the shore. A single excursion into the tidelands with a flashlight will often reveal more of an animal's habits than a score of daylight trips. Seen at low tide in the daytime, a limpet seems as immobile and inactive as a dry barnacle; but at night the foot extends, the shell lifts from the rock, and the animal goes cruising in search of food, which it scrapes from the rock with a filelike ribbon, the radula. It is useful to remind oneself periodically of the limpet's mobility and behavior, for its position in the littoral zone is related not only to tide levels. These, like other mobile animals, may seek out a protective crevice or an especially favored food; they may flee from predators; or they may move up or down with changes in light or turbulence, and migrate with the seasons. In the daytime one could hardly guess that these are mobile ani-

*The surface area of the foot obviously has an influence on the tenacity of an animal, but even after surface area has been accounted for, there is considerable variation among species: measured forces of adhesion range from 484 to 6,900 grams per cm² for different species. Interested readers should consult Grenon and Walker (1981) for a recent view of the factors influencing the tenacity of a limpet.

mals with well-developed behaviors. Characteristically, however, they are inactive during daytime low tides, when in contrast we observers of the shore are most active.

The commonest species on these high Pacific rocks is *Collisella digitalis* (Figs. 17, 21), which ranges from southern Alaska to Cabo San Lucas, Baja California. This is the exceedingly common form that has been listed as *Acmaea persona* in some older accounts. Great aggregations occur even where (or especially where) the surf is fairly brisk, occasionally extending clear up to the splash line in areas sheltered from direct sunlight. The aggregations, evident at low tide, are not static, but are reformed with each tide. When splash from advancing waves stimulates individuals to move out in search of food, the aggregation begins to disperse. Then as the tide again recedes, the clusters reform, not always in exactly the same spot nor with exactly the same component members, but at least with some measure of consistency. The average, small, dingy-brown specimen found in clusters high on the shore will measure under 25 mm, even under 15 mm; but solitary individuals may often be half again as large and more brilliantly marked. Some spawning has been reported during winter, spring, and summer months, although recruitment is highest in the spring and fall. On steep vertical surfaces, limpets tend to occupy the lower levels when young, then migrate up the shore in fall and winter, possibly stimulated by increased wave ac-

Fig. 17. A group of limpets, *Collisella digitalis* (§6), with periwinkles and barnacles.

tion. In spring, they migrate down again, but a smaller distance, the net effect being that older, and larger, animals on the average occupy higher positions on the shore. *C. digitalis*, apparently preferring rough rocks, notably tolerates conditions such as swirling sand, mud, debris, and even sewage or industrial pollution (most of which are disastrous to other limpets), and is therefore possibly the most abundant acmaeid on the whole outer coast north of Point Conception.

A heavily eroded form, the former subspecies *textilis* Gould, is now known to be simply an ordinary *Collisella digitalis* whose shell has been pitted by the parasitic fungus *Pharcidia balani* (also known as *Didymella conchae*) also found on other limpets inhabiting successively lower levels (§§11, 25). The reader should refer to accounts of these other forms, since some of them are occasionally found on the upper beach; *C. scabra* in particular occurs as high as *C. digitalis* at many sites, and the two may compete for food.

It has been recognized for some years that some of the limpets superficially resembling *Collisella digitalis* are in fact another species. Though similar in external shape, and sometimes in color pattern, these limpets always lack the large solid-brown spot inside the apex of the shell (which can be seen only if the animal is removed) and the ridges on the outside of the shell that are characteristic of *C. digitalis*; the bluish-white interior of this related species, however, may have small brown spots in the center. This species, which turns out to be common from southern Alaska to northern Baja California, is *C. paradigitalis*.

§7. The giant owl limpet, *Lottia gigantea* (Fig. 18), ranges from Neah Bay, Washington, to Bahía Tortugas, Baja California; it is found in the high intertidal, especially in the southern part of its range. The largest specimens, however, more than 7 cm long, are found in the middle zone, especially on surf-swept rocks. These large individuals are nearly always females because *Lottia* changes sex, from male to female, as it grows. At low tide, most individuals of *Lottia* on open rock occupy a characteristic home scar that fits precisely the margin of the shell (the scar can be seen beside the animal in the lower figure). Surrounding the home scar is an area of algal film, 1,000 cm² or so, which is kept free of would-be settlers and intruders. This territory is actively defended by the resident *Lottia*, intruders such as *Collisella digitalis* and other *Lottia* being bulldozed off if necessary. It is rarely necessary, however, for as in most territorial systems, the interactants here seldom come to damaging blows. Instead, intruders typically flee from a resident *Lottia*, either immediately or after the resident delivers a few (ritualized?) thrusts. *Lottia* only occasionally pursues an intruder to the edge of the territory, somehow knowing that the message has been received. Within its territory of algal film, the limpet does all of its grazing, returning to its home scar at low tide. Interestingly, the outward route taken by a foraging animal is not retraced on the return trip; rather, the path of the journey describes a circle or figure 8. Occasionally, one *Lottia* has been observed to take over the home site of another.

Fig. 18. *Above*, several giant owl limpets, *Lottia gigantea* (§7), with the territories of some appearing as cleared areas in the dark algal turf. *Below*, an owl limpet that has been removed from its home scar, visible to the right of the limpet.

The Mexicans justly prize the owl limpet as food. When properly prepared, it is delicious, having finer meat and a more delicate flavor than abalone. Each animal provides one steak the size of a silver dollar, which must be pounded between two blocks of wood before it is rolled in egg and flour and fried. Unfortunately, *Lottia* is not specifically protected under our fish-and-game regulations—a pity, for its edibility has been its downfall in some locations. It is much too interesting an animal to be carelessly tossed down a gourmet's gullet.*

§8. In localities that have spray pools at or above the high-tide line, the vivid, yellow-green strands of the plant *Enteromorpha* are likely to appear, especially where there is an admixture of fresh water. In these brackish pools, which often have a high temperature because of long exposure to the sun, very tiny red "bugs"—crustaceans—may often be seen darting about. These animals, easily visible because of their color, are specimens of the copepod *Tigriopus californicus*. This hardy animal, which ranges from Alaska to Baja California, breeds easily in captivity and may be raised at home in old coffee cups or cheese glasses on a diet of pablum or the like. In tide pools they feed primarily on algal film and detritus rasped from the rocks with specialized mouthparts. The animal's reproductive capacity is considerable, and under optimum conditions a population can double in numbers every 4 to 6 days depending on the temperature. Eggs are laid in a single mass attached to the ventral surface of the female.

Tigriopus is one of the harpacticoid copepods, most of which live near the bottom, or among detritus and other accumulations everywhere on the seashore. *Tigriopus*, however, seems to be the most successful colonizer of the isolated high pools. Some of these pools may contain only a liter or two of water, and are subject to changes in salinity from rain and evaporation. At home in conditions fatal to most intertidal animals, *Tigriopus* remains active in water up to three times normal salinity and can even tolerate immersion in saturated brine for 12 days if then returned to normal seawater. The copepods are carried from one pool to another by wave splash and by hitching rides on the legs of shore crabs (*Pachygrapsus*, §14) as they dip in and out of the high pools.

Where pools have an accumulation of sand, clumps of small algae, and the like, the waving tentacles of a spionid worm, *Boccardia proboscidea* (§152), may be seen emerging from the bottom, although the rest of the worm is often difficult to find.

§9. A few other animals may occur in the high intertidal zone: occasional small clusters of leaf, or goose, barnacles (§160) and even small mussels (§158). These, however, achieve their greatest development in other zones and are therefore treated elsewhere.

*Current (1984) California law sets a daily limit of 35 for any invertebrate, including *Lottia*, not specifically covered by other regulations. Would-be collectors are also cautioned that laws are subject to change at any time, and it is advisable to check the latest regulations before collecting any marine life under the authority of the required sport-fishing license.

All the above are abundant and characteristic; they are fairly well restricted to the uppermost zone, and we should say that all except *Alloioplana* (which takes searching for) and the uncertain *Ligia* can be found and identified by the neophyte. They should be a pleasant-enough introduction for whoever cares to proceed farther into the tidelands.

Zone 2. High Intertidal

In this belt there is a greater variety of species and a tremendously greater number of individuals of many of the species represented than in the highest zone. Here there are some more or less permanent tide pools, with their active fauna of hermit crabs, snails, and crabs. Barnacles occur here also, certainly larger and possibly in greater abundance. Plants begin to appear in the uppermost part of this zone, and grow lush in its lower part. Most of the blue-green algae and some of the greens begin here and extend on down into the lower zones. The slim, brown rockweed *Pelvetia* begins at about the middle of Zone 2, and cuts off fairly sharply at the lower boundary. In mildly surf-beaten areas the first alga to appear may be the brown, crinkly *Endocladia*, taxonomically a red alga, which sparsely covers the granite ledges with tufted clusters (Fig. 19). Below this the more massive *Pelvetia* and *Fucus* appear (Fig. 20). These plants not only provide bases of attachment for sessile animals, but furnish shelter for others that adhere to the rock and protect them from desiccation (See Figs. 329–37 for several of the more common Pacific coast algae.)

In the highest zone one can collect dry-shod, but in this zone and downwards wet feet are likely to penalize the un-rubber-booted observer. However, it is sometimes safer to wear sneakers, or even old leather shoes, in an area heavily overgrown with slippery seaweeds, since rubber boots are somewhat clumsy. Even the coldest ocean seems to become warmer after a bit of immersion. It is a good rule, when exploring an unfamiliar seashore, to use one hand and both feet to assure stability.

§10. *Littorina scutulata* (Fig. 11), one of the periwinkles, takes up the burden here, where *L. keenae* leaves off, and covers the territory down to the lower reaches of the upper tide-pool zone, where *Tegula* becomes the dominant snail. The two species of *Littorina* may be found side by side; but of a thousand individuals, more than 80 percent will indicate this depth zoning. *L. scutulata*, for instance, never occurs more than a meter or so above high-tide line, whereas in surfy regions, *L. keenae* may be found 6 meters above the water. The two may be differentiated easily. *L. keenae* has an aperture bounded at the inner margin by a large flattened area; it is dingy gray in color, chunky in outline, and comparatively large (up to 2 cm high). The smaller, daintier *L. scutulata* is slim, has one more whorl to the shell, lacks the flattened area, and may be prettily checkered with white on shining brown or black. The range of *L. scutulata* is from Alaska, and that of *L. keenae* from Oregon, both south to Baja California.

Fig. 19. A high intertidal assembly: tufts of *Endocladia*; California mussels, *Mytilus californianus*; barnacles, *Semibalanus cariosus*; and scattered limpets, *Collisella digitalis* (§6). Some periwinkles (*Littorina* sp.), rock snails (*Nucella* sp.), and chitons (*Nuttallina*) can be seen along the bottom of the picture.

Fig. 20. The characteristic Zone 2 alga *Pelvetia*, with a dash of *Fucus*.

Smaller *Littorina scutulata* tend to occur lower on the shore, where certain of the rockweeds (*Pelvetia* or *Fucus*) apparently serve as a sort of nursery, for it is on their fronds and stems that the young are often found. Like *L. keenae*, *L. scutulata* feeds mainly on the film of diatoms and microscopic algae that coats the rocks, but since macroscopic algae are also present within its range, *L. scutulata* readily eats these too. In turn, *L. scutulata* is eaten by several predators, including gastropods of the genus *Acanthina*, the seastar *Leptasterias*, and shore crabs. Breeding has been reported during all seasons except summer, and, as in *L. keenae*, fertilized eggs are packaged in capsules; in this species, however, several eggs are present in each capsule.*

§11. A number of limpets find their optimum conditions in this zone, and scattered individuals may be found of others like *Collisella digitalis* (§6; Figs. 17, 19, 21) or types more abundant lower down (§25). A large, high-peaked, spectacular form, *Notoacmea persona* (Fig. 21), occurs high up, sometimes even in the uppermost zone. It avoids light and occurs in dark places—in deep crevices under boulders or (particularly) on the roofs of caverns—emerging to feed only at night. The exterior surface of the shell is olive-green and typically variegated with white dots sprinkled over the surface. Those on the anterior and anterolateral areas are particularly large and make a "butterfly" pattern in anterior view. These large spots are also translucent, permitting light to penetrate the shell, and there is evidence that the illumination so received may mediate the animal's tendency to avoid bright sunlight. An average specimen is 34 × 25 × 16 mm, and the life span probably exceeds 6 years. It ranges from the Aleutians to Morro Bay.

The dainty and relatively flat *Collisella scabra* (Fig. 21), the ribbed limpet, ranging from Cape Arago, Oregon, to southern Baja California, is regularly and deeply scalloped. Its average dimensions are 19 × 15 × 7 mm. Individuals occupy specific home sites at low tide, to which they return consistently after foraging on nearby rocks during periods of immersion. The food of *C. scabra*, microscopic algae and diatoms, is similar to that of *C. digitalis*, which also occurs on these high rocks, and competition for food likely exists when they co-occur. Competition is apparently lessened, however, by slight differences in the habitats that characterize the two species: *C. scabra* is found most often on horizontal or gently sloping surfaces, whereas *C. digitalis* favors vertical or overhanging rock faces.

Populations of *Collisella scabra* high in the intertidal differ in several respects from those lower on the shore. Sutherland (1970, 1972), working near Bodega Bay, found that in the high zone the density of limpets was

---

*This seems as good a place as any to note that the snails long known as *L. scutulata* have been recognized recently as two separate species, representatives of which are nearly identical in morphology. The two can be distinguished by the morphology of the egg capsules and penes, by electrophoresis of body proteins, and to a limited extent by shell morphometrics. The name *L. scutulata* has been retained for one of the species, and the name *L. plena* Gould revived for the second.

Fig. 21. Four common limpets of high intertidal rocks, all shown close to natural size: *left, Notoacmea persona* (§11); *below left, Collisella digitalis* (§6); *below center, C. scabra* (§11); *below right, C. pelta* (§11).

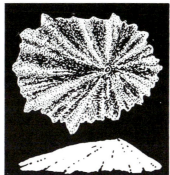

low, but that individuals there grew rapidly and reached a relatively large size; spawning of these limpets was distinctly seasonal, from January to March. In the low zone, limpet density was high. New recruits settled in greater numbers lower on the shore, and owing to the homing habit of *C. scabra*, densities there stayed high. Individuals grew slowly, presumably owing to competition for food, and reached a smaller maximum size than those few individuals that settled in the high zone; spawning activity in the low zone was less seasonal, occurring over a longer period.

The brown-and-white shield limpet, *Collisella pelta* (Fig. 21), also occurs in this zone. One of the most widely distributed limpets, its vertical distribution probably includes nearly the entire littoral. North of Crescent City it occurs high up with *C. digitalis*, but in many localities, such as around Monterey, it occurs farther down on the shore; it is a common inhabitant of mussel beds. The strong shell is highly variable in color pattern—exteriors of brown, green, or black all are common—and in texture

as well—ribs are usually present, but some shells are smooth. The range of
C. pelta is from northern Japan and the Aleutians to Punta Rompiente, Baja
California. A typical northern specimen measures 38 × 27 × 17 mm (to 40
mm in length); specimens in southern California are more commonly about
25 mm in length. The largest individuals are said to occur on the large, im-
movable rocks of reefs frequently covered by masses of *Ulva*, the sea lettuce, and on the surf-swept sea palm, *Postelsia*. Small black specimens frequently occur on brown algae such as *Egregia* where they may be confused
with *Notoacmea insessa*. When about 10 mm long, these specimens usually
move off the algae onto nearby rocks, change their color pattern, and develop ribbing. Unlike many limpets that graze primarily on microalgae or
encrusting forms, *C. pelta* usually eats larger, erect algae when they are
present, although it eats microalgae as well.

All the limpets illustrated are typical and representative specimens of
their species, and show such marked characteristics of form and texture
that identification, from the illustrations, of equally typical specimens seen
in life should be easy. The field observer is quite likely, however, to find
specimens with somewhat intermediate traits, and it is possible to arrange
the shells of a large number of limpets such that it is difficult to tell where
one species ends and the other begins.

§12. The pleasant and absurd hermit crabs are the clowns of the tide
pools. They rush about on the floors and sides of the rock pools, withdrawing instantly into their borrowed or stolen shells and dropping to the bottom at the least sign of danger. Picked up, they remain in the shell; but if
allowed to rest quietly in the palm of one's hand, they quickly protrude and
walk about. Among themselves, when they are not busy scavenging or
love-making, the gregarious "hermits" fight with tireless enthusiasm tem-
pered with caution. Despite the seeming viciousness of their battles, none,
apparently, are ever injured. When the vanquished has been surprised or
frightened into withdrawing its soft body from its shell, it is allowed to dart
back into it, or at least to snap its hindquarters into the shell discarded by
its conqueror.

It is a moot question whether or not hermit crabs have the grace to
wait until a snail is overcome by some fatal calamity before making off with
its shell. Many observers suppose that the house-hunting hermit may be
the very calamity responsible for the snail's demise, in which case the hermit would obtain a meal and a home by one master stroke. The British
naturalist J. H. Orton (1927), however, determined conclusively that the
common British hermits are unable to ingest large food, and that their fare
consists entirely of very small particles of food and debris. In the Monterey
Bay region a great many small hermits use the shells of *Tegula* of two species, snails that are strong, tough, and very ready to retreat inside their
shells, closing the entrance with the horny operculum. It would seem that
ingress to a *Tegula* shell occupied by a living animal determined to sell out
as dearly as possible would be very difficult, even if the hermit ate its way
in. Attacks have actually been observed in which the snail, by sawing the

rough edge of its shell back and forth across the hermit's claw, convinced the hermit that that particular shell was not for rent.*

Even when the hermit finds a suitable house, its troubles are far from over, for at intervals it must find larger shells to accommodate its growing body. It is a lifelong job and one in which it seems never to lose interest. A few hermits and some spare shells in an aquarium will provide the watcher with hours of amusement. Inspecting a new shell—and every shell or similar object is a prospect—involves an unvarying sacred ritual: touch it, grasp it, rotate it until the orifice is in a position to be explored with the antennae, and then, if it seems satisfactory, move in. The actual move is made so quickly that only a sharp eye can follow it.

Precisely what kind of shell is satisfactory, or at least preferred, may vary with the species of hermit. In California, *Pagurus samuelis* (see below) is partial to *Tegula* shells, whereas *P. hirsutiusculus* (§201) occurs more often in the shells of *Nucella* or *Olivella*. Of course, which shells are available also plays a role. At Coyote Point in San Francisco Bay, for example, what is now available to the local hermits are the shells of several introduced snails that were not present in the Bay a century ago: as a result, these alien, but available, shells have become the most common homes of the native hermit crabs. In this general connection, it should be noted too that in some locations a lack of suitable shells may limit the population density or the size distribution of hermits.

Along the California coast the commonest hermit in the upper tide pools is *Pagurus samuelis* (Fig. 22), a small crab with bright-red antennae and brilliant-blue bands around the tips of its feet. At Pacific Grove there are ten of these to every one of the granular-clawed *P. granosimanus* (§196), which tend to occupy a somewhat lower position on the shore. In the north, however, the latter is larger and more abundant, apparently having a dominant range in Puget Sound and to the north, although its total range is from Alaska to Mexico. *P. samuelis*, on the other hand, having about the same total range (in this case, British Columbia to Punta Eugenia, Baja California), is most numerous along the California coast. We have taken ovigerous (egg-bearing) specimens of *P. samuelis* from early May through July, and of *P. granosimanus* in late May, both in central California; in southern California, egg-bearing females of *P. samuelis* have been found throughout much of the year. During the breeding season, a courting male may carry about a mature female for a day or more at a time, periodically butting shells with her repeatedly, before mating. With the courtship complete, actual consummation takes less than a second, during which time the animals extend most of the way out of their shells. Eggs, as shown in the photograph of another hermit (Fig. 23), are carried on the female's abdomen—a region always protected by the shell, but provided with a continuous current of water for aeration by the action of the abdominal appendages.

---

*Although there exist rare reports of hermit crabs killing large snails (see Hazlett 1981), our species do not attempt this method of acquiring a home.

Fig. 22. *Pagurus samuelis* (§12) in the shell of *Calliostoma annulatum* (§182); about natural size. This hermit, the most common in upper tide pools of California, has bright-red antennae and blue bands around the tips of its legs.

Fig. 23. An egg-bearing female hermit crab, *Pagurus hemphilli*, removed from her borrowed shell; about twice natural size. This brownish-red hermit (§28) is characteristic of middle and lower tide-pool regions.

§13. The original residents of the homes so coveted by hermit crabs, the turban snails of the genus *Tegula*, often congregate in great clusters in crevices or on the sides of small boulders. The shells are pretty enough when wet; dry, they have a dingy-gray color that blends very well with the dry rocks. Older specimens usually have shells that are badly eroded, owing to mechanical abrasion and the activities of an endolithic marine fungus. The black turban, *Tegula funebralis* (Figs. 24, 41), ranging from Vancouver Island to central Baja California, is the dominant type of the upper level. Lower down, *T. funebralis* is replaced by the brown turban (§29).

Samples of *Tegula funebralis* taken at successive tide levels reveal a definite size gradient, with larger snails being found lower on the shore. According to work done in Washington by Paine (1969), juvenile *T. funebralis* apparently settle high on the shore, where they live for 5 to 7 years; then, as the snails age, they slowly migrate to lower levels. The distribution is maintained, other workers have found, by behavioral means. If groups of snails collected from various shore levels are released together on a rock face, they will sort themselves out, reestablishing a size gradient within one complete tidal cycle.

Growth is fairly slow, although the rate varies with geographic location, being generally faster to the south. In Oregon, snails take 2 years to reach a shell diameter of 10 mm and about 6 years to reach 20 mm. The rate declines rapidly in older snails, and specimens attaining the maximum size of 30 mm may be 30 years old.

Attaining this age is not easy, or common, since *Tegula* has many predators. Among these are sea otters, octopuses, the red rock crab (*Cancer antennarius*, §105), and seastars, especially *Pisaster ochraceus* (§157). Yet the snail is not without its defenses. Minute quantities of chemicals emanating

Fig. 24. Black turban snails, *Tegula funebralis* (§13). The wet shells at the far left are a brilliant black, the others an inconspicuous gray.

from predatory seastars and crabs are detected by the snail and stimulate it to seek higher ground, above the predators' foraging range; in tide pools, snails often leave the water entirely to avoid contact with a predator. If contact is made, the snail flees the premises immediately, its many tentacles lashing wildly as it attempts to escape. The attempts are not always successful, of course, and in many locations a heavy toll is taken, with the black turban the most important item in the predator's diet. *Tegula* in turn eats many species of algae, harvested with its elaborate radula (Fig. 25). Beach wrack, various microalgae, and the microscopic film that covers the rocks are all included in the snail's catholic diet. Given a choice, however, *Tegula* seems to prefer soft, fleshy algae such as *Gigartina* and *Macrocystis*.

Italians consider *Tegula* fine food. The animals are cooked in oil and served in the shell, the bodies removed with a pin as they are eaten. The lucky Americans who can overcome their food prejudices in this and other respects will at least achieve a greater gastronomic independence and may even develop an epicurean appreciation of many of the intertidal delicacies. They will also no doubt contribute to the declining fauna in such places as Moss Beach. Certainly a long-lived, slow-growing snail such as *Tegula* cannot withstand an intensive fishery; the current (1984) bag limit for *Tegula* is 35 in California.

Fig. 25. The radula of *Tegula funebralis* as it would appear when protruded from the snail's mouth, showing the elaborate arrangement of hooked teeth; this scanning electron micrograph is 55 times natural size.

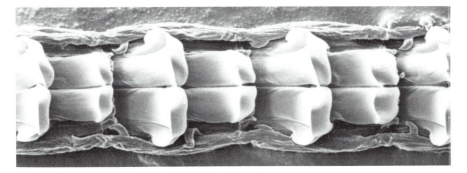

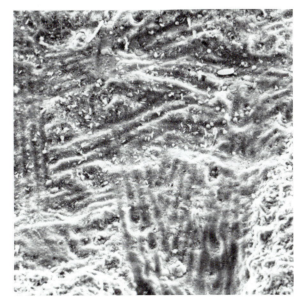

Fig. 26. *Left,* graze marks on the shell of *Tegula* produced by the radula of *Collisella asmi,* a commensal limpet occurring on turban snails. *Above,* the radula of *C. asmi.* Both scanning electron micrographs are 325 times natural size. The striking differences between the radulae of *Tegula* and *Collisella* indicate the importance of this structure as a taxonomic character (see also Figs. 9, 148, 185).

The brown turban (*Tegula brunnea*), more often than the black, carries specimens of the commensal slipper limpet *Crepidula*, whose shell apex is hooked in contrast to that of the true limpets. Both *Tegula* species frequently also carry small specimens of several true limpets, including *Collisella asmi*, *C. digitalis*, *C. pelta*, and *C. paradigitalis*. All except *C. asmi* are juveniles that soon move off the turban snail, presumably when food becomes scarce or when their increasing size makes a tight fit impossible on the host's curved shell. Of the true limpets only *C. asmi* is specifically associated with *Tegula* and reaches adult size on its host. This small, tall limpet (shell to 11 mm long, to 8 mm high), with a dark-brown or black shell, feeds exclusively on the film of microscopic plants that coat the host's shell (Fig. 26). Although largely restricted to *Tegula* (but occurring occasionally on the bivalve *Mytilus californianus*), this limpet is not averse to transferring from one snail to another, which it likely does when aggregations of the host form during low tide. The range of *C. asmi* is Alaska to Baja California.

§14. A third obvious and distinctly "visible" animal occurring in this zone is the pugnacious little rock crab *Pachygrapsus crassipes*, without which any rocky beach must seem lonesome and quiet (Fig. 27). These dark-red or green square-shelled crabs run sideways or backward when alarmed— their eyesight is excellent—scurrying away before the intruder or rearing up and offering to fight all comers. Common in the rocky intertidal from Charleston, Oregon, to the Gulf of California, they may be found on top of the rocks at night and hiding in crevices and pools during the day. Their diet consists mainly of minute algae and diatoms scraped from the surface with their chelae, but the variety of foods consumed is great: larger algae, dead animal matter, and occasionally living prey, including limpets, hermit crabs, and isopods—even an unwary beach fly that settles too close. Certainly, to see a group of them attack a discarded apple core is to understand one method by which the rock pools are kept clear of all foreign matter that is to any degree edible. They and the beach hoppers are the most active scavengers in this particular ecological association.

Since *Pachygrapsus* spends half or more of its time out of water, it is, as expected, well adapted to semiterrestrial life. These crabs tolerate wide ranges of temperature and salinity, and can survive nearly 3 days out of water, drawing on water retained in the gill chamber.

Fig. 27. The rock crab *Pachygrapsus crassipes* (§14), a squarish dark-red, green, or striped crab, is a common and pugnacious inhabitant of the high intertidal; about natural size.

Females carry their eggs, nearly 50,000 of them in an average brood, attached to the underside of the recurved tail. The peak reproductive season seems to be in the summer, although ovigerous females have been reported in California from February to October.

§15. Shore observers in the La Jolla region will find, on and under fairly bare rocks on the upper shore, the small, clean-cut, and rather beautifully marked volcano-shell limpet, *Fissurella volcano*. For some obscure reason this animal moves down into the low intertidal as it ranges northward. It has been reported from Crescent City to Bahía Magdalena, Baja California. The shell of *Fissurella* is pink between the red or purple ridges that radiate from the small opening at the apex. The opening, or "keyhole," serves as an exit for wastes and for water that has passed over the gills in the mantle cavity. This general arrangement of plumbing, shared with the abalones but different from that of the true limpets, is considered a primitive feature.

§16. In the muddy or sandy layer between closely adjacent rocks, or in the substratum, the upper tide pools from central California to Baja California are characterized by the presence of the hairy-gilled worm *Cirriformia luxuriosa* (Fig. 28). Also common in the middle and low intertidal zones, *Cirriformia* may be found in bits of the blackest and foulest mud. The gills of this worm, which resemble thin, reddish roundworms, will often be noticed on the substratum, as will the orange-red feeding tentacles. While the gills handle respiratory chores, the tentacles gather minute organisms and organic detritus for food, which is passed along a whitish, ciliated groove in each tentacle to the anterior of the buried worm. Both structures are withdrawn when the animal is disturbed. This deployment of respiratory and feeding structures is particularly efficient under the circumstances; it gives the worm an opportunity to burrow in the mud for protection, while leaving its "lungs" and tentacles behind, at the surface. *Cirriformia* is thus one of the few animals that can live in an environment of foul mud; the noxious substratum that repels others protects this animal from enemies and desiccation. The tentacles are exposed to the many species of intertidal fish that roam the surface, of course, and one would think they would provide a feast for the interested diner. However, only one fish, the high cockscomb (*Anoplarchus*, §240), apparently partakes regularly of the banquet, suggesting that the exposed tentacles may contain some distasteful chemical included for defense.

Our collecting records show that sexually mature specimens of *Cirriformia* have been observed in June under interesting conditions. We quote from a report dated June 21, 1927, from 9:30 to 10:30 P.M.: "For the past several evenings I have noticed that the *Cirriformia* were very active in mud-bottom pools of the upper tide flats. Some of them lie half exposed, whereas only the tentacles are protruded normally. Tonight I have noticed further that the shallow pools containing them were murky and milky. Then I noted some of the worms almost fully out of the substratum, exud-

Fig. 28. The hairy-gilled *Cirriformia luxuriosa* (§16), with red gills and orange-red feeding tentacles that extend to the surface when the animal is buried; a bit greater than natural size.

ing milky fluid or more solid white 'castings.' I took two worms back to the laboratory, examined the exudate, and found it to be spermatic fluid." Eggs were not specifically observed at this time, but others have documented the spawning of this species in southern California in June and July.

§17. Under flat rocks in the upper tide-pool region (but possibly as abundant in the middle and low tide pools) may be found the polyclad flatworm *Notoplana acticola* (Fig. 29), one of the most common polyclads of rocky shores, at least throughout California. It is tan in color, with darker markings around the midline. Large specimens may exceed 2 cm in length, although the average will be nearer to half this. By examining the pharynx contents, Heath found that the food of many flatworms consists of one-celled plants, spores of algae, microcrustaceans, and worm larvae. Boone found that captive *N. acticola* fed on the red nudibranch *Rostanga*.

These worms (and other species) crawl about on the damp undersides of freshly moved rocks, appearing to move much as a drop of glycerine flows down the side of a glass dish. This unique motion is achieved both by muscular creeping and by cilia, which propel the animal on a secreted layer of mucus. The animal's grip is so effective that it is difficult to remove the soft-bodied worms from the rock without damage. We use a camel's-hair brush, to which they adhere as firmly as to the rock unless they are immediately cast into a jar of water.

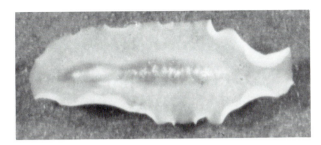

Fig. 29. The flatworm *Notoplana acticola* (§17), a tan polyclad usually about 1 cm long, is common under rocks from the high zone to low intertidal pools.

On the California coast near Dillon Beach, Thum (1974) found that the majority of mature *N. acticola* were functional hermaphrodites (eggs and sperm both present in one individual) throughout the year. Eggs were laid chiefly during the summer.

§18. From northern British Columbia to Point Conception *Petrolisthes cinctipes*, a porcelain crab (Fig. 30), is also a common and characteristic inhabitant of the under-rock associations, although, like *Notoplana*, it is almost as characteristic of the middle zone. *Petrolisthes* is a filter feeder, diatoms and organic debris being strained from the water by fanning motions of the crab's hairy maxillipeds. A large specimen will measure 14 mm across the carapace, and it is these small, flat crabs that scurry about so feverishly for cover when a stone is upended. Their pronounced flatness makes it easy for them to creep into crevices, and they avoid imprisonment by throwing off, or autotomizing, a claw or walking leg at the slightest provocation. Indeed, *Petrolisthes* is one of the champion autotomizers of this coast—no mean distinction, for many animals have this peculiar ability.

When autotomizing, a crab casts off its limb voluntarily; an automatic reaction instantly closes the severed blood vessels to prevent bleeding to death. This process takes place at a prearranged breaking point near the base of the limb, marked externally by a ringlike groove. The muscles and tendons there are adapted to facilitate breakage; and after autotomy a membrane consisting of two flaps is forced across the stump by the congestion of blood at that point. Regeneration begins at once, and a miniature limb is soon formed, though the new limb undergoes no further growth until the next molt, when it grows rapidly to several times its previous size.

Female crabs will often be found with large egg masses under their curved and extended abdomens; ovigerous *Petrolisthes* in central California have been taken in all months except November. These eggs do not hatch into young crabs, but into fantastic larval forms that shore visitors will never see except in photographs (Fig. 31). Animals in the first larval stages, called zoeae, are minute, transparent organisms swimming at the surface of the sea. They seem so totally unrelated to their parents and to their own later stages that zoologists long mistook them for an independent species of animal. The pelagic zoea of *Petrolisthes*, somewhat resembling a preposterous unicorn, is a wondrous sight under a microscope. Projecting

Fig. 30. *Petrolisthes cinctipes*, a porcelain crab (§18), the most abundant crab under rocks in Zones 2 and 3. This flat reddish-brown crab is a champion at autotomizing its limbs, which may become pinned under shifting rocks.

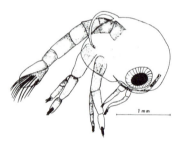

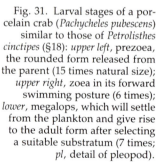

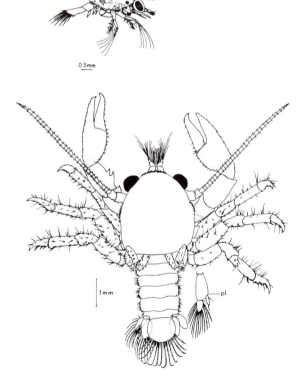

Fig. 31. Larval stages of a porcelain crab (*Pachycheles pubescens*) similar to those of *Petrolisthes cinctipes* (§18): *upper left,* prezoea, the rounded form released from the parent (15 times natural size); *upper right,* zoea in its forward swimming posture (6 times); *lower,* megalops, which will settle from the plankton and give rise to the adult form after selecting a suitable substratum (7 times; *pl,* detail of pleopod).

from the front of the carapace is an enormous spine, often longer than all the rest of the animal. If this spine has any function, it is, presumably, to make ingestion by the zoea's enemies difficult. An enterprising pup that had just attacked its first porcupine would understand the principle involved. There are also two long posterior spines on each side, which protect the larva from behind. After casting its external shell several times as it grows in size, the zoea changes to still another form, the megalops (also formerly regarded as a separate genus, *Megalopa*). The megalops, related to a crab in much the same way that a tadpole is related to a frog, molts several times, finally emerging as a recognizable young crab.

   Petrolisthes has in recent years become a favorite in the curio business, since its flat shape makes it an attractive subject for embedding in blocks of plastic for key rings, pendants, and the like. The bright-red color of these embalmed specimens is due to a preservative or other treatment; the color in life is closer to brown.

   §19. Of the common and shoreward-extending brittle stars, the dainty black-and-white (sometimes gray) *Amphipholis pugetana* (Fig. 32) is perhaps the highest, being characteristically an upper tide-pool animal. Since large adults measure only about 2 cm in arm spread, it is further distinguished by being one of the smallest of all Pacific coast ophiurans, although it is a rather hardy form that does not autotomize readily, unlike many of the other brittle stars. *Amphipholis* occurs in great beds, so commonly that it would be difficult to find a suitable rock in this zone that hid no specimens; but it is never found with intertwined arms, as is its larger relative *Amphiodia* (§44; see the discussion there for a possible significance of this concentration of population).

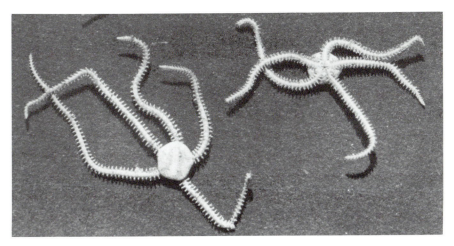

Fig. 32. Dorsal and ventral views of *Amphipholis pugetana* (§19), one of the smallest brittle stars of the Pacific coast.

The recorded range of *Amphipholis pugetana* is from Alaska to San Diego; but it extends farther south, for we have also taken it, as well as the similar and cosmopolitan *A. squamata*, in northern Baja California. The two species of *Amphipholis* may usually be distinguished by examining the length of the arms: those of *A. pugetana* are 7 to 8 times the diameter of the disk, whereas in *A. squamata* they are 3 to 4 times the disk diameter. In addition, the latter, which is by far the more common species in the Monterey Bay area, broods its young; the former does not.

§20. Representatives of two related groups of crustaceans, the amphipods and the isopods, are found under practically every embedded rock and boulder in the upper tide-pool zone, and farther down also. The amphipods, a group containing the sand hoppers, or beach fleas, are, as their popular names imply, lively animals. Being compressed laterally, so that they seem to stand on edge, they are structurally adapted to the jumping careers that they follow with such seeming enthusiasm. *Melita sulca*, with a recorded range of Friday Harbor to Isla Cedros, Baja California, is almost certain to be present under most of the suitable rocks down as far as the middle tide pools. But then, over 150 species of amphipods have been described from California alone; and, although detailed keys are available for the serious collector, identification of most of our species is best left to a specialist. Some examples of amphipod genera common at various tide levels on Pacific rocky shores are shown in Figure 33.

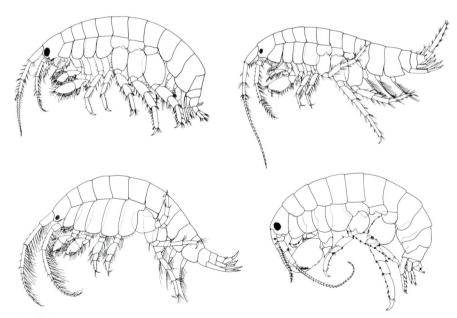

Fig. 33. Representative genera of gammarid amphipods on rocky shores (§20). *Clockwise from top left:* Elasmopus, Melita, Hyale, and Photis. Bodies usually are 1 cm long or less.

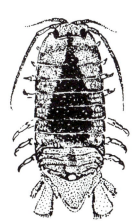

Fig. 34. Drawing and photograph of *Cirolana harfordi* (§21), an isopod common along the entire Pacific coast. This drab pale-gray to nearly black crustacean is found under rocks at all tide levels from the high zone down.

§21. The isopods, being flattened bellywise, or dorsoventrally, are crawlers, usually slow ones. These are close relatives of the terrestrial pill bugs, so called from their habit of curling up into little balls when disturbed; they are also known as sow bugs because of their resemblance to swine. Representatives of the group are common on land, where they are found in damp places—under boards lying on the ground, for example. The drab *Cirolana harfordi* (Fig. 34), a common inhabitant of the entire coastline from British Columbia to at least Ensenada, occurs most abundantly under rocks in the high intertidal but will be found in other zones as well. This isopod, up to 2 cm in length, is a voracious carnivore, feeding on polychaete worms, amphipods, and any dead animals encountered. Its food, located over a distance by chemoreception, is consumed in great chunks, and hordes of *Cirolana* attracted to a dead fish have stripped the bones clean in a short time. Once gorged with food, the animal retires to a protected spot and may subsist on this one meal for a month or more if no other opportunity arises; digestion is slow. This combination of habit and physiology seems adaptive in that it maximizes food intake while minimizing the risk of potentially dangerous situations, such as exposure to the several species of predatory fish that feed on this and other isopods. *Cirolana* females bearing clusters of 18 to 68 young in a ventral brood pouch have been reported during the spring in central California.

§22. Small, round boulders that are always immersed in the pools are very likely to be covered with the tiny, tightly coiled calcareous tubes of one or more species of a cosmopolitan group of serpulid worms known as spirorbids (Fig. 35), which includes species of *Spirorbis* and several other genera. The almost microscopic red gills protrude from the tubes to form

ones commonly cover themselves with bits of gravel or shell, so that when contracted they are part of the background. Many a weary shore visitor has rested on a bit of innocent-looking rock ledge, only to discover that a couple of centimeters or so of extreme and very contagious wetness separated him from the bare rock (Fig. 37). Those bits of adherent material are not there to deceive us, of course, but rather to shield the anemone's body from solar radiation at low tide and to reduce desiccation, a factor implicated in limiting the distribution of this animal in some locations.

These animals, which occur from Alaska well into Baja California, are not restricted to pure seawater, for they live well where there is industrial pollution or sewage, their hardihood belying their apparent delicacy. They can be kept in marine aquaria, even where there is no running seawater, provided only that the water is aerated or changed once or twice a day. Contracted, *Anthopleura* is rarely more than 4 cm in diameter; expanded, the disk may be more than twice that size and may display rather dainty colors over a groundwork of solid green. The tentacles are often banded with pink or lavender, and lines of color frequently radiate from the mouth to the edges of the transparent disk, marking the position of the partitions (mesenteries) inside.

Fig. 37. A dense cover of the aggregated anemone *Anthopleura elegantissima* (§24) on a large rock. This anemone, which is green with tinges of pink on the tentacles, contains thousands of photosynthetic symbionts.

Most anemones, including *Anthopleura*, are voracious feeders, and when hungry they will ingest almost anything offered. Their tentacles are armed with minute, stinging structures called nematocysts, which paralyze any small animals so unfortunate as to touch them. We have many times seen small shore crabs disappearing down the anemone's gullet, and as often have seen an anemone spew out bits of crabs' shells. The surprising rapidity of an anemone's digestion was illustrated by specimens of another species, *Metridium* (§321), with which we experimented at Pender Harbor, British Columbia. One half of a small chiton that had been cut in two was placed on the disk of an anemone in a glass dish. The disk was immediately depressed at the center, so that the food disappeared into the body cavity. Within 15 minutes the cleaned shell of the chiton, all the meat completely digested away, was disgorged through the mouth—a striking tribute to the potency of anemone enzymes.

Although *Anthopleura* consumes food in a characteristically animal fashion, it also has a more productive, plantlike side, for living within the anemone's tissues are thousands of photosynthetic, plantlike symbionts, either dinoflagellates (zooxanthellae), unicellular algae (zoochlorellae), or both. These symbionts, which contribute to the anemone's nutrition, place *Anthopleura* in an unusual dual role in the budgetary affairs of intertidal communities—that of producer as well as consumer. Indeed, the photosynthetic rate of its symbionts is such that *Anthopleura* contributes to primary productivity at a rate comparable to that of many intertidal algae. In keeping with the symbionts' need for illumination, anemones with symbiotic algae respond to lighting conditions: they move either toward or away from light, until some optimum intensity is found. Such anemones also expand in moderate light, exposing more photosynthetic surface, and they contract in darkness. Anemones without symbionts, pale forms collected from deeply shaded habitats, seem indifferent to light.

Anthopleura reproduces sexually, by means of gametes released separately by males and females during the summer and fall, and also asexually, by longitudinal fission. Both reproductive methods have their benefits. Sexual reproduction results in new combinations of genes and in larvae dispersed away from the parents, perhaps to succeed in new and different territories. If the animal is successful in finding a suitable habitat, asexual reproduction then turns out multiple copies of the individual and spreads the successful animal rapidly across the rocks. In this manner, a single settling larva may give rise to great aggregations of individuals, all genetically the same, and thus all having the same sex and patterns of coloration. On a large boulder, one can often tell where one of these aggregations of genetically identical individuals (a clone) has expanded until its boundary has touched that of another. Between the two clones is a corridor of bare rock, which, as Francis (1973a,b, 1976) has described, is a war zone (see Fig. 344). Anemones on the periphery of the neighboring clones exhibit aggressive behavior toward non-clonemates, employing knoblike swellings (acrorhagi) loaded with nematocysts to sting their opponents. Considerable damage can be inflicted, and a cost is exacted from anemones on the periphery of both clones: they are usually smaller than anemones from the center and they are without gonads when other anemones are ripe. The clone as a whole, however, presumably benefits from the sacrifices of these "warriors" who share the common genotype.

The growth rate of *Anthopleura*, like that of many other anemones, varies directly with food supply, within limits. At Friday Harbor, specimens of *A. elegantissima* grow fastest from April to August and may actually decrease in size from September to March. Interestingly, asexual reproduction is most frequent during the lean months and may be triggered by a reduction in food supply. (See Sebens 1982a,b for details of the interplay of size, growth, and reproduction.)

Although not much is known specifically about the life span of *Anthopleura*, several examples of extreme longevity have been observed in this group. Professor Ashworth tells of several anemones that were donated to the University of Edinburgh in 1900 by a woman who had collected them, already fully grown, 30 years before. She had kept them throughout that period in a round glass aquarium, strictly observing a daily rite of aerating the water with a dipper and a weekly rite of feeding them on fresh liver, which she believed they preferred to anything else. When they came into the possession of the University they were fed, possibly not so regularly, any scraps of protein that came to hand, such as shredded crab meat or even beefsteak. Nevertheless, these anemones continued on in the best of health, annually producing clouds of sperm and eggs. Unfortunately, this experiment in longevity was brought to an untimely end after some 80 years by the ineptitude of (we understand) a botanist; but there seemed to be no reason why the animals might not have lived a century or more.

§25. Also occurring more or less exposed on the tops and sides of rocks are a number of forms more properly treated elsewhere. Occasional acorn barnacles extend their range down into this middle zone when competition and predation are not too keen, and such open-coast forms as the red barnacle (§168) and the common tough-skinned seastar (§157) may be locally plentiful, depending on the exposure. There may be clusters of mussels and goose barnacles, and the small chiton *Nuttallina* is occasionally found.

Three common limpets may be found at this level on the protected outer coast. The large *Notoacmea scutum* (Figs. 38, 360), the plate limpet, is the flattest of the tribe. The shell color is variable, but clean specimens are often brownish with white spots or lines. Frequently the shell has a greenish cast, and sometimes long trailers of the stringy green algae *Enteromorpha* and *Ulva* are attached. This is an active limpet, which moves up and down the shore with the tide, some individuals traveling nearly 2 meters between successive low tides; as with several of its brethren, activity is greater at night. The plate limpet's food consists of microscopic algae when on otherwise barren rocks, but when encrusting red algae are available,

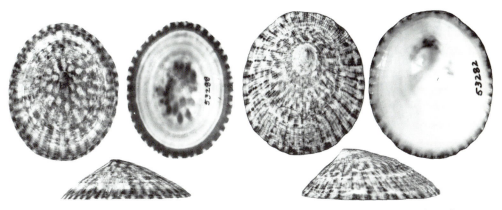

Fig. 38. Limpets of the middle zone: *left*, Notoacmea scutum (§25), the plate limpet, whose flat shell is often brownish with white spots; *right*, *Collisella limatula* (§25), the file limpet, with coarse imbricate ribs on the shell and black on the sides of the foot.

these are preferred. Growth is highly seasonal in central California, highest from the late spring through early summer and lowest in the winter; in contrast, spawning apparently occurs throughout the year. Near Bodega Bay, limpets reach maturity and grow to approximately 17 mm in length by the end of their first year, reaching 26 mm by the end of their second year. Predators include crabs, shorebirds (e.g., oystercatchers), and seastars; like *Tegula* (§13), *N. scutum* flees in a vigorous, running escape response upon contact with seastars. Even before contact, the limpet may sense the predator and respond, most commonly by moving up the nearest vertical surface and away from the seastar.

Notoacmea scutum ranges north from San Pedro in southern California all the way to Alaska and on to northern Japan. It is one of the commonest large limpets in northern California. South of Monterey, however, it becomes increasingly rare; and south of Point Conception it is replaced almost entirely by *Collisella limatula*, former literary references to the contrary notwithstanding (most southern *N. scutum* identifications were incorrect). Shells of both species are thin and flat, but *N. scutum* particularly has a knife-edge shell easily broken in removal, and it lacks the black sides of the foot so characteristic of the other. An average specimen of *N. scutum* measures 33 × 25 × 9 mm.

Collisella limatula (Fig. 38), frequently mistakenly called *Acmaea scabra* in the past, is sculptured with coarse and imbricate ribs (hence the name "file limpet"), which are sometimes obscured by algae. A diagnostic feature of this species is the characteristic deep-black color around the side of the foot, the bottom being contrastingly white. Inactive when exposed, these limpets begin to move when splashed with water by the rising tide. During their excursions they feed on microscopic algae and on larger encrusting forms, but rarely on erect algae, an important food of *Collisella pelta* (§11), which often occurs in the same area. *C. limatula* is reported to range from Newport, Oregon, to southern Baja California, becoming increasingly common south of San Francisco. The average specimen is 29 × 23 × 9 mm, and maximum size is attained in 2 to 3 years, assuming, of course, that the limpet successfully avoids its predators, to which it responds in ways similar to *N. scutum*.

A southern species, similar but not closely related to *Collisella scabra*, was described by A. R. (Grant) Test as *C. conus* (Test 1945, 1946). It is ivory-colored to brown and, with *C. limatula*, replaces *N. scutum* on most reefs south of Point Conception. Its average dimensions are 15 × 12 × 6 mm.

Occasional specimens of any of the limpets previously mentioned (§§6, 11) may turn up here; *Notoacmea fenestrata* and *Acmaea mitra* of the lowest zone (§117) are also found.

Protected-Rock Habitat

Animals here are attached to, or living on, protected rocks and rockweed. Many of the commonest inhabitants of this subregion can scarcely be considered obvious, since in order to see them advantageously one must almost lie down in the pools, looking up into the crevices and under ledges.

§26. The large red, yellow, or purple seastar *Patiria miniata* (Fig. 39), locally called "sea bat" because of its webbed rays, is quite possibly the most obvious animal in this association, extending downward to deep water. Usually with five rays (but sometimes with four to nine), this seastar is typically an omnivore or scavenger, feeding by extending its voluminous stomach over a great variety of sessile or dead plants and animals. Because it is readily available, because it has an unusually long breeding season, and because it extrudes its sexual products so obligingly, *Patiria* is used extensively for embryological experimentation. Some male and female specimens, when laid out on wet seaweed, discharge their ripe sperm and eggs at almost any time of the year, but especially during the late winter and spring. With ordinary aseptic precautions, the ova can be fertilized by fresh sperm; and if spread out in scrupulously clean glass dishes, they will develop overnight into motile embryos and later into minute larvae, which swim by vibrating their cilia. The young seastars develop inside these larvae, gradually absorbing the larval form as they grow. After the adult specimens have shed their sexual products, they should be put back alive into the tide pools to prevent depletion of the race. In the Monterey region, and probably because of collecting activities, this interesting animal is no longer found in the great congregations that formerly occupied the middle tide pools, although it still occurs in large numbers in the outermost pools and reefs. The recorded range is from Sitka to Islas de Revillagigedo, Mexico; but south of Point Conception the specimens are few, small, and stunted, and are almost invariably found under rocks. The bat star is rare intertidally on the Oregon coast.

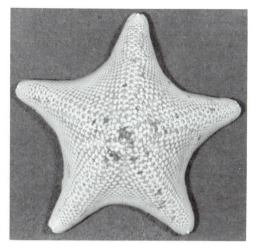

Fig. 39. The bat star, *Patiria miniata* (§26), a common resident of protected-rock habitats. *Above,* two specimens reflect some of the great variation in pattern and color, from yellow to red to purple, characteristic of this species; natural size. *Left,* the oral surface, showing the commensal polychaete *Ophiodromus pugettensis* (§26).

A polychaete, *Ophiodromus pugettensis* (formerly *Podarke*), often lives commensally on the oral (or under) surface of *Patiria* (Fig. 39). It usually frequents the ambulacral grooves, but may crawl actively over the whole oral surface. There may be several worms, up to 20, on an individual seastar, and worms move regularly between host individuals. The hosts are apparently located by means of scent, some waterborne chemical "essence" of *Patiria*. Although most likely to be encountered first as a commensal on *Patiria*, *Ophiodromus* also occurs as a free-living polychaete among fouling organisms in the lower intertidal and on mud bottoms in quiet water. Such free-living specimens collected from mud in Garrison Bay, San Juan Island, had a diet consisting primarily of harpacticoid copepods, but also of a wide variety of other small benthic invertebrates. Curiously, free-living *Ophiodromus* specimens show no interest, at least in the laboratory, in mounting host seastars and becoming commensals.

Fig. 40. The purple shore crab, *Hemigrapsus nudus* (§27), a dominant crab of the middle intertidal; natural size. Purple spots on the claws differentiate *H. nudus* from the smaller and slimmer *Pachygrapsus* (§14) of higher rocks.

§27. The little square-shelled rock crab *Pachygrapsus* is still abundant, but the purple shore crab, *Hemigrapsus nudus* (also §203), is the dominant representative of this group in the middle tide-pool region, and it also extends into the next lower zone. *Hemigrapsus* (Fig. 40) is as characteristic of the rockweed zone as *Pachygrapsus* is of the naked-rock zone higher up, or as *Cancer antennarius* (§105) is of the lower tide-pool zone. The purple spots on the claws of *Hemigrapsus*, particularly noticeable on their white undersides, differentiate it from the smaller and slimmer *Pachygrapsus*. Egg-bearing females of *Hemigrapsus* have been taken from November to April in Monterey Bay, but in the Puget Sound area the females are ovigerous in early summer. The number of eggs in a brood increases with the size of the female, 13,000 being average for the one brood produced each year. The purple shore crab's range is from Yakobi Island, Alaska, to Bahía Tortugas, Baja California. It, like *Pachygrapsus*, is primarily herbivorous.

§28. Hermit crabs are also abundant in this zone. Specimens of *Pagurus samuelis* larger than those encountered in the upper tide-pool region (§12) can be found, but even the large ones are small in comparison with the average *P. hemphilli* (Fig. 23), a species characteristic of the middle and lower tide-pool regions. The stalwart and clean-cut *P. hemphilli*, up to 5 cm long and colored a brownish red dotted with tiny blue granules, can be recognized by the laterally compressed wrist of its big claw, which subtends a sharp angle with the upper surface. Other hermits have the top of the wrist more or less rounded. Also, this is the only reddish hermit common intertidally, and the tips of the walking legs are yellow. Egg-bearing females of *P. hemphilli* have been taken in February and March. This hermit ranges from Alaska to the Channel Islands, occupying the shells of large *Tegula* almost to the exclusion of anything else.

Hermit crabs are sometimes afflicted with rhizocephalan parasites (most likely *Peltogasterella gracilis*) that produce long white sacs on the abdomen. These are actually specialized barnacles that gain access to their host as free-swimming cyprid larvae scarcely distinguishable from those of other barnacles. Once attached, the animal loses all its appendages and becomes a sac of generative organs devoted solely to the production of eggs and sperm, food and respiration being provided by the host. Hermit crabs so afflicted are not common on the Pacific coast of North America, and they are not easy to find, since the hermit's shell must be crushed in a vise to determine whether or not the animal is affected.

§29. In the middle zone the brown turban snail, *Tegula brunnea* (Fig. 41), ranging from Cape Arago, Oregon, to southern California, takes the place of *T. funebralis* (§13), its black brother in the upper zone, although each overlaps the territory of the other to some extent. Both species are definitely larger in the middle zone, however, and both are more often found carrying specimens of another snail, *Crepidula adunca* (Fig. 42), the hooked slipper snail. This small, dark-brown form, with its sharply recurved hook at the shell apex, looks not unlike a limpet, and it affixes itself with limpetlike strength to the *Tegula* that carries it. Mature *Crepidula* snails lead an entirely sedentary life, straining food from the water that passes over the gill. Often a small specimen will be seen on top of a larger one. Almost without exception, the smaller individual will be a male and the larger a female, for these snails are protandric hermaphrodites, undergoing sex changes from male to bisexual to female as they grow. Breeding females, which can be found year-round in California, carry eggs in the shell cavity, where they can be seen if the animal is pried from its support. With a hand lens the embryos of *Crepidula* can sometimes be seen whirling around in their envelopes in the egg packets. The hooked slipper snail is known to occur from the Queen Charlotte Islands, British Columbia, to northern Baja California.

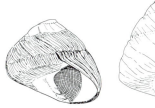 

Fig. 41. Turban snails of the genus *Tegula*: *left*, *T. funebralis* (§13), the black turban; *right*, *T. brunnea* (§29), the brown turban. *T. brunnea* is the more common in Zone 3. Both are slightly reduced.

Fig. 42. Two slipper snails, *Crepidula adunca* (§29), with their sharply hooked apexes; 4½ times natural size. This species is commonly found on *Tegula*.

An eastern form, *Crepidula fornicata*, has been introduced on our coast, along with oysters, and is now abundant in Puget Sound, occurring as well in Tomales Bay. It was also introduced on the English coast about 1880, and its habits and natural history, somewhat similar to those of our native *C. adunca*, have been studied there by Orton (1912a,b). It is considered an enemy of the oyster beds on the Essex coast because it uses the same kind of food in the same way. "Water is drawn in and expelled at the front end of the shell, the in-going current entering on the left side, passing over the back of the animal and out at the right side. . . . Between the in-going and out-going current the gill of the animal acts as a strainer, which collects all the food material that occurs floating in the water." On its way toward the mouth, food is divided into two main batches, coarse and fine. The fine material, consisting of diatoms, etc., is collected in a cylindrical mass in a groove and ingested at intervals. The coarse particles are stored in a pouch in front of the mouth, so that the animal can feed whenever it wishes.

Crepidula fornicata, like *C. adunca*, is a protandric hermaphrodite, and Orton has found the adults sticking together in long chains, one on the back of the other, with the shells partially interlocked by means of projections and grooves. Old individuals are sometimes permanently attached to their support or to each other by a calcareous secretion of the foot. As many as 13 individuals lie in a chain—the lower ones old and female, the middle ones hermaphroditic, and the upper ones young and hence male. Orton allows a year to the individual; thus in such a chain the oldest would be 14 years old.

§30. Specimens of *Leptasterias pusilla* (Fig. 43), a dainty little six-rayed seastar with a total arm spread usually under 2 cm, may be very numerous in this zone, crawling about in the sea lettuce or dotting the tide pools. At night, specimens are likely to be seen moving around on top of the rocks that have hidden them during the day. Unlike their larger relative *L. hexactis*

Fig. 43. *Leptasterias pusilla* (§30), a small six-rayed seastar.

(§68), of the lower tide pools, these seastars are delicate and clean-cut, usually light gray in color. *L. pusilla* feeds mainly on small gastropods, and its breeding habits are famous. The mother broods the eggs and larvae in clusters around her mouth region until the larvae have attained adult form. Ovigerous females may be found in January and February, and the very minute, liberated offspring are seen in the tide pools during February and March. This species is found in central California, including San Mateo, Santa Cruz, and Monterey counties, where it is very common.

§31. Encrusting the sides of rocks with gay colors are sponges and tunicates of several types. A species of *Haliclona*, probably *H. permollis* (Fig. 44; formerly *Reniera cinerea*), is the vivid-purple, fairly soft, but not slimy encrustation common in this association during some years. It is recognizable as a sponge by the regularly spaced, volcano-like "craters," the oscula, that produce the porousness characteristic of all the animals of this group, including the common bath sponge. *H. permollis* is known from British Columbia to central California. A similar species occurs in Puget Sound (§198).*

Elvin (1976) has studied the growth and reproduction of *Haliclona permollis* on the coast of central Oregon, and found both to be highly seasonal. Growth rates of the sponge reached a maximum in the fall and were minimal from December to April. The production of eggs was initiated during

_____

*The genus *Haliclona* is in need of extensive taxonomic revision. It is complex, containing more than 200 species that are frequently difficult to distinguish. In California, there are probably three or more species, but specific names await further critical study.

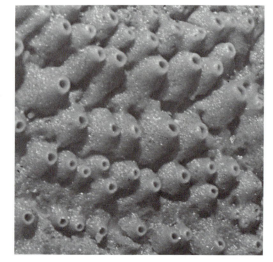

Fig. 44. *Haliclona* sp. (§31), a purple
sponge with volcano-like oscula.

early March, and embryo production was maximal in the late spring, apparently related to an abundance of particulate food.

§32. There are likely to be several types of encrusting red sponges growing in narrow crevices and on the undersides of overhanging ledges. Indeed, there are about ten intertidal species of red to orange encrusting sponges along the Pacific coast. *Ophlitaspongia pennata* (Fig. 45) is a beautifully coral-red form characterized, especially after drying, by starry oscula; its surface is velvety. De Laubenfels remarked (1932) that it occurs clear up to the half-tide mark (higher up than any other sponge), especially on vertical rocks under pendant seaweed, hence shaded from direct sunlight. *O. pennata* is recorded from Sooke (Vancouver Island), British Columbia, to near Puertocitos, Baja California.

Plocamia karykina (Fig. 46), which grows in bright-red layers up to 2 cm thick, is firm and woody, with a smooth surface. The oscula are usually large, irregular, and rather far apart. De Laubenfels (1932) noted that it "emits copious quantities of a colorless slime not conspicuous before injury." A common neighbor is the slightly different-looking *Antho lithophoenix* (formerly *Isociona*), which is more of a vermilion-red, softer, and with a very lumpy, almost papillate, surface. Its oscula are very small and appear in depressions. The microscopic spicules of *A. lithophoenix* are straight, spiny rods, instead of short, thick, double-headed affairs like those of *P. karykina*. The brilliantly splashed colors of these red sponges are a feature of rocky caverns and granite rock faces in this zone, particularly in the Monterey Bay region. *P. karykina* is known to range from Vancouver to Laguna Beach. Other sponges occasionally occurring are mentioned in §140.

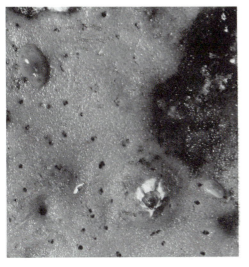

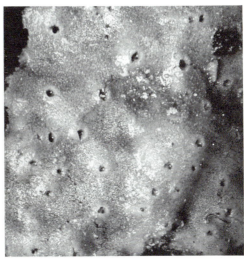

Fig. 45. The red sponge *Ophlitaspongia pennata* (§32), characterized by star-shaped oscula (most evident in dried specimens); slightly reduced. Two small red nudibranchs, *Rostanga pulchra* (§33), are visible.

Fig. 46. The red sponge *Plocamia karykina* (§32), with large irregular oscula.

§33. A vivid-red nudibranch (shell-less snail), *Rostanga pulchra*, averaging only 12 mm in length, is very commonly found feeding on red sponges—an excellent example of protective coloration or camouflage (Fig. 47). Even *Rostanga's* egg ribbons, which may be found throughout the year attached to, or near, the sponge in a loose coil, are vividly red. Laboratory studies have revealed that the nudibranch's close association with red sponges (primarily *Ophlitaspongia, Axocielita,* and *Plocamia,* §§32, 140) is initiated and maintained by behavioral means. Its free-swimming larvae are induced to settle when a piece of *Ophlitaspongia* is added to the culture dish, and if an adult should move off one sponge, it can locate another, by scent, some distance away. *Rostanga pulchra* ranges from Vancouver Island to Baja California.

§34. Two other nudibranchs occur this high in the tide pools. One, *Triopha catalinae* (Fig. 48; formerly *T. carpenteri*), which is found in the Monterey Bay region and as far north as Alaska, can be seen crawling upside down suspended from the underside of the air-water surface film of pools. As it moves along, the animal spins a path of slime, which prevents its weight from distorting and finally breaking the surface film. If the film is once broken, the animal sinks to the bottom of the pool, crawls up the side, and climbs to the surface film again—a long, slow labor that may be watched to advantage in an aquarium. A good many nudibranchs, even the large ones, have this habit. *T. catalinae* is seldom more than 25 mm long (occa-

Fig. 47. The nudibranch *Rostanga pulchra* (§33) with its loosely coiled egg ribbons attached to, or near, one of several species of red sponge that provide food, and presumably camouflage, for this shell-less snail.

Fig. 48. The clown nudibranch *Triopha catalinae* (§34), a striking form, whose orange-red nipple-like appendages contrast with the white body.

sionally 100 mm), but the brilliant orange-red of its nipple-like appendages, contrasted with its white body, makes even small specimens very
noticeable. Dumpy and saclike in outline, it has none of the firmness of
many other sea slugs and collapses readily. It might be supposed that
so tender-looking a morsel, apparently defenseless, would not last long
among the voracious tide-pool animals, but for some reason as yet unknown the nudibranchs are avoided. Obviously this challenge to all enterprising human tasters could not go unanswered indefinitely. Professor
Herdman took up the gauntlet by eating a vividly colored nudibranch alive.
He reported that it had a pleasant oysterlike flavor, so the question remains
open. Frank MacFarland said that the bright colors seem to serve as a warning, but more tasters are needed to find out why that warning is heeded.
Certain special, well-documented cases of inedibility among nudibranchs
will be mentioned later. *T. catalinae*, itself a carnivore, eats several species of
bryozoans, and apparently prefers those with a bushy, arborescent form
over encrusting species. The white, loosely coiled egg ribbons of this animal have been observed in the spring and early summer in Washington.

§35. *Diaulula sandiegensis* (Fig. 49) occurs more frequently on the sides
of boulders, or sometimes even under them. The colors are dainty rather
than spectacular—consisting of as many as a dozen dark circles or spots on
a light-brown-gray background. Like other nudibranchs, *Diaulula* is hermaphroditic. At Monterey it has been observed depositing eggs, in broad
white spiral bands, almost throughout the year, records being at hand for

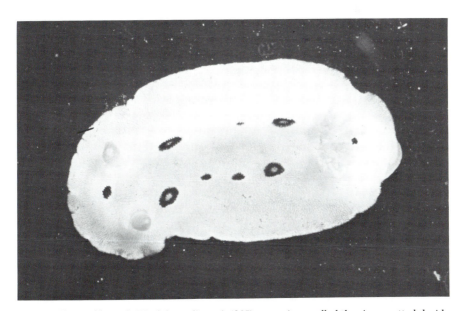

Fig. 49. The nudibranch *Diaulula sandiegensis* (§35), sometimes called the ring-spotted dorid.

the summer, early October, January, and February. This continuous breeding, common among central Pacific coast invertebrates, is very possibly a result of the even temperature throughout the year; such a cause seems especially likely when it is remembered that on the northern Atlantic coast, where there is great seasonal change, the animals have definite and short reproductive periods.

Diaulula and other dorids respire by means of a circlet of retractile gills at the posterior end; if the animal is left undisturbed, these flower out very beautifully. Never occurring in great numbers, but found steadily and regularly in the proper associations, *Diaulula* ranges from Alaska (and Japan) to the southern tip of Baja California. The specimens at Sitka are darker in color, gray-brown, and larger than the average Monterey examples. Specimens 10 to 12 cm long are occasionally seen on the flats south of Dillon Beach. This species feeds on sponges, apparently preferring the purple sponge *Haliclona permollis* (§31), to which it is attracted chemically.

§36. Some of the bolder chitons, or sea cradles, can be considered characteristic members of this zone. Most of them live under rocks, coming out to forage at night; but *Tonicella lineata*, the lined chiton (Fig. 50), is often in plain sight during the day. Although this is a small species, usually not over 2 cm in length, it is the most strikingly beautiful chiton found in the Pacific coast intertidal. In semisheltered areas it maintains the high station

Fig. 50. *Tonicella lineata* (§36), the lined chiton, generally found on rocks covered with coralline algae. This small species is one of the most distinctively marked on the Pacific coast.

(that is, high for a chiton) that characterizes *Nuttallina* (§169) on the surf-swept open coast. In southeastern Alaska, *T. lineata* is the most abundant shore chiton, but it occurs there with the leather chiton, *Katharina*, which on the California coast is restricted to surf-swept areas.

Tonicella is generally found on rocks having a substantial growth of coralline algae, and these algae, along with the epiphytic diatoms coating them, are the chiton's major food source. The chiton's association with corallines begins almost immediately, as the larvae of *Tonicella* have been found to settle selectively on corallines in laboratory experiments. Spawning occurs in the spring or early summer, and in some localities, it is correlated with (triggered by?) the spring bloom of phytoplankton. *Tonicella* ranges from the Aleutian Islands to the Channel Islands.

Another chiton, *Lepidochitona hartwegii* (formerly *Cyanoplax*; see §237 for the related *L. dentiens*), is an oval, olive-green form that may be found almost at will, although not in great numbers, by lifting up the clusters of *Pelvetia* on vertical rocks. In such places it spends its days feeding on the algae, while being protected from sunlight, drying winds, and foraging shorebirds.

The natural history of chitons is discussed in §76.

§37. The small, squat anemone *Epiactis prolifera* (Fig. 51) is locally abundant; it is usually found on the protected sides of small, smooth boulders and sometimes on the fronds and stipes of algae, especially *Cystoseira* and the sea lettuce *Ulva*. This anemone is notable for its breeding habits. Young individuals, which are most of those seen in the field, are nearly all functional females. As they grow, testes begin to develop, so that fully mature individuals possess both eggs and sperm. A breeding system of this sort, in which the population consists only of females and hermaphrodites,

Fig. 51. *Epiactis prolifera* (§37), a small striped anemone of the protected coast. Young in various stages of development are attached to the brooding area of this adult. A small specimen of *Tonicella* lies above and to the left of the anemone.

is unusual for animals, although it is well known among plants. Another notable feature, and one more obvious to the shore visitor, is that the eggs, instead of being discharged freely into the water, are retained in brooding areas at the base of the body wall, where they develop into young anemones without leaving the parent, almost forming an accretionary colony. A fully expanded adult, with half a dozen flowering young, is a lovely thing to see.

Epiactis is capable of considerable independent movement; if a stipe of alga bearing several specimens is moved to an aquarium, the anemones will leave the plant for the smooth walls of the tank. In the same way, the young, when they are sufficiently mature (after about 3 months), glide away from the parent. California specimens average perhaps 2 cm in diameter; Puget Sound specimens are larger. The animal may be brown, green, or red with the stripes similar in color but darker. It ranges from southern Alaska to La Jolla.

§38. The solitary hydroid *Tubularia*, treated (in §80) as an outer tidepool inhabitant, is occasionally found here also, as are small, fernlike colonies of *Abietinaria* (§83).

§39. The vividly orange-red solitary coral *Balanophyllia elegans* (Fig. 52) is abundant, but it must be well sheltered from desiccation and direct sunlight. In February and March at Pacific Grove, the transparent and beautiful cadmium-yellow or orange polyps may be seen extending out of their stony, cuplike bases until the base constitutes only a third of each animal's bulk. Large individuals may be slightly over 1 cm in diameter unexpanded. Eggs are fertilized within the parent's gastrovascular cavity, and wormlike planula larvae, already orange, are released through the mouth to settle nearby. *Balanophyllia's* range is from Puget Sound to southern California.

Another coral, *Caryophyllia alaskensis*, occurs intertidally in the Puget Sound region; according to J. Wyatt Durham, it may be similar in appearance when alive to *Balanophyllia elegans*. Perhaps some northern records for *B. elegans* should apply to *C. alaskensis*.

Astrangia lajollaensis, smaller and with light-orange polyps, is said to occur from Point Conception south; it is comparatively abundant at La Jolla.*

In fact as well as in appearance these corals are related to the anemones, and their method of food-getting is similar. Working with a related east-coast species, *Astrangia danae*, Boschma (1925) found that in the natural state the coral fed on diatoms, small crustacean larvae, etc., and that in captivity it would take almost anything offered, including coarse sand. In an average of less than an hour, however, the sand, covered with mucus, was regurgitated onto the disk. The tentacles then bent down on one side,

*A good-sized population of *Astrangia lajollaensis* lives subtidally off Hopkins Marine Station at Pacific Grove, apparently far disjunct from the main population in southern California (Fadlallah 1982).

Fig. 52. The orange solitary coral *Balanophyllia elegans* (§39): *left,* unexpanded specimens, each animal in its own theca and separated from its neighbors; *above,* an expanded specimen, showing the nearly transparent tentacles of the polyp.

allowing the sand to slip off. Living copepods coming into contact with a tentacle remained affixed, harpooned and probably paralyzed by stinging nematocysts, and were then carried to the center of the mouth disk by the tentacle, the disk rising at the same time to meet the morsel. Larger animals usually struggled when they came in contact with the tentacles, but other tentacles then bent over to help hold the captive until it quieted.

§40. A form similar in appearance to *Balanophyllia,* but having white-tipped tentacles and lacking the hard skeleton, is the small red anemone *Corynactis californica,* which often clusters under the overhang of rocks (Fig. 53). The clusters, some covering a square meter or more of substratum, may each represent a single clone, since this anemone reproduces asexually by fission. *Corynactis* does not like sunshine; hence the most fully expanded specimens are seen at night. It occurs intertidally from Sonoma County to Santa Barbara, and subtidally to at least the Coronado Islands near San Diego.

§41. The purple sea urchin is found frequently here, but not in the great beds that characterize the outer pools and reefs (see §174).

§42. Living on and in the rockweed itself, especially in *Pelvetia,* are several diminutive forms rarely found elsewhere. By shaking a mass of freshly gathered rockweed over a dishpan, one can easily dislodge and capture hosts of tiny amphipods (beach hoppers), whose colors perfectly match the slaty green of the weed. Several forms are common here, among them some species of *Ampithoe.* In general appearance they are scarcely distinguishable from *Melita* (§20).

Fig. 53. The small red or pink anemone *Corynactis californica* (§40); about 1½ times natural size. The clubbed white-tipped tentacles are clearly visible in these expanded specimens.

To enable them to cling to plants and other objects, both amphipods and isopods have the last joint of the legs incurved to form a claw. The shape of isopods (flattened horizontally) adapts them for crawling over rocks rather than over weeds, whereas amphipods (flattened vertically, or laterally) seem better adapted for balancing on vegetation. The leaps of amphipods are made with a sort of "flapping" motion. Driving power for the leap is obtained by suddenly snapping backward the bent posterior of the abdomen—a spring strong enough to propel the animal a meter at a leap and make it adept at avoiding the specimen jar.

§43. It is in this area also that one is likely to first see examples of the roundworms (nematodes), a group of animals that include such well-known forms as the hookworm and the eelworm. Referring to the abundance and widespread distribution of these worms, it has been said that if everything else in the world were to be removed, there would still remain a ghost or shadowy outline, made up of roundworms, of mountains, rivers, plants, soil, and forests, and even of animals in which roundworms are parasitic. Their characteristically wiggling bodies are very noticeable to the keen-eyed observer who pulls loose a bit of alga, especially where decay has set in. Some forms writhe so rapidly that they seem almost to snap themselves back and forth.

Under-Rock Habitat

A characteristic group of animals is found here on the underside of loosely embedded rocks (but not in the substratum) or between adjacent rocks and boulders where there is no accumulation of substratum. This area offers the maximum protection from wave shock, sunlight, and desiccation. Some of the brittle stars (*Ophioplocus*, §45) are diagnostic of this association and rarely found elsewhere, but such widely ranging and generalized forms as the flatworms (*Notoplana*, §17) may be the most abundant with respect to numbers of individuals. Members of the under-rock group are even less obvious than those of the protected-rock habitat, but they are easily found by anyone willing to turn over a few rocks.

§44. The fragile and beautiful brittle stars, whose delicacy requires the protection of the under-rock habitat, are the most striking of all animals found here. They are bountifully represented, both in number of individuals and in number of species. In Monterey Bay at least six shore species occur in this habitat, and there are still others in the south, most of them restricted to the low intertidal.

The tiny *Amphipholis pugetana*, already treated (§19) as an inhabitant of the upper tide-pool region, is almost as plentiful here in the middle zone. A much larger and hence more conspicuous form is *Amphiodia occidentalis* (Fig. 54), ranging from Alaska to San Diego, but not reported intertidally south of Monterey Bay. This is an annoyingly "brittle" brittle star with long snaky arms bearing spines that take off at right angles. An interested observer who places a specimen in the palm of his hand to examine it

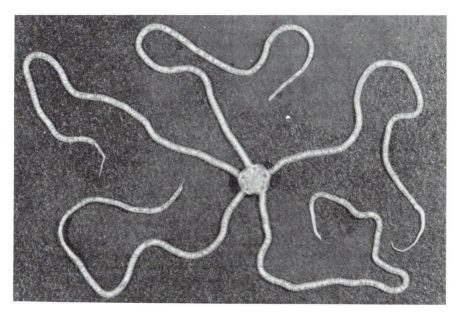

Fig. 54. *Amphiodia occidentalis* (§44), a common long-armed brittle star of the under-rock habitat in Zone 3.

is likely to see the arms shed, piece by piece, until nothing is left but the disk, possibly bearing a few short stumps; normally, regeneration takes place quickly. During October the disk of the female, up to 13 mm in diameter, is greatly swollen with eggs; at such times it is detached on the slightest provocation, possibly in a kind of autotomy. Whether the remaining part of the body, which is little more than the intersection of the arms, will regenerate a new disk is not known.

Although closely restricted to the underside of rocks (especially where the rocks are embedded in fine sand with detritus) and even occurring in the substratum, the brittle stars, like many other animals that hide away during the day, come out on calm nights and crawl about on top of the rocks at ebb tide. In their under-rock retreat, they are almost invariably found in aggregations of from several to several dozen, so closely associated that their arms are intertwined; studies of this intertwining habit lead us to the borderline of the metaphysical. Working with Atlantic brittle stars, isopods, and planarians, Professor W. C. Allee found (1931) that aggregations have a distinct survival value for their members, bringing about a degree of resistance to untoward conditions that is not attainable by isolated individuals. By treating individual animals and also naturally and spontaneously formed aggregations with toxic substances, he found not only that the mass had greater resistance to the action of the poisons (partly because of absorption by secreted slime and the bodies of the outermost animals) but also that a specialized protective material was given off by the aggregations.

This rather astounding discovery opens new and unexplored vistas to students of biology. It will certainly throw some interesting light on animal communities in general, and may conceivably clarify the evolutionary background behind the gregariousness of animals, even behind that of human beings.* At the moment, it explains a phenomenon regarding anemones that had puzzled us for some time. A single anemone may easily be anesthetized for preservation, but a large number in a tray show a tremendous degree of resistance, even though the amount of anesthetic employed per individual is greater for the group. Furthermore, the pans in which anemones are anesthetized become increasingly unsuitable; that is, each successive batch of animals is more resistant than the batch before. We finally took to scrubbing out the pans with hot water and soap powder and then revarnishing them.

Several animals that do not form aggregations in their natural habitat, such as *Nereis* and *Neoamphitrite*, do so under artificial conditions, as when placed in trays. Since *Amphiodia* aggregates in its natural state, however, it

*Indeed, there is a burgeoning field of "chemical ecology," at present far better developed for terrestrial systems than for marine systems, owing to the importance of insects and their relationship to agriculture and forestry. Even so, considerable work is in progress on marine animals, and quite a number of investigators are exploring the various attractants, repellents, toxins, and the like that are important parts of the daily lives of marine animals. The field is enormous, and encompasses nearly all the activities of animals in behavioral, ecological, and evolutionary time. Interesting problems remain everywhere.

may be assumed that it does so for protection against some usual environmental factor. It is not known whether the aggregations disband when the tide is in, but this is probably the case.

Some more-tangible aspects of brittle stars in general are their methods of locomotion and feeding habits. The locomotion over a firm surface is peculiar—quite different from that of seastars, and very rapid compared with the snail-like pace of even the fastest seastar. Most brittle stars that we have observed pull with two arms and push with the other three, apparently exerting the power in jerky muscular movements. Their very flexible arms take advantage of even slight inequalities in the surface. Burrowing is equally unusual. When placed on fine sediment the tube feet go busily to work moving sand from below the arms and disk and piling it on top. As a result of this concerted effort, the animal sinks straight down, slowly disappearing from sight beneath the sand.

A good deal is known of the brittle star's feeding habits. The hard jaws surrounding the mouth can be rotated downward, thus greatly enlarging the mouth, and then rotated upward, so that they form a strainer. The digestive tract is restricted to the disk and does not extend into the arms as is the case with seastars. Moreover, a brittle star, again unlike the seastars, cannot project its stomach outside its body to envelop food. It has been observed that many brittle stars feed by accumulating food particles on their tube feet and passing the material from foot to foot toward the mouth. Some feed on detritus on the bottom. Others, often those in deeper water and in great numbers, hold their arms upward, and are thus suspension feeders. *Amphiodia* employs both methods: feeding on deposits with the arms extended horizontally over the substratum, and suspension feeding with the arms held vertically in the water.

Any area colonized by animals is full of organic debris, which is continually being sorted over by animals of various capacities and with varying requirements, with the Victorian "economy of nature" factor always at work. The whole thing calls to mind the sorting effect of a set of sieves. Certain animals, like crabs, dispose of sizable chunks of food; others, like beach hoppers, eat minute particles and may, in time, reduce large pieces of edible matter; certain of the sea cucumbers and brittle stars, buried in the substratum, sweep the surface with their tentacles for adherent particles, an office that is performed for rock surfaces by limpets and periwinkles; and other animals, including some brittle stars, sea cucumbers, and annelid worms, pass quantities of dirt and sand through their alimentary canals in order to extract whatever nourishment these may contain. Everything is accounted for: particles so small as to be overlooked by everything else are attacked and reduced by the last sieve of all, the bacteria.

In a Monterey Bay population of *Amphiodia*, most spawning occurred in the late spring and early summer. Sexual products were released into the water, with the eggs developing into swimming larvae.

§45. *Ophioplocus esmarki* (Fig. 55) is a larger, brown or red-brown brittle star, common in some regions on a sandy-mud substratum under flat rocks,

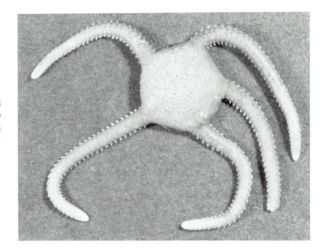

Fig. 55. *Ophioplocus esmarki* (§45), a large brown brittle star with short stubby arms.

especially disintegrating·granite. It averages 13 to 19 mm in disk diameter and has short stubby arms with a spread of 7 to 9 cm. It is relatively hardy and maintains its form well when preserved. In central California, specimens noted during January had the genital bursae, or sacs, swollen with eggs, and young were found in the bursae in July. *Ophioplocus* and *Ophioderma* (§121) differ from most brittle stars in having spines on the arms that extend outward at an acute angle to the arms instead of being at right angles. *Ophioplocus* ranges from Tomales Bay to San Diego. In this southern range *Ophioderma* and *Ophionereis* are common in the lower tide-pool zone but often stray up into the middle zone; both (§121) are large and striking, but neither is found above Santa Rosa Island.

§46. The porcelain crab discussed with the upper-tide-pool animals (*Petrolisthes*, §18) is also a common and easily noticed member of the middle-tide-pool under-rock association. A transparent tunicate (*Clavelina*, §101) is fairly common here, but belongs more characteristically to the lower intertidal. Some serpulid worms occur (§§129, 213), but they are also more common farther down.

Belonging to this zone is a large, slow-crawling isopod, *Idotea urotoma* (Fig. 56), which will be found under rocks, especially in the south. It may be recognized by its pill-bug appearance, paddle-shaped tail, and somewhat rectangular outline. It is highly variable in color (but often brown) and about 16 mm in length (see also §205).

Betaeus longidactylus (Fig. 57), the "long-fingered" shrimp, may be found under rocks or in pools. In adult males the nearly equal large claws are extraordinarily long, sometimes more than half as long as the animal's body. The elongate carapace projects somewhat over the animal's eyes, like a porch roof—a peculiarity that is still more exaggerated in a related shrimp that lives on wharf piling (§326). The legs of *B. longidactylus* are red, and

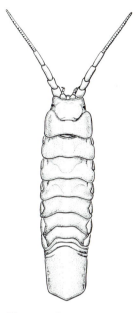

Fig. 56. The isopod, or pill bug, *Idotea urotoma* (§46). The paddle-shaped tail and rectangular body are characteristics of this variously colored species.

Fig. 57. *Betaeus longidactylus* (§46), the long-fingered shrimp, most plentiful in southern California on the outer coast, but occurring as far north as Monterey in quiet water; twice natural size.

there is a fringe of yellow hairs on the last segment of the abdomen. The body and the big claws vary in color from blue and blue-green to olive-brown. It occurs from Monterey Bay to the Gulf of California, but is most plentiful in the south, where it occurs on the outer coast, being restricted in the north to quiet-water conditions, such as in Elkhorn Slough.

§47. A common feature of the under-rock fauna below Santa Barbara is a great twisted mass of the attached and intertwined tubes of a remarkable sessile snail, *Serpulorbis squamigerus* (Fig. 58; formerly *Aletes*), which occurs at times even on the sides of rocks. For a reason not readily apparent, this animal lost or gave up the power of independent locomotion and in the process evolved a new method of securing food. Mucus from glands in the foot, formerly employed to lubricate the snail's path, is now used to entangle minute animals and bits of detritus. The mucus expands in water to form a delicate net some 50 cm² in area. The net is left suspended for 10 to 30 minutes, and then retrieved, the radula and jaws delivering the net and its catch to the mouth.

Serpulorbis is ovoviviparous (that is, it lays eggs but retains them within the shell until the young are hatched), and produces offspring that have snail-like, coiled shells. It is only after the young gastropod has at-

Fig. 58. The fixed snail *Serpulorbis squamigerus* (§47), with an encrusting sponge; ¼ natural size. The snail in the upper left center is relatively free of epibionts and shows the characteristically furrowed and ridged tube.

tached to a rock and started to grow that its shell loses all resemblance to a snail's. Calcareous matter is added to the mouth of the shell to produce a long wormlike tube that is irregularly coiled or nearly straight. The hinder part of the body is black, the rest mottled with white. When disturbed, *Serpulorbis* retracts slowly, like an anemone—a trait that distinguishes it instantly from tube worms, which snap back in a flash, often with nothing more than a shadow to alarm them. Even experienced zoologists, if unfamiliar with the Pacific fauna, may mistake this tubicolous mollusk for a serpulid worm. *Serpulorbis* ranges from Monterey to Baja California, but it is only occasional in the Monterey Bay region, occurring singly or in small clusters of tubes. The specimens shown were photographed in Newport Bay.

The empty tubes of *Serpulorbis* provide hiding places for a variety of small crabs, amphipods, ophiuroids, polychaetes, and others seeking temporary shelter.

§48. More than a hundred species of ribbon worms (nemerteans) are found on the Pacific coast. One of them, *Emplectonema* (§165), is abundant here; but it is treated elsewhere, since it occurs even more commonly under mussel beds. The similar but larger *Amphiporus bimaculatus*, however, has possibly its chief center of distribution in this zone. Recorded from Sitka to Ensenada, it may be found occasionally as a twisted brown cord, capable of almost indefinite expansion and contraction, but averaging 6 to 10 cm in length when undisturbed. As suggested by its specific name, *Amphiporus* has two dark spots on its head, not unlike great staring eyes but having no connection with the light-perceptive organs. The "real" eyes are the ocelli—dozens of minute black dots centered in two clusters that are marginal to the dark spots and extend forward from them. Otherwise the worm's head is light in color, a striking contrast to the rather beautiful dark orange of the rest of the body.

Nemerteans are ribbon worms, and they have a physical organization much more primitive than that of segmented worms like *Cirriformia* and the serpulids. They are unique in the possession of a remarkable eversible organ, the proboscis, which, in the case of some mud-living forms (*Cerebratulus*, §312), may be longer than the animal bearing it. The proboscis of *Amphiporus* is armed with a poisonous barb or stylet, not large enough to penetrate human skin but very effective when directed against smaller worms.

It is almost impossible to preserve some of the nemerteans because of their habit of breaking up at the slightest provocation (*Amphiporus*, however, is only moderately fragile). Often they cannot even be picked up whole. If the head fragment of an autotomized specimen is left in the animal's natural environment, or undisturbed in a well-aerated aquarium, it will regenerate a complete new animal. Some interesting experiments in this connection are discussed in §312.

§49. A very common, insignificant-looking inhabitant of this region is the scale worm, or polynoid. *Halosydna brevisetosa* (Fig. 59), whose geographic range extends from Kodiak Island to Baja California, is the most common species from Alaska to Monterey. South of Monterey it begins to be replaced by a form described as *H. johnsoni* (referred to in earlier edi-

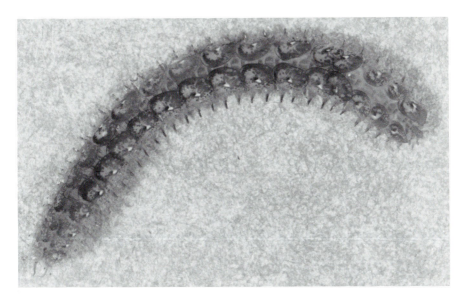

Fig. 59. The common scale worm *Halosydna brevisetosa* (§49), often found as a parasite or in commensal association. The number of scales, 18 on each side, is a diagnostic feature.

tions as *H. californica*). *H. johnsoni*, however, is probably not a separate species, but rather a form of *H. brevisetosa*. Free-living specimens seldom exceed 2 cm or so in length. Many, however, exchange their freedom for a symbiotic but doubtless comfortable residence in the tubes of other worms. In this case, contrary to the usual run of things, the symbiotic, dependent form is considerably larger and more richly colored than its self-supporting relatives. Perhaps the commensal habit is adopted later in life.

The polynoid worms are pugnacious fighters; and in captivity they will attack one another with their strongly developed, four-jawed proboscises, displacing scales or removing entire segments from the after ends of their companions' bodies. *H. brevisetosa*, like allied annelids, bears its scales, called elytra, in two longitudinal rows on the upper (dorsal) surface and may autotomize them when roughly handled. The elytra of *H. brevisetosa* shield the ciliated, dorsal surface and form a tunnel through which a respiratory current of water can flow, even when the worms are wedged tightly into crevices. This animal has 18 pairs of mottled scales with white centers.

Many other worms, such as *Neanthes virens* (§311), are occasional in the under-rock habitat.

§50. Here, as in the upper tide pools, small isopods and amphipods, pill bugs and beach hoppers, are more numerous than any other animals found under rocks. *Melita sulca* (§20) is the commonest hopper in this zone, but others may occur, including the very active, lustrous, slaty-green *Traskorchestia* (of undetermined species), which is closely related to the tiny freshwater "shrimps." It should be mentioned in extenuation of the hoppers that their variant name of "beach flea" is a libelous misnomer. None of the beach hoppers has the slightest use for man or other warm-blooded animals.

The isopod *Cirolana harfordi* (§21), likewise not restricted to any one habitat, occurs also, as do several others.

§51. A long, eel-like fish, one of the blennies (Fig. 60), is very common under boulders in this association from southeastern Alaska to Baja California. The young are greenish black; the adults (up to 30 cm long) are nearly black, with traces of dusky-yellowish or whitish mottling toward the

Fig. 60. The black prickleback, *Xiphister atropurpureus* (§51), a representative tide-pool fish and a common under-rock inhabitant of middle and lower intertidal regions.

tail fin. This species, *Xiphister atropurpureus* (formerly *Epigeichthys*), lives even as an adult in the under-rock habitat, frequently in places that are merely damp when the tide runs out. Most specimens have three stripes of light brown or white, which are bordered by darker colors and radiate backward and upward from the eyes. Not all *X. atropurpureus* have this character, and other blennies may have it also, but the markings are fairly diagnostic. During the winter on protected rocky beaches and during the spring on more-exposed shores, females lay masses of eggs under boulders set on substrata of pebbles, small rocks, or shells. Such a substratum provides the many channels and pockets this fish prefers for a snug-fitting, yet well-aerated, living space. The pockets also serve to hold the nonadhesive egg masses, which are usually tended by a single male.

Another common species is the larger *Xiphister mucosus* (up to 58 cm long). Contrary to previous reports that only the young of this species are found in the intertidal zone, adults have been found commonly in the intertidal of California. The young are often a pale, translucent olive, and the adults, a greenish black, gray, or brownish. The range is from Alaska to near Point Conception.

Three other less-common blennies represent an entirely different group,* which is usually tropical. They are more compressed laterally than those described above and not so elongate, and they are often brilliantly striped. Two of these, *Gibbonsia montereyensis* and the usually striped *G. metzi*, range from British Columbia to Baja California; the third, *G. elegans*, is common only south of Point Conception. Many people regard the blennies as excellent eating. However, some that we tried had green flesh, whose cooking smell was so reminiscent of defunct kelp that our research was discontinued.

With the blennies, or in similar situations, the little clingfish *Gobiesox maeandricus* is very common, slithering over smooth, damp surfaces or clinging to a rock with limpetlike strength.

Burrowing Habitat

Animals of the burrowing habitat occur in the mud, sand, or gravel substratum underlying tide-pool rocks. A small, stout trowel is a good tool here, for the surface boulders often rest on a matrix of muddy gravel bound together by chunks of rock of varying sizes. Ordinarily, however, the prying up of a large boulder will disturb the substratum, and any animals present will be visible. The normal expectancy would be that the sand or mud under rocks would harbor forms similar to those burrowing in the sand or mud of estuaries. This proves true to some extent, although the under-rock burrowing animals actually seem to form a fairly specific assemblage.

---

*The common term blenny (or blennioid) refers to fishes in several distinct families, including the Clinidae (e.g., *Gibbonsia*) and the Stichaeidae (e.g., *Xiphister* and *Anoplarchus*) as well as the Blenniidae.

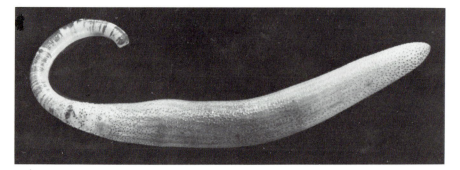

Fig. 61. The peanut worm, or sipunculan, *Phascolosoma agassizii* (§52), a common burrower in the sediment under tide-pool rocks.

§52. Foremost of these substratum animals is the sipunculid worm *Phascolosoma agassizii* (Fig. 61), which ranges from Kodiak Island, Alaska, to Bahía de San Quintín, Baja California. In the south, *Phascolosoma* tends to be replaced by the more beautifully flowering *Themiste* (§145). These rough-skinned worms are related to the annelids, although themselves devoid of segmentation. They have been popularly termed peanut worms, a name that really describes the specimens only in their contracted state. *Phascolosoma* is abundant in the muddy crevices between rocks, and is often associated with terebellids (see the next section). A contracted specimen 5 cm long (larger than the average) is capable of extending its introvert, which bears a mouth surrounded by stubs of tentacles, to an extreme length of 15 cm, much as one would blow out the inverted finger of a rubber glove.

In Monterey Bay, *Phascolosoma* spawns in March, April, and May, whereas in the San Juan Islands it breeds from the middle of June to early September. The larvae are apparently quite long-lived, and have been kept alive in the laboratory for as long as 7 months. If this length of larval life is also possible in the natural habitat, the offspring of an adult could be dispersed a considerable distance by currents. In the quiet channels of upper British Columbia large representatives of this form are among the most characteristic substratum animals.

§53. The brown or flesh-colored terebellid worms (Fig. 62), commonly up to 15 cm long and occasionally to 28 cm, build their parchmentlike, sand-encrusted tubes under rocks or in the substratum between rocks. Several species of these polychaetes are involved. *Thelepus crispus*, *Eupolymnia heterobranchia*,* and possibly some species of *Terebella* seem to prefer the cool oceanic waters of open shores but may be found also in the

---

*Hartman (1968) concluded that this form, described as *Lanice heterobranchia*, is "too incompletely known to refer to any known species." Another species of *Eupolymnia*, *E. crescentis* Chamberlin, is fairly common in sandy-mud sediments of bays.

quiet waters of Puget Sound. Another form, *Neoamphitrite robusta* (§232), which reaches great size in Puget Sound, may also occur on outer coasts. All of these worms superficially resemble *Cirriformia* (Fig. 28), but the cirratulids have gills extending almost to the middle of the body, which is tapered at both ends, whereas the terebellids have long, delicate tentacles that are restricted to the enlarged head region.

Differentiating the many species of terebellid worms is a task for the specialist. We have had little success in identifying them by observations in the field, but the following suggestions supplied by Olga Hartman will be found useful: *Thelepus* has threadlike and unbranched (cirriform) tentacles, with gills of about the same length. The tentacles of *Neoamphitrite* and *Terebella* are branching (dendritic), dark, and longer than the white gills; but *Neoamphitrite* has only 17 thoracic segments, whereas *Terebella* has more—23 to 28.

A specimen of any of these tube worms, if deprived of its home and placed in an aquarium with bits of gravel, will quickly set about remedying the deficiency. In a few hours a makeshift burrow will have been constructed of the gravel, cemented together with mucus and detritus, and the worm's transparent, stringy tentacles will be fully extended on the bottom of the container, writhing slowly in search of small edible particles, which are moved toward the mouth along ciliated grooves on the tentacles.

Fig. 62. *Thelepus crispus* (§53), a terebellid polychaete of the middle zone, with its sand-encrusted tube.

In Bermuda, Welsh (1934), primarily interested in nerve structure, investigated the autonomy of the tentacles of a similar form. He found that they would live, squirm about, and react normally in most ways up to 12 days after they had been removed from the body. He also determined that new tentacles (in experimentally denuded worms) grew at the rate of 1 to 1.5 cm per day, and that the fully developed tentacle, up to 50 cm long, acted as a food-getting organ.

Most of the tubes of our local forms, when carefully examined, will be found to contain presumably uninvited guests who share the substantial protection of the tubes, profiting from the food-laden water. Commensal polynoid worms (§49) are often present, and with them a minute pea crab, *Pinnixa tubicola*. Only males of this crab have ever been noted free, and then only once, in Elkhorn Slough. The range is Alaska to San Diego.

§54. *Neoamphitrite* is one of several animals whose breeding time is known to be correlated with the spring tides and hence bears some sort of relationship to the phases of the moon. In this case one investigator, Scott (1911), believed that both moon and tides are entirely secondary factors, and that the egg-laying is induced by the higher temperatures at the time of spring tides, caused by long exposure to sunlight and possibly by the changes of pressure brought about by the alternate increase and decrease in the depth of water. The increased food supply may be another eliciting factor.

At any rate, both males and females discharge their ripe sexual products into their tubes during spring lows and expel them from the tubes into the water via the respiratory currents. A possible reason for discharging the eggs and sperm at this time is that these products have a much greater opportunity to mix than if they were voided into the tons of water moving at flood tide—a reasoning that applies equally well, of course, to other animals. At the time of egg-laying the female's wavelike body contractions become stronger and faster as the ripe eggs that are to be released are separated from the immature eggs that are to be retained.

§55. In sandy and gravelly substrata, extending downward from the middle tide pools, the dirty-white sea cucumber *Leptosynapta albicans* (Fig. 63) is common. This and similar forms (the synaptids) lack the respiratory trees that are common to other cucumbers; they have, instead, a delicate, semitransparent skin that lets through enough dissolved oxygen for purposes of respiration. The skin, through which the internal anatomy can be seen vaguely, is slippery, but it is armed with numerous white particles, calcareous anchors that stiffen the body and give the animal better purchase in burrowing; these anchors often come off on one's fingers. Although it lacks the tube feet characteristic of most of the other cucumbers and of echinoderms in general, this remarkably efficient little wormlike form is at home from Puget Sound to San Diego. When disturbed, it has the habit, also wormlike, of autotomizing by constricting itself into several portions—a trait that makes it difficult to collect and difficult to keep in an

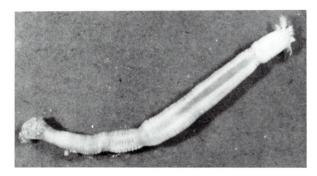

Fig. 63. *Leptosynapta albi-cans* (§55), a dirty-white sea cucumber that lacks tube feet. This burrower uses its 10 to 12 feathery tentacles to collect sedi-ment and detritus, which it gleans for food.

aquarium unless sand for burrowing is provided. Experiments on a related species have demonstrated that only the head portion will regenerate a complete animal.

When the tide is in, *Leptosynapta* lies fairly well buried but stays close enough to the surface to protrude its 10 to 12 feathery tentacles. The animal feeds by ingesting fine sediment and debris, assisted by the tentacles. Here on the Pacific coast this species is rarely more than 3 to 5 cm in length, but specimens up to 25 cm long have been found in other parts of the world; and a related form from southeastern Alaska, mostly from quiet water, may be 15 cm long or more (*Chiridota*, §245).

§56. The brittle star *Amphiodia* (§44) is possibly more commonly found buried in the substratum than crawling on the underside of rocks, but it is certainly not so easy to see there. In little pockets under rocks, where the bottom is soft and oozy but with coarse particles, one may sometimes sift out with the fingers a dozen or more of these snaky, long-armed specimens.

§57. In southern California and northern Mexico the muddy bottom under tide-pool rocks is often riddled with the burrows of a small, white, crayfishlike animal, the ghost shrimp *Callianassa affinis* (Fig. 64). A large specimen may be 6 to 7 cm long. Although not so feeble and delicate as its relatives of the mud flats, *C. affinis* is one of the softest animals found in the tide pools. There it uses the protection of rocks, and then burrows deeply into the substratum so as to obtain the maximum shelter. From the protec-tion of its permanent home, *C. affinis*, often found in pairs, feeds on sus-pended plankton and surface detritus brought in with water pumped through the burrow by means of the abdominal appendages (pleopods). The ghost shrimp's recorded range is from Goleta to Bahía de San Quintín, and we have taken specimens (identified by the National Museum) from Bahía de Todos Santos. Like its larger mud-flat relatives (in connection with which the natural history of the ghost shrimps is discussed in more detail, §292), this *Callianassa* has the crayfish habit of flipping its tail suddenly so as to swim backward in a rapid and disconcerting manner. Ovigerous specimens have been taken in early May.

Fig. 64. *Callianassa affinis* (§57), a small white ghost shrimp of southern California and northern Mexico.

Fig. 65. *Typhlogobius californiensis* (§58), a blind goby that occurs in ghost-shrimp burrows; slightly reduced.

§58. Associated with the above is the blind goby *Typhlogobius californiensis* (Fig. 65). It lives in the burrows made by the ghost shrimp and is about 6 cm long, somewhat resembling a very stubby, pink eel. Its eyes are rudimentary and apparently functionless.

Pool Habitat

§59. A minute, red-orange flatworm, *Polychoerus carmelensis*, only slightly longer than it is wide, may be so abundant locally during the spring and summer as to dot the Monterey tide pools with color, particularly during early-morning low tides. However, any disturbance of the tide pool, by water entering on a rising tide or by a would-be collector mucking about, will cause the worms to retreat under stones on the pool bottom. This form (a turbellarian) is interesting in that it is one of the few free-living flatworms to have no intestine and to possess a tail. These animals are hermaphroditic, and the male organs of one fertilize the female organs of another. (Self-fertilization rarely takes place in hermaphroditic animals except possibly by accident.) *Polychoerus* deposits eggs primarily at night in transparent, gelatinous capsules, which may be found attached to pebbles and stones or in folds of the alga *Ulva* during later summer. Relatives in the same order (Acoela) reportedly feed on copepods, other flatworms, or diatoms; little is known of the feeding habits of our local form, except that it will capture and consume copepods in the laboratory.

§60. Those who in the early evening, about the time total darkness sets in, examine tide pools bordered with red algae will probably be rewarded with an exhibition of phosphorescence by another flatworm, *Monocelis* (probably *Monocelis cincta*), an alloeocoel. These little dots of light can be seen crawling rapidly along the underside of the air-water film or swimming about at the surface. For those not wanting to wait for nightfall, *M. cincta* may be collected from the sediment on the bottom of these middle tide pools. Transferred to a petri dish, the slender worms move about actively in short darts with frequent changes of direction. Between darts, the worms pause, attached by adhesive glands at the posterior end of the ventral surface; these glands are also used in turning and stopping. Like *Polychoerus*, *Monocelis* is hermaphroditic, with cross-fertilization. It forms hard-shelled eggs.

Certain small annelid worms, likewise luminescent, may be found swimming in pools with *Monocelis*; these swim even more rapidly than *Monocelis* and are generally somewhat larger.

§61. During the late winter or early spring (in February 1930, at Laguna Beach, for example), but unfortunately not every year, the middle and upper tide pools are temporarily inhabited by hordes of a very small crustacean, *Acanthomysis costata*. These are mysids, called opossum shrimps because the eggs are carried in thoracic pouches; they are very transparent and delicate, with a slender body up to 13 mm long, enormous eyes, and long, delicate legs on the thorax. The thoracic legs have outer branches, or exopodites. *A. costata* seems to be a neritic (near-shore) species restricted to southern California, but it is overlapped by the similar *A. sculpta*, which is often abundant in Puget Sound.

§62. Especially at night, the transparent shrimp *Heptacarpus pictus* (formerly *Spirontocaris*), a pale-green form often banded with red, is likely to be found, darting about in the pools or quiescent in bits of sea lettuce or in rocky crevices. There is such a fairylike beauty to this ephemeral creature that the inexperienced observer will be certain that he is seeing a rare form. The elusive animal is likely to lead him a lively chase, too: when at rest, it is difficult to see against the vivid colors of a tide pool; and when discovered, it darts backward in a disconcerting manner by suddenly flexing its tail. Once captured, the living specimen should be confined in a glass vial not much larger than itself and examined with a hand lens. The beating heart and all the other internal organs can be seen very plainly through the transparent body.

About 95 percent of the transparent shrimps found in rocky pools will be of the species *Heptacarpus pictus*. The very similar *H. paludicola* (Fig. 66) is much more common on mud flats and in eelgrass beds, but it does occur in rock pools.

In the lower tide pools other species of *Heptacarpus* will be found; these are not transparent, but all are striking and all have the tail bent under the body in a characteristic manner.

Fig. 66. The transparent shrimp *Heptacarpus paludicola* (§§62, 278) of mud flats and eelgrass beds; about twice natural size. *H. pictus* (§62), of mid-zone tide pools, is similar in appearance.

Zone 4. Low Intertidal

It is here that the field observer finds his most treasured prizes, since this zone, bared only by the minus tides (from 0 to −1.8 ft, or −0.6 meters, at Pacific Grove), is not often available for examination (Fig. 67). In its lower ranges, the region of the outer tide pools is uncovered only a few times each month, sometimes not at all. Many of the middle-intertidal animals occur here also, but much of the population consists of animals unable to exist higher up and not so accustomed to the rhythm of the tides. This tremendously crowded environment contributes a greater number of species than all the other tide levels. Triton is a fruitful deity, and here, if anywhere, is his shrine. Not only do the waters teem, but there is no rock too small to harbor some living thing, and no single cluster of algae without its inhabitants. Since these creatures live and thrive in an environment that seems utterly strange to us, it is no wonder that we find interest in their ways of feeding, of breathing, of holding on, of ensuring the continuity of their kind—and in their strangely different weapons and methods of attack and escape.

Rock Habitat

In the low intertidal, only a few animals on unprotected rock faces are exposed to wave action, sun, and wind, except on open coasts where the surf is high. (The animals often found on bare rocks—purple seastars, mussels, and goose barnacles—are treated in Part II with the fauna of the surf-swept open coast.) Rather, almost every square centimeter in the lower tide-pool area is covered with protecting rockweeds, corallines, or laminarians. Though few of the animals are completely exposed, a rich and varied population can be seen without disturbing the rocks.

Fig. 67. A rich area of protected outer coast at Carmel; Ed Ricketts collecting in the low zone.

§63. A black, tube-building cirratulid worm may form calcareous masses on bare rocky shores, particularly where there are limestone reefs. Reported from British Columbia to southern California, this is *Dodecaceria fewkesi*, formerly known as *D. fistulicola* (and including *D. pacifica*). Colonies of *Dodecaceria* start with the settlement of a single worm. This "founder" worm divides itself in two, and both halves regenerate the missing parts. The process is repeated over and over again until eventually colonies may measure more than a meter across. Compact colonies of these worms may also occasionally be found on the shells of the red abalone (§74) or the barnacle *Balanus nubilus*.

   In the south there is another bare-rock form, *Chama arcana* (§126; formerly *C. pellucida*). In some regions this clam is practically a reef builder, but it achieves its maximum development in quiet waters.

   §64. The other zones exhibit a few characteristic and striking animals that occur in great numbers, but it would be hard to pick any highly dominant forms out of the lower tide-pool life. Perhaps the solitary great green anemone, *Anthopleura xanthogrammica* (Fig. 68), is the most obvious; certainly it is the form most frequently observed by the layman. This surf-loving animal might have been included as justly with the open-coast fauna, but since it is here in the tide-pool region that the casual observer is

Fig. 68. The giant green anemone, *Anthopleura xanthogrammica* (§64), common in deep pools and channels of Zone 4; about ¼ maximum size. The specimen at bottom center is contracted.

first likely to see it, the inductive method of this book justifies its treatment here. So far as is known, the huge specimens of *A. xanthogrammica* that are found in the deep pools and channels are exceeded in size by few others. Some Antarctic anemones are larger, as are the giant anemones of the Australian Barrier Reef, specimens of which have been reported as over 30 cm in diameter, with stinging capacities as great as the stinging nettle. Although our form sometimes attains a diameter of 25 cm, a bare hand placed in contact with its "petals" will feel no more than a slight tingling sensation and a highly disagreeable stickiness.

Typically a west-coast animal, this magnificent anemone has the enormous range of Unalaska and Sitka to Panama, and it has been the subject of observation and experimentation since Brandt first observed the Alaskan species in 1835. It is a more or less solitary animal, typically of great size and with a uniformly green color, and it is restricted to the lowest tide zone, or lower in areas where surf and currents continually provide a fresh supply of water. Unlike the smaller, aggregated *Anthopleura elegantissima* (§24), which occurs in great beds high in the intertidal zone, *A. xanthogrammica* cannot survive where there is sewage, industrial pollution, or sludgy water.

Those specimens of *Anthopleura xanthogrammica* that live in sunlight are a vivid green, whereas specimens found in caves and other places sheltered from direct sunlight are paler. The amount of green pigment produced by the anemone apparently relates at least in part to the health and efficiency of photosynthetic symbionts (dinoflagellates and unicellular green algae), which live in the anemone's tissues in a mutually beneficial arrangement, as mosses and algae live together to form lichens. The anemone's bright-green pigment is thought to screen the symbionts from excessively bright light that might be damaging. In dim light, however, the symbionts require all the light they can get, and the anemones oblige by producing little or none of the green pigment. In its preference for bright (but not the brightest) light, as in many other physiological traits, this giant green anemone is similar to the aggregated species of *Anthopleura*.

The diet of *Anthopleura xanthogrammica* consists primarily of mussels, sea urchins, and crabs that have become detached from the substratum. As if anticipating this future source of food, the larvae of *A. xanthogrammica* apparently settle preferentially in mussel beds (at least we know that young anemones are found here more often than lower down); later they migrate downward to take up a position characteristic of large anemones, in pools and channels below the bed, where they wait for a tasty morsel to appear from above. The spawning period is late spring to summer.

A sea spider, *Pycnogonum stearnsi* (§319), will often be found associated with the giant green anemone in central California, apparently forming some such relationship as most of the other sea spiders form with the hydroids on which their young live. Nor are sea spiders the only animals appreciative of *Anthopleura*'s society: a giant amoeba, *Trichamoeba schaefferi*, which is visible to the unaided eye if the specimen is in strong light, has been taken from the base of the anemone.

Around the bases of the green anemone at La Jolla may be found a predatory or quasi-parasitic snail, *Opalia funiculata* (Fig. 69; formerly *O. crenimarginata*). This snail occupies at San Diego, and presumably elsewhere in its recorded range of Santa Barbara County to Peru, the ecological position that *Pycnogonum stearnsi* does at places like Carmel and Duxbury Reef. A somewhat desultory study of this matter, lasting several years, has failed to reveal the presence of *Pycnogonum* and *Opalia* living together on the same anemones in southern California; but at Pacific Grove and at Shell Beach (Sonoma County) there were several finds of *Pycnogonum stearnsi* and *Epitonium tinctum*, together on the same anemones. *Epitonium*, which feeds on the tentacles of *Anthopleura*, is another of the wentletraps (Fig. 70), as the small, long-spired snails like *Opalia* are called. *Opalia* has been found

Fig. 69. The snail *Opalia funiculata* (§64), a wentletrap: *above*, feeding on the anemone *Anthopleura elegantissima* (Gunnar Thorson's Christmas card for 1956); *right*, apertural view of the shell.

Fig. 70. Shells of the wentletrap *Epitonium tinctum* (§64), with the granular egg capsules of the species; about 7 times natural size.

in association with *Anthopleura* as far north as Bodega. In Pliocene times, where La Jolla is today, there were *Opalia* almost 8 cm long, which would suggest, if proportions mean anything, that there could have been anemones the size of washtubs in those days.

The wentletraps, many of which have been identified as feeders upon coelenterates, produce a violet or purple toxic fluid that is thought to be an anesthetic, permitting the animal to pierce the body of an otherwise resisting anemone with its proboscis. *Pycnogonum*, with its strong claws, needs only to dig and poke its way in, and requires no tranquilizer (if that indeed is the function of the wentletrap's purple juice).

§65. *Tealia crassicornis* is another large anemone, beautifully red in color, usually with green blotches, and up to 10 cm in diameter, that may be found only at the most extreme of the low tides, attached to the sides and undersurfaces of rocks. Another *Tealia*, *T. coriacea*, is also common here, often half buried in the sand or gravel between large rocks. The two species may be most readily distinguished by examining their columns. That of *T. coriacea* usually has large tubercles to which bits of shell, gravel, or sand are attached; the column of *T. crassicornis* is generally smooth and without adhering material. *T. crassicornis* is also common on the coasts of Europe and, through circumpolar distribution, occurs on the east coast of America as far south as Maine, where it is called the thick-petaled rose anemone. Thus we have, as far south as La Jolla at least, another cosmopolitan species to keep company with the sponge *Haliclona*, the cucumber *Leptosynapta*, and the flatworm *Polychoerus*. A third species of *Tealia*, *T. lofotensis*, is a bright-cherry-red anemone with distinctive white warts on the column; this predominantly subtidal form occurs from Alaska to San Diego, and is also circumpolar.*

Other anemones, *Epiactis prolifera* and *Corynactis* (§§37, 40), and the solitary coral *Balanophyllia* (§39) may be found abundantly in this zone.

§66. Seastars are likely to be common and noticeable in this prolific zone. In most well-known regions of the Pacific (except Oregon), *Patiria* (§26) may be larger and more plentiful here than in the middle tide-pool region, but an examination of virgin areas indicates that the center of distribution is probably subtidal. Where surf keeps the rocks bare of large algae, *Pisaster* (§157), the common seastar, may also be abundant. There are a number of seastars, however, that are characteristic inhabitants of the low intertidal.

The many-rayed sunflower star, *Pycnopodia helianthoides* (Fig. 71), is possibly the largest seastar known, examples having a spread of more than 60 cm being not uncommon. Certainly it is the most active of our Pacific asteroids, and specimens have been known to travel 50 cm in a minute; though all seastars can and do move, they are scarcely to be classed in general as active animals. Probably they possess the most highly special-

*A fourth common species of *Tealia*, *T. piscivora*, occurs from Alaska to the Channel Islands, but it seems to be exclusively subtidal.

Fig. 71. *Above*, the under (oral) surface of the sunflower star, *Pycnopodia helianthoides* (§66), showing the tube feet. *Right*, *Pycnopodia* righting itself. This specimen is about 20 cm in diameter, but individuals three times this size have been found subtidally, making *Pycnopodia* one of the largest seastars; it is also one of the most active.

ized form of locomotion to be found in the intertidal zones. Locomotion is effected by the tube feet, which cling so strongly that many will be broken when a seastar is forcibly removed from a rock. On horizontal surfaces and on sand, the tube feet seem to be used as levers in a "stepping" motion, not unlike the legs of other animals. Traction is increased in many forms by the secretion of a sticky mucus and by a muscle-operated suction cup at the tip of each tube foot. These provide the necessary foothold when the seastar is climbing a steep surface or hanging inverted. They also secure the animal against wave shock, and allow it to move while remaining firmly attached.

Pycnopodia is readily distinguished from all other seastars in this zone by its soft, delicate skin, often colored in lively pinks and purples, and by the bunches of minute pincers (pedicellariae) on its upper surface. It also has the greatest number of rays—up to 24. It is often popularly called "the 21-pointer," but the number of rays is apparently dependent on its age, since it starts out in life with only six. Unless carefully supported when lifted from its tide pool, *Pycnopodia* indicates its attitude toward the human race by shedding an arm or two, and persistent ill-treatment will leave it in a very dilapidated condition.

The eating habits of *Pycnopodia* vary greatly from one location to another, as is the case for many seastars, depending on the availability of certain prey. The recorded diet is broad and includes various bivalves, gastropods, chitons, crabs, sea cucumbers, seastars, and potentially any dead animal matter encountered. A favorite item in the sunflower star's diet is the sea urchin, and many of the nicely cleaned tests of *Strongylocentrotus* to be found in tide pools indicate that *Pycnopodia* has recently fed well. It evidently engulfs its prey, spines and all, with its ample stomach, and leaves the well-cleaned remains behind. On occasion, the anemone *Anthopleura xanthogrammica* (§64) may be an inadvertent beneficiary of *Pycnopodia*'s foraging in sea urchin beds. Sea urchins respond to the presence of *Pycnopodia* by beating a hasty retreat, and in dense aggregations, this means climbing over the backs of their neighbors, who promptly attempt to dislodge them. Such a "stampeded" urchin will likely as not be swept away by the next wave, and, as misfortune often has it, directly into the waiting mouth of a hungry *A. xanthogrammica*. The range of *Pycnopodia* is from Unalaska to San Diego, but it is found less frequently south of Monterey.

§67. A less-common multi-rayed seastar, rare in the intertidal of central California but slightly more common to the north, is the sun star *Solaster dawsoni*. The number of arms in this species is usually 12 but varies from 8 to 15. A similar form, *S. stimpsoni* (Fig. 72), usually has 10 rays, very slim and uniformly tapering to a point. Both have a less-webbed appearance than *Pycnopodia*, having disks smaller in proportion to the size of the animal, and both are found only at or below the extreme low-tide mark. The colors range from purple-gray to orange, and the skin, which is harder than *Pycnopodia*'s and softer than *Pisaster*'s, is rough to the touch. *S. dawsoni*, ranging from the Aleutian Islands to Monterey (at the latter place found in

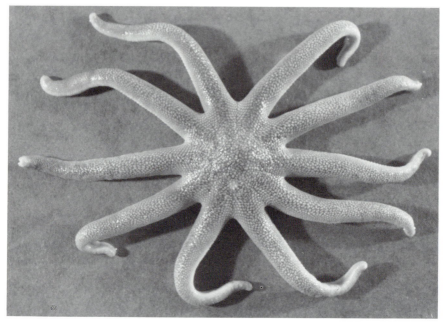

Fig. 72. The sun star *Solaster stimpsoni* (§67), a ten-armed orange to purple-gray seastar of the low intertidal and subtidal; usually 25 cm or less in diameter.

deep water only), is larger than *S. stimpsoni* (up to 36 cm) but very rarely occurs inshore. *S. stimpsoni*, ranging from the Bering Sea to Trinidad Head (Humboldt County), is not infrequent in the low intertidal. Both feed almost exclusively on other echinoderms: *S. stimpsoni* feeds on sea cucumbers, and *S. dawsoni* feeds primarily on seastars, including, and indeed preferring, its cousin *S. stimpsoni*. The dining plans of *S. dawsoni* are well known, however, and many of its kindred seastars leave immediately when contacted by this notorious predator.

§68. Two ovigerous seastars occur in this zone, and one of them, *Leptasterias hexactis* (Fig. 73), is very common. This little six-rayed form, up to 6 cm in spread and hence considerably larger than the delicate *L. pusilla* (Fig. 43) of the middle pools, is a feature of the low intertidal fauna throughout its range of Alaska to Newport Bay. The madreporite (a perforated plate through which water is admitted to the water-vascular system) is conspicuous in this species, and the rays are slightly swollen at the base, giving this dark-brown or olive species (less often dull pink or green) a bulky, unwieldy appearance. In Puget Sound, the breeding season begins in November (February in California) and terminates in April. The number of eggs in a brood varies from 52 to 1,491 depending on the size of the female, and the presence of the egg mass precludes the female from feeding. During this time, incubating females are likely to be humped up over a brood of eggs

Fig. 73. Three specimens of the small six-armed seastar *Leptasterias hexactis* (§68), slightly enlarged, showing some of the variety in shape and color pattern. This small but active seastar is an important predator on mollusks in the low zone.

and to keep fairly well hidden in rock crevices. Broods are carried for 6 to 8 weeks before minute seastars are discharged, most commonly in April or May; there is no pelagic larval stage.

The diet of *L. hexactis* has been extensively studied in Washington. As is the case for most seastars, this one too is a carnivore, and a culinary generalist at that, feeding on barnacles, snails, limpets, small mussels, chitons, and other animals. To some extent, its diet overlaps that of the larger seastar *Pisaster ochraceus*, and these two may often be in competition for available food. Some differences in diet exist, however, with *Leptasterias* apparently selecting hard-to-capture, calorie-rich prey (such as limpets), and *Pisaster* seeming to concentrate on sessile prey (such as mussels) whenever available.

§69. The other ovigerous seastar, *Henricia leviuscula* (Fig. 74), is most frequently found where rocks are encrusted with sponges and bryozoans. Although not nearly as common as the above, *Henricia* resembles *Leptasterias* in its usually small size (although 13-cm individuals have been seen) but differs in its usually vivid, blood-red color. The rays are long and sharply tapering. Smaller females of this species brood their eggs in January, keeping absolutely in the dark, and hence well hidden, during this period. Interestingly, larger females apparently do not brood but discharge their eggs into the sea. *Henricia* is a ciliary, particle feeder that may also feed

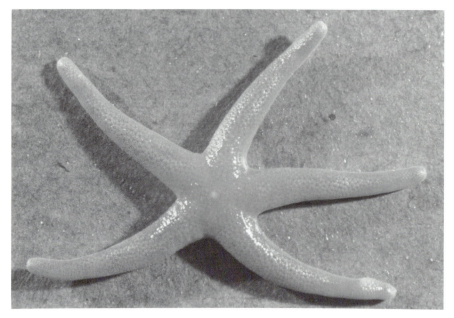

Fig. 74. A red seastar, *Henricia leviuscula* (§§69, 211), most frequently found on protected rocks covered with sponges and bryozoans.

on sponges and bryozoans. It ranges from the Aleutian Islands to Baja California; a related east-coast species (*H. sanguinolenta*) is circumpolar in distribution but does not range as far south as Monterey. The distinctive, gawky shape and vivid color of this animal make it a form that will stand out in the mind of the collector who finds it.

§70. The leather star, *Dermasterias imbricata* (§211), may be found occasionally in the lowest rock pools of the protected outer coast, but it is far more common in completely sheltered bays and sounds.

§71. Southern California has two characteristic rocky-shore seastars, both semitropical forms: *Linckia columbiae* (Fig. 75) and *Astrometis*. A symmetrical *Linckia* might be mistaken for *Henricia* were it not for the former's mottled gray and red colors and for the difference in the range. *Linckia*, however, a typical nonconformist, rarely falls into the bourgeois error of remaining symmetrical. This grotesque little species, which may have a spread of 10 cm, is famous for autotomy and variability. Monks (1904) said of them: "In over 400 specimens examined not more than four were symmetrical, and no two were alike. . . . The normal number of rays is five, but some specimens have only one, while others have four, six, seven, or even nine."

Referring to autotomy she says: "The cause of breaking is obscure. . . . Whatever may be the stimulus, the animal can and does break of itself. . . . The ordinary method is for the main portion of the starfish to re-

Fig. 75. Three specimens of the highly variable *Linckia columbiae* (§71), a southern form noted for autotomy.

main fixed and passive with the tube feet set on the side of the departing ray, and for this ray to walk slowly away at right angles to the body, to change position, twist, and do all the active labor necessary to the breakage. . . . I have found that rays cut at various distances from the disk make disks, mouths and new rays in about six months.''

"Comet" forms are a common occurrence, in which the end of one large ray bears a minute disk with from one to six tiny rays. Here is polyvitality (if one is permitted to use such a word) with a vengeance. How the single arm of an animal that normally has a disk and five arms can live, regenerate, and grow offers a striking example of the flexibility and persistence of what might have been termed "vital purpose" a generation or two ago. In the matter of autotomy, it would seem that in an animal that deliberately pulls itself apart we have the very acme of something or other.

Linckia columbiae reaches its farthest northward extension in the southernmost part of our territory, at San Pedro and San Diego, extending from there to the Galápagos Islands.

§72. *Astrometis sertulifera* (Fig. 76), compensating in symmetry for *Linckia*'s temperamental ways, still clings to a mild autotomy. Specimens that are roughly handled reward their collector by making him the custodian of a flexible three- or four-rayed animal. The rich ground color of the slightly slimy skin is brown or green-brown. The spines are purple, orange, or blue, with red tips, and the conspicuous tube feet are white or yellow. Surrounding the bases of the large spines are numerous tiny pincers, pedicellariae, which *Astrometis* can employ most remarkably to capture active

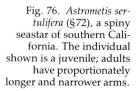

Fig. 76. *Astrometis sertulifera* (§72), a spiny seastar of southern California. The individual shown is a juvenile; adults have proportionately longer and narrower arms.

prey—even fish as large as the seastar itself can be captured and held by the combined force of the many pincers. This active and noticeable seastar, having an arm spread of up to 15 cm, ranges from Santa Barbara to the Gulf of California but never occurs in central or northern California.

§73. *Strongylocentrotus franciscanus* (Fig. 77), radiating spines like a porcupine, is the largest sea urchin found in this section, which is fitting in view of the length of its name. This giant brick-red or purple form often has a total diameter exceeding 18 cm. Its gonads are considered a delicacy by many, and now form an important export product for the Japanese market. We have sampled these gonads, eaten à l'Italienne (raw) with French bread, and found them very good—extremely rich, and possibly more subtle than caviar. If the race were not already being depleted by appreciative gourmets, and more recently by biologists, urchins could be highly recommended as a table delicacy.

Sea urchins are closely related to seastars and brittle stars. Whoever examines the test or shell of an urchin that has been denuded of spines will recognize this relationship in the pentamerous design and in the holes through which the tube feet protrude. Urchins and seastars have three kinds of projecting appendages: spines, tube feet, and pedicellariae. In the urchins the movable spines are the most conspicuous of the three. When an urchin's test or shell is prodded with a sharp instrument, the spines converge toward the point touched so as to offer a strictly mechanical defense. If a blunt instrument or the arm of a seastar is used, however, the spines turn away from the point of attack and give the pedicellariae free play. These peculiar appendages are thin, flexible stalks armed with three jaws apiece; each jaw is provided with a poison gland and a stiff sensory hair.

The urchin can maintain a stout defense with its pedicellariae against the attack of, say, a predatory seastar; but if the attack is continued for long enough, the urchin is likely to succumb because of the loss of its weapons, since each pedicellaria is sacrificed after inflicting one wound.

Like the seastars, urchins make auxiliary use of their movable spines. Those on the lower, or oral, surface may behave like stilt-like legs, and together with the tube feet permit the animals to walk about. Locomotion is relatively slow, but the fixed algae on which the urchin feeds are not famous for their speed of retreat. Movable jaw parts, forming a structure called, from its design, "Aristotle's lantern," cut the seaweed into portions small enough for ingestion in the urchin's huge, coiled intestine. The kelps *Macrocystis* and *Nereocystis* are preferred foods, but a variety of brown, red, and green algae are eaten as well. Urchins are also scavengers and will apparently eat almost anything they can manage, including dead fish.

An interesting feature of the digestive tract of *Strongylocentrotus franciscanus* is the frequent presence of a small rhabdocoel flatworm, *Syndesmis franciscana*. This commensal (or parasitic) worm, up to 6 mm in length, usually reveals itself by sluggish movements when the intestine of its host is opened. Sometime in the past few million years *Syndesmis* presumably dis-

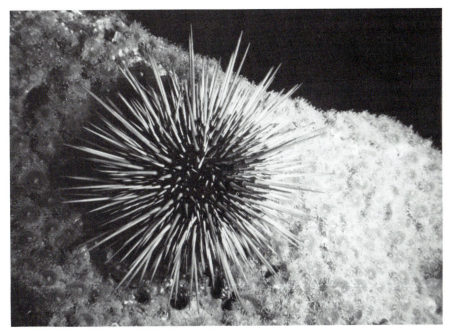

Fig. 77. *Strongylocentrotus franciscanus* (§73), the giant red urchin, an herbivore on low intertidal and subtidal rocks all along the coast. This individual is surrounded by a bed of the sea anemone *Corynactis* (§40).

covered that food-getting conditions on the inside of urchins were much superior to those on the outside and forthwith gave up free-living habits for good. How it gets inside is still a puzzle, however.

Urchins reproduce much like the seastar *Patiria* (§26), by pouring out eggs and sperm into the water for chance fertilization. Embryonic development of the free-swimming stages is also very similar in the two species, and *Strongylocentrotus*, even more than *Patiria*, has become an important object of embryological research. The sexes are separate, and in central California the main spawning period occurs in April and May, somewhat earlier than this to the south and later to the north.

The range of *Strongylocentrotus franciscanus* is from Alaska to Isla Cedros, off the coast of Baja California. Characteristically, this species inhabits only the deeper pools and the rocky shores extending downward from the low-tide line. Juveniles are further restricted in distribution, being found almost exclusively under the spiny canopy of adults. Here the juveniles probably gain protection from predators and also benefit from large pieces of food captured by the adults.

Often occurring with the giant red urchin, but preferring strongly aerated waters with violent surf (and hence treated in that habitat), is a smaller urchin, *Strongylocentrotus purpuratus* (§174). This species has a larger number of spines, which are vividly purple and more closely cropped than those in the bristling panoply of *S. franciscanus*.

§74. The very word "abalone" conjures up a host of associations for the Californian who has accustomed himself to steaks from this delicious shellfish. Still more vivid memories of foggy dawns will be recalled by the sportsman who has captured this huge, limpetlike snail on the low-tide rocks and reefs. Alas, they are no longer common intertidally; abalones for market today are taken by divers.

Haliotis rufescens, the red abalone (Fig. 78), may grow to be 30 cm long, but 17.8 cm (7 inches) is the minimum size that may legally be taken, and larger specimens are rare near shore. Abalone shells have been valued since their discovery by human beings with an eye for iridescent colors; and abalone pearls, formed when the animal secretes a covering of concentric layers of pearly shell over parasites or irritating particles of gravel, have in times past made fashionable jewelry. One of the common instigators of abalone pearls is a small boring clam, *Penitella conradi* (formerly *Pholadidea parva*), a relative of the shipworm *Teredo*, which except for the great difference in size, looks much like *Penitella penita* (Fig. 242). *P. conradi* also bores into mussel shells and occasionally into shale.

Haliotis occurs most frequently on the undersides of rock ledges, where it clings, limpet-fashion, with its great muscular foot. If taken unawares, individuals may be loosened from their support easily; but once they have taken hold, it requires the leverage of a pinch bar to dislodge them. Stories of abalones that hold people until the incoming tide drowns them are probably fictitious, but it is nevertheless inadvisable to

Fig. 78. *Above*, two red abalones, *Haliotis rufescens* (§74), attached to rocks encrusted with coralline algae; their shells are covered with fouling organisms, among them a stalked tunicate. *Below*, a red abalone that has been turned on its back and is extending its great foot preparatory to righting itself.

try to take them by slipping one's fingers under the shell and giving a sudden pull. We have captured them in this manner when no bar was available, but there is always the danger of a severe pinch.

The life history of the abalone has been carefully investigated, chiefly in connection with its commercial-fishery role. Growth is highly variable, but a red abalone reaches a length of 10 cm in about 6 years; under ideal conditions, rates nearly twice this have been recorded. Growth slows with increasing size and age, and animals approaching full size may be well over 20 years old. Sexual maturity is reached while the animals are relatively young, probably within 4 years, and fecundity is high. Animals in their first breeding season may produce only a few thousand gametes, but a specimen 15 cm in length may produce several million. The spawning of gametes, white sperm or gray-green eggs, has been recorded throughout the year, but tends to be greater in the spring and summer. Spawning may be precipitated by several conditions: a sudden change in water temperature, exposure to air for 1 to 2 hours, or rough handling. Recently, it has been shown that the addition of a small amount of hydrogen peroxide to seawater will cause spawning in both males and females, and another chemical (γ-aminobutyric acid; see Morse et al. 1979) has been found that will induce abalone larvae to settle and begin metamorphosis. Both findings may prove to be important for the mass rearing of abalone in mariculture.

The animal's diet is strictly vegetarian. Young abalone graze on surface films of microscopic plants, whereas larger animals eat mainly larger algae. These larger abalone, although capable of considerable movement, tend to be rather sedentary, occupying a crevice or other suitable spot and using the foot to capture loose plants as they drift by. The abalone's food strongly influences the color of its shell. When individuals eat only brown or green algae, the shell of the red abalone is actually greenish or white. Only when red algae are a significant part of the diet is the shell red (or brownish), and then, only the part of the shell being secreted at the time. Since some red algae are being consumed much of the time, the overall cast of the shell is usually pinkish to red; but in some individuals, distinct bands of color reflect seasonal or other changes in diet.

The abalone is eaten by a variety of predators, including crabs, octopuses, fishes, sea otters, man, and seastars. The last elicits a twisting, galloping escape response by the abalone, which departs amid a dense cloud of viscous mucus from the respiratory pores. Predation by sea otters and man is notable, too, both for the substantial impact of each and for the conflict that arises when both have aims on the same resource. Certainly, abalone are scarce now in places where they once were common, and it seems clear that the otter (in some localities) and overharvesting (in general) have both contributed. There is hope, however. The technical expertise to raise abalone to 1 to 2 years old in mariculture is now available, and transplanting these "hatchery" abalone to depleted grounds may replenish natural populations. In the meantime, the commercial catch of abalone continues to decline, as it has for the past several decades.

When an abalone is to be eaten, its shell is removed, the entrails are cut away, and the huge foot is cut into several steaks about 1 cm thick. Commercial fish houses pound these steaks with heavy wooden mallets; amateurs use rolling pins or short pieces of two-by-four. Whatever the instrument, the pounding must be thorough enough to break up the tough muscle fibers. Preferably, the meat should then be kept in a cool place for 24 hours before cooking. When properly prepared and cooked, abalone steak can be cut with a fork; otherwise it is scarcely fit to eat.

Abalones are known to have existed as long ago as the Upper Cretaceous age. They now occur in the Mediterranean, and even on the coast of England, but it is only in the Pacific that they attain great size, most being found off California, Japan, New Zealand, and Australia. The red abalone is distinguished by its great size, by the (usually) three or four open and elevated apertures in its shell (through which water used in the gills is discharged), and by the whole forests of hydroids, bryozoans, and plants that frequently grow on its shell. It ranges from Sunset Bay, Oregon, to northern Baja California, and it is most abundant subtidally on the central California coast between Mendocino County and San Luis Obispo. Specimens are now rare in the intertidal zone.

On the outside Alaska coast, as at Sitka, the small and dainty *Haliotis kamtschatkana*, commonly known as the pinto abalone, may sometimes be taken in abundance, since it, like *H. rufescens*, occurs in aggregations. It has rarely been taken as far south as Point Conception. Our specimens from Sitka Sound were up to 11 cm long, with butterscotch-colored flesh and brown-and-gray sides. Steaks from this small abalone are said to require no pounding, but it is probably well to work over any abalone before frying it. In British Columbia, *H. kamtschatkana* takes about 2 years to reach a length of 35 mm and another 3 to 4 years to reach 75 mm. Sexual maturity comes at about 65 mm in length (in Alaska), and spawning occurs during the spring.

There are eight species of abalone on the Pacific coast. For details concerning their ranges and habits, as well as a series of excellent color plates of the shells, the reader is referred to Cox's bulletin (1962).

§75. Below Point Conception, and as far south as Bahía Magdalena, Baja California, *H. fulgens*, the smaller green abalone, is the common form. It may be distinguished by its flat shell, which has fine corrugations and 5 to 7 open holes with the rims only slightly elevated. This abalone reaches sexual maturity in from 5 to 7 years, and spawns from early summer through early fall. Some green abalones are taken commercially, and we have often seen Mexicans gathering them for their own use. Besides humans, the other main predator is the octopus, which eats small as well as large specimens.

Still another *Haliotis*, the black abalone (§179), may occur with either of the above, but it is more common in surf-swept regions.

Underneath these and other species of abalone, a dark, large-clawed shrimp, *Betaeus harfordi*, may be found living commensally among the mollusk's mantle folds. *Betaeus* rarely leaves the abalone, but when removed experimentally, the shrimp relocates its host by means of a chemical scent emanating from the abalone's tissues. *B. harfordi* ranges from Fort Bragg to Bahía Magdalena.

§76. The seashore observer who, from seeing *Tonicella* (§36) and an occasional *Nuttallina*, has become familiar with the chitons for which the Pacific is famous will surprise himself some day by turning up a perfectly enormous brick-red sea cradle with no apparent shell. This giant, *Cryptochiton stelleri* (Fig. 79), sometimes called the "gumboot," is the largest chiton in the world, growing up to 33 cm long. It is reputed to have been used for food by the coast Indians and was eaten by the Russian settlers in southeastern Alaska. After one experiment we decided to reserve the animals for times of famine; one tough, paper-thin steak was all that could be obtained from a large *Cryptochiton*, and it radiated such a penetrating fishy odor that it was discarded before it reached the frying pan.

Although this giant form is sensitive to light and feeds mostly at night, as do the other chitons, it may remain out in the tide pools and on

Fig. 79. *Cryptochiton stelleri* (§76), the gumboot chiton. *Above*, a chiton partially detached from a rock at low tide, exposing the ventral surface; the many dark gills are evident along its slitlike mantle cavity. *Right*, the dorsal surface cut away to show the valves, or "butterfly shells," commonly found on the beaches north of Point Conception.

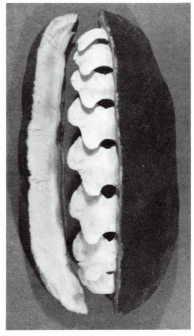

the rocks all day when there is fog. *Cryptochiton* feeds on fixed algae, rasping its food into small particles, snail-fashion, with a large radula, which may be examined by reaching into the chiton's mouth with a pair of blunt forceps and drawing out the filelike ribbon. (This practice should not be indulged in without good reason, since it undoubtedly leads to the demise of this magnificent animal.)

Chitons, although related to snails, are placed in a separate class of mollusks. Like snails, they have a long, flat foot, with the internal organs between it and the shell; but chitons have many gills in a row along each side of the foot, whereas most snails have only one or two. A chiton's shell is formed of eight articulated plates, which allow the animal to curl up almost into a ball when disturbed. In *Cryptochiton*, however, the plates are not visible externally, the fleshy girdle having completely overgrown them. The hard, white, disarticulated plates of dead specimens are often cast up on the beach as "butterfly shells."

Cryptochiton ranges from Alaska west to Japan and south to San Nicolas Island off southern California, though specimens are not numerous below Monterey Bay. The animal apparently has few enemies, grows slowly, and may live for more than 20 years. Sometimes in the spring great congregations of them gather on rocky beaches, having presumably come in from deep water to spawn. An account of the breeding habits of a related chiton (*Stenoplax heathiana*) is given in §127.

More than 25 percent of the cryptochitons examined have a commensal scale worm, *Arctonoe vittata* (formerly *Halosydna lordi*), living on the gills. These worms, with 25 or more pairs of scales, sometimes shed them when disturbed. *Arctonoe* grows up to 8 to 10 cm long and is light yellow in color. The same worm may be found in the gill groove of the keyhole limpet *Diodora aspera* (§181) and in the ambulacral groove of the seastar *Dermasterias imbricata* (§211).

Another guest that may be found clinging to the gills of *Cryptochiton* is the pea crab *Opisthopus transversus* (Fig. 80). This tiny crab, likewise never free-living, takes to a variety of homes, and in the San Diego area alone it has been found in association with 13 different species. Among others, it is found regularly in the California cucumber (below), frequently in the mantle cavities of the quiet-water mussel and the gaper (*Tresus*), occasionally in the giant keyhole limpet (*Megathura*), and even in the siphon tubes of rock-boring clams.

§77. Tourists at Monterey who go out in the glass-bottomed sightseeing boats are often shown "sea slugs," which are actually sea cucumbers. These rather spectacular animals, *Parastichopus californicus* (Fig. 81), are usually subtidal, but they often come to the attention of the shore visitor. Visitors to San Francisco's Chinatown (and probably visitors to other American Chinatowns) have also seen *Parastichopus* and a similar cucumber in the food shops. The body wall, first boiled and then dried and smoked, is a great delicacy. It also has the reputation, along with ginseng and bird's-nest soup, of being an efficient aphrodisiac. In the South Seas the production of this trepang, or *bêche-de-mer*, is an important fisheries industry, and there is now a small commercial fishery for *Parastichopus* in southern California.

P. californicus occurs from British Columbia to Baja California, and in the south—from Monterey to Baja—there is a similar-looking *Parastichopus*, *P. parvimensis*, which is definitely intertidal, at least south of Point Conception, living as high as the middle tide pools. The cylindrical, highly

Fig. 80. The pea crab *Opisthopus transversus* (§76), a common commensal of the gumboot chiton and the California sea cucumber; about 2½ times maximum size.

Fig. 81. The California sea cucumber *Parastichopus californicus* (§77), usually found in deep water but sometimes intertidally. In the background is a branched gorgonian, a colonial anthozoan.

contractile bodies of these cucumbers, black, dull brown, or red, are up to 46 cm long. The animals are covered above with elongate warts and below with the tube feet with which they attach themselves and crawl. Cucumbers may be flaccid when unmolested; but when annoyed, they immediately become stiff and turgid, shorter in length, and very thick.

We have pointed out that the lay observer must tax his credulity before grasping the fact that urchins are closely related to seastars and brittle stars. Sea cucumbers tax one's credulity still further, for they are related to all three of these, despite their wormlike bodies, their possession of tentacles, and their apparent lack of a skeleton. Their pentamerous symmetry is much less obvious than in urchins. It can be seen readily, however, in *Cucumaria miniata* (Fig. 239). The water-vascular system characteristic of echinoderms is manifest in the cucumber's tube feet, which, with wormlike wrigglings of the body wall, serve for locomotion. The tentacles around the mouth are actually modified tube feet. Like other echinoderms ("spiny skins"), cucumbers have a calcareous skeleton; but in their case it is only vestigial, composed of plates and spicules of lime buried in the skin and serving merely to stiffen the body wall.

Cucumbers in general have a specialized form of respiration that is unique among the echinoderms. Water is pumped in and out of the anus, distending two great water lungs (the respiratory trees) that extend almost the full length of the body. As would be expected, this hollow space, protected from the inclemencies of a survival-of-the-fittest habitat, attracts commensals and parasites. The respiratory tree is the common home of two microscopic, one-celled animals, *Licnophora macfarlandi* and *Boveria subcylindrica*, each of which clings by means of a ciliated sucking disk at the end of a posterior fleshy stalk. In aquaria the same pea crab that occurs with *Cryptochiton*, *Opisthopus transversus*, may be found in and about the posterior end of *Parastichopus*. When cucumbers are being relaxed for preservation, the crab will often leave his heretofore dependable home in the interior to cling to the outside.

A scale worm, *Arctonoe pulchra* (formerly *Acholoe*, *Halosydna*, or *Polynoe*), distinguishable by the dark spot on each scale (*A. vittata* scales are without spots), may often occur with the crab. This worm, and its association with other hosts including several seastars, mollusks, and polychaetes, has been the subject of considerable study. Worms removed from a host apparently use the chemical scent diffusing from the host's body to find their way home. Furthermore, given the choice between the original host species and an alternative species, the displaced worm returns to the original host with impressive regularity. Interestingly, if forced to remain on a new host species for 2 to 3 weeks, worms begin to prefer their new host over the original. *A. pulchra* adults are aggressively territorial, and although several small specimens may reside on one host, usually only one large worm (20 mm or more in length) is present even on a large host. Would-be newcomers are immediately confronted, and the ensuing combat frequently leads to major damage or even to death of one of the combatants. In this connection, the attractants issuing from hosts presumably assist the vanquished, or otherwise displaced individuals, in finding suitable new homes.

Many cucumbers lie half buried in the soft substratum, passing through the intestinal tract quantities of sand and mud from which their food is extracted. *Parastichopus parvimensis* is such a form. *P. californicus*, however, is a rock-loving cucumber. Considering its habitat and the nature of its tentacles, it seems likely that this cucumber brushes its stumpy appendages along the surface of the substratum as a related English form is known to do, sweeping minute organisms and bits of debris into its mouth.

The collector who insists on taking living specimens of these animals away from the tide pools for subsequent observation is sure to be provided, if the water gets stale, with firsthand information on the general subject of evisceration, a cucumber trait that has been termed "disgusting" by unacclimated collectors. When annoyed, *Parastichopus* spews out its internal anatomy in a kind of autotomy. The organs thus lost are regenerated if the animal is put back into the ocean. In the usual evisceration, the hindgut just inside the anus is ruptured by the pressure of water caused by a sudden contraction of body-wall muscles. This contraction voids first the respiratory trees and subsequently the remainder of the internal organs. Evisceration also occurs spontaneously (or at least without human interference) on a seasonal basis in the field. Natural populations of both *P. californicus* (in Puget Sound) and *P. parvimensis* (in southern California) have been shown to eviscerate more commonly in October and November. In one or more well-known examples the trait has protective value. An English form is called "the cotton spinner" from its habit of spewing out tubes covered with a mucus so sticky when mixed with seawater that it entangles predatory animals as large as lobsters.

The sexes are separate. As with many other marine animals, hosts of ova and sperm are discharged into the water, fertilization being pretty much a matter of chance. In Puget Sound, *P. californicus* adults migrate into shallow water as the gametes ripen, and spawning takes place in late July and August.

§78. In rocky clefts and tide pools where currents of pure water sweep by, delicate sprays of hydroids will be found, often in conspicuous abundance. Until their peculiarities are pointed out to him, the uninitiated is quite likely to consider them seaweeds, for only on very close inspection, preferably with a microscope, does their animal nature become apparent. They may occur, like *Eudendrium californicum* (§79), in great bushy colonies sometimes 15 cm tall, or, like *Eucopella everta* and various species of *Halecium* and *Campanularia*, as seemingly independent animals about 2 mm high, forming a fuzz about the stems of larger hydroids (see Figs. 82–89).

Whether large or small, the hydroids are likely to be first noticed for their delicate beauty and often exquisite design. Few things in the exotic tide-pool regions will bring forth more ecstatic "oh's" and "ah's" when first

examined than these hydroids, and one cannot but wonder why their fragile patterns have never been used as the motif for conventional designs.
Plumularia setacea (Fig. 82), one of the plume hydroids, is possibly the most
delicate of all intertidal animals. Underwater, and against many backgrounds, it positively cannot be seen, so perfect is the glassy consistency of
the living specimen. This is all the more remarkable when it is remembered
that there is a covering of skeletal material over the stems and over the
budlike individual animals, or hydranths. When removed from the water,
or seen against certain backgrounds, the tiny sprays, a centimeter or so
long, are plainly visible and well worth examination. *P. setacea* extends
from Vancouver Island to California. In the north, the larger and more
robust *P. lagenifera*, which occurs sparsely on the central California coast,
largely replaces *P. setacea*, until, north of Vancouver Island and clear up at
least to Sitka, it becomes the commonest plume hydroid.

Hydroids have a rather startling life history. It is a grotesque business,
as bewildering to the average person as if one were asked to believe that
rose bushes give birth to hummingbirds, and that the hummingbirds' progeny become rose bushes again. The plantlike hydroid that the shore collector sees gives rise by budding to male and female jellyfish, whose united
sexual products develop into free-swimming larvae, the planulae, which
attach and become hydroids, like their grandparents. The life cycle, then, is
hydroid-jellyfish-larva-hydroid.

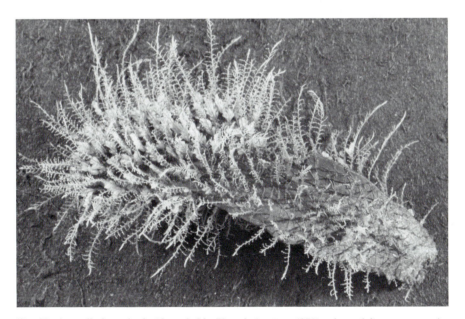

Fig. 82. A small plume hydroid, probably *Plumularia setacea* (§78), whose delicate sprays of
hydranths fur various objects in low-zone pools.

Unfortunately for the efficiency of this account, few of the rocky tide-pool hydroids produce free-swimming jellyfish. We may speculate on the reason for this: a jellyfish budded off from a rock-pool hydroid would begin its independent life in such a dangerous neighborhood that its chances of survival would be almost nil, compared to its chances of being dashed against the rocks and destroyed. Accordingly, the jellyfish generation, among the rock-pool species, is usually passed in saclike gonangia distributed about the hydroid stem. In some cases, as with the sertularians and the plumularians, the gonangia have degenerated into little more than testes and ovaries; but the medusae of *Tubularia marina*, although attached, actually have tentacles and look like the tiny jellyfish that they are. Species like *Obelia* (§§271, 316), which inhabit more protected waters or the piling of wharves, commonly have a conspicuous, free-swimming jellyfish generation. It should be mentioned that not all jellyfish are born of hydroids. All the hydromedusae come from hydroids, however; and most of the scypho-medusae, or large jellyfish, are budded from a stalked, attached, vaselike animal (the scyphistoma), which is the equivalent of the hydroid stage. The hydroid stage of one of our most common hydromedusae, *Polyorchis*, is still unknown, however.

Hydroids are interesting for another reason—they provide a protective forest for many other small animals, some of them, like the skeleton shrimp, absurd beyond belief. These dependents will be mentioned in detail later (§§88, 89, 90).

The accompanying photographs and drawings should aid in the identification of the hydroids listed in the following paragraphs. These are, we believe, the commonest forms, but there are innumerable others, and an attempt to list them would only lead us into confusion worse confounded.

§79. *Eudendrium californicum* (Fig. 83), mentioned above, is one of the most conspicuous of the hydroids, with bushy colonies that are sometimes 15 cm tall. The brown, slender stem by which the spirally branching colony attaches to the rock is stiff and surprisingly strong, and the roots are very firmly attached. The zooids—the individual animals in the composite group—look like bits of pink cotton on some lovely sea plant. Their rings of minute, white tentacles, each ring surrounding a central mouth, can be seen readily if the colony is held to the light in a jar of seawater. They retract slowly if touched, and lack the protective hydranth cups (thecae) into which so many of the hydroids snap back their tentacles.

Hydroids belong in the same great group as jellyfishes, anemones, and corals; in all of these the main internal body cavity is the gut, and ingested food is dumped there for digestion. The stem must be regarded as part of the colony, since through it runs a tubular extension of the gut that connects with all the zooids and transmits digested food. A good-sized cluster of *Eudendrium* may have thousands of living zooids, all with tentacles outstretched for the capture of passersby smaller than themselves. The prey is stung by the nematocysts and carried into the gut through the central mouth. During the winter this hydroid bears attached jellyfish in

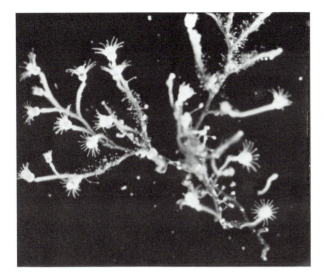

Fig. 83. A small colony of the hydroid *Eudendrium californicum* (§79); enlarged. The spirally branching colony, which may reach 15 cm tall, attaches to the substratum with a stiff brown stem.

the form of orange (female) and green-centered pink (male) gonophores. The female gonangia, borne at the base of the cuplike hydranths, may be turgid, and may invest the zooid so completely as to hide all but the tips of the tentacles.

Except on very low spring tides, colonies of *Eudendrium* are rarely exposed. They occur, even in regions of high surf, from Sitka, Alaska, to San Diego.*

§80. *Tubularia marina*, differing from its common and clustered brother *T. crocea* of boat bottoms and wharf piling (§333), appears to be strictly solitary. Individuals may be spaced evenly a few centimeters apart in rocky crevices or under ledges. The stem, usually not more than 2 cm long, bears a relatively large but dainty hydranth, often vividly orange (or pink). The medusoids are red, with pink centers, and have two to five very noticeable, long tentacles. Specimens that are sexually mature, that is, with sessile medusae present, have been taken in Monterey Bay in February, June, and July. The range is from Trinidad (near Eureka) to Pacific Grove.

§81. The hydroid *Garveia annulata* (Fig. 84), sometimes 5 cm high, with 20 to 30 zooids similar in size and appearance to those of *Eudendrium*, is recognizable by its brilliant and uniform color—orange root, stem, and hydranth, with lighter-orange tentacles. The colonies are likely to be seen at their best during the winter and spring months, when they commonly appear growing through or on a sponge. They also grow on coralline algae or at the base of other hydroids, and they themselves are likely to be over-

---

*Species in the genus *Eudendrium* are in need of review and revision. *E. californicum* ranges from British Columbia to Monterey, but other species are easily confused with it. The range of the genus extends from at least Sitka, Alaska, to San Diego.

Fig. 84. The orange-colored hydroid *Garveia annulata* (§81). Although similar in organization and appearance to *Eudendrium*, *Garveia* is easily recognized by its brilliant orange color.

Fig. 85. A colony of *Hydractinia milleri* (§82), which occurs as fuzzy pink masses on rocks in Zone 4.

grown with still other hydroids. *Garveia* ranges from Sitka, Alaska, to Santa Catalina Island.

§82. *Hydractinia milleri* (Fig. 85), which occurs in fuzzy pink masses on the sides of rocks, is interesting because it illustrates one of the most primitive divisions of labor—nutrition, defense, and reproduction are each carried on by specialized zooids. *H. milleri* is known from the outside coast of Vancouver Island, and it ranges at least as far south as Carmel. In Monterey Bay, recently settled colonies have been observed in August.

§83. The hydroids considered above are characterized by naked hydranths. Those that follow have hydranths housed in tough skeletal cups (the thecae), and most of them withdraw their tentacles into these protective cups very rapidly at the slightest sign of danger. The fernlike colonies of *Abietinaria anguina* (Fig. 86) have well-developed thecae, and the obvious skeleton is all that one ordinarily sees. During the winter or spring months one of the clean-cut sprays should be placed in a dish of seawater, left unmolested for a few minutes, and then observed. The transparent zooids will be seen popping out of the hard cups, forming a total pattern that is amazingly beautiful. At other times, the dead branches will be furred with other, smaller hydroids and with the suctorian protozoan *Ephelota gemmipara*, and will be covered with brown particles, which are diatoms.

Abietinaria anguina sometimes carpets the undersides of ledges or loosely embedded rocks, together with a smaller, undetermined relative, possibly the *A. amphora* also reported from the outer shores of Vancouver Island. The bushy-colonied *A. greenei* (Fig. 87), another of the numerous Pacific intertidal species of *Abietinaria*, will be taken for a totally different animal. The various species range well up the coast and south as far as San Diego, *A. anguina* and *A. greenei* being fairly well restricted to oceanic waters.

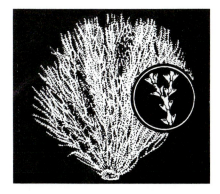

Fig. 86. *Abietinaria anguina* (§83), a hydrozoan with fernlike sprays; natural size.

Fig. 87. *Abietinaria greenei* (§83), a bushy-colonied hydrozoan not at all similar in overall appearance to its close relative *A. anguina*; about natural size.

Fig. 88. An ostrich-plume hydroid, *Aglaophenia latirostris* (§84), whose tan to dark-brown plumes form clusters attached to rocks and large algae on semiprotected shores.

Fig. 89. *Sertularia furcata* (§86), a small hydroid forming furry growths to about 1 cm tall, here on a blade of the surfgrass *Phyllospadix*.

§84. The Pacific coast is famous for its ostrich-plume hydroids, which belong to the genus *Aglaophenia*. Giant specimens (*A. struthionides*) are more common in violently surf-swept clefts of rock (§173) or in deep water, but they may be found here also. In the habitat considered here, however, there are at least two species so similar in appearance that only the larger form is illustrated. *A. inconspicua* occurs as 2- to 3-cm bristling plumes foxtailing over surfgrass or corallines, and is sometimes found at the base of larger species of the same genus. *A. latirostris* (Fig. 88) is a clustered form with dainty, dark-brown plumes as long as 7 to 8 cm. Surprisingly enough, *A. latirostris* was originally described from Brazil. It has since been recorded from Puget Sound and as far north as Alaska, and it is common in California. *A. inconspicua* is also common in central and southern California, and extends into Oregon.

§85. Several other skeletal-cupped forms, all beautiful, turn up very frequently in the lower tide pools. *Eucopella caliculata* forms a creeping network on the "stems" and fronds of algae that grow just at or below the extreme low-tide line. This network and the stalked solitary zooids that arise from it are opaque white and easily seen; even the extended tentacles are visible against the proper background. This hydroid produces a short-lived, free-swimming medusa stage. Several species of *Eucopella* are widely distributed on the Pacific coast, and *E. caliculata* is almost cosmopolitan in both hemispheres.

§86. *Sertularia furcata* (Fig. 89) is a small hydroid, known from the outer coast of British Columbia to San Diego, that forms a furry growth on the blades of surfgrass and occasionally on algae. The related *Sertularella turgida* is one of the most common and widespread hydroids on the coast. It is recorded from many points between Vancouver Island and San Diego, from both the protected outer coast and the fully protected inner coast, and from deep water as well as the intertidal. The stout and robust stalks of *S. turgida* have the zooids, which are swollen to turgidity, arranged alternately on the stem, rather than opposite each other as in *Sertularia furcata*.

The coppinia masses (compact, fuzzy encrustations of gonangia concentrated about the stems of the hydranths) and the strangely irregular creeping or mildly erect hydroid colonies of *Lafoea* species (*L. dumosa* and *L. gracillima*) will often be seen in the north; these are very characteristic, and once recognized, they are not likely to be mistaken for anything else. Both species range from Alaska to San Diego but will be found intertidally only in the north.

§87. The hydrocoral *Allopora porphyra* (Fig. 90) is more closely related to the hydroids than to the corals it resembles. Distinct and vividly purple colonies may be found encrusting rocky ledges at very low tide levels where the surf is fairly powerful but where no sediment occurs. In some of the semiprotected regions south of Monterey, and on the Oregon coast, *Allopora* colonies cover several square meters of continuous rock surface. On the Pacific coast and elsewhere, deep-water colonies of similar forms,

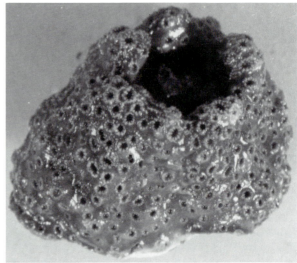

Fig. 90. *Allopora porphyra* (§87), a vividly purple hydrocoral of the very low tide levels, sometimes covering several square meters of rock. This specimen is encrusting the shell of a dead barnacle.

including the red *A. californica,* form huge erect masses, bristling with branches and the sharp points of the colorful lime skeleton. In the Caribbean they form some of the "coral reefs."

Many hydrocorals expand only at night, but our California shore form is even more conservative—it never expands at all, at least not while being watched. Occasionally, however, one gets a glimpse of a white polyp down in the tiny craters that protect the feeding zooids.

§88. As would be expected in a haunt so prolific as the lower tide-pool zone, the shelter of the hydroid forests attracts a great many smaller animals, both sessile and active. In this work, it is difficult to say at just what point animals become too inconspicuous to be considered, for in the tidelands it is almost literally true that

> Great fleas have little fleas
> Upon their backs to bite 'em
> And little fleas have lesser fleas
> And so *ad infinitum.*

A hydroid colony no bigger than can be contained in one's cupped hands may be almost a whole universe in itself—a complete unit of life, with possibly dozens of units in one tide pool. Each little universe includes amphipods, isopods, sea spiders (grotesque arthropods that are spiders only in appearance), roundworms, other hydroids, bryozoans, and attached protozoa, with a possible number of individuals that is almost uncountable. Though all of these animals are at least visible to the naked eye, and are abundant, characteristic, and certainly not lacking in interest, most of them are too small to be seen in detail without a hand lens or microscope and hence cannot be included in this handbook. It is impossible, however, to pass over some of the largest of these fantastic creatures without more detailed mention.

§89. Caprellids, amphipods very aptly known as skeleton shrimps, may be present in such multitudes as to transform a hydroid colony into a writhing mass, and the tyro will insist that it is the hydroids themselves that do the wiggling, so perfect is the resemblance between the thin, gangly crustaceans and the stems they inhabit (Fig. 91). If caprellids were a meter tall instead of around 25 mm, as in the case of the relatively gigantic *Metacaprella kennerlyi* (Fig. 92), no zoo would be without them, and their quar-

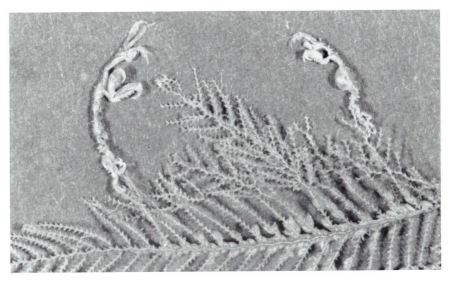

Fig. 91. Two caprellids less than 2 cm long on an *Abietinaria* colony. The female, on the right, bears a centrally located brood pouch.

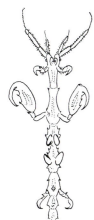

Fig. 92. *Metacaprella kinnerlyi* (§89), a pink-banded skeleton shrimp (caprellid) most commonly found on *Abietinaria* colonies (§83).

ters would surpass those of the monkeys in popularity. Specimens seen under a hand lens actually seem to be bowing slowly, with ceremonial dignity; clasping their palmlike claws, they strike an attitude of prayer. Often they sway from side to side without any apparent reason, attached to the hydroid stem by the clinging hooks that terminate their bodies, scraping off diatoms and bits of debris and filtering particles from the water. As a counterbalance to the animal's preposterous appearance, it evidences great maternal solicitude; the female carries her eggs and larvae in a brood pouch on her thorax. *M. kennerlyi*, a prettily pink-banded amphipod ranging from Alaska to southern California, is most commonly found in *Abietinaria* colonies. A substantial number of smaller species (e.g., *Caprella equilibra*) will be found in *Aglaophenia*, bryozoan colonies, and elsewhere. *Caprella californica* (§272), for example, occurs on eelgrass.

Caprellids can, when they wish, climb rather actively about on the branches of their hydroid homes in a manner suggestive of measuring worms. They take hold with their front appendages and, releasing the hold of their hind legs, bend the body into a loop, moving the clinging hind legs forward for a new hold. The body is extended forward again and the motion repeated until the desired destination is reached.

§90. Among the pycnogonids, or sea spiders, the male carries the eggs on an extra pair of appendages lacking in many females. Sometimes the thin, gangly-legged animal will be seen weighed down with a white mass of eggs that must be as heavy as he is. The bodies of sea spiders are not large enough to contain all of the internal anatomy, so that, as with seastars, the stomach extends into the legs. Few large clusters of *Aglaophenia* (§§84, 173) are without these weird creatures, and usually the comparatively large *Ammothea hilgendorfi* (Fig. 93), with a leg spread up to 25 mm, is the most conspicuous species. *A. hilgendorfi* (long known as *Lecythorhynchus marginatus*) is well disguised by its color—amber, with darker bands on the legs—and by its sluggish habits, so that specimens often may not be seen until the hydroids on which they live are placed in a tray and carefully examined. One of its larval stages, not even faintly resembling the adult (if the life history of this form resembles that of similar types that have been investigated), actually lives within the hydroid, entering by means of sharply pointed appendages and feeding on the body juices of the host. The legs of sea spiders, which end in recurved hooks, are admirably adapted to crawling about on vegetation and hydroids. *A. hilgendorfi* has been reported from the vicinity of Vladivostok (on *Zostera*), and it occurs on the California coast south to Isla Cedros, Baja California.

Tanystylum californicum, a much smaller sea spider, occurs on *Aglaophenia*. It is rounder and more symmetrical than its larger fellow lodger, and its legs are banded with white instead of light yellow, in a color pattern that matches its host.

Aglaophenia latirostris is for some reason a favored pycnogonid habitat, and most of the sea spiders mentioned here have been collected at one time or another on this hydroid. Particularly common along the California coast

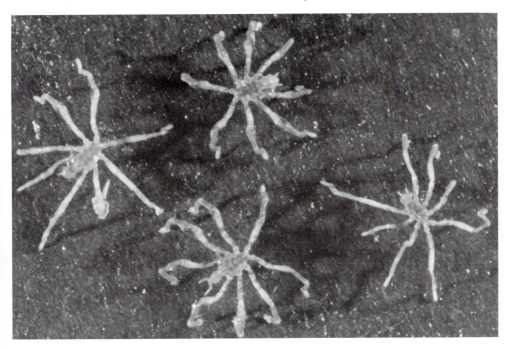

Fig. 93. *Ammothea hilgendorfi* (§90), a sea spider whose color (amber, with darker bands on the gangly legs) blends with that of the *Aglaophenia* colonies on which it lives.

during some years (possibly alternating in abundance with *Ammothea hilgendorfi*) is the thin and gangly *Phoxichilidium femoratum*, reminiscent of the *Anoplodactylus* (§333) taken with *Tubularia* and *Corymorpha*. *P. femoratum* is known to range from Dutch Harbor, Alaska, to Laguna Beach.

§91. Sponges of many kinds are common in the low intertidal landscape. Far simpler in organization than any of the hydroids, and near the bottom of the complexity scale in the animal kingdom, some of the sponges seem to be more or less loose aggregations of one-celled animals—colonies of protozoa, one might say—banded together for mutual advantage. This loose organization and the factors that promote cell aggregations have been the subject of many experiments. If the cells of *Microciona prolifera*, a common sponge of the Atlantic, are pressed through a fine cloth and sustained under proper conditions, the dissociated cells migrate actively and aggregate instantly upon contact. After 2 to 3 weeks, they will have formed a functional group. *Microciona* is not alone in this ability, and many other sponges will form such aggregations. Often if cell suspensions of two different sponges are mixed, for instance *Haliclona oculata* (purple) and *M. prolifera* (red), the cells will re-sort themselves in a species-specific manner: small purple *Haliclona* and small red *Microciona* will emerge from the cell mixture.

Although the sponges have no mouths, stomachs, or other specialized internal organs, they do have cells with flagella, or lashing "tails," which pull a continuous current of water through the colony, bringing in food particles and expelling waste products. The incoming, food-bearing current passes through innumerable fine pores, and the outgoing, waste-bearing current through characteristic craterlike vents (oscula). Two food-capture systems apparently operate in many sponges. Large particles are captured by cells lining the incurrent canals, whereas small particles, in the size range of bacteria, are captured by the special flagellated cells called choanocytes. Each choanocyte bears a filtering collar of closely spaced protoplasmic fingers, which surrounds the single, lashing flagellum.

The flagella responsible for driving the water current are microscopic, but the observer who would see for himself the currents of water that mark these plantlike encrustations as animals may do so easily by transferring one of the large pored sponges from its tide pool to a clean dish containing fresh seawater and adding a little carmine to the water. Notwithstanding the ease of making this simple experiment, there has long been a popular belief that sponges are plants. English observers once supposed, the *Cambridge Natural History* noted, that sponge colonies were the homes of worms, which built them as wasps build nests and as mud wasps build craterlike holes. The sharp-eyed and sharper-witted Ellis disproved this theory when he stated, in 1765, that sponges must be alive, since they sucked in and threw out currents of water. It is pleasant to note that the mighty intellect of Aristotle, far on the other side of the foggy Middle Ages, knew them for animals.

The volume of water passing through a sponge is impressive. Reiswig (1975a) estimated that in less than 8 seconds *Haliclona permollis* (§31) cycles through its body a volume of water equal to its own tissue volume. Another investigator, Parker (1914), remarked that in the vicinity of large tropical sponges the "water often wells up so abundantly from the sponges as to deform the surface of the sea much as a vigorous spring deforms that of a pool into which it issues." Laboratory tests indicate that a single such sponge colony, with a score of "fingers," will circulate more than 1,500 liters of water in a day.

In addition to the vital role played by the thousands of lashing flagella, recent studies have shown that sponges are assisted in their water-circulating endeavors by a simple phenomenon of physics, also put to good use in fireplace chimneys and prairie-dog tunnels. The oscula of sponges are often elevated slightly, and tidal or other water currents directed across the oscula create a "draft" through the sponge; water will even flow through the interior of a dead sponge if a water current is present outside the sponge (see Vogel 1977). To take full advantage of the benefits provided by the external water flow, many sponges adapt their shapes and orient themselves with respect to the external current.

Sponges reproduce both by an asexual method (budding) and by a

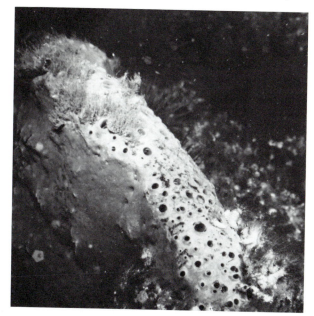

Fig. 94. *Speciospongia confoederata* (§92), a giant gray sponge at the lowest tide level; greatly reduced.

sexual method. In sexual reproduction, only sperm are released into the water; they exit via the excurrent stream and are taken in by the incurrent stream of another individual. Once inside, the sperm are trapped by choanocytes and transferred to the eggs. After fusion of the gametes, free-swimming, flagellated larvae are usually produced.

§92. A giant sponge, *Spheciospongia confoederata* (Fig. 94), occurs at the lowest tide level in great, slaty clusters on which several people could be seated. On close inspection these clusters are recognizable as sponges by the presence of numerous pores and oscula. At a distance of a meter, however, there is nothing to distinguish them from a hundred boulders that may be scattered about in the same goodly stretch of tide flat. *Spheciospongia* is known from Monterey Bay, and it has a recorded range of Queen Charlotte Islands (British Columbia) to Cabo San Quintín (Baja California).

§93. Masses of the cream-colored *Leucosolenia eleanor* (Fig. 95), one of the simplest and perhaps the most primitive of calcareous sponges, are common at the lowest intertidal level in crevices and at the bases of rocks. A similar "antler sponge" at Plymouth, England, has been observed to attain its complete growth as a massive cluster, several centimeters in diameter, in 6 to 8 months. Growth of the Pacific species seems to start in the spring with delicate colonies of slender, branching tubes, 2 mm or less in diameter, which become a three-dimensional lattice of connecting growth by late fall; hence, it is presumed to be an annual here also. It is common in Monterey Bay, and a similar species occurs in Puget Sound (§229).

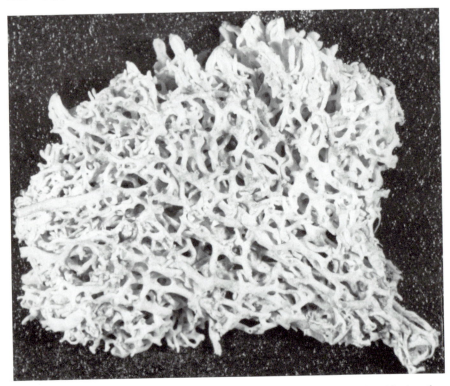

Fig. 95. The sponge *Leucosolenia eleanor* (§93), a cream-colored three-dimensional lattice of slender branching tubes.

§94. The urn-shaped *Leucilla nuttingi* (Fig. 96), recorded from Sansum Narrows, British Columbia, to Cabo San Quintín, Baja California, may be found growing suspended in crevices and under ledges at the extreme low-tide mark, where it is seldom exposed to the air. Large specimens, creamy white in color, may reach a length of 5 cm. Visitors from other regions commonly mistake this form for *Grantia*, a simple sponge that it closely resembles but to which it is not intimately related. Like *Leucosolenia*, the *Grantia* at Plymouth, England, is known to become adult in 6 to 8 months. At Pacific Grove large specimens of *Leucilla* (formerly *Rhabdodermella*) have been taken at all times of the year; only a few kilometers north of San Francisco, however, this sponge appears to be more seasonal, being most obvious in early spring and scarce by late summer.

§95. *Leucandra heathi* (Fig. 97; formerly *Leuconia*), a very sharp-spined, cream-colored sponge with a large, volcano-like central osculum, typically occurs as a dome- or pear-shaped colony, but it accommodates its shape to that of the crevice it occupies. The average size is small, but large specimens may be several centimeters in diameter. The large osculum, which acts as a

Fig. 96. *Leucilla nuttingi* (§94), an urn-shaped sponge found in protected microhabitats at the extreme low-tide level.

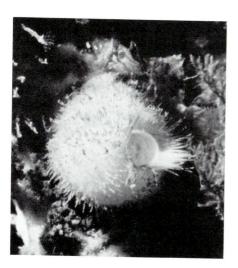

Fig. 97. *Leucandra heathi* (§95), a dome-shaped sponge from 2 to 10 cm in height with long spines surrounding the large single osculum.

centralized excurrent pore, is fringed with spines longer than those that bristle formidably over the rest of the body. *Leucandra*'s dangerous-looking spicules are calcareous, however, and so crumble easily—quite the opposite of the really dangerous siliceous spicules found in *Stelletta* (§96).

A large, shelled protozoan (1 to 3 mm in diameter), *Gromia oviformis*, may often be found on *Leucandra*. The spherical shell of this protozoan is transparent and glassy, exposing the ivory to light-brown interior protoplasm and leading many to mistake them for fecal pellets. A nearly cosmopolitan species, *G. oviformis* also occurs commonly on the holdfasts of surfgrass and algae and in sheltered crevices, from Canada to Mexico and on the Atlantic coasts of Europe and America.

§96. It is amusing to note the controversies that centered on sponges a few hundred years ago, when their structure was still imperfectly known. The herbalist John Gerard wrote in 1636: "There is found growing upon the rocks neer unto the sea, a certaine matter wrought togither, of the forme [foam] or froth of the sea, which we call spunges . . . whereof to write at large woulde greatly increase our volume, and little profite the Reader." The tactful Gerard might have been describing some of our white sponges. Possibly the most obvious of these are species of *Stelletta*, a feltlike, stinging sponge that encrusts the sides of low-tide caves. Certainly, this form is the most obvious to the collector who handles it carelessly and spends the rest of the day extracting the spicules from his hands with a pair of fine forceps. The spicules are glassy and have about the same effect on one's fingers as so much finely splintered glass. There are other white sponges, most of them smooth and harmless, but any uniformly white sponge with bristling spicules had best be pried from its support with knife and forceps. On one of the harmless varieties (*Leucilla nuttingi*, §94) a small white nudibranch, *Aegires albopunctatus*, may be found, matching the color of its home and its food source, just as the red *Rostanga* (§33) matches its red sponge.

§97. A good many other sponges occur in the low rocky intertidal, but we cannot consider them all in detail. Several others, however, are common enough to deserve mention.

Lissodendoryx firma, a strong and foul-smelling cluster up to 8 cm thick and 15 cm in diameter, has a lumpy, yellow or tan surface like a dried bath sponge, and is malodorously familiar to all prowlers in the Monterey Bay tide pools. When the sponge is broken open, its interior is found to be semicavernous and to harbor hosts of nematodes, annelids, and amphipods. This is one of the few sponges that may be found actually in contact with the substratum, on the undersides of rocks.

The famous and cosmopolitan "crumb of bread" sponge, *Halichondria panicea*, may also occur spottily up and down the coast; occasionally it is abundant in central California. Although most characteristically a low-intertidal form, it may extend well into the mid-zone in some locations. Encrustations are orange to green, amorphous, up to 6 mm thick and 3 cm in diameter, and of fragile consistency; the oscula are usually raised 1 mm above the superficially smooth to tuberculate surface. A variety of nudibranchs prey on this sponge.

Purple colonies of *Haliclona* (§31) occasionally extend into the low-tide zone from the middle tide pools, but these are rare compared to the red sponges. *Plocamia*, *Ophlitaspongia* (§32), and other red encrusting forms are often common here, splashing the sides of rocks with delightful bits of color. (See also §140, on under-rock forms, in this connection.) Near relatives of the familiar bath sponges are represented here only by small and rare specimens; the bath sponges grow luxuriantly only in warm to tropical waters.

§98. Under overhanging ledges in the La Jolla–Laguna region, and on the sides of rocks in the completely sheltered Newport Bay, occur clusters of the coarse-textured yellow sponge *Aplysina fistularis* (formerly *Verongia thiona*), known as the sulfur sponge. It is shown in Figure 58 growing pretty well up in the tide pools, with the tubed snail *Serpulorbis*. Clusters that fit one's cupped hands have from two to four raised oscula. This animal turns purple or black after being removed from the water, even if immediately preserved. This trait, and the texture, distinguishes it from *Hymeniacidon* (§183), another sponge occasionally abundant in the same habitat and similar superficially, although very different internally.

Living on *Aplysina* clusters is another animal, actually related to the sea hares but more nearly resembling a soft-shelled limpet. This little-known creature, *Tylodina fungina* (Fig. 98), is colored the same yellow as the sponge—another of the many examples of apparent protective coloration. Such examples ought never to be accepted without reservations, for very often animals that are "protectively colored" in one environment turn up in other habitats where their coloring makes them conspicuous. In this case the color may be due to the diet and little more.

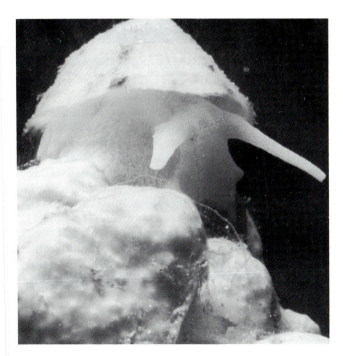

Fig. 98. *Tylodina fungina* (§98), crawling on a sponge. The snail, typically about 3 cm long, is yellow, which matches its sponge background.

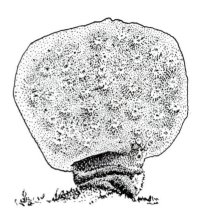

Fig. 102. *Polyclinum planum* (§103), a stalked, flattened tunicate; natural size. A live specimen can be seen at the right in Figure 101.

(Fig. 102). This animal consists of a definite, usually somewhat flattened, brown bulb some 5 to 7 cm long by two-thirds as wide, mounted on a thick stalk, or peduncle. The zooids, arranged in clusters, give the bulbous colony a flowered-wallpaper effect.

As was the case with hydroids and sponges, many more species of tunicates are likely to be turned up in the lower tide pools than can profitably be treated here. At the risk of being misunderstood, we might again quote the excellent Gerard and say, "to write at large woulde greatly increase our volume, and little profite the Reader." Another semicompound tunicate, *Perophora annectans*, must be considered, however, because it illustrates another organization of bodies into colonies, a looser one than the thickly embedded arrangements of the encrusting *Aplidium* or flaplike *Polyclinum*. The globular, green zooids of *Perophora* colonies adhere closely to the substratum but can be easily told apart, like so many tightly packed, tiny grapes. The brilliant-red colonial tunicate *Metandrocarpa taylori* shows this same habit and will occasionally be found in this low zone attached to undersurfaces, especially along granitic stretches of the coast. Like *Perophora*, *Metandrocarpa* ranges from British Columbia to San Diego.

§104. The sea hare *Aplysia californica* (Fig. 103), a tectibranch,* occurs in various environments—from completely sheltered mud flats to fairly exposed rocky shores. It occurs below the low-tide line and is also a common feature of the intertidal zone almost up to the upper pools. We have found great numbers of small specimens with the under-rock fauna in northern Baja California in February, and hosts of medium-sized specimens—about 13 cm long—at Laguna Beach in May.

Aplysia is hermaphroditic, having both testes and ovaries and both male and female organs of copulation. It may thus play female during one

---

*The term "tectibranch" is rarely used now. It refers to those opisthobranchs, of several orders, that have "covered" gills, as opposed to the "naked" gills of nudibranchs.

Fig. 103. Two views of the sea hare *Aplysia californica* (§104), a large herbivorous opisthobranch. Its color, an olive-green or brown, often with darker splotches, makes *Aplysia* inconspicuous in an algal forest.

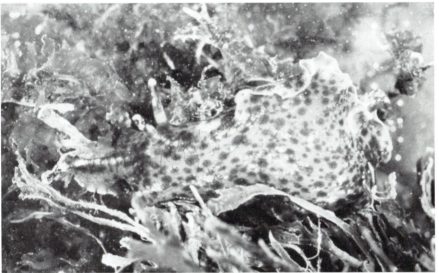

copulation and male during the next, as its whims happen to dictate; or it may play both roles at the same time. At Elkhorn Slough in the spring of 1931, seven or eight *Aplysia* were seen copulating in a "Roman circle," each animal having its penis inserted in the vaginal orifice of the animal ahead. Copulation lasts from several hours to several days.

Mated adults may be found throughout much of the year, but especially from the late spring to late fall, depositing their eggs in yellow, stringy masses larger than one's two fists. From a test count of a portion of such a mass MacGinitie estimated that the total number of eggs in the single mass was in the neighborhood of 86 million. MacGinitie (1934) also recorded a ±5 percent computation to the effect that a captive sea hare weighing almost 2.7 kilograms (under 6 pounds) laid 478 million eggs in less than 5 months. The total egg mass, when untangled, was 60,565 cm in length—about ⅓ mile! The largest of the 27 layings amounted to 17,520 cm, at an average rate of 5.9 cm or 41,000 eggs per minute; and each egg averaged 55 microns in diameter (a micron is 0.001 mm). The larvae become free-swimming in 12 days and contribute their mite of food to a marine world of hungry, plankton-seeking forms. Presumably, few more than the biblical two from one of these litters can run the gauntlet into adulthood, long before which their flesh will have become distasteful to predaceous neighbors.

The planktonic period lasts for a month or so, and then the larvae settle, primarily on the red algae that usually form an important part of the sea hare's diet. *Aplysia* grows very rapidly and may attain nearly maximum size within a year, after which many of the mature animals apparently die. Such a short life span is quite unusual for so large an animal and deserves more detailed study.

There are other miscellaneous points of interest. *Aplysia* has an internal shell, which it has tremendously overgrown, and a more complicated digestive system than almost any other algae-feeding invertebrate. The food, which can be sensed at a distance, is first cut up by the radula and then passed through a series of three stomachs, the second and third of which are lined with teeth that continue the grinding process begun by the radula. Breathing is assisted by two flaps—extensions of the fleshy mantle— that extend up over the back and are used to create currents of water for the gills, so that the animal may be said to breathe by "flapping its wings." When disturbed, *Aplysia* extrudes a fluid comparable to that of the octopus, except that it is deep purple instead of sepia. A handkerchief dipped into a diluted solution of this fluid will be dyed a beautiful purple, but the color rinses out readily; possibly the addition of lemon juice or some other fixative would make the color fast. The ink, whose color is derived from pigments contained in the red algae consumed by *Aplysia*, seems to be distasteful to some predators, as are various parts of the sea hare's body. In any event, very few animals are known to prey on adult *Aplysia*. The color of the sea hare itself is an inconspicuous olive-green or olive-brown, often with darker blotches. *Aplysia* is common as far north as Monterey Bay, and occasional specimens show up at Bodega in Sonoma County and Humboldt Bay.*

---

*Aplysia has been used extensively as a laboratory subject and, at least in terms of its behavior and physiology, is one of the best-known marine animals. Readers interested in cellular aspects of behavior are referred to the excellent book on *Aplysia* by Kandel (1979), which also contains a useful bibliography.

Fig. 104. *Cancer antennarius* (§105), a conspicuous large crab of low intertidal rocks, especially common where some sand has gathered.

§105. Another obvious animal in this zone, with hosts of less-obvious relatives, is the large crab *Cancer antennarius* (Fig. 104), which snaps and bubbles at the visitor who disturbs it. The carapace of *C. antennarius* is rarely more than 13 cm wide, but its large claws provide food quite as good from the human point of view as those of its big brother *C. magister*, the "edible" or Dungeness crab of the Pacific fish markets. (Except to the north during winter months when it occurs on shallow sandy beaches, *C. magister* rarely occurs in the intertidal in California as an adult, but it is common on the Oregon coasts in summer, when it comes into shallow water to molt.) *C. antennarius* can be identified easily by the red spots on its light undersurface, especially in front.

Here another word anent popular names is necessary, for to many people *Cancer antennarius* is known as "the rock crab," although the present writers have given that name to the bustling little *Pachygrapsus*. The reason is that the vastly more numerous and obvious *Pachygrapsus* seemed to be in dire need of a nontechnical name. It has been called "the striped shore crab," but except in occasional specimens the stripes are not particularly conspicuous. *Pachygrapsus*'s predominant green color is a more striking characteristic, but the name "green shore crab" designates a European form.

To return to *Cancer antennarius*: female specimens carrying eggs are most often seen from November to January (a few can be found at other times), with most embryos well advanced by the end of that period. In central California, sexual maturity is attained in about 2 years, and maximum size, over 15 cm across the carapace, is reached in 5 to 6 years. The crab is both a scavenger and a predator, hermit crabs and snails such as *Tegula* being among the prey consumed. The prey's calcareous defenses are penetrated either by a methodical chipping away of bits of shell from around the aperture, eventually exposing the occupant, or by a simple crushing of the shell in the powerful claws. *C. antennarius* is a rather delicate animal that does not live well in an aquarium—a direct corollary of its habitat. It is interesting to contrast it in this respect with the much hardier purple shore crab (*Hemigrapsus*) of the middle zone and with the extremely tough and resistant *Pachygrapsus* of the upper zone. It is an almost invariable rule that the lower down in the tidal area an animal lives, the less resistance it has to unfavorable conditions.* The range is from Coos Bay, Oregon, to Baja California.

§106. *Cancer productus* (Fig. 105), related to *C. antennarius*, may be considerably larger and therefore more useful for food, but it is not plentiful enough to be commercially important. Adults are uniformly brick red above. The young exhibit striking color patterns, in which stripes are very common. Like *C. antennarius*, *C. productus* may often be found half buried in the sandy substratum under rocks. At night it stalks about in the tide pools, so large and powerful that it dominates its immediate vicinity unless predatory man happens along. Like the other *Cancer* species, this one too is a carnivore, consuming live prey (e.g., barnacles and smaller crabs) as well as dead fish and such. *C. productus* is a good example of an animal that is pretty well restricted by its physiological makeup to this rocky habitat, for it lacks equipment for straining the fine debris encountered on bottoms of mud or pure sand out of its respiratory water.† It ranges from Kodiak Island to at least San Diego, and probably into Baja California.

§107. A dark-olive-green spider crab, *Pugettia producta* (Fig. 106), occurs so frequently on strands of the seaweed *Egregia* and others that it is commonly called the kelp crab. The points on *Pugettia*'s carapace and the

*Although the general correlation of durability and vertical position on the shore is probably still a good one, some would question whether *Cancer antennarius* is "a rather delicate animal that does not live well in aquaria." Jon Geller, for example, reports that it can be maintained in aquaria without much difficulty (pers. comm.). It is, however, not very tolerant of brackish conditions.

†As is perhaps usual, some individuals will be found under conditions that seem physiologically less than optimal. *Cancer productus* can be found occasionally on the mud flats of Bodega Harbor, and it is common in deeper waters offshore in the Morro Bay region (also from Point Conception to Santa Barbara), where it supports a small commercial fishery.

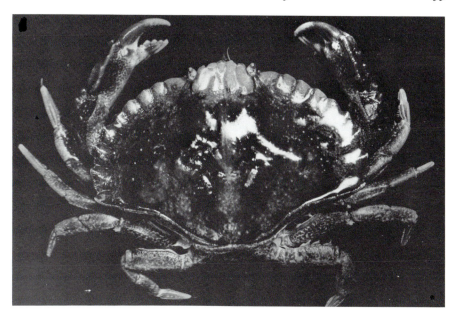

Fig. 105. *Cancer productus* (§106): *above*, an adult, uniformly brick red above and up to 16 cm in carapace width; *left*, a striped juvenile.

spines on its legs are sharp (the latter adapt the animal to holding to seaweed in the face of wave shock), and the strong claws are rather versatile. The collector who is experienced in the ways of kelp crabs will catch and hold them very cautiously, and the inexperienced collector is likely to find that he has caught a Tartar. Even a little fellow will cling to one's finger so strongly as to puncture the skin, and a kelp crab large enough to wrap itself about a bare forearm had best be left to follow the natural course of its life.

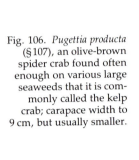

Fig. 106. *Pugettia producta* (§107), an olive-brown spider crab found often enough on various large seaweeds that it is commonly called the kelp crab; carapace width to 9 cm, but usually smaller.

Most spider crabs, which rest or move slowly during the day, pile their carapaces with bits of sponges, tunicates, and hydroids that effectively hide them, but *Pugettia* is moderately active and keeps its carapace relatively clean. Limited observations on this point suggest that specimens from rocky tide pools have less foreign growth than specimens from wharf piling, where the animals are very common. We have examined many wharf-piling crabs that had barnacles and anemones growing on their backs, or even on their legs and claws. In this connection we should note that like other spider crabs, *P. producta* undergoes a terminal molt upon reaching sexual maturity, after which individuals may become heavily encrusted with growths of various uninvited organisms that have settled on their surfaces.

The kelp crab is a prodigious breeder. In California, females may be found carrying eggs throughout the year, and some have been observed mating while still carrying the eggs of a previous brood. Under maximum conditions, a female seems capable of producing a new crop of offspring about every 30 days.

Pugettia, which is often parasitized by a sacculinid (§327), is reported as ranging from British Columbia well down into Baja California. Below Point Conception it is joined by the southern kelp crab *Taliepus nuttallii*, which ranges from there to Bahía Magdalena, Baja California. The carapace of *Taliepus*, usually fairly clean, is rounder than that of *Pugettia*, and often purple with blotches of lighter colors. Both crabs are primarily herbivorous, although *Pugettia* is known to consume sessile animals (e.g., barnacles, hydroids, and bryozoans) when its normal plant foods are not available.

§108. Our friends *Hemigrapsus nudus* and *Pagurus hemphilli*, the pugnacious purple shore crab and the retiring hermit, are frequent visitors in the lower tide-pool region, *P. hemphilli* being the only hermit found this far down, at least in the California area. Many other crabs, some of which are almost certain to be turned up by the ambitious observer, are not sufficiently common here to justify their inclusion in this account.

§109. The northern Pacific coast is noted for its nudibranchs, for there are probably greater numbers here, of both species and individuals, than in any other several regions combined. Several nudibranchs have already been mentioned, but it is in the lower zone that these brilliantly colored naked snails come into their own. It would be difficult for a careful observer to visit any good rocky area, especially in central California, without seeing several specimens. Their reds, purples, and golds belie their prosaic name, "sea slugs," so often imposed by people who have seen, presumably, only the dead and dingy remains. No animals will retain their color when exposed to the light after preservation, and a colorless, collapsed nudibranch is certainly not an inspiring object. Their appearance "before and after" is one of the strongest arguments against preserving animals unless they are to be used specifically for study or dissection.

Nudibranchs breathe through "crepe paper" appendages at the rear end (actually a rosette of gills around the anus) or through delicate, finger-like protuberances along the back. Nudibranchs with centralized gills in a single retractable cluster are generally grouped together as dorids. Two common dorids have already been mentioned: the red *Rostanga pulchra*, found on sponges (§33); and *Diaulula sandiegensis*, which characteristically has ringlike markings on the back (§35). *Anisodoris nobilis*, the sea lemon, is a bright-yellow nudibranch with a white gill plume, sometimes mottled with light brown (§178). A similar sponge-eating dorid is *Archidoris montereyensis* (Fig. 107), which is paler yellow and has dark, almost black, spots scattered irregularly on its back. (The dark spots of *Archidoris* extend to cover some of the tubercles on this nudibranch's back; those of *Anisodoris* do not, and occur only between the tubercles.) The coiled, yellow or cream-colored egg ribbons of *A. montereyensis*, each containing as many as 2 million eggs, may be found throughout the year. The range of this species is Alaska to San Diego. There are several species of *Cadlina*, characterized by short, rounded tubercles scattered over the back. Probably the most common is *C. luteomarginata* (Fig. 108), whose whitish back is edged by a yellow line and bears yellow-tipped tubercles. Specimens average about 40 mm long, occasionally twice this, and the range is from Vancouver Island to Punta Eugenia, Baja California. Many nudibranchs, but especially the dorids, have a penetrating, fruity odor that is pleasant when mild but nauseating when concentrated. Undoubtedly, this odor is one of the reasons why nudibranchs seem to be left strictly alone by most predatory animals; recent studies have indicated, for instance, that fish often quickly reject a nudibranch if it is accidentally ingested.

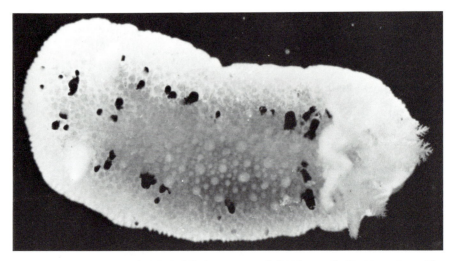

Fig. 107. A sponge-eating dorid, *Archidoris montereyensis* (§109); usually 25–50 mm long. The dark spots covering some of the tubercles serve to distinguish *Archidoris* from a similar yellow nudibranch, *Anisodoris* (Fig. 197).

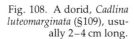

Fig. 108. A dorid, *Cadlina luteomarginata* (§109), usually 2–4 cm long.

Most of the dorids so far mentioned are prettily but not brilliantly colored (except for *Rostanga*). One of the brightest dorids on the California coast, sometimes found as far north as Coos Bay, is *Hopkinsia rosacea* (Fig. 109), which is a bright, uniform magenta or cerise ("rose pink," according to MacFarland 1966). A pink bryozoan (*Eurystomella*, §140) is the main food of *Hopkinsia*. More characteristic of southern California waters are such gaudy, blue-and-yellow creatures as *Hypselodoris californiensis* (with sepa-

Fig. 109. *Hopkinsia rosacea* (§109), a small dorid, 2–3 cm long, that feeds primarily on the pink bryozoan *Eurystomella* (§140).

rate, yellow spots), *Mexichromis porterae* (with a continuous, yellow band around the margin of the back), and the lovely violet *Chromodoris mac-farlandi* (with three yellow stripes on the back).* Two brightly colored dor-ids are common in central California: *Triopha catalinae* (formerly *T. carpen-teri*), with a snow-white body and brilliant-orange protuberances, occurs as far north as Alaska; *T. maculata* (Fig. 110), which usually has a brown, yellowish-brown, or red body speckled with light-blue spots (and often bearing reddish-orange processes and gill plume), ranges from Humboldt County to Cabo San Quintín, Baja California. The two species of *Triopha* differ somewhat in their feeding habits: *T. catalinae* prefers arborescent bryozoans, whereas *T. maculata* tends to eat encrusting forms during most of the year (some arborescent forms are included in the diet of large *T. maculata*).

Another group of nudibranchs commonly encountered is the eolids, characterized by rows of cerata on the back. Cerata are dorsal appendages containing a central core that is a branch of the "liver," or digestive gland. The eolids characteristically prey on cnidarians, and their cerata may be packed with nematocysts from the hydroids and anemones consumed. Since the nematocysts are "independent effectors," little mechanical de-vices that work by a trigger or by exposure to some chemical, it is possible

---

*All three of these chromodorid nudibranchs were listed as species of *Glossodoris* in the Fourth Edition. On the basis of work by Bertsch (1977, 1978a,b), however, the three species belong to three other genera.

Fig. 110. *Triopha maculata* (§109), a brown-bodied dorid sprinkled with light-blue spots; the body is usually 2–4 cm long.

that they may be used by the nudibranch against predators or in aggressive encounters with potential competitors. Perhaps the commonest, and loveliest, of the eolids is *Hermissenda crassicornis* (Fig. 111). Although variable in color, *Hermissenda* always has orange areas on the back and a clear-blue line around the sides. This eolid is more or less common throughout its range from Alaska to Baja California. According to MacFarland (1966), *Hermissenda* is "voracious and irritable, attacking other nudibranchs and members of its own species indiscriminately." More-recent quantitative studies, confirming its aggressive nature, have shown that encounters between two individuals result in fights involving lunging and biting about 20 percent of the time. Food deprivation increases the proportion of aggressive encounters, and presumably also increases the incidence of cannibalism. Even when protected from other members of its own species, the life span of *Hermissenda*, which may grow to a maximum length of 8 cm, is short, probably not exceeding much more than 4 months. Year-round reproduction and a short generation time of about 2.5 months (from egg to the time of first reproduction) apparently keep the species going (see also §295).

A striking contrast in coloration to *Hermissenda* is provided by the uniformly dull-grayish-mauve or pinkish-gray *Aeolidia papillosa* (Fig. 112), which is cosmopolitan in cold and temperate seas. Another contrast occurs in the arrangement of the cerata: clustered in *Hermissenda*, more or less uni-

Fig. 111. The eolid *Hermissenda crassicornis* (§109), whose white cerata tips contain nematocysts from hydroid prey.

Fig. 112. The eolid *Aeolidia papillosa* (§109), often 4–6 cm long, is cosmopolitan in temperate and cold waters. A colonial tunicate (*Perophora*) and a sertularian hydroid are also present.

form along both sides in *Aeolidia*. *Aeolidia* is a voracious predator on sea anemones, consuming 50 to 100 percent of its body weight each time it feeds, which is at least once a day. Laboratory tests suggest that adult *Aeolidia* prefer to eat the aggregated anemone *Anthopleura elegantissima* (§24), and are often found with it. However, this anemone possesses a battery of defensive mechanisms (including clone formation, its high position on the shore, and various behavioral responses), which combine to reduce the effectiveness of the predator, although they by no means deter individual attacks. The result, it seems, is that *Aeolidia* is more frequently found in association with one of its "least preferred" prey, *Metridium senile*.

Another of the widely distributed North Pacific nudibranchs, found from Vancouver Island to San Diego, is the uniformly gray *Dirona albolineata*. As the specific name implies, this species has white lines: they are vividly white, adding a conspicuous trim to the large, arrowhead-shaped cerata (Fig. 113).

Many of the nudibranchs are sporadic or seasonal in occurrence; some apparently come near shore only during the summer. In some years certain species may not be seen at all. Good samplings of our rich opisthobranch fauna (nudibranchs, sea slugs, and their allies) are available in the MacFarland monograph (1966) and in books by Behrens (1980), McDonald

Fig. 113. *Dirona albolineata* (§109), a gray nudibranch with white lines on the foot and cerata; body typically 2–4 cm long.

and Nybakken (1980), and Morris, Abbott, and Haderlie (1980), all with many color plates.

§110. On the "leaves" and "stems" of outer-tide-pool kelps, one almost always finds an encrusting white tracery delicate enough to be attributable to our childhood friend Jack Frost. But a hand lens reveals a beauty of design more intricate than any ever etched on frosty windowpanes. These encrustations are usually formed by colonies of the bryozoan, or ectoproct, *Membranipora membranacea* (Fig. 114), so named in the middle of the eighteenth century by Linnaeus, regarded as the founder of modern classification. *M. membranacea* is found in temperate regions all over the world and is common on this coast at least as far south as northern Baja California.* The minute, calcareous cells, visible to the keen naked eye but seen to better advantage with a lens, radiate in irregular rows from the center of the colony. The colony may be irregular and may cover many square centimeters, or it may be fairly small and round. The minute crowns of retractile tentacles (not unlike those of hydroids) by which the animals feed can be seen only with a microscope. As is usually the case with sessile animals, bryozoans have free-swimming larvae, and the process of larval settlement has been described for several species. The process can be watched by any careful observer with a good lens. For example, when a larva of *Membranipora* comes in contact with a bit of kelp, it tests the substratum briefly, opens its shell, and settles down; the shell flattens over the larva; and the bit of tissue enlarges rapidly to form a colony of several cells, which soon reaches visible size.

Living on *Membranipora* colonies, and feeding on them, are two small nudibranchs (Fig. 115). The first, *Corambe pacifica*, reaches a maximum length of 10 mm, but is nearly as wide as it is long. *Corambe*'s ground color is a pale, translucent gray that makes it almost indistinguishable from its bryozoan background; its yellowish liver shows through the center of its back, however, and surrounding this area the white of the foot may be seen. The rest of the animal's back, especially toward the margins, is flecked with irregular, sometimes broken, lines of yellow. There is a deep notch in the rear margin of the dorsum. *Corambe* occurs from Vancouver Island to Punta Eugenia, Baja California. The second nudibranch to be found on *Membranipora* is *Doridella steinbergae*, similar to *Corambe* in size and color but lacking the posterior notch. *Doridella* occurs from the Canadian San Juan Islands to Isla Coronados, Baja California.

*At least three species of *Membranipora* have been reported fairly commonly from the Pacific coast: *M. membranacea*, *M. tuberculata*, and *M. villosa*. They have often been confused, and in particular, many Pacific identifications of *M. membranacea* (primarily an Atlantic species) may actually be *M. villosa*. Furthermore, recent studies (Yoshioka, 1982b) have suggested that *M. villosa* (and *M. serrilamella*) may be only a phenotypic variant of *M. membranacea*, induced to change by the presence of nudibranch predators. It now seems that predation by *Corambe pacifica* and *Doridella steinbergae* Lance induces the formation of protective spines on the bryozoan; these spines had earlier been considered an important diagnostic feature.

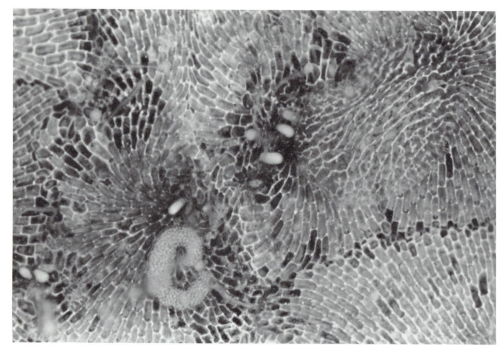

Fig. 114. The lacy white encrusting bryozoan *Membranipora* sp. (§110); greatly enlarged. Present but remarkably well camouflaged in the upper right quadrant is the nudibranch *Doridella steinbergae* (§110), with its egg mass below it to the left.

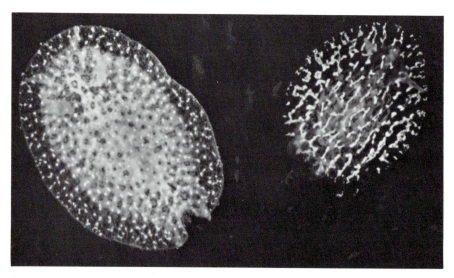

Fig. 115. Two tiny nudibranchs (greatly enlarged) that live and feed on *Membranipora: left, Corambe pacifica; right, Doridella steinbergae.* The larger specimen is 5 mm long.

§111. The lower tide pools contain many other bryozoans. Possibly the most obvious, though certainly not the most common, is *Flustrellidra corniculata* (Fig. 116; formerly *Flustrella cervicornis*), reported from the coast of Alaska to Point Buchon (San Luis Obispo County). The colonies are encrusting, but soft rather than calcareous; they form great, dull, gray-brown masses over the erect seaweeds. Their surface is covered by erect, horny, branching spines more than 6 mm tall, giving the colony a distinctly fuzzy appearance.

§112. Other bryozoans are certain to be mistaken for hydroids—the delicate *Tricellaria occidentalis* (Fig. 117), for example, whose erect, branching colonies resemble those of *Obelia*, although the two belong to entirely unrelated groups. Probably some lusty forebear of *Tricellaria* made the happy discovery, back in an age geologically remote, that the branching pattern of growth was successful in resisting wave shock and provided plenty of area for waving tentacles—a discovery that the hydroids made somewhat earlier. Bushy sprays of the white *Tricellaria* will often be found in rock crevices, with all the minute, tentacle-bearing colonies facing in one direction, so that the "stem" will be curved slightly downward. The animal ranges south from the Queen Charlotte Islands (and probably Alaska) to Isla Cedros, Baja California; it is abundant in the San Francisco Bay area.

§113. Branching bryozoans of the genus *Bugula* are found all over the world, and several species occur on the Pacific. A particularly handsome form is *Bugula californica* (Fig. 118). The whitish or yellowish colonies are large and definite, 7 cm or more in height, and the individual branches are arranged in a distinctly spiral fashion around the axis. In Monterey Bay, larvae of *B. californica* settle in all but the winter months, and in the warmer waters of Los Angeles Harbor, settlement may occur throughout the year. This species has been reported from British Columbia to the Galápagos Islands, and lush growths have been seen at Fort Ross, Pacific Grove, and the Channel Islands.

The avicularia ("bird beaks") of *Bugula*, thought to be primarily defensive in function, are classic objects of interest to the invertebrate zoologist. It is a pity that these, like so many other structural features of marine animals, can be seen only with a microscope. If the movable beaks of avicularia were a half-meter or so long, instead of a fraction of a millimeter, newspaper photographers and reporters would flock to see them. The snapping process would be observed excitedly, and some enterprising cub would certainly have one of his fingers snipped off. Avicularia and similar appendages, situated around the stems that support the tentacled zooids, probably have functions similar to those of the pedicellariae of urchins and seastars. Whatever else they do, they certainly keep bryozoan stems clean, as anyone will grant who has observed their vicious action under the microscope.

§114. Another, and perhaps more common, *Bugula*, occurring along the Pacific coast from the Bering Sea to Santa Cruz Island, is *B. pacifica* (Fig. 119); it is one of the most common bryozoans at Monterey Bay. It is slightly

Fig. 120. An entoproct, *Barentsia ramosa* (§115), forming thick-matted colonies in caves and on rocky underhangs; greatly enlarged.

§115. In the rocky caves of Santa Cruz, rank clusters of *Barentsia ramosa* (Fig. 120), an entoproct, form thick-matted colonies with headlike bunches at the end of 1- to 2-cm stalks. *Barentsia*'s tentacles contract unlike those of any hydroids or ectoprocts herein considered, being pulled in toward the center like the diaphragm of a camera shutter. The similar *Pedicellina cernua* may also be found furring the stems of plants, hydroids, and ectoprocts. The entoprocts are superficially similar to ectoprocts, and in older classifications were considered to be closely related; they are now classified as a distinct phylum by most zoologists.

§116. There are a good many small crustaceans in this zone, mostly amphipods and isopods. For the most part, they look pretty much alike superficially, and there seems to be no need for enumerating them here, for anyone interested in their differentiation will already have passed beyond the unpretentious scope of this handbook. We should mention, however, the hosts of half-centimeter beach hoppers that populate each cluster of the coralline alga *Amphiroa*. Whoever likes to deal with numbers of astronomical proportions should estimate the number of these amphipods in a handful of corallines, multiply by a factor large enough to account for the specimens present in one particular tide pool, estimate the probable number of tide pools per kilometer of coast, and, finally, consider that this association extends for something more than two thousand kilometers. One who has done this will have great respect, numerically speaking, for the humble amphipod, and will be willing to agree with A. E. Verrill that "these small crustacea are of great importance in connection with our fisheries, for we have found that they, together with the shrimps, constitute a very large part of the food of most of our more valuable fishes," and that they "occur in such immense numbers in their favorite localities that they can nearly always be obtained by the fishes who eat them, for even the smallest of them are by no means despised or overlooked even by large and powerful fishes, that could easily capture larger game."

§117. There are a number of shelled snails in this zone, *Nucella* and *Tegula* (§§165, 29) being fairly common. Larger and more spectacular than either, but not so common, is the brick-red top snail *Astraea gibberosa*, which ranges from the Queen Charlotte Islands, British Columbia, to Bahía Magdalena, Baja California. Taken intertidally in the north, it is usually subtidal in California. *A. gibberosa* may be 7 cm or more in diameter, and the larger, tan-colored *A. undosa* (Fig. 121) of southern California (Point Conception to central Baja California) may reach a diameter of 11 cm. Large shells of this type are quite likely to be overgrown by algae or hydroids, so that they are not conspicuous unless the living animal is present to move them along. The cone-shaped shell of *Astraea* is so squat, and the whorls so wide and regular, that identification is very easy.

A murex, known as *Ceratostoma foliatum* (Fig. 122) and related to the richly ornamented tropical shells so familiar to the conchologist, occurs on this coast. Mature, uneroded specimens are distinctive, bearing three projecting flanges on the largest whorl of the shell and a tooth on the outer lip of the aperture. The functions of the flanges, which develop gradually in young snails but then may disappear in old eroded specimens, have been the subject of considerable speculation. Suggestions include strengthening the shell, providing large size and a sharp weapon to discourage predatory fish, and increasing the likelihood that a snail detached by waves will land with its foot down. All may be true, of course, but the last suggestion has some experimental support: when dropped into water, specimens with all three flanges intact landed foot down more than half the time, whereas those with the dorsal flange filed off usually landed with the aperture facing up. *Ceratostoma* is an active and carnivorous snail, closely related to the oyster drill that plays such havoc with commercial oyster beds on the east coast. Its diet consists primarily of bivalves (e.g., *Protothaca*), which are preferred, and barnacles. In Washington, the snail reaches maturity in 4 to 5 years at a shell length of about 8 cm. Shortly thereafter, shell length actually decreases each year as the rate of erosion exceeds that of growth, and old snails (16 years or more) may be only 5 to 6 cm long. Mature individuals, scattered for most of the year, aggregate in the early spring to form small breeding groups. After mating, all the females in a group deposit their egg capsules side by side in one common mass on the rocks. Each female contributes approximately 40 egg capsules containing a total of 2,000 eggs. Young snails emerge 4 months later. The recorded range

Fig. 121. *Astraea undosa* (§117),
the wavy top shell,
a large gastropod of
low intertidal rocks
in southern California.

Fig. 122. The active and carnivorous snail *Ceratostoma foliatum* (§117), with its characteristic egg capsules.

of *C. foliatum* is Sitka to San Diego, but it is seldom found south of Point Conception.

Neither *Astraea* nor *Ceratostoma foliatum* will be found frequently, and both occur only in the lowest of low-tide areas.

Some of the limpets occur here, and at least two are characteristic. The white-cap or dunce-cap limpet, *Acmaea mitra* (Fig. 123), is rarely found living above the line of extreme low tides, however common its empty shells may be on the beach. Its height and size are diagnostic; the average specimen will be 26 × 23 × 16 mm, making it the tallest of the limpets. Individuals up to 43 mm long have been taken; in bulk these are probably the largest limpets, although some of the flat species may cover more sur-

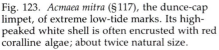

Fig. 123. *Acmaea mitra* (§117), the dunce-cap
limpet, of extreme low-tide marks. Its high-
peaked white shell is often encrusted with red
coralline algae; about twice natural size.

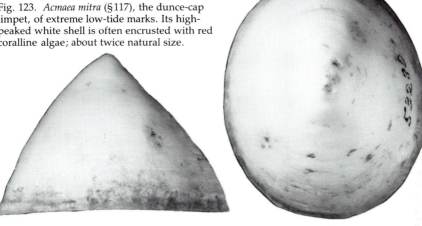

face. The shell is thick and white, but it is frequently obscured by growths
of the encrusting red corallines that constitute the main food of this limpet.
A. mitra's range is from the Aleutian Islands to Isla San Martín, northern
Baja California, and the animals are ordinarily solitary and fairly uncommon; but Mrs. Grant (see annotation under A. R. G. Test 1946) has observed thousands of specimens south of the Point Arena lighthouse, where
they occupy pits of the purple urchin. It is also fairly common far down in
the quiet waters of Puget Sound (§195). In California, this species spawns
in the winter.

The flatter *Notoacmea fenestrata* (Fig. 124), formerly considered a subspecies of *N. scutum* (Fig. 38; §25), is now considered a separate species. It
occurs from the Aleutians to northern Baja California, and is said to be the
commonest limpet at Jenner. The apex of *N. fenestrata's* shell is lower than
in *A. mitra*, the faces are convex, and the aperture is more nearly circular
than in any other limpet. Individuals may be identified readily by their
fairly consistent size (25 × 23 × 11 mm for a standard shell) and by the
ecological distribution recorded by Grant—smooth, bare rocks in sand.
When the tide recedes, the animals usually crawl down to sand and bury
themselves, so that often one must dig to find them. At Dillon Beach, however, *N. fenestrata* is known to occur fairly high in the intertidal.

As in the zones of previous sections, limpets more characteristic of
other levels may be found occasionally, and reference should be made to
§§11 and 25.

§118. The last form to be mentioned as belonging strictly to the
protected-rock habitat is a tiny colonial chaetopterid worm, whose membranous tubes mat large areas of vertical or slightly overhanging rock faces,
especially where the boulders rest on a sandy substratum. The tubes are
light gray, about 1 mm in diameter, and 5 to 6 cm long; they are intertwined

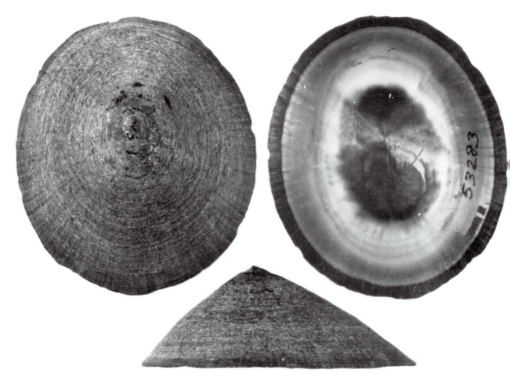

Fig. 124. *Notoacmea fenestrata* (§117), a limpet commonly found near the sand line on boulders set in sand; about twice natural size.

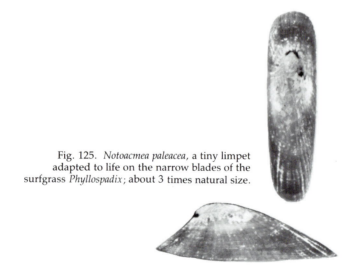

Fig. 125. *Notoacmea paleacea*, a tiny limpet adapted to life on the narrow blades of the surfgrass *Phyllospadix*; about 3 times natural size.

at the base, and encrusted with sand. A cluster the size of one's fist will contain, at a guess, several hundred tubes. As is so often the case, this form is easily overlooked until the observer has it brought to his attention and identifies it. After that it is obvious and seems to occur everywhere.*

§119. A variety of surfgrass (*Phyllospadix*) grows in this habitat on the protected outer coast; on its delicate stalks occurs a limpet, ill adapted as limpets would seem to be to such an attachment site. Even in the face of considerable surf, *Notoacmea paleacea* (Figs. 125, 349), called the chaffy limpet, clings to its blade of surfgrass. Perhaps the feat is not as difficult as might be supposed, since the flexible grass streams out in the water, offering a minimum of resistance. Also, the size and shape of the limpet adapt it to this environment. A large specimen, usually less than 13 mm long, is higher than it is wide, and its width (about a sixth of its length) matches closely that of the blade of grass. An average specimen is 6 × 1 × 3 mm; on the right anterior margin, and sometimes on the left, the shell is notched. The ground color of *N. paleacea*'s shell is brown, with white at the apex where the brown, ridged covering has been worn away. The surfgrass provides not only a home but also food for this limpet, which feeds on the microalgae coating the blades and on the epithelial layers of the host plant. Indeed, some of the plant's unique chemicals find their way into the limpet's shell, where they may possibly serve to camouflage the limpet against predators such as the seastar *Leptasterias hexactis*, which frequents surfgrass beds and hunts by means of chemical senses. The surfgrass limpet is fairly common in its recorded range of Vancouver Island to northern Baja California; it is abundant at Carmel.

Another common marine plant, *Egregia*, provides on the midrib of its elongated leathery stipe (sometimes 6 meters long) attachment sites for another limpet, *Notoacmea insessa*, limited specifically to this habitat (Fig. 126). The shell is dark brown, shiny, and smooth, or nearly so; but no description or illustration should be needed to identify an animal so obvious and so limited in its habitat. A typical shell is 16 × 14 × 9 mm. The life history of *N. insessa* is intimately connected with that of the host plant upon which it feeds. Although capable of spawning year-round, the limpet's main spawning period is in the spring and summer, with settlement timed to occur during the summer months when *Egregia* is growing rapidly. With abundant food, the new recruits themselves grow rapidly, and apparently mature at 5 to 7 mm in length, about 3 months after settlement. As the limpet feeds on the plant's tissue, it excavates deep depressions; these fit the limpet comfortably and presumably provide some protection against desiccation, abrasion, and wave action. The limpet's activities, however, weaken

*This species has not been encountered "everywhere" for many years, and its identity is still uncertain. Another (perhaps the same?) species of chaetopterid, which resembles the above in some ways, is *Phyllochaetopterus prolifica*. *P. prolifica* inhabits membranous tubes, measures 1 to 1.5 mm in diameter and 10 to 15 cm long, and occurs in large clusters on piling and in protected areas of the low intertidal zone from British Columbia to southern California.

Fig. 126. The specialist limpet *Notoacmea insessa* (§119), on the large brown alga *Egregia* upon which it feeds, forming deep depressions.

the plant, and during winter storms many limpets are lost when fronds break loose and are cast ashore. Many others survive, of course, to propagate the species, but few live more than a year. The range is Alaska to Bahía Magdalena, Baja California.

A third limpet associated with plants is *Collisella triangularis*, a small form that occurs on coralline algae or on the shells of *Tegula brunnea* encrusted with such algae. The high, elongate shell, to 7 mm long, is white with brownish spots, but this too is likely to be covered by encrusting algae. The range is Port Dick, Alaska, to Santa Barbara, and possibly into Baja California.

§120. A good many small or obscure animals, listed elsewhere as referable to other environments, may turn up here occasionally. Common in the north, and rare to occasional at Monterey, is the sessile jellyfish *Haliclystus*, thought to be more characteristic of quiet waters (§273). From Puget Sound north, but rarely as far south as the Washington outer coast, the hydrozoan jellyfish *Gonionemus* (§283) may be locally abundant in semi-protected areas, in coves off the open coast, or on the lee shore of protecting islands. Wave shock, although possibly regulating the distribution of both *Gonionemus* and *Haliclystus*, is thought to be of secondary importance to the need for pure and clear oceanic water, although *Gonionemus*, at least, can tolerate some reduction in salinity. Both species may be successful competitors in semisheltered and entirely quiet waters; but, especially in the case of *Haliclystus*, there must be strong currents of oceanic water from neighboring channels.

Under-Rock Habitat

The fixed clams, the great tubed worms, the brittle stars, and the crabs are the most noticeable under-rock animals. The newly ordained naturalist will perhaps be attracted first of all by the brilliant and active brittle stars. He is likely to find most of the brittle stars that occur under the middle-tide-pool rocks, and, especially in the south, he will surely find others. A reminder is perhaps in order here. This is a rich habitat, but for it to remain so, rocks turned over should be replaced right side up.

§121. Two of the large and striking brittle stars not found north of Santa Rosa Island stray with considerable frequency into the middle zone, but never occur there in the great numbers that characterize this low zone. One of them, *Ophioderma panamense* (Fig. 127), has a superficial resemblance to *Ophioplocus* (Fig. 55), but it is larger, richer brown in color, and smoother in texture, and it has banded arms. This species rarely autotomizes, and should prove to be a fair aquarium inhabitant. We have taken specimens as large as 2.5 cm in disk diameter with an arm spread of 18 cm from the Laguna region, which is near the upper limit of its recorded range (Santa Rosa Island to Peru), and others have collected specimens with disk diameters nearly twice this size. *Ophioderma* is an active species, emerging at night from its hiding place to roam over the bottom. Small living prey or other suitable food in its path are captured by a loop of the arm and brought to the mouth. As in *Ophioplocus*, *Ophioderma*'s short spines extend outward at an acute angle to its arms.

The other species commonly found here, *Ophionereis annulata* (Fig. 128), seems to be the southern counterpart of *Amphiodia* (Fig. 54), although it does not autotomize so readily. This very snaky-armed echinoderm, generally having a disk 1.3 cm or less in diameter and arms up to 10 cm in spread, ranges from San Pedro to Ecuador. Unlike many brittle stars, *O. annulata* moves about by the stepping motions of the podia rather than by movements of the arm.

§122. *Ophiopholis aculeata* (Fig. 129, upper left), a dainty, variably colored, and well-proportioned animal—according to human aesthetic standards—is notable for an astonishing range, both geographically and bathymetrically. The same species not only extends throughout the world in northern temperate regions but occurs at various depths from the low intertidal zone to the sea bottom at 600 fathoms. The physical conditions in the tide pools at Pacific Grove may not differ markedly from those at Sitka or on the coasts of Maine, England, or Spitsbergen, but there is certainly a marked difference between conditions at the surface and those on the

found on this coast in Alaska, British Columbia, and Puget Sound, and it has been taken as far south as Santa Barbara.

§123. The ultra-spiny brittle star *Ophiothrix spiculata* (Fig. 129, upper right) has its disk so thickly covered with minute spicules that it appears fuzzy. The arm spines are longer, thinner, and more numerous than those of *Ophiopholis*, with which it is sometimes found. *O. spiculata*'s color is so highly variable that it furnishes little clue to the animal's identity, but it is often greenish-brown, with orange bands on the arms and orange specks on the disk. The disk may be 17 mm in diameter, and the arm spread several times this. The egg sacs of ovigerous specimens may be seen protruding from the disk between the arms; in Monterey Bay, spawning has been noted in July. Although this is a southern form, ranging from Moss Beach (San Mateo County) to Peru, specimens are not at all uncommon at the northern extremity of the range. In some locations, population densities may be very high, on the order of 80 individuals per 0.1 square meter. Food gathering is accomplished by projecting several arms into the water and trapping particles in mucus secretions on the spines and tube feet. Adherent particles are then cleaned from the spines by the prehensile tube feet and passed to the mouth.

Associated with *Ophiothrix*, but treated with the more open-coast forms, may be found any of several "sea scorpions" (isopods), including the large *Idotea stenops*, also found clinging to brown algae (especially the boa kelp *Egregia*), and *I. schmitti* (§172).

§124. *Ophiopteris papillosa* (Fig. 129, lower) occurs with other brittle stars in the south, where it is not uncommon, but there it seldom reaches the great size and spectacular appearance that characterize the rare Pacific Grove specimens. In the south, too, *O. papillosa* is a relatively hardy animal, whereas at Pacific Grove it is so fragile that only once have we succeeded in taking an unbroken specimen. The spines are large and blunt, almost square-cut, and relatively few, although the temperamental animal is nonetheless distinctly bristly in appearance. The arms reach a spread of 15 to 18 cm and are brown, banded with darker colors. This brittle star is never found in great numbers, and is never accessible except on the lowest of low tides. Its recorded range is from Barkley Sound on Vancouver Island to Isla Cedros, Baja California.

§125. The avid turner of rocks will see many such fixed bivalves as *Chama* (§126) and the rock scallop *Hinnites giganteus* (Fig. 130). *Hinnites*, when young, is not to be distinguished from other scallops, and it swims about by flapping its two equal shells. But the ways of placid old age creep on it rapidly. The half-buried undersurface of some great rock offers the appeal of a fireside nook to a sedentary scholar, and there the scallop settles. One valve attaches to the rock and becomes distorted to fit it; the other valve adapts itself to the shape of the attached one, and there *Hinnites* lives to an alert old age, with its half-open shell ready to snap shut and send a

Fig. 130. The rock scallop *Hinnites giganteus* (§125). The animal's right valve (the light one) attaches when the scallop settles down, and becomes distorted to fit substratum contours.

spurt of water into the face of some imprudent investigator. The largest intertidal specimen of this species, which ranges from the Queen Charlotte Islands, British Columbia, to Baja California, may have a diameter of 15 cm, with subtidal specimens reaching 25 cm. In Puget Sound, spawning occurs during June.

The rock scallop is highly prized as food, but restraint should be exercised in its collection, for this is a slow-growing animal, taking perhaps 25 years to reach full size. Even medium-sized specimens may be several years or a decade old, and heavy collecting by clammers and divers has already depleted populations in many areas. The current bag limit on this species in California is 10 per day.

§126. *Hinnites* is fairly flat, without any sudden curves, but *Chama arcana* (Fig. 131; formerly *C. pellucida*) is often taller than it is long. Nearly hemispherical, and only 2 cm or so high, the translucent, white shell of this species is like nothing else to be found in the tide pools.* Except that it snaps its shell shut at times most inconvenient for the observer, we have little to record of its natural history; nothing is known about its breeding habits. Subtidal populations may fall prey to the seastar *Pisaster giganteus*,

---

*There is one bivalve with which *Chama arcana* may be initially confused. *Pseudochama exogyra* (Conrad) closely resembles *C. arcana* except that in the former, the beaks of the shell coil to the left as viewed from above, whereas in *Chama* they coil to the right. The two species occupy similar habitats and have the same geographical range.

Fig. 131. The fixed clam *Chama arcana* (§126), which is attached by the larger, deeply concave left valve.

and in this situation, the dense growth of epibionts that normally covers the shell seems to make detection, or attack, more difficult for the seastar; removal of the epibionts results in increased mortality in the field. In the Baja California region *Chama* grows so thickly on the tops and sides of rocks as to be practically a reef-building form; in the Pacific Grove region, however, it is usually found under rocks at the lowest of low tides. It is reported to range northward from Baja to Oregon, but it is rarely if ever seen in Oregon.

Most of the species of *Chama*, which share with the unrelated *Pododesmus* the name "rock oyster," are tropical forms, and are found attached to coral reefs. The few species in the world today are only a fraction of the number of extinct forms. Many are highly frilled and ornamented—taken by some as an indication that the species is reaching its old age and is likely to become extinct soon.

The rock oyster or "jingle," *Pododesmus*, is occasional here, but it is more common on rocks in quiet bays or on wharf piling (§217).

§127. The other molluscan inhabitants of the under-rock areas are chitons ("sea cradles"), several species of which have already been mentioned (see Figs. 132–34 for illustrations of various specimens). Most of the 110–125 species of chitons that inhabit the west coast are light-sensitive and will be found underneath the rocks in this zone, carefully protected from the sun.

Most chitons are vegetable feeders; but some are actively carnivorous, and others apparently prefer a mixed diet. In most cases the sexes are separate, and in some species, the females seem to retain their eggs until the males have liberated their sperm into the water. Spawning has been reported at various times from February to November, depending on the species, location, and year.

Stenoplax heathiana (Fig. 132) is a rather beautifully marbled, elongate

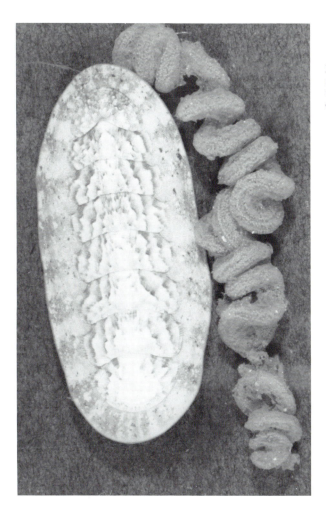

Fig. 132. *Stenoplax heathiana* (§127), a light-gray to white chiton, with its egg string.

(up to 8 cm long), light-gray to white chiton, ranging from Fort Bragg to Bahía Magdalena, Baja California; it is particularly common in Monterey Bay. Hidden during the day, this chiton emerges at night to feed on drift algae and may often be found exposed at dawn. Along a considerable stretch of coast, all mature *S. heathiana* individuals spawn on the same day in May or June, and at almost the same hour. Usually each female lays two long, jellylike spirals of eggs, which average 79 cm long and contain, together, from 100,000 to nearly 200,000 eggs. These jellylike egg spirals are unusual for chitons, most of which spawn eggs unattached to each other. The *Stenoplax* larvae begin moving within 24 hours, and in 6 days the young break through, to swim freely for from 15 minutes to 2 hours before they settle down. After settling, they remain inactive for about 2 days, but within 10 to 12 days they undergo several metamorphoses and become miniature chitons. The same sequence of events appears to apply to the other species of *Stenoplax* and to *Cryptochiton* (§76).

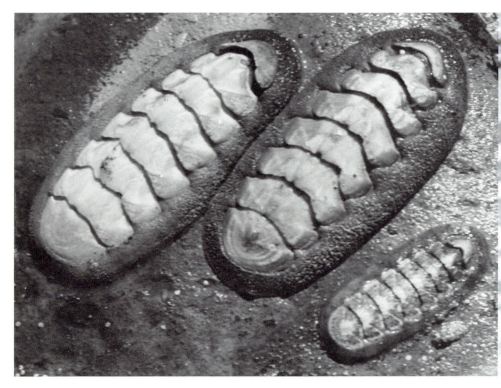

Fig. 133. *Stenoplax conspicua* (§127), an inhabitant of the undersides of fairly smooth rocks in southern California.

In the La Jolla tide pools the most abundant large chiton to be found at present is *Stenoplax conspicua* (Fig. 133), whose girdle is covered with fine, velvety bristles. Ten-centimeter specimens are common there, in the San Diego area, and in northern Baja California, the extreme range being Santa Barbara to the Gulf. Both of these large and noticeable chitons occur under fairly smooth and round rocks buried in types of substratum of which sand is the dominant constituent.

§128. Other common chitons (Fig. 134) are readily distinguished by their shape or color. *Ischnochiton regularis*, reported from Mendocino to Monterey only, is a uniform and beautiful slaty blue, up to 5 cm long. Specimens of *Callistochiton crassicostatus*, a mottled green or brown, have a noticeable median keel to the plates, are less than 2.5 cm long, and range from Trinidad to Isla Cedros, Baja California. *Placiphorella velata* is recognized by its almost circular outline; it is 5 cm or less in length, and has been reported from Alaska to Baja California. The food-getting habits of *Placiphorella* are noteworthy—it can graze the rocks like other chitons, but it is also an active predator. Prey like amphipods and worms are trapped under the chiton's normally uplifted head flap when the flap is quickly lowered onto the victims. *Lepidozona mertensii*, up to 4 cm long, with angular ridges and straight sides, is usually a mottled red and ranges from Alaska to Baja California. The similar but dull-brown *L. cooperi* occurs here also, and from Neah Bay, Washington, to Baja California. All of these are exceedingly shy animals, presumably because they find sunlight highly injurious. But the several species of *Mopalia*—broad, chunky chitons that have the mantle haired to the point of furriness—are more tolerant of daylight. We have found fine, healthy specimens living on the walls and ceilings of caves, and during foggy weather they may often be found on rocks. Ordinarily, however, they are under-rock animals like their relatives. Some *Mopalia*, however, live in sheltered bays; and *M. muscosa* is evidently tolerant of surprisingly muddy conditions on scattered rocks in such locations as Yaquina Bay, Oregon, and Tomales Bay.

The four common species of *Mopalia* may be identified according to the following field characters. *M. muscosa* (§221) is unmistakable: the thick, stiff hair is diagnostic. Of the other three, *M. hindsii* and *M. ciliata* have a definite notch at the posterior end of the girdle, and the sculpturing is very evident; *M. hindsii* has fine hairs, whereas those of *M. ciliata* are denser and flatter. The fourth species, *M. lignosa*, lacks both the girdle notch and prominent sculpturing, but it has prominent color lines and its posterior valve has a decided central beak. All four *Mopalia* range at least from British Columbia to southern California, occurring between tides on shores of low to moderate surf; *M. hindsii* and *M. lignosa* can also be found where surf is strong on the open coast.

§129. The large tube worms are of two sorts: serpulids, in which the tube is white, hard, and calcareous; and sabellids, which are larger,

Placiphorella velata

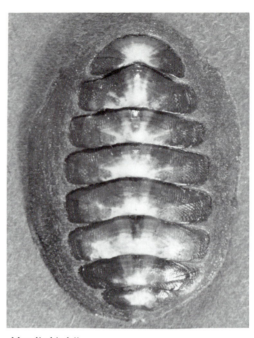

Mopalia hindsii

Lepidozona cooperi

Mopalia ciliata

Mopalia muscosa *Mopalia lignosa*

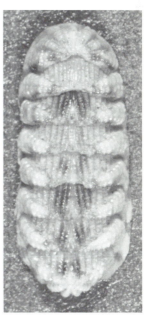

Ischnochiton regularis *Lepidozona mertensii* *Callistochiton crassicostatus*

Fig. 134. Under-rock chitons (§128) of Zone 4.

plumed worms with membranous tubes. The sabellid *Eudistylia polymorpha* (Fig. 135) is one of the vivid sights of the north Pacific tide pools. The dull-yellow or dull-gray, parchmentlike tube is very tough and may be 46 cm long, extending far down into the crevices of rocks; nothing short of prolonged labor with a pick will dislodge the complete animal. When the worm is undisturbed, its lovely gills protrude out of the tube into the waters of the tide pool, where they look like a delicate flower or like the tentacles of an anemone. Touch them, and they snap back into the tube with such rapidity as to leave the observer rubbing his eyes and wondering if there really could have been such a thing. So sensitive are the light-perceiving spots on the tentacles that the mere shadow of one's hand passing over them is enough to cause the tentacles to be snapped back. Three distinct color varieties are known (maroon, orange, and brown); the most common is maroon with orange tips. This worm ranges from Alaska to San Pedro but is not common south of Pacific Grove.

Eudistylia vancouveri is another sabellid of the northwest. These worms are gregarious, forming large, shrublike masses on piling, although they appear to be susceptible to reduced salinity; surprisingly enough, *E. vancouveri* is also found on vertical rock faces in heavy surf, where it would seem in danger of being washed away. As with *E. polymorpha*, the color is variable, but often the tentacles are crossed by alternating bands of red and white. Specimens exhibiting characteristics of both *E. polymorpha* and *E. vancouveri* have been found, suggesting the possibility of hybridization. The range of *E. vancouveri* is from Alaska to central California.

A solitary serpulid worm, *Serpula vermicularis* (§213), may be locally common. It has a sinuous to nearly straight stony tube several centimeters in length, which is stoppered up with a red operculum when the tentacles are withdrawn.

A smaller tube worm, *Marphysa stylobranchiata*, occurs in under-rock chitinous tubes much like those of *Eudistylia* but considerably smaller, and has two big head palps projecting from the tube in place of the great sabellid's tentacular crown. *Marphysa* is a eunicid, a very near relative of the palolo worm of lunar spawning habits (§151), but it apparently lacks that spectacular trait on this coast.

§130. The worm *Glycera americana* should be mentioned in passing; although commoner in sloughs, particularly in eelgrass (§280), it will sometimes be found under rocks on the protected outer coast. These slender worms may be mistaken for *Nereis* (§163; *Neanthes virens*, §311, is occasional here also), but they differ in that the head tapers to a point and in that the animal can shoot out a startling introvert. The long, slim, and fragile *Lumbrineris zonata* will be taken here occasionally, as well as the ubiquitous scale worm *Halosydna* (§49) and a good many smaller forms. Other worms frequently found are mentioned in the pool-and-holdfast habitat (§153), but of the enumerating of worms there is no end; certainly 20 species could be taken in this environment at Monterey alone. Their differentiation offers a challenge even to the specialist; field determinations are rarely conclusive.

Fig. 135. The sabellid worm *Eudistylia poly-morpha* (§129): *top*, clustered tubes; *center*, a cluster of expanded plumes, amid a bed of *Corynactis*; *left*, one expanded plume, surrounded by hydroids and bryozoans.

Fig. 136. *Loxorhynchus crispatus* (§131), a large masking crab, occasionally intertidal on the protected outer coast but more often subtidal on piling, rocks, and kelp holdfasts.

§131. Of the masking crabs, *Loxorhynchus crispatus* (Figs. 136, 351) is easily the champion. Although it occurs so far down in the intertidal as to be more justly considered a subtidal form, it will be found often enough in the intertidal between northern California and Baja California to justify mention, and it is certain to occasion interest whenever found. *Loxorhynchus* is generally a slow-moving crab, at least during the day and when left alone. But at night it is more active, and if harassed it is quite capable of scuttling across the bottom at a respectable pace. Until it moves, however, one never suspects that a crab is there.

To some animals the accidental growth of algae or sessile animals on their shells seems to be a source of danger, presumably because the weight and the water resistance of this growth might hamper their movements or ability to cling to the substratum. Such animals keep their shells scrupulously clean. Others tolerate foreign growths. Still others, notably some of the spider crabs, go to the extreme of augmenting the natural growths by planting algae, hydroids, sponges, etc., on their backs. This masking may

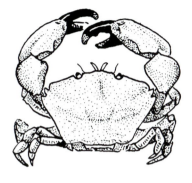

Fig. 137. *Lophopanopeus leucomanus heathii* (§132), a black-clawed crab; slightly enlarged.

serve a double purpose: first, to make the animal inconspicuous to its ene-mies; and second, to enable it to stalk its prey without detection. In the case of *Loxorhynchus*, masking is presumed to be primarily defensive, since the diet of this crab consists mainly of algae and sessile, or slow-moving, invertebrates such as sponges, bryozoans, and sea urchins.

The masking activities of *Loxorhynchus* have been studied in detail by Wicksten (1975, 1977, 1978), who has found that these crabs attach masking materials to their bodies by wedging or impaling them among hooked setae on the back and legs. Crabs that have had these hooks removed experi-mentally are unable to attach materials until their next molt, when a new set of hooks is produced. Masking materials are normally removed from the cast-off exuvia of the molt and recycled in redecorating the crab's new exterior. Large crabs, especially males over 8 cm in carapace width, ap-parently no longer actively decorate their backs, but, as with many non-decorating crabs, these too may nevertheless be covered by various growths that have settled and grown there without assistance.

A more common and more truly intertidal masking crab is *Scyra acu-tifrons*. It is not more than half the size of *Loxorhynchus*, and it is not so effi-cient a masker, although some larger specimens may be completely cov-ered with encrustations of sponges, bryozoans, tunicates, and barnacles. It has been reported from Cook Inlet, Alaska, to Punta San Carlos, Baja Cali-fornia, but it is uncommon below Pacific Grove.

§132. Under-rock investigations will turn up a number of other crabs, which may be differentiated by reference to the illustrations. *Pachycheles* (§318), a common wharf-piling inhabitant, will occasionally be routed out of an unaccustomed home here.

Lophopanopeus frontalis is a common southern California form, its place being taken in central California by *L. leucomanus* ssp. *heathii* (Fig. 137).*

*Lophopanopeus frontalis is perhaps more common in harbors and quiet bays than on the outer coast. It is now apparently rare in the northern part of its range, which is Santa Monica Bay to Bahía Magdalena, Baja California.

Fig. 138. *Mimulus foliatus* (§132), a spider crab; maximum size is about 4 cm in carapace width.

Mimulus foliatus (Fig. 138), of which ovigerous females have been seen throughout the year, may have a carapace of bright red or yellow or some pattern of these. On adults, growths of sponges or bryozoans may add to the coloration. The range of *Mimulus*, which is more common offshore on kelp than intertidally, is Unalaska to San Diego (a record from Mazatlán is in error).

The queer-looking anomuran crab *Cryptolithodes sitchensis* (Fig. 139) has so large a carapace that none of its appendages is visible from above, and it resembles nothing so much as a timeworn fossil. This slow-moving and variably colored species ranges from Alaska to Point Loma in San Diego County, and full-sized specimens (carapace to 7 cm wide) are not uncommon even in the southern part of the range.

The fuzziest of the crabs is the anomuran *Hapalogaster cavicauda* (Fig. 140), which ranges from Cape Mendocino to Isla San Gerónimo, Baja California. In life the animal presents an interesting sight, especially when, as during the winter months, the already swollen tail is distended with eggs; the male also shows the characteristic redundance of tail. This species and the more northern *H. mertensii* (§231) are omnivorous forms, straining water for plankton with their hairy outer maxillipeds and scraping bits of algae from the rocks.

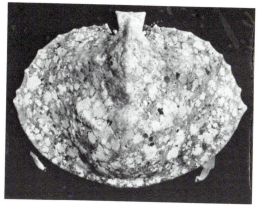 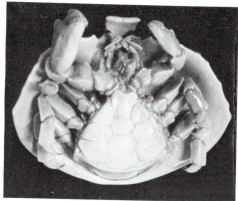

Fig. 139. *Cryptolithodes sitchensis* (§132), the umbrella-backed crab, an anomuran: *left*, dorsal view; *right*, ventral view.

Fig. 140. The furry *Hapalogaster cavicauda* (§132), an anomuran crab; about 1½ times natural size.

Not so much under rocks as in shallow crevices, abandoned sea-urchin pits, and other holes will be found *Oedignathus inermis*, a slow-moving, stout, clawed anomuran that superficially resembles a fiddler crab because of its large claw. It has a large, somewhat bulbous abdomen, and body hair is sparse or lacking. This crab occurs from Alaska to Monterey, but it is rare in central California. It is often encountered on the Oregon coast.

§133. Conditions of semitropical heat and aridity being what they are, it is quite in the order of things for the intertidal animals of the south to be restricted to the under-rock habitat. From Santa Barbara to as far down into Baja California as we have collected, there is a prolific and interesting fauna, but only the upturning of rocks will reveal it. Such rock-beach inhabitants as *Ischnochiton*, octopuses, brittle stars, ghost shrimps, and blind gobies (the last two restricted to the substratum) form an association in which a common top snail and several crabs also figure.

The crab *Cycloxanthops novemdentatus* (Fig. 141), a dull-reddish-brown animal with black chelae, will be taken for a small *Cancer*. But the dark-red *Paraxanthias taylori* (Fig. 142) will be readily identified by the obvious bumps on its big claws. The diet of both species consists chiefly of algae, especially corallines, but some animal matter as well. The larger *Cycloxanthops* has been observed breaking open and eating purple urchins. Both of these crabs extend down into Baja California from Monterey; but though they are found in the intertidal at La Jolla, for instance, neither is common in the northern part of their range.

Pilumnus spinohirsutus (Fig. 143), recognizable immediately by its hairiness, is another of the retiring, probably light-avoiding crabs of the southern coast. It has been recorded from San Pedro to Bahía Magdalena, Baja California (and erroneously to Ecuador), but the dredging of southern harbors has severely reduced its numbers in recent years.

Any of these three crabs may be found under rocks at Laguna, La Jolla, or Ensenada—sometimes all of them under a single large rock, possibly with *Lophopanopeus* and a small octopus thrown in for good measure.

Fig. 141. *Cycloxanthops novemdentatus* (§133), a black-clawed reddish-brown crab about natural size for a female; males may be nearly twice this size.

Fig. 142. *Paraxanthias taylori* (§133), a dark-red crab with conspicuous bumps on its massive claws. Natural size for a female; males somewhat more than half this size.

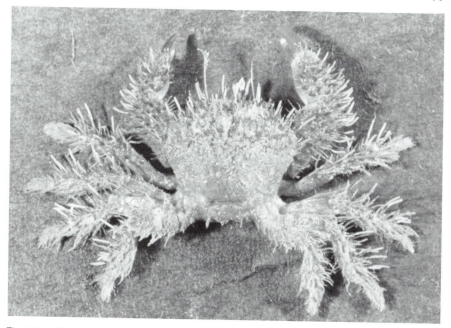

Fig. 143. *Pilumnus spinohirsutus* (§133), a retiring and exceptionally hairy crab of the southern California coast; slightly reduced.

§134. *Lysmata californica* (Fig. 144; formerly *Hippolysmata*) is a conspicuous, transparent shrimp up to 7 cm long (more commonly to 4 cm), with broken, red stripes running fore and aft. A dozen may occur on a single undersurface, but the collector who can capture more than one of them is doing well; for when the rock is moved, they hop around at a great rate and disappear very quickly. Clustered in protected spots during the day, they emerge at night in great numbers from along jetties and other rocky areas. The numbers are such that a bait fishery exists for these "red rock shrimp" in some locations (e.g., Los Angeles Harbor). *Lysmata* is one of the "cleaning" shrimps—so called because it removes parasites and decaying tissue from the surfaces of large animals that happen by, in this case especially moray eels. It is one of the least specialized of the cleaners, making only part of its living in this manner and sometimes falling prey to a larger animal that was to have been cleaned. *Lysmata* has been recorded from Santa Barbara to Bahía Sebastián Vizcaíno, Baja California.

§135. The much-maligned octopus—which, in the vernacular, shares the name of devilfish with the giant manta ray of tropical waters—is not often found in the intertidal regions of the north, but from Santa Barbara south a small species, *Octopus bimaculoides* (Fig. 145), is a common under-rock inhabitant of the outer tidelands. The octopus has eyes as highly developed as ours, and a larger and better-functioning brain than any other

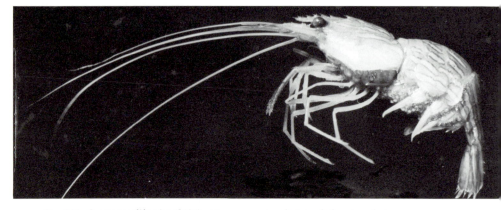

Fig. 144. *Lysmata californica* (§134), a transparent red-striped cleaning shrimp.

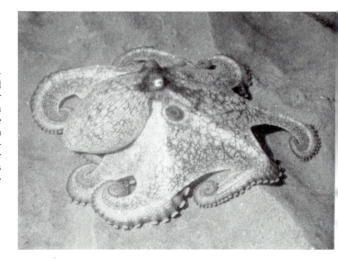

Fig. 145. *Octopus bimaculoides* (§135), a small octopus common under rocks and on mud flats in southern California. The specific name comes from the two dark round eye-like spots, set farther apart than the animal's real eyes (white in the picture).

invertebrate animal. Its relatively high degree of intelligence is quite likely a factor in its survival in the face of persistent collecting as food. Italians and Chinese justly relish the animals as food, and so does an occasional American. Delicious as octopus is, however, it is decidedly tough and rubbery, and thin slicing or the meat-grinder treatment is recommended.

In the metropolitan areas around Los Angeles and San Diego, octopuses are no longer to be had in their former abundance, for, besides the many people who hunt them specifically, hunters of abalones and spiny lobsters capture them incidentally. Nevertheless, a good many are still found, even in the areas mentioned, where it is probably safe to say that several thousand are taken for food every year. This observation is based on personal experience. For several years we have collected, observed, and photographed along the Corona del Mar shore at Newport Bay and in the region north of Laguna. On one visit to Corona del Mar we questioned one of several crews who we had supposed were collecting bait. It developed that they were capturing octopuses, and by a rather pernicious method: they poisoned a likely looking pool to force out the mollusk, which they promptly hooked and deposited in their gunnysack. These two men had captured 13 octopuses, totaling possibly 14 kilograms, and several other crews appeared to have equally good "luck." The whole procedure was approximately duplicated on each of the dozen or more times we visited the spot. So devastating a method of collecting as poisoning pools would be regrettable anywhere, unless the collectors were in genuine need of food, for it causes wholesale destruction of much of the other life in the vicinity. (Not only is this old method regrettable, it is now strictly forbidden by law.) Along the coast of Baja California, the octopuses are more fortunate, for the Mexicans we have talked with along the shores are interested in abalones and spiny lobsters only and regard the eating of octopuses with the same horror that most Americans do.

Between Tijuana and Ensenada we have found octopuses very numerous in April, May, and December, in several different years. In December 1930 they were so abundant that one could count on finding a specimen under at least every fourth rock overturned. Their clever hiding and escaping strategies make them difficult to see, difficult to capture after they are seen, and difficult to hold after they are captured. Two animals may be found under the same rock, but to take them both with bare hands is almost an impossibility. A captured specimen will sometimes cling to one's hand and, if the animal is large enough, to one's forearm; but unless the collector has acquired some skill in handling the wily animals, the specimen will at once let go, shoot out of his hand like a bar of wet soap, and disappear. A little observation will convince one that in a given area probably half of the specimens escape notice despite the most careful searching—a highly desirable situation from the points of view of the conservationist, the "regular" octopus fisherman, and the octopus.

This octopus is known as *Octopus bimaculoides* because of its two large, round, eyelike spots. These are wider apart than the animal's real eyes, which are raised above the body surface in knobs reminiscent of the light towers on sailing ships. Each of the eight arms has two rows of suction disks, so powerful that when even a small specimen decides to fight it out by main strength it is very difficult to detach him from his rock. In an aquarium, very small specimens survive well, withstanding inclemencies of temperature and stale water most amazingly for animals so delicate; but large specimens (an arm spread of 60 cm is large) must be kept in cool and well-oxygenated water to survive.

Octopus bimaculoides is inferior to the rest of its tribe in the ability to change color, a trait employed by many for social signaling and camouflage; even so, it can produce startling and often beautiful effects and harmonize its color so perfectly with its surroundings as to be almost invisible until it moves. The usually larger *O. dofleini*, which is occasionally found intertidally in Monterey Bay and northward to Alaska and Asia, is a more versatile color-change artist. Another species found occasionally in the intertidal region of Monterey is *O. rubescens* (Fig. 146). The usually small intertidal specimens of *O. dofleini* and *O. rubescens* are difficult to distinguish (and it is not always clear to which species earlier reports refer). *O. rubescens*, primarily a subtidal species, ranges from Alaska to Baja California.

The octopus, together with the squid and *Nautilus*, belongs to the molluscan class Cephalopoda. Though the octopus has no shell, an internal vestige is to be found in most squids—in some forming the "cuttlebone"—and *Nautilus* has a well-developed external shell. Like all the octopods (and the related decapods, such as the squid), the octopus has an ink sac, opening near the anus, from which it can discharge a dense, sepia-colored fluid, creating a "smoke screen" that should be the envy of the navy. Jack Calvin once spent hours changing water in a 20-liter jar so that he might better observe the half-dozen octopuses therein, but his arms were exhausted before the animals' ink sacs.

Fig. 146. *Octopus rubescens* (§135), probably the most common shallow-subtidal and intertidal octopus from Alaska to Baja California.

The octopus can move rapidly over sand or rocks by the use of its arms and suckers; but in open water its arms trail away from the direction of motion with an efficient-looking, streamlined effect as it propels itself swiftly backward with powerful jets of water from its siphon tube. One common method this animal uses for hunting is to lie quietly under rocks and then to dart out to capture passing fish or crustaceans (crabs seem to be a favorite food). Captured prey are quickly subdued by a potent, paralyzing venom and then dispatched with the strong, beaklike jaws, which are normally concealed inside the mouth opening. Shelled mollusks, such as various limpets, abalones, and bivalves, as well as hermit crabs occupying snail shells, also commonly fall prey to the octopus. In this case, the octopus first rasps a small hole in the shell with its radula, and then paralyzes the victim by injecting venom into the opening. All of this should serve as fair warning to collectors inclined to handle the octopus. Although bites are uncommon, they are painful and long remembered.

The octopus's method of reproduction is decidedly unique. At breeding time one arm of the male enlarges and is modified (hectocotylized) as a copulatory organ. From the generative orifice he charges this arm with a packet of spermatozoa, which he deposits under the mantle skirt of the female. In some squids a portion of the arm is detached and carried in the mantle cavity of the female until fertilization takes place—often a matter of several days. This detached arm was formerly thought to be a separate animal that was parasitic in the female cephalopod. In dissecting male squid, the senior writer has seen the packet of spermatozoa explode when the air reached it, shooting out a long arrowlike streamer that was attached with a cord to the spermatophore.

The act of fertilization accomplished, the male octopus has but to retire to repair his damaged arm. The female, however, has before her a long vigil, for a description of which we quote from a paper by Fisher (1923). A small octopus had been captured, late in June 1922, near the Hopkins Marine Station at Pacific Grove. It was kept alive in an aquarium for 3½ months, where it fed irregularly, and usually at night, on pieces of fish and abalone. Dr. Fisher writes:

"For reference in conversation, the octopus was named 'Mephisto,' later changed for obvious reasons to 'Mephista.' She had a little shelter of stones in a front corner of the aquarium. A movable board was so arranged as to exclude all but dim light, since in the language of animal behaviorists an octopus is negatively phototactic (unless, perchance, it is too hungry to care).

"During the night of July 4 two or three festoons of eggs were deposited on the sloping underside of the granite roof of Mephista's retreat. These eggs were nearly transparent, and resembled a long bunch of sultana grapes in miniature. The central axis, consisting of the twisted peduncles of the eggs, was brown. During the following week about forty other clusters were deposited, always at night. The process was never observed.

"In the meantime she had assumed a brooding position. The masses of eggs rested against the dorsal surface of the body and the arms were curled backwards to form a sort of basket or receptacle. . . . [During one period of observation] the ends of the arms were in frequent gentle movement over the clusters and among them—almost a combing motion.

"Mephista was never observed away from the eggs during the day, unless badly irritated. If a morsel of fish or abalone was dropped close to her she endeavored to blow it away with water from the funnel. If this was not sufficient she caught the food in one or two arms and dropped it as far as she could reach. If then the fish was returned she flushed reddish brown, seized the offending morsel, crawled irritably out of her little cell, and, with a curious straightening movement of all the tentacles, fairly hurled the food from her. She then quickly returned to the eggs. In the morning the rejected food would be found partly eaten.

"Although the eggs were laid in early July they showed no signs of development for nearly a month. By August 20 the embryos were well advanced and their chromatophores were contracting and expanding. From time to time bunches of eggs had been removed for preservation, without causing unusual reactions on Mephista's part. About ten bunches were left. They hatched during my absence from the laboratory near the middle of September. On my return, October 1, Mephista was still covering the remains of her brood—the dilapidated clusters of empty egg-capsules. She refused to eat either fish or abalone, and had become noticeably smaller. On October 11 she was found dead at her post. . . . Possibly during the post-brooding period the food reaction is specialized—that is, limited to relatively few things. I find it hard to believe that in *Polypus* death is a normal sequel to egg-laying.''*

Two years later another specimen, named Mephista II, repeated the performance, with some variation in details, and provided further opportunities for observing the embryos and newly hatched young (Fisher 1925). The latter, a shade less than 2 mm long, were so translucent that some of the internal organs were visible, and "the beating of the systemic and branchial hearts may be observed." The body form was more like that of a squid than that of an adult octopus. The young of Mephista II swam as soon as they were hatched; but they would not eat any food that was offered, even bits of yolk from their own eggs, and none lived more than 3 to 4 days.

The process of hatching usually involves a violent struggle. First the embryo squirms until it splits the capsule at one end; then "by vigorous expansions and contractions the body is forced out through the opening. . . . The escape from the shell proceeds rather rapidly until the visceral sac and funnel are free. The somewhat elastic egg-shell seems to close around the head, behind the eyes, and the animal has to struggle violently

---

*It turns out that death after spawning is indeed a common occurrence among cephalopods, for both males and females.

to extricate itself. This final stage may occupy from ten to forty minutes."
Once, in what might be interpreted as a flurry of temper at the difficulty of
being born, one of the tiny animals discharged ink into its capsule.

§136. Conchologists have celebrated the richness of the southern
California shores, and the observer interested in living animals as well as
the collector of shells will be impressed by it. A smooth brown cowry,
Cypraea spadicea (Fig. 147), and a cone snail, *Conus californicus* (Fig. 148), are
likely to be found, both representative of families well developed in the
tropics and featured in shell-collectors' cabinets. The California cone, like
its numerous tropical relatives, is an active predator that immobilizes its
prey with venom. Its radula is reduced to a series of harpoonlike teeth that
are used one at a time (Fig. 148). The snail's armament is carried on a long,
prehensile proboscis that is cautiously sneaked near enough to the prey for
a quick jab from the poison-bearing tooth. The California species feeds on a
wide variety of living prey, including other gastropods (e.g., *Nassarius* and
Olivella), bivalves, and polychaetes; its feeding habits are less specific than
those of cones found on tropical reefs. Some tropical cones paralyze small

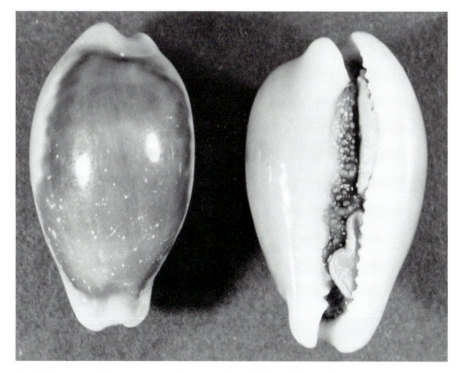

Fig. 147. A smooth brown cowry, *Cypraea spadicea* (§136); shell usually less than 6 cm in
length, although rarely to 12 cm.

 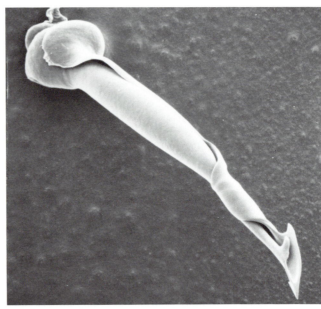

Fig. 148. *Left, Conus californicus* (§136); about 1½ times natural size. *Right,* one of its harpoon-like radular teeth; 133 times natural size. These teeth are used singly, and only once, to puncture prey and deliver a venom.

fish, and some are dangerous to man. The recorded range of *C. californicus* is from the Farallon Islands near San Francisco to Bahía Magdalena, Baja California.

The top snails and purple snails are amply represented. The smooth turban *Norrisia norrisi* (Fig. 149) will be noticed because of its size and numbers, and enjoyed because of the pleasing contrast that the vivid-red flesh of the living animal forms with the lustrous brown of the shell. It is a satisfying experience to find crawling about in the tide pools living specimens of a type that one has known from empty shells only—especially since, for some of these mollusks, there are no known methods of anesthetization that will permit preservation of the relaxed animal. The center of *Norrisia's* distribution is offshore in kelp beds where the animals feed on a variety of brown algae. The recorded range is Point Conception to Isla Asunción, central Baja California.

More often than not, individuals of *Norrisia* will have a pinkish slipper limpet attached to them. As its name implies, *Crepidula norrisiarum* (Fig. 150) is characteristically associated with *Norrisia*; but it may be found on other snails, or occasionally on a crab.

Septifer bifurcatus (Fig. 151), a ribbed and slightly hairy mussel not unlike *Mytilus*, is found in these parts, but only under rocks or nestled in crevices, often well up in the intertidal zone. Specimens may be more than

45 mm long, but the average is smaller. Another common bivalve is *Glans carpenteri* (Fig. 152; formerly *Cardita*), a very small, rectangular, cockle-like form. The northern extent of *Septifer* is Crescent City, northern California; that of *Glans* is the Queen Charlotte Islands, British Columbia. Both species extend into Baja California.

Fig. 149. The smooth turban snail
Norrisia norrisi (§136); natural size.

Fig. 150. Several specimens of the slipper limpet
Crepidula norrisiarum (§136) on the turban snail
Norrisia.

Fig. 151. *Septifer bifurcatus* (§136),
a prominently ribbed mussel.

Fig. 152. *Glans carpenteri* (§136),
a very small (to about 15 mm long)
rectangular bivalve.

Fig. 153. *Tegula eiseni* (§137), a beaded turban, common in rubble and on rocks in southern California; shell to 2.5 cm in diameter.

Fig. 154. *Amphissa versicolor* (§137), an active snail of variable coloration, including white, yellow, and brown; about 6 times natural size.

§137. *Tegula eiseni* (Fig. 153), a rusty-brown turban with raised, beaded bands, is gaily colored by comparison with *T. funebralis*, its drab relative of the more shoreward rocks; and it is more retiring, hiding under rocks during the day. South of the Los Angeles area the two may occur on the same beach, but north of that region *T. eiseni* is rarely found. It is recorded, however, as far north as Monterey.*

No northern snail fills the place held by *Tegula eiseni* at, for instance, La Jolla. *Amphissa versicolor* (Fig. 154), although differing in feeding habits, possibly comes nearest in terms of abundance. Restricted to the underside of rocks loosely buried in a gravel substratum, this excessively active form sports a proboscis longer than the average. Most snails retract into and extend from their shells very deliberately. *Amphissa* emulates its enemy the hermit crab. No sudden danger finds it napping, and it can retreat and advance a dozen times while *Tegula* is doing so once. This animal and expanded specimens of the giant moon snails (§252) are good for exhibition to whoever still believes that snails are mostly shell. The recorded range is Fort Bragg to Isla San Martín, Baja California.

§138. A polyclad flatworm, *Notoplana acticola*, has already been mentioned (§17) as an inhabitant of the upper tide pools. Along the central California coast, this polyclad may be abundant in the lower tide pools as well.

*This seems as appropriate a place as any to point out that many records for Monterey may indicate shells found perhaps but once or twice. Since the days of Josiah Keep, Monterey and vicinity have been a favorite collecting ground for conchologists, and they have assiduously recorded their findings and preserved their rarities in cabinets. This has resulted in a certain skewness for many distribution records, for species north as well as south.

In the south under rocks in this lowest zone two additional forms, both probably unidentified, will be found very commonly, and others occasionally. One of the common ones we think of, for lack of any other name, as the "pepper-and-salt" flatworm. The other is an orange-red form. Two others, somewhat less common but more striking, have papillate, furry backs and affix themselves to the rocks with suckers on their undersides. The first of them, also unidentified, is a soft brown form up to 4 cm long. The second, probably a new species of *Thysanozoon*, the "scarf dancer," is not an impressive sight when found attached by its suckers and huddled up on the underside of a rock; but when it stretches itself to get under way in the clear water of a tide pool, it presents a different picture. It is brilliantly colored and swims gracefully, undulating its margins delicately but effectively and justifying its name. Italians living near Naples call the *Thysanozoon* there the "skirt dancer."

§139. The most common habitat of nemerteans, or ribbon worms, is in the mud of bays and estuaries, so they are considered in some detail in that section (§312). However, a good many forms, some of them large and common, will be found at home under rocks on fairly open coast, and therefore require mention here. The most conspicuous of these, and indeed one of the most striking of all intertidal animals, is *Micrura verrilli* (Fig. 155), whose body is vividly banded with sparkling lavender and white. It may be up to 30 cm long, and ranges from Alaska to Monterey Bay, but unfortunately it occurs infrequently. *Paranemertes peregrina* (§312), purple or brown with a white underside, is commoner, ranging from the Bering Sea to Ensenada, Baja California; it is up to 15 cm in length. Near Big Sur, and

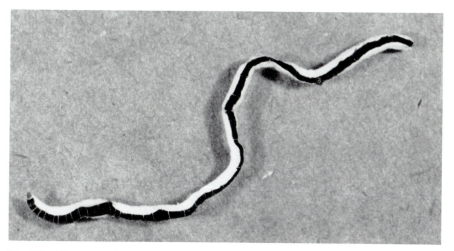

Fig. 155. *Micrura verrilli* (§139), a common and striking nemertean banded with lavender and white.

at Laguna, La Jolla, and Ensenada, we have found a long, stringy, pinkish nemertean we take to be the remarkable *Procephalothrix major*. Its slightly flattened body is only about 2 mm wide, but it reaches a length of over a meter and is comparatively strong. Its mouth is set back some 5 cm from the tip of the head, and this 5-cm portion is cylindrical. Still another rocky-shore form is *Lineus vegetus*, known from Pacific Grove and La Jolla, a slender animal up to 15 cm long, brown in color, with numerous encircling lighter lines. In contraction the body coils in a close spiral. This nemertean has been the subject of some interesting regeneration experiments (§312). A bright-reddish-orange, soft-bodied nemertean, capable of considerable attenuation, is often common in the spring and summer on Oregon beaches; found also under rocks throughout California and as far north as Friday Harbor, this is *Tubulanus polymorphus* (which includes *Carinella rubra* of earlier editions). A good many other forms may occur, most of which are treated in Coe's papers (1905, 1940, 1944).

§140. Some of the bryozoans have been treated previously, so that the shore visitor who has come with us this far will recognize several common encrustations under the rocks as belonging to this group. *Eurystomella bilabiata* (Fig. 156) is characteristically old-rose-colored and always encrusts flatly on stones and discarded shells. A close examination, even with the naked eye, reveals much loveliness, but the aesthetic appeal is much enhanced, as with other bryozoans, by the use of a binocular microscope, or even a ten-power hand lens. So aided, the food-gathering tentacles, the lophophore, can be clearly seen. This species is preyed upon by the nudibranch *Hopkinsia rosacea* (§109), whose color matches that of its bryozoan prey. *Hippodiplosia insculpta* (Fig. 157) is deep buff, and sometimes mildly erect in a double-layered leaflike formation on sticks. When encrusting under rocks, it may have an apparent thickness of several layers, rolled over at the margin. The living zooids of both these bryozoans can be seen, in fresh colonies taken during the winter or spring, with the lowest power of a compound microscope. One of the so-called corallines, *Phidolopora pacifica* (Fig. 158), is a bryozoan colony. Small, lacy clusters of this orange calcareous form are not uncommon under rocks in the south. The large colonies, up to a size that would make a double handful, can be collected only by dredging. The latticed form of this species provides less resistance to the force of moving water, and provides for a more efficient routing of feeding currents, which exit through the holes. *E. bilabiata* ranges from Alaska to Mexico (most abundantly north of California), *H. insculpta* from Alaska to Costa Rica (primarily as a subtidal form south of Pacific Grove), and *P. pacifica* from British Columbia to Peru.

Attached forms also include the encrusting red sponges mentioned in §32, some of which are more at home in this deep littoral than farther up. The brownish-red *Axocielita originalis* (formerly *Esperiopsis*) rarely occurs higher up. It encrusts the undersides of stones, but also occurs in crevices.

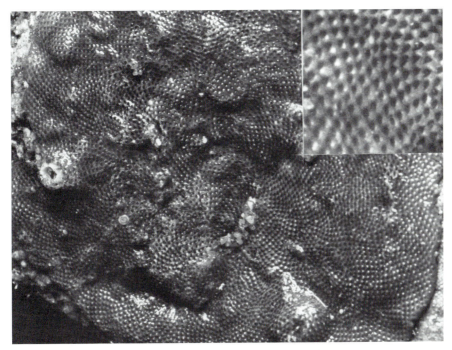

Fig. 156. The pink bryozoan *Eurystomella bilabiata* (§140); *above*, a portion of the animal, slightly enlarged, with a portion inset; *below*, expanded feeding tentacles, or lophophores, greatly enlarged.

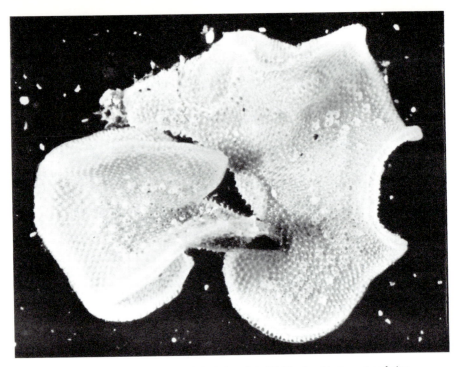

Fig. 157. The leaflike bryozoan *Hippodiplosia insculpta* (§140); about twice natural size.

Fig. 158. *Phidolopora pacifica* (§140), a lacy orange bryozoan.

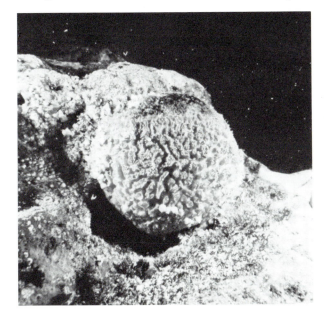

Fig. 159. *Tethya aurantia* (§140), a round or hemispherical orange sponge from low intertidal and subtidal rocks; about ½ natural size.

A hard, pearly-white encrusting form, *Xestospongia vanilla*, characteristic of the entire Pacific coast, looks like thin, smooth cake frosting. The very lumpy, hemispherical, bright, orange or yellow, woody-fibered *Tethya aurantia* var. *californiana* (Fig. 159) may be taken at Carmel, far down in rocky crevices or on the undersides of rocks. *T. aurantia's* base, with its radiating core structure, is visible if the colony is detached; it is diagnostic, and there is nothing else even slightly like it in the tide pools. This is a cosmopolitan species, almost worldwide in distribution, and more abundant than might be supposed at first glance, since many specimens are overgrown with green algae.

§141. Although the southern California cucumber *Parastichopus parvimensis* has been treated (§77) as a protected-rock animal, it often occurs, especially when a low tide comes on in the heat of the day, in the under-rock habitat. North of central California the white *Eupentacta quinquesemita* occurs in this habitat, too, but it is much more common in sheltered waters like Puget Sound (§235).

Lissothuria nutriens, a small, red, flattened cucumber that is remarkable for carrying its young on its back, is occasionally found between Monterey and Santa Barbara. Strangely, however, central California, a region otherwise very fecund, has no shore cucumbers comparable in abundance with *Parastichopus* in the south or with the orange *Cucumaria miniata* in the north, though there are occasional patches of *C. miniata* in the Bodega region. Presumably, more-efficient animals just happen to have crowded them out. Another viviparous holothurian (cucumber), red with a white

Fig. 160. Two subspecies of the seastar *Pisaster giganteus*: left, *P. giganteus capitatus* (§142); right, *P. giganteus giganteus* (§§142, 157).

underside, is the small, cylindrical *Pachythyone rubra. L. nutriens* and *P. rubra* may also be found among kelp holdfasts and surfgrass roots.

§142. A form of the seastar *Pisaster giganteus*, called *capitatus* (Fig. 160), is a not uncommon under-rock member of the southern California fauna. It is by no means gigantic, despite its name, being usually smaller than the common *P. ochraceus*. It is certain to be taken for a different species, even by those familiar with the typical *P. giganteus* of Monterey Bay and north (Fig. 160), for it differs in habitat, shape, and color, as well as size. The few spines are large, stumpy, and vividly colored a slaty purple on a background of ocher. The northern limit of its range is about San Luis Obispo, and we have taken it from as far south as Ensenada. With it may be taken an occasional *Astrometis* (§72) or small *Patiria* (§26), most of the southern representatives of the latter species being stunted and limited to the under-rock habitat.

§143. The noisy pistol shrimp *Alpheus clamator* (Fig. 161; formerly *Crangon dentipes*) may be taken from under rocks in the lower extremity of the rocky intertidal zone and from the sponges and bryozoans growing in kelp holdfasts. Unlike the good children of a past generation, it is more often heard than seen. Very often the shore visitor will hear the metallic clicking of these shrimps all around him without being able to see a single specimen and, unless he is quick and industrious, without being able to take one. Captured and transferred to a jar, the animal is certain to cause excitement if his captor is inexperienced: it is hard to believe, on hearing the inevitable sound for the first few times, that the jar has not been violently cracked. This startling sound is made by the animal's snapping a thumb against the palm of its big claw; the suddenness of the motion is made possible by a triggerlike device at the joint. Any attempt at a close investigation of the operation is likely to result in the investigator's being left in possession of the detached claw, while the denuded shrimp retires to grow himself a new snapper. Studies on related species, such as *A. californiensis* of southern mud flats (§278), suggest that the snapping claw is used both offensively and defensively. The explosive popping may be directed offensively toward prey; small prey some distance from the shrimp are stunned by the concussion and then easily captured. These shrimp are also territorial animals. They defend their burrows or home crevices against intruders, and the distinctive snapping likely expresses a part of their territorial nature, either as a form of communication or, if necessary, as a weapon of defense. As with most aggressively territorial species, of course, some provision must be made for the acceptance of mates. Precisely how this is accomplished by *Alpheus* is unknown, but both *A. clamator* and *A. californiensis* are often found in sexual pairs within their burrows. *A. clamator* has been recorded from San Francisco to Bahía San Bartolomé, central Baja California.

In similar environments the colorful shrimp *Heptacarpus* (§150), of several species, may be seen. Amphipods and isopods of several types occur, but none seems to be specifically characteristic here or of particular interest to anyone but a specialist. Sea spiders (pycnogonids) are seen occasionally. One, *Ammothella biunguiculata* var. *californica*, is common and fairly characteristic under stones at Laguna Beach and elsewhere in southern California, with specimens also being taken off hydroids.

Fig. 161. The pistol, or snapping, shrimp *Alpheus clamator* (§143); maximum length less than 4 cm. The large snapping claw is shown with the thumb in the cocked position.

Fig. 168. An aggregation of the California spiny lobster, *Panulirus interruptus* (§148); these animals are uncommon in the north.

Spiny lobsters are omnivorous feeders—almost anything will do, plant or animal, fresh or decayed—and they are themselves preyed upon by octopuses and large fish. They move by walking—forward, sideways, or backward—or by propelling themselves stern first through the water by rapid flips of their powerful tails.

Panulirus is an entirely southern form, not authentically known north of Cayucos but occurring down into central Baja California. A similar lobster occurs off the Florida coast, where it is known as the Florida crayfish.

§149. Another enlivener of southern tide pools is the vividly gold or orange garibaldi, a fish known to science as *Hypsypops rubicundus*. Vertebrates have scant place in this account, since an adequate treatment of fish alone would require a separate book, but the garibaldi is too common and too obvious to omit entirely; because of *H. rubicundus*'s abundance at La Jolla, one locality is known as Goldfish Point. Sessile animals (e.g., sponges, cnidarians, and bryozoans) are apparently the preferred food, although algae, worms, and small crustaceans are taken as well. The young garibaldi has vivid-blue splotches that are lost as the fish grows older. We should also note that the garibaldi is by law fully protected off the coast of California and may not be taken.

A second fish that merits mention is variously known as the tide-pool sculpin, the rockpool johnny, and *Oligocottus maculosus*. It is a small fish with a sharply tapering body, a large ugly head, and large pectoral fins. It is red-brown and prettily marked. These sculpins have well-developed homing instincts and are endowed with navigational abilities to match.

They regularly leave their pools during high tide, and return before low tide to the same pool or group of pools; sculpins experimentally displaced some distance away find their way back home. Both the distribution and behavior of this fish are influenced by the degree of wave exposure. On fairly protected rocky shores, *O. maculosus* has a broad vertical range, occupying pools in the high and middle zones as well as lower down. On shores exposed to violent wave action, it is essentially restricted to the upper intertidal, and under these conditions, individuals rarely venture outside the protection of their home pools.

§150. Several species of shrimps may sometimes be common in the lower pools. At one time they were all placed in the genus *Spirontocaris*, but now most are referred to as *Heptacarpus*, although one of our common low-intertidal species remains as *Spirontocaris*. These are the "broken-back" shrimps, which characteristically bend their tails suddenly under and forward and thus swim backward. Those found in the higher tide pools are small and transparent; those centering here (Fig. 169) are opaque and relatively massive. *S. prionota* ranges from the Bering Sea to Monterey. *H. palpator* ranges from San Francisco to Bahía Magdalena, and *H. stimpsoni*

Fig. 169. Two "broken-back" shrimps of the lower tide pools (§150): *upper, Spirontocaris prionota; lower, Heptacarpus palpator.* About twice natural size.

(§278) from Sitka to San Diego, but almost invariably subtidally or in eelgrass in quiet water. Less commonly, any of half a dozen others may be found, including *H. taylori* in the south. A large and beautiful brown (sometimes white) shrimp, *H. brevirostris*, has been reported from Alaska, and ranges at least as far south as the area between Carmel and Big Sur; it has been found there in some quantity (under rocks in pools where the water is very pure) and also around Coos Bay, Oregon. The typical littoral specimens, however, are smaller than the large and robust "bastard" shrimps of the Alaska shrimp dredgers.

These shrimps show a great variety of brilliant coloring, which makes them conspicuous in a white tray but very difficult to see against the colorful background of a tide pool. There is no quick way of distinguishing the various species, reference to exact anatomical descriptions being the only method.

To the north very much larger shrimps of the edible species of *Crangon* and *Pandalus* (§283) may occasionally be trapped in the pools, but these must be regarded as rare visitors from another habitat.

§151. Some of the rhythms in nature are as completely baffling today as they have been since they were first observed. For instance, the palolo worm of the tropical seas near Samoa, living the year round in coral burrows at the sea bottom, is an entirely reliable calendar. To quote from an excellent English work, *The Seas*, by Russell and Yonge (1975):

"But true to the very day, each year the worms come to the surface of the sea in vast swarms for their wedding dance. This occurs at dawn just for two days in each of the months, October and November, the day before, and the day on which the moon is in its last quarter; the worms are most numerous on the second day, when the surface of the ocean appears covered with them. Actually it is not the whole worm that joins in the spawning swarm. The hinder portion of the worm becomes specially modified to carry the sexual products. On the morning of the great day each worm creeps backwards out of its burrow, and when the modified half is fully protruded it breaks off and wriggles to the surface, while the head end of the worm shrinks back into its hole. The worms are several centimetres in length, the males being light brown and ochre in colour and the females greyish indigo and green. At the time of spawning the sea becomes discoloured all around by the countless floating eggs.

"The natives are always ready for the spawning swarms as they relish the worms as food. . . . The worms are eaten either cooked and wrapt up in bread-fruit leaves, or quite un-dressed. When cooked they are said to resemble spinach, and taste and smell not unlike fresh fish's roe."

All of which is far from home but not entirely disconnected, for we have at least two similar annelid worms on the Pacific coast. One is *Odontosyllis phosphorea*, whose swarming periods have been plotted at Departure Bay. Only a very persistent or a very lucky observer will see these worms, for except in their brief periods of sexual maturity, when they appear at night in countless hordes, specimens are rarely found.

In the tide pools of the California coast, another syllid, less than a centimeter in length, is more common than *O. phosphorea*. At dusk they may be seen swimming actively at the surface of pools, appearing as vivid pinpoints of phosphorescence.

The inevitable question with regard to these animals is: How do they know when the moon has reached a certain stage in a certain one of the 13 lunar months? The obvious answer would be that they do not know, but govern their swarming time according to tides, which in turn are governed by the moon. Unfortunately, this answer will not hold. The tides are governed not by the moon alone but by the sun and more than half a hundred additional factors. To suppose that the worms can take all these factors into consideration passes the bounds of credulity. Furthermore, worms placed experimentally in floating tanks have spawned naturally at the usual times, although they had no means of sensing what the state of the tide might be. We are driven, then, back to the moon as the undoubted stimulus for their spawning, and confronted with a mystery beside which any of Sherlock Holmes's problems seem pale and insipid.

Nor are these worms the only sea animals that follow amazing cycles. The grunion (§192) is, if possible, an even more startling example. More and more it is apparent that one element in these cyclic patterns of behavior is a physiological timing mechanism or "biological clock" of some sort.

§152. The occasional sandy pools found rarely between rock outcroppings a meter or so apart—at Carmel, more frequently along the outer coast of Washington, as at Queets and Mora, and all along the Oregon coast—may be characterized by the spionid worms *Boccardia proboscidea* and *B. columbiana*. The two species are very much alike, and have rather catholic distributions on the shore. Both extend from British Columbia to southern California, penetrating clay or shale rocks with vertical, U-shaped burrows and protruding their two minute tentacles from one end of the burrow. *B. columbiana* also bores into coralline algae and the shells of mollusks.

In these pools will also be found a few young edible crabs, *Cancer magister*, which stray up into the intertidal from deeper sandy bottoms. In the northern latitudes of Oregon and in Washington, where it is known as the Dungeness crab, *C. magister* often comes near shore or into bays to molt, and the resulting windrows of ecdysed "skeletons" inspire undue public worry that the crabs are dying off because of some catastrophe.

The exoskeleton of a crab presents a formidable barrier to love-making in some species. Although it seems to be the general rule that the mating of crabs requiring internal fertilization takes place when the female is still soft from molting, mating in several of the xanthid crabs (e.g., *Cycloxanthops*, *Lophopanopeus*, *Paraxanthias*) and others occurs when the female's shell is hard. Snow and Neilsen (1966) of the Oregon Fish Commission were able to watch the mating behavior of the Dungeness crab in an aquarium (Fig. 170). They followed the process for 192 hours and have provided a minute-by-minute description.

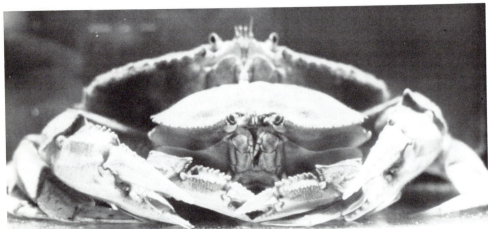

Fig. 170. Mating behavior in *Cancer magister* (§152), the Dungeness crab: *top*, premating embrace, the male on top; *center*, the female molting her old carapace; *bottom*, position of the abdominal flaps during copulation.

A few days before the female is ready to molt, the male, who does not molt at this time, clasps the female in a belly-to-belly premating embrace, and holds her in this way for several days. If the female becomes too restless, the male holds her more firmly and strokes her carapace with his chelipeds "in an up and down motion that seemed to pacify her." When the female is ready to molt, she nibbles at her partner's eyestalks and is permitted to turn over. The female begins to molt while still confined in the basket formed by the male's appendages. During actual shedding of the old exoskeleton the female is not held tightly; and when the ecdysis is complete, the male pushes the cast-off "shell" away. Copulation does not take place immediately; possibly it is necessary for the new carapace to harden somewhat. Finally, at the right moment, the female allows herself to be turned over to the proper position, and extends her abdominal flap to receive the male, who inserts his gonopods into her spermathecae. As observed in the aquarium, this was not quite the end of the process; a postmating embrace lasted for two days. Perhaps this was associated with the close quarters of aquarium life. As a result of mating, the male accumulates scars or marks on his chelipeds from the incessant stroking of the female before she molts; excessive wear has been attributed to polygamy, but it may be the result of a single, protracted mating.

Snow and Neilsen are not sure how the male knows when the female is about to molt; but since molting may be controlled by hormones, it does not seem improbable that some hormones may be released in the stages preliminary to molting, thus alerting the male. Such a chemical signal has been reported for the rock crab *Pachygrapsus* (§14). Alternatively, the male may stimulate or trigger molting by his actions.

The average life span of *C. magister* is about 6 years, the maximum about 8. Some become sexually mature as early as 2 years, but most probably reproduce first when they are 3 years old. Males, the only sex that may be harvested, reach the legal size of 6¼ inches (15.9 cm) or more across the carapace in about 4 years. After reaching maturity, molting and mating occur only once each year, generally during the early summer. Eggs are not fertilized at this time, however; rather, sperm are stored by the female for several months until she extrudes her eggs in the late fall or early winter. Hatching in California generally occurs in January and February, but farther north, cooler temperatures require somewhat longer brooding periods. The larval period, from hatching through seven pelagic larval stages, lasts about 4 months.

The commercial catch of *Cancer magister*, which ranges from Alaska to southern California, fluctuates greatly from year to year, and in the late 1950's there was a decline in the central California catch that seemed at first to be only another oscillation in the fishery cycle. That region, however, has yet to recover. Since the crab fishery to the north has in general prospered (with the usual fluctuations), the problem seems to be a regional one. Several possible explanations, none universally accepted, have been advanced for the precipitous decline in central California. Pesticides, sewage, and industrial pollutants may have increased larval and adult mortality. High levels of microbial fouling have been noted on egg masses in the area and have been correlated with increased mortality. Several fishes that prey on pelagic crab larvae have been successfully introduced into nearby rivers, and these may be the culprits. Then there is the nemertean worm *Carcinonemertes errans*.

Carcinonemertes errans is a predator that feeds on the eggs of *Cancer magister*. It is present on both males and females, but during copulation it will quickly transfer from a male to a female. Once on a female, the worms wait in protected spots on the crab's exoskeleton until oviposition, when they migrate to the egg mass and begin to feed. As many as 100,000 worms have been recorded from a single crab, and since each worm consumes an average of 70 eggs during the host's brooding period, the potential effect is apparent: an entire brood, ranging from 700,000 to 2.5 million eggs in *Cancer magister*, could theoretically be wiped out. Complete destruction apparently does not happen though, because an increase in the density of worms somehow inhibits their feeding rate. Nevertheless, it has been estimated that *Carcinonemertes* may annually destroy an average of more than 50 percent of the eggs of the central California *C. magister* population. A disturbing observation, but one that may produce a test of the worm's impact, is that *Carcinonemertes* densities have been increasing to the north in the last few years.

Although *Carcinonemertes errans* occurs only on *Cancer magister*, a closely related species, *C. epialti*, has been found on nine different species of crabs, including *Hemigrapsus oregonensis*, *H. nudus*, *Pachygrapsus crassipes*, and *Pugettia producta*. This nemertean is also an egg predator, but it occurs in much lower densities.

§153. An investigator with sufficient strength and enthusiasm will find much of interest in sampling, at extreme low tide, the holdfasts of various algae or modest clumps of surfgrass, roots and all. A typical large holdfast may contain hundreds of individuals of dozens of different species. Nemerteans and occasional nestling clams live in the inner parts, and bryozoans, hydroids, sponges, tunicates, small anemones, holothurians (§141), and chitons cover the outside. Crawling over and through this living mass are hordes of snails, hermits, crabs, brittle stars, seastars, amphipods, and isopods.

Here is a perfect little self-sufficient universe, all within a volume less than that of a collecting bucket. A census of the holdfast inhabitants from various regions has been suggested as an index of richness for comparing different areas. Most of the forms occurring thus in the Pacific have been considered in connection with other habitats. The few that do not occur elsewhere are forms so small as to fall outside the scope of this work; but, if only for the sake of renewing a large number of old acquaintances, the observer will find holdfasts and roots worth investigating. Holdfasts cast up on the beach are usually rather barren; because they have been torn off the rocks by wave action and tumbled about in the surf, most of the inhabitants have left for other shelter or have been washed out. However, holdfasts cast up from deeper water may occasionally contain such comparative rarities as the brachiopod *Terebratalia transversa* (Fig. 235).

The holdfasts, particularly in the interstices, shelter many small worms, most of which frequent other habitats as well. A *Lumbrineris*, possibly *L. erecta* (= *L. heteropoda*), green, and stouter than the attenuated *L. zonata* of §130, occurs in *Phyllospadix* roots from Monterey well into Baja California. A similar *Lumbrineris*-like form, very slim and breakable, green, and highly iridescent, is the cosmopolitan *Arabella iricolor*. The carnivorous adults of this species, which are also to be found burrowing in sand or in rock crevices on protected beaches, roam about in the holdfasts, probably scavenging dead animal matter. The juvenile worms apparently have adopted a different and intriguing life-style: they have been found as parasites inside the body of the tube-dwelling polychaete *Diopatra ornata*. *Anaitides medipapillata*, one of the "paddle worms" (phyllodocids), so called because of its broad, leaf-shaped appendages, will be turned up in this assemblage or under rocks—not frequently, but with admiration and enthusiasm on the part of the fortunate observer, who will admire its rich purple-blue with brown or cream trimmings and iridescent glints. Specimens of this sort may be many centimeters long. The range is central California to Mexico, with specimens more common in southern California than farther north.

Also in clumps of *Phyllospadix*, sometimes in quantity sufficient to bind their roots together, will be found tubes of the small *Platynereis bicanaliculata*, similar in appearance to *Nereis* but lacking the broadness of the posterior appendages. Specimens experimentally deprived of their white tubes will construct others very quickly in aquaria, collecting and cementing together with mucus any bits of plant tissue available. The pieces of plant, some of which attach permanently and grow, apparently provide a protective cover for the tube, besides supplying food for these primarily herbivorous worms. The valued tubes, with their attached gardens, are vigorously defended against intruders, with the polychaete's jaws providing the primary weapon. Contrary to many examples throughout the animal kingdom in which territorial behavior seems largely a matter of bluff or communication, here the fights are quite real indeed. The jaws of polychaetes can be formidable weapons, and smaller individuals are sometimes eaten unceremoniously by larger ones.

Sea spiders (pycnogonids) have also been taken in the holdfast. One is characteristic of *Phyllospadix* roots, but it has also been taken on hydroids, or even walking about on the pitted rocks; this is *Ammothella tuberculata*, which extends from northern California to Laguna Beach. Cole (1904, Figure 7, Plate 12) illustrates this form, which has a spread of under 13 mm.

The amphipods occur here in profusion, and among the more than 70 different species (many of them undescribed) that we have turned up in Monterey Bay the following might be considered characteristic of kelp holdfasts and similar habitats: *Eurystheus thompsoni*, from Puget Sound to Baja California, especially common in the south; *Aoroides columbiae*, known from Alaska to Baja California and from many habitats, including coralline algae, clusters of *Leucosolenia*, and even, higher up, *Pelvetia*; *Hyale*, of several species, possibly the commonest low-tide hoppers of the Monterey Bay region and north at least to Sonoma County.

2 🦌 Outer-Coast Sandy Beaches

The sandy beaches of the protected outer coast are more barren than any of the various intertidal regions of the rocky shores discussed so far. They are distinct from the typical open Pacific beaches, which are exposed and subjected to strong surf but nevertheless support a highly specialized but limited fauna of a few abundant species, and from the sheltered sandy beaches of bays and estuaries, which support a rich fauna with many species. The protected sandy beaches are usually at most a few hundred meters long, often between rocky outcrops (Fig. 171). Some of the animals of the surf-swept beaches do not tolerate these sheltered conditions, though these beaches are not quiet enough for the animals of the sheltered bay flats.

In more-northern regions along the coasts of Oregon and Washington there are occasional small beaches of gravel and cobbles (there are also some cobble beaches in California and Baja California). These are the most barren of all shores because they can be sustained only in regions of heavy wave action (for example, a relatively "sheltered" situation where wave refraction may be heavy against the apparent lee of a headland), and there is no refuge for life among the wide spaces between gravel and cobbles. Sea lions and elephant seals may haul out on such beaches.

On sandy protected beaches of this type in many parts of the world a varied microfauna lives among the sand grains. This so-called interstitial fauna includes, in addition to tiny polychaetes and nematodes, such oddities as a bryozoan reduced to a single polypide, marine tardigrades, and odd, copepod-like animals called mystacocarids. Although not all of the oddities have been found on our shores, several representatives of the interstitial fauna may be found among the sand grains of the protected outer coast; and in the richer sands of protected bay shores, this fauna is both diverse and abundant.

§154. Often the difference between a protected beach and an exposed beach is a matter of degree. This degree is in some ways subtly marked by the beach hoppers characteristic of most sandy beaches on our coast. The

Fig. 171. A small sandy beach of the protected outer coast.

large, "short-horned" beach hopper *Megalorchestia corniculata* (Fig. 172; formerly *Orchestoidea*) frequents the smaller, short, steep beaches, and may occur at the sheltered ends of long beaches, where the "long-horned" beach hopper *M. californiana* (Fig. 173; see also §185) is the most abundant hopper; but according to Bowers (1964), the reverse situation does not occur. *M. corniculata*, then, is a creature of steeper, coarser beaches, where sand sorting is poorer but where the sand is perhaps damper and richer in oxygen. It does not retreat far up the beach, as does *M. californiana*, perhaps because its preferred beaches do not have much backshore. On some beaches the two species overlap, but their requirements and occurrences are complemental rather than competitive. Thus, they are good examples of Gause's principle: two species of the same requirements should not (or cannot) occupy the same ecological situation, or "niche." The range of *M. corniculata* is Humboldt County to southern California; that of *M. californiana*, Vancouver Island to Laguna Beach.

During the daytime the hoppers are usually in shallow burrows, and they are best observed at night or on cloudy days, when they come out to feed on seaweed or picnic leavings. Sometimes large hoppers may be found under piles of seaweed. During the night the hoppers leave their burrows open, and evidently are not too particular about finding a "home" burrow. Often they can be observed fighting for a burrow just before dawn, perhaps because it is easier to obtain a hole by fighting a weaker hopper than to dig a new one. The hoppers will also fight for bits of food and, presumably, for females.

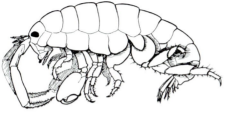

Fig. 172. The beach hopper *Megalorchestia corniculata* (§154); about 2½ times adult size.

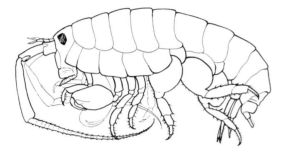

Fig. 173. The large beach hopper *Megalorchestia californiana* (§154); about 2½ times adult size.

Another beach hopper, *Megalorchestia benedicti* (formerly *Orchestoidea*), occurs on both open and protected beaches and is often seen during the daytime, apparently oblivious of its vulnerability to birds. It is smaller than the two big hoppers, and is perhaps protected from some predation by its size.

There are at least five common species of beach hoppers on our various beaches (and a sixth in southern California). Anyone who consults the standard systematic literature will find the descriptions too intricate for an untutored mind, but Bowers has noted (1963) that the species have characteristic patterns of pigmented spots in life and may be distinguished on that basis (Fig. 174).

When beach hoppers come out at night, they move down the beach to feed near the water's edge, and toward dawn they return to their own—or the other fellows'—burrows on the upper beach. Precisely how they navigate up and down the shore is unknown. Early observations suggested that the animals orient by the moon, but later studies, unable to confirm a lunar role, suggest that bearings may be taken on certain fixed landmarks, perhaps a prominent rock face or some other element of the backshore. The animal's basic activity pattern seems to be governed by internal biological clocks, with a circadian rhythm (of 24 hours) as well as a tidal rhythm (of 24 hours and 50 minutes). When kept in continuous darkness in the laboratory and isolated from the shore, these animals continue to display a diurnal rhythm of activity for several days: each night they emerge, without an external cue, and later rebury themselves. The rhythm of the tides is evident as well, for each day the animals burrow into the sand 50 minutes later than the day before, as though somehow anticipating the next higher high water.

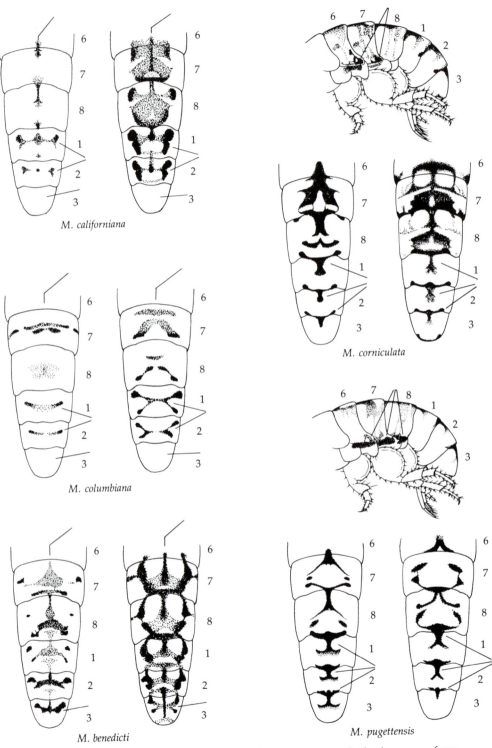

Fig. 174. Identifying pigment patterns of the five species of *Megalorchestia* common from central California to Canada (§154). The patterns are present only in living animals.

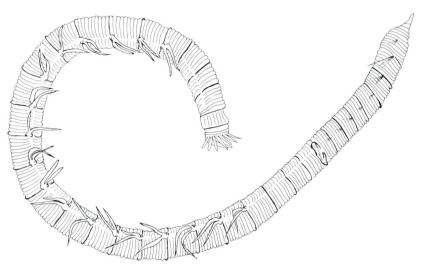

Fig. 175. The bloodworm *Euzonus mucronata* (§155), abundant at about the mid-tide line on many protected sandy beaches.

§155. The inhabitant *par excellence* of these protected beaches is the bloodworm *Euzonus mucronata* (Fig. 175; formerly *Thoracophelia*). This bright-red opheliid worm, about 4 to 5 cm long, lives in a narrow band at about mid-tide level (from near the surface at high tide to 25 or 30 cm down at low tide). The bright-red color of the worm is due to the hemoglobin in its blood plasma. Its range is from Vancouver Island to Punta Banda. The position of the most densely populated band of worms will change somewhat with the season and the tide cycle, but usually the worms are found where the sand is damp but not mushy. Sometimes, in particularly sheltered situations, the presence of the worms will be revealed by small, close-set holes at the surface, as at the La Jolla beach studied by McConnaughey and Fox (1949). In these places they may attain a density of 25 to 30 per cm². *Euzonus* evidently feeds on the nutrient material found on the surface of sand grains, since it passes sand through its alimentary tract. The worms do not swim or crawl, but burrow. When turned out of the sand, they curl up in a circle or coil like a watch spring.

The abundance of bloodworms on some beaches indicates the rich supply of nutrient material in sand that seems barren to us. McConnaughey and Fox estimated that a worm bed a mile long (1.6 km) might contain 158 million worms, or 7 tons of them, which would cycle 14,000 tons of sand a year for a possible yield of 146 tons of organic matter.

A greenish, nereid-like polychaete, up to 13 or 15 cm long, is often found in the bloodworm zone. This is *Nephtys californiensis*, reported from

British Columbia to Baja California. *N. californiensis* is suspected of preying upon the bloodworm.

§156. So deep down as to be chiefly a subtidal form, especially where the sand is firm with an admixture of detritus, but sometimes to be seen from the shore, is a lugworm that belongs more properly in the mud flats of protected bays, *Arenicola* (§310). It is a burrowing animal that breathes by keeping two entrances open to the surface. About one of them, characteristic castings of sand will be noticed. On these beaches it occurs, so far as we have observed, only where rocky outcrops divert the surf in such a way as to induce the deposition of some silt—the situation to be expected, since the worm feeds by eating the substratum and extracting the contained organic matter.

PART II

Open Coast

It may seem amazing that in this surf-swept environment the same animals and plants should occur almost unchanged all the way from Sitka to Point Conception (a few kilometers west and north of Santa Barbara), for the long intervening stretch of coast works an irregular traverse across 22½ degrees of latitude—across most of the North Temperate Zone. But this range seems less remarkable when we consider that conditions are actually very uniform. Along the California coast there is an upwelling of relatively cold water, such that the resulting water temperature is lower than would be expected considering the latitude; and in the north the Kuroshio, the warm Japan Current, sweeps inshore in direct proportion to the increase in latitude, with the result that the waters along the outside coast of British Columbia and southeastern Alaska are very nearly as warm as those along the California coast as far south as Point Conception. Even the variation in water temperature from summer to winter is only a few degrees, for the water is almost constantly agitated. As if conspiring with these powerful oceanic currents, there are great depths close offshore, throughout the range from Sitka to Point Conception, that yield not only uniform oceanic conditions, but also relatively uniform weather conditions. All along this immense coastline, fogs and cool weather are the rule during summer, and there are no great peaks of heat or cold at any time of year.

Close inshore, however, at the shifting interface of the tide level, seashore plants and animals can be exposed to a surprisingly wide range of temperatures. The ocean may be cold and relatively constant in temperature a few meters offshore, but the variation of water temperature at the sea's edge, especially in tide pools, may exceed within a day the annual range of offshore extremes. When this fluctuation is combined with the varying air temperatures (and sometimes there are sunny summer days, even at Pacific Grove!), the range may equal that experienced by terrestrial beings. Too often we have jumped to conclusions about the temperature relations of seashore animals from the data for seawater gathered at the end of a dock or at weather stations kilometers away. But the seashore is a rough, turbulent part of the world, and it is difficult to obtain good records of temperature and humidity in this active zone. Nevertheless, the similarity of the common fauna and flora suggests that these ranges must be of about the same order of magnitude all along the coast, however they may differ seasonally and from place to place.

3 🐟 Open-Coast Rocky Shores

Along the surf-swept open coast, the rocky cliffs of the inaptly named Pacific (Fig. 176) have developed associations of animals with phenomenal staying power and resistance to wave shock. It is not the fact that these animals will not live elsewhere that makes them characteristic, but the fact that no other animals can tolerate the rigorous conditions of heavy surf. Obviously, the prime requisites here are the ability to hold on in the face of breaking waves and the possession of structures fit to resist the sudden impact of tons of water. These necessities have produced the tough skins and heavy shells generally present, the strong tube feet of seastars, and the horny threads of natural plastic by which mussels are attached. Short squat bodies, or ones that bend to minimize sheer stress, are the norm, and most animals that move, such as seastars and gastropods, do so while remaining firmly attached. Many can be found in crevices or otherwise situated to take advantage of any protrusion that might break a wave's force.

Perhaps a word of warning is in order here. Especially from central California northward, completely exposed rocky points like those considered in the first part of this chapter are very dangerous places. It has been stated that in some regions any person within 6 meters of the water, vertically, is in constant danger of losing his life; and every year newspapers, with monotonous regularity, report the deaths of people who have been swept from the rocks by unexpected waves of great size. This loss of life is usually as unnecessary as it is regrettable. To lie down, cling to the rocks like a seastar, and let a great wave pour over one takes nerve and a cool head, but more often than not it is the only sane course of action. To run, unless the distance is short, can be fatal. Like most mundane hazards, the danger from surf largely disappears when its force is recognized and respected; hence no one need be deterred by it from examining the animals on exposed points, always bearing in mind that a brief wetting is better than a permanent one. When collecting at the base of cliffs, we always keep handy a rope made fast at the top of the cliff. It is best to avoid dangerous-looking places, and in any event it is advisable to watch from a safe vantage

Fig. 176. Surf-swept rocks on the open coast, characterized by organisms adapted to withstand tremendous wave shock. Although in general the surf is high, protected microhabitats abound in crevices and on the lee of large rocks.

point before venturing into an unfamiliar place, to observe the wave action and study the lay of the sea. Above all, do not trust the old notion about every seventh or ninth wave being higher, with the implication that the high one may be followed by half a dozen "safe" ones. Sometimes there may be several high waves in succession. And, finally, never go collecting or observing alone in any place where there is the slightest danger.

Interzonal Animals

The force of the waves on stretches of completely open coast is so great that animals up near the high-tide line are wetted by spray on all but the lowest tides. Consequently, there is not the relatively sharp zonation that we find elsewhere, and some of the most obvious animals are fairly well distributed throughout the intertidal zone. Accordingly, we shall depart somewhat from our usual method and shall treat those wide-ranging animals first, taking up the more definitely zoned animals later.

§157. The most conspicuous of the open-coast animals is *Pisaster ochraceus* (Figs. 177, 221), sometimes popularly called the purple or ocher seastar but more often the common seastar—an animal distinctly different from the common seastar of the Atlantic. Since these animals are not "fish" in any sense, "seastar" is perhaps a better name for them than starfish, and has been used throughout this book. Specimens varying from 15 to 36 cm in diameter and having three color phases—brown, purple, and yellow— are commonly seen on exposed rocks from Prince William Sound, Alaska, to Point Sal (near Point Conception). In the warmer waters below that point a subspecies, *P. ochraceus* ssp. *segnis*, extends at least to Ensenada.

Pisaster neither has nor seems to need protective coloration. Anything that can damage this thoroughly tough animal, short of the "acts of God" referred to in insurance policies, deserves respectful mention. To detach a specimen from the rocks one heaves more or less mightily on a small crowbar, necessarily sacrificing a good many of the animal's tube feet but doing it no permanent injury thereby, since it will soon grow others to replace the loss. The detached tube feet continue to cling to the rock for an indefinite period.

Fig. 177. The common seastar *Pisaster ochraceus* (§157), also called the purple or ocher seastar, the most conspicuous of the open-coast animals.

The action of the pedicellariae, minute pincerlike appendages that keep the skin of seastars and urchins free of parasitic growths, can be demonstrated particularly easily with *Pisaster*: allow the upper surface of a husky specimen to rest against the skin of your forearm for a few seconds; then jerk it away. The almost microscopic pincers will have attached themselves to the epidermis or hair, and you will feel a distinct series of sharp nips. A classical experiment to demonstrate the function of pedicellariae is to drop crushed chalk on the exposed surface of a living seastar. The chalk is immediately ground to a powder so fine that it can be washed away by the action of the waving, microscopic cilia on the skin. Considering that sessile animals must struggle for even sufficient space for a foothold, and that seastars live in a region where the water may be filled with the minute larvae of barnacles seeking an attachment site, the value of these cleansing pedicellariae is very apparent. It is even possible that they have some value in obtaining food, for despite their small size, they have been observed to catch and hold very small crabs, which may eventually be transferred to within reach of the tube feet.

The tube feet also play an important part in the seastar's breathing. In addition, delicate, fingerlike extensions of the body wall are scattered over the animal's upper surface. These extensions are lined, both inside and out, with cilia, and they assist in respiration. Many small shore animals have not even this degree of specialization of breathing apparatus—sponges, for example, get what oxygen they need through their skins. In connection with the general subject of respiration, it is interesting to note the great number of marine animals, including some fishes, that are able to get oxygen from the air as well as from the water. This is necessarily the case with a large proportion of the animals herein considered. Some, in fact, like some of the pill bugs (isopods) and a fish that lives on the Great Barrier Reef, will drown if they cannot get to the air. Among the shore crabs the gill structure shows considerable progress toward air-breathing—a process that reaches its climax in some of the tropical land crabs, which have developed the gill chamber into a lung and return to the sea only to breed.

Pisaster will eat almost anything it can get its stomach onto, or around; the stomach is everted and inserted into the shell of a clam or mussel, or around a chiton or snail. Digestion, therefore, takes place outside the body of the diner, but the stomach must be in contact with the meal. Seastars are often observed humped up over mussels, apparently pulling the shells apart in order to intrude their stomachs. This is not entirely necessary, however, since the stomach can squeeze through a very small space (a slit as narrow as only 0.1 mm will suffice). A seastar may well be strong enough to pull a mussel or oyster apart, but it may not often need to do so. The food of *Pisaster* is governed to a considerable extent by circumstance: where there are mussels, it eats mostly mussels; and where mussels are absent, it eats mostly barnacles, snails, limpets, and chitons. For many of the smaller food items, *Pisaster* seems to be in competition with *Leptasterias* (§68).

Many of the molluscan prey eaten by *Pisaster* have evolved defensive behaviors that reduce the seastar's rate of predation on them. "Running responses" of California limpets to the presence of *Pisaster* were first noted (but not published) in 1947 by Eugene C. Haderlie, and since then have received much study. Gastropods such as the limpet *Notoacmea scutum* and the snail *Tegula funebralis*, for instance, respond to *Pisaster* and several other seastars by moving up a vertical surface or by moving directly away, usually with as much speed and vigor as they can muster. For many prey, it is not necessary to be actually touched by the seastar; the mere presence of the predator nearby (and its scent in the water) is often enough to set them in motion. Interestingly, one common shore limpet, *Collisella scabra*, does not attempt to evade *Pisaster*. Whatever the reason, it is worth noting that this particular limpet is more frequently eaten by *Pisaster* than would be expected from its relatively rare occurrence within the seastar's normal foraging range.

Spawning of *Pisaster* occurs in April and May in Monterey Bay, and a month or so later in Puget Sound. Unfortunately, little is known about the life of very small specimens, which tend to be uncommon at most sites.* Adult *Pisaster* grow in proportion to their food supply: with abundant food, they grow faster, and if starved, they actually decrease in size. This, of course, makes the animal's life span a matter of some debate. Perhaps an estimate of 20 years or so is close.

As far as we can tell, the principal enemies of adult *Pisaster* are people: hordes of schoolchildren on field trips who must each bring back one, tourists and casual visitors from inland who hope the star will dry nicely in its natural colors, collectors for curio stores, and sport fishermen who think that all seastars are some sort of marine vermin to be removed for the betterment of whatever it is they are after. Though it will probably be impossible to exterminate such a robust and abundant animal as *Pisaster ochraceus*, there are indications that it will cease to frequent heavily utilized beaches. One can almost tell where people have been by the scarcity of *Pisaster*. This mistreatment of the humble *Pisaster*, as abundant and inexhaustible as it may seem to be, is unfortunately only one of the more obvious of many wasteful and destructive collecting practices that have inspired laws protecting intertidal invertebrates in the states of California, Oregon, and Washington.

Other enemies of the common seastar are few, and somehow more respectful. Sea otters eat only part of a seastar, leaving the remainder alive and able to regenerate lost components. Sea gulls also eat a few, but they are easily forgiven; it must be a hungry gull indeed to tackle such an unrewarding chore.

Up to 1 percent of the *Pisaster* seen between northern California and

---

*Competition for food between small specimens of *Pisaster* and adults of the generally smaller seastar *Leptasterias hexactis* might be expected to be intense. Adults of the two species feed on similar kinds of prey.

Monterey Bay will be the more delicate and symmetrical *P. giganteus* (Fig. 160), with beautifully contrasting colors (which are lacking in *P. ochraceus*) on and surrounding the spines. This seastar, which occurs from Vancouver Island to Baja California, would seem to be improperly named, since the average intertidal specimen is smaller than the average *P. ochraceus*. Subtidal specimens, however, may exceed 60 cm in diameter. *Pisaster giganteus*, like its close relative, is an active predator, and experiments indicate that the seastar chooses mussels over several other types of prey if given a choice in the laboratory. Its diet in the field, however, is broad and includes barnacles, chitons, mobile snails, and vermetid gastropods, as well as mussels. Undoubtedly, its diet depends to a great extent on which prey are present and on the defensive prowess of its intended prey. Many mobile prey treat *P. giganteus* as they do *P. ochraceus*, dissociating themselves from its presence as rapidly as possible, and leaving the predator to "select" from prey that it may deem less desirable but that in practice are eminently more catchable. Spawning in this species occurs in March or April in the Monterey Bay area.

§158. The common seastar is the most obvious member of the triumvirate characteristic of this long stretch of surf-swept rocky coast. The others are the California mussel and the Pacific goose, or leaf, barnacle. The mussel *Mytilus californianus* (Fig. 178), ranging from Alaska to southern Baja California, forms great beds that extend, in favorable localities, from above the half-tide line to well below extreme low water. Here is another animal that is distinctly at home in crashing waves. Indeed, it occurs only where there is surf, whereas *Pisaster* may be found elsewhere. Each animal is anchored to the rock by byssal hairs, tough threads extruded by a gland in the foot. Contrary to the natural expectation, however, *Mytilus* is actually capable of limited locomotion, particularly when young and provided that it is not too tightly wedged into position by other mussels. It can achieve a slight downward motion by relaxing the muscle that controls the byssal hairs, but it can go still farther than this by bringing its foot into play. A young mussel, by extending its foot and gripping the rock with it, can exert a pull sufficient to break the byssal hairs a few at a time, or even pull the byssus out by the roots. Once the hairs are all broken, it can move ahead the length of its foot, attach new hairs, and repeat the process.

Small mussels fall prey to shorebirds, several species of crabs, and predatory snails, which drill neat holes in the shell and extract the soft body bit by bit. The chief enemies of large mussels are *Pisaster*, winter storms, man, and along some stretches of coast, the sea otter. The predatory activities of *Pisaster* are of such magnitude that this seastar may actually determine the lower extent of the mussel bed in some locations, a relationship that has been demonstrated experimentally on the coast of Washington by Paine (1974): he removed all the *Pisaster* from a section of shore, and found that the lower border of the mussel bed gradually extended seaward. Occasionally, however, clumps of very large mussels will

Fig. 178. California mussels, *Mytilus californianus* (§158), with the goose barnacle *Pollicipes polymerus* (§160) and a clump of the red alga *Endocladia muricata*.

be found in the low zone, even when *Pisaster* is present. These clumps represent young mussels that, by chance, have escaped predation long enough to attain a size larger than *Pisaster* can successfully attack—they are now immune to further predatory advances. Another enemy of mussels is storm waves, which often rip loose large patches of mussels from the beds. Here, too, a second detrimental influence of *Pisaster* may be seen. The seastar, by eating mussels in a small patch, may so weaken the structural integrity of the larger bed that great sections may be torn away by subsequent waves. An effect similar to this is produced, at least to the north, by logs battering against the shore. Small holes are punched in the fabric of the bed, later to be enlarged by storm waves.

The California mussel is fine eating. Coastal Indians have known this for millennia, and modern residents still collect them for food. But mussels are to be strictly avoided during summer months, when they may accumulate toxins produced by minute organisms (dinoflagellates) in the genus *Gonyaulax* (Fig. 179). A great abundance of these organisms gives a red cast to the sea, and excessive numbers of them may cause the death of fishes and invertebrates. There are several kinds of dinoflagellates associated with

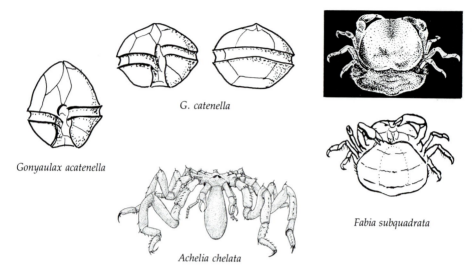

Fig. 179. Associates of the California mussel: *above left*: microscopic dinoflagellates of the genus *Gonyaulax* (§158), the cause of mussel poisoning; *below left*, the pycnogonid *Achelia chelata* (§159), enlarged; *right*, dorsal and ventral views of *Fabia subquadrata* (§159), the mussel crab (natural size).

red tides in various parts of the world, but *Gonyaulax* is responsible for the fatalities associated with mussel poisoning on this coast. In California, mussels are quarantined between May 1 and October 31, and public health officials regularly take samples of mussels to determine their toxicity. *Gonyaulax* generally becomes dangerously abundant only along the open coast. It apparently rarely invades bays, so oysters, *Protothaca* (rock cockles), *Mya*, and most of the gapers and Washington clams (§§300, 301) are usually safe, being restricted to quiet waters; but in inlets near the open sea they too may be toxic and should be avoided in the summer. Other open-coast bivalves besides mussels may also become toxic, especially razor clams (§189); the sand crab *Emerita* (§186), a plankton feeder, also accumulates dangerous amounts of *Gonyaulax* toxin. Regardless of the time of year and collecting location, if after a hearty meal of mussels or another bivalve, your lips begin to feel numb, as though the dentist had just injected novocaine to work on your front teeth, and your fingertips begin to tingle, you had better take measures to rid yourself of your meal as soon as possible and, of course, head for the nearest medical facility.

The breeding season of *Mytilus californianus* extends throughout the year, although peaks have been noted during July and December in California. The sexes are separate. In the warm waters of southern California, mussels may reach a length of 86 mm one year after settlement and 150 mm by the end of 3 years; growth rate varies inversely with water temperature in this form, being considerably slower to the north. The maximum length of specimens in most intertidal populations is about 130 mm, but subtidal mussels may greatly exceed this. The largest, collected subtidally off Baja California, was nearly twice this length.

§159. In perhaps 1 to 3 percent of the California mussels at various places along the central California coast will be found a rather large pea crab, *Fabia subquadrata* (Fig. 179). In the waters of Puget Sound this crab infests by preference (or necessity?) the horse mussel *Modiolus modiolus* (§175). Pearce (1966), who studied the life cycle of this pea crab in Puget Sound, found evidence that *Fabia* is not a harmless or benign commensal, but a true parasite that robs food from its host, sometimes damaging one of the bivalve's gills in the process. The mature female crab is almost as large as the end of a thumb, and this is the one usually found in the mussel. Many consider *Fabia* as tasty as the oyster crab that infests the eastern oyster. The eastern oyster crab is considered such a delicacy that an exception is made in the public health codes, which usually prohibit the marketing or serving of parasitized animals as food, to permit the sale of infested oysters.

In the mature stages of *Fabia* the female is so much larger than and so different from the male that they were originally described as separate species. Both sexes in the first crab stages (just after the megalops) are rather similar in size and appearance, and both infest mussels. Some young crabs infest various other bivalves, but apparently adults are found only in mussels. At the end of the first crab stage the males and females leave their hosts for a sort of mating swarm in the open water. The female then returns to a mussel and goes through five more stages in about 21 to 26 weeks before producing eggs. In Puget Sound, mating takes place in late May, and the eggs appear in November. A few males may return to a host also, but most apparently disappear, having accomplished their essential task. The mating swarm is thus the only time mature males and females are together, and the sperm transferred apparently lasts the female her remaining lifetime of seclusion, usually a year.

Pea crabs have not been found in either *Mytilus edulis* or *M. californianus* at San Juan Island, but on Vancouver Island 18 percent of an intertidal population of *Modiolus modiolus* was infested.

Another, and probably equally unwelcome, guest of the California mussel is a pycnogonid, *Achelia chelata* (Fig. 179). This pycnogonid is recorded from various places along the central California coast, but has been found parasitizing mussels only just south of the Golden Gate in San Francisco and at Duxbury Reef in Marin County. Individual infestation may be as high as 20 to 25 pycnogonids in a large mussel. The damage to the host in these cases may be so extensive that the gills are almost completely destroyed, and even the gonads may be attacked. Another case of pycnogonid infestation of a bivalve has been reported from Japan; both host and guest are different species. This parasitism has not been observed a few kilometers north of Duxbury Reef at Tomales Point, although *Fabia* is frequently encountered in mussels there. So far, no case of parasitism by both pycnogonids and pea crabs in the same mussel has been found, and it may be that the two parasites are mutually exclusive; but this may not be the case in the Japanese situation.

§160. The Pacific goose barnacle, *Pollicipes polymerus* (Figs. 178, 180; formerly *Mitella*), the third member of an assemblage so common that we may call it the *Mytilus-Pollicipes-Pisaster* association, is fairly well restricted to the upper two-thirds of the intertidal zone. These three animals, "horizon markers" in marine ecology if ever there were any, are almost certain to be associated wherever there is a stretch of rocky cliff exposed to the open Pacific. The goose barnacle has been recorded at least from Sitka to the middle of Baja California; specimens on the open coast of Alaska, however, are comparatively few and scattered.

Pollicipes is commonly chalky-colored when dry. A vividly colored form of this species is restricted to caves or rocks sheltered from direct sunlight. In the Santa Cruz region, clusters of this form increase in the brilliancy of markings as they range deeper into the darkness of the cave. Although the adult goose barnacles resemble but vaguely the squat and heavy-shelled acorn barnacles, the two are closely related, and both have substantially the same life history.

At San Juan Island, *Pollicipes polymerus* reaches sexual maturity during its first year and breeds from late April to October with a peak in July. Similarly, at Monterey, according to the studies of Hilgard (1960), this barnacle is engaged in reproductive activity much of the year; animals brooding embryos were found over an 8-month period from spring to fall in 1957, while the water temperature varied from 12.3° to 17°C. Both sexes in each hermaphroditic individual mature at the same time, but cross-fertilization is thought to be necessary. A mature goose barnacle produces 3 to 7 broods per year, with perhaps 100,000 to 240,000 larvae being liberated from each brood. Though there are a great many *Pollicipes* on our shores, it would seem that there would be no room for anything else if all this reproduction resulted in adult barnacles.

The goose barnacle often feeds on amphipods and other creatures up to the size of house flies, rather than on the finer fare of the acorn barnacles. To capture food, it holds its cupped feeding appendages into the current, and one may notice a tendency of groups of this barnacle to line up with the direction of the wave or current movement (Fig. 180). Although it is apparently immobile at low tide, *Pollicipes* is capable of a slow, stately sort of motion, which may be elicited in sunlight by passing a hand over a cluster of barnacles. As the shadow moves across them, they will be observed (if they have not been tired out by some previous shadow) to bend or twist their stalks so that their orientation is shifted. The individuals of a cluster all move in the same direction. The significance of this shadow reaction is not clear, at least at low tide; perhaps it has some value in capturing food when *Pollicipes* is covered with water.

The *California Fish and Game Commission Publication* for 1916 gave a recipe for preparing goose barnacles, the "neck" or fleshy stalk of which the Spanish and Italians consider a choice food.

Fig. 180. Clusters of goose barnacles, *Pollicipes polymerus* (§160), showing the uniform alignment of individuals.

On the plates of *Pollicipes* one may often find a form of the common upper-zone limpet *Collisella digitalis* (§6). Those individuals residing on the goose barnacles are usually lighter-colored and more lightly striped than their counterparts on adjacent rocks; in each case, however, the color pattern blends with the respective background.

The name "goose barnacle" comes to us from the sixteenth-century writings of that amiable liar John Gerard, who ended his large volume on plants with ". . . this woonder of England, the Goosetree, Barnakle tree, or the tree bearing Geese." After declining to vouch for the authenticity of another man's report of a similar marvel, Gerard continued: "Moreover, it should seem that there is another sort heerof; the Historie of which is true, and of mine owne knowledge: for travelling upon the shores of our English coast between Dover and Rumney, I founde the trunke of an olde rotten tree, which (with some helpe that I procured by fishermens wives that were there attending on their husbandes returne from the sea) we drewe out of the water upon dry lande: on this rotten tree I founde growing many thousands of long crimson bladders, in shape like unto puddings newly filled before they be sodden, which were verie cleere and shining, at the neather end whereof did grow a shell fish, fashioned somewhat like a small Muskle, but much whiter, resembling a shell fish that groweth upon the rocks about Garnsey and Garsey, called a Lympit: many of these shels I brought with me to London, which after I had opened, I founde in them living things without forme or shape; in others which were neerer come to ripeness, I founde living things that were very naked, in shape like a Birde; in others, the Birds covered with soft downe, the shell halfe open, and the Birde readie to fall out, which no doubt were the foules called Barnakles. I dare not absolutely avouch every circumstance of the first part of this Historie concerning the tree that beareth those buds aforesaide, but will leave it to a further consideration: howbeit that which I have seen with mine eies, and handled with mine handes, I dare confidently avouch, and boldly put downe for veritie."

Gerard was quite in accord with the modern tendency to popularize science. However, he perceived that pure scientists would desire certain definite information, so he added: "They spawne as it were in March and April; the Geese are formed in Maie and Iune, and come to fulnesse of feathers in the moneth after." Also, he catered to the modern feeling that pictures cannot lie: his graphic representation of the birth of barnacle-geese is reproduced in Figure 181.*

The remainder of the wave-swept rocky-shore animals lend themselves better than the foregoing to treatment according to tidal zones that correspond to the upper, middle, and lower intertidal regions of the protected outer coast.

---

*Adding an interesting twist is a finding of more-modern observers. Cheng and Lewin (1976) reported that goose barnacles (*Lepas* spp. in this case) washed ashore at La Jolla were most often attached to large feathers!

Britannica Concha anatiferæ.
The breede of Barnakles.

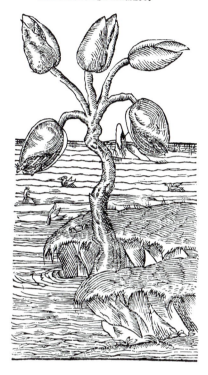

Fig. 181. The barnacle tree (§160); from John Gerard's *Herball*, 1597.

Zone 1. Uppermost Horizon

The unbroken force of waves on the open coast extends this zone of splash and spray far up the shore. The more frequent wetting experienced by inhabitants here reduces the threat of desiccation and enables organisms to live higher on the shore than in more protected situations.

§161. As in other regions, the first animal to be encountered will be the dingy little snail *Littorina*; but whereas it occurs a little above the high-tide line on protected outer shores, it will be found here 6 to 7 meters above the water. *Littorina* is famous for its independence of the ocean. Tropical species have taken to grass and herbage at the top of low oceanic cliffs, and even to trees bordering the water.

Next in the downward progression, but still meters above the water, come very small specimens of the barnacle *Balanus glandula* (§4), whose sharp, encrusting shells are likely to exact a toll of flesh and blood from the careless visitor. In regions of excessively high surf, however, they may be absent, not because they lack the ability to hold on once they have colonized the region, but because the delicate, free-swimming larvae may never be able to get a foothold if there is high surf during the season when they should attach. A tremendous and epic struggle for existence constitutes the daily life of the animals of this region, where the only uncontested advantage is an abundance of oxygen. Barnacles and mussels begin to feed within 60 seconds after immersion, and their great haste is understandable when one considers that they are covered, in this highest zone, only when a wave dashes up over them. They must fight for a foothold, fight to keep it, and fight for their food.

Limpets occur here also, but not in the great numbers that characterize the highest zone of more-protected shores. The commonest form is *Collisella digitalis* (Figs. 17, 21), but others will be found, including *C. pelta* and the giant *Lottia*, which, on these open shores, ranges high up. *Lottia* was once common at Moss Beach and along most of central and southern California; it is not so abundant now but is still fairly common on offshore islands.

§162. On vertical faces of rock, often well above tide level and wave action but in situations where the surface is damp from spray, especially in broad fissures and caves, the rock louse *Ligia* is often abundant. The animals are a general, dark, rock color, and are not easily seen until they move, especially in shadowed places. A visitor to such places becomes aware that the rock surface seems to be crawling as the *Ligia* scuttle away, usually upward, in fits and starts. Often the animals will evade capture by dropping from the rock surface when reached for, then running for safety among the jumble of boulders and cobbles at the base of the rock face.

There are two species of *Ligia*, the slender, spotted *L. occidentalis* (§2), common from Sonoma County southward, and the much broader, more uniformly colored *L. pallasii* that occurs from Sonoma County northward to Alaska (Fig. 182). The larger, northern species occurs sporadically as far south as Santa Cruz, but *L. occidentalis* is the common one at Monterey and in Marin County. The species differ somewhat in their habits and habitat

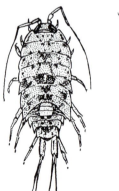
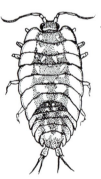
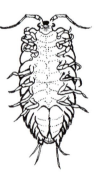

Fig. 182. Two rock lice of the genus *Ligia* (§162): *left, L. occidentalis* (see also §2 and Fig. 12); *center and right, L. pallasii*. All drawings are about natural size, but *L. pallasii* is usually much larger north of Bodega Bay.

preferences. The faster-moving *L. occidentalis* tends to occur on drier, rocky beaches, where it forages in more nearly terrestrial situations and then returns to damp microhabitats to replenish body moisture. *L. pallasii* is slower and more of a stay-at-home, spending most of its time in cool, moist sections of sea cliffs, especially where there is some freshwater seepage. Both feed on the algal film coating the rocks and scavenge such dead plant and animal material as they can find. Sexual dimorphism is marked in *L. pallasii*. Adult males are considerably wider (and often longer) than females, and on a crowded surface, females and juveniles may be found under the shield provided by the larger males. The smaller size of the female apparently makes it easier for the male to hold her beneath him during copulation. It also provides him with the unusual ability to run during copulation (taking her with him) if necessary to avoid predators or waves. The main breeding season is from the spring to early summer. During this time, females can be found with broods, averaging 48 young, held beneath the thorax in an incubation pouch.

The beginnings of social order have been observed in a Japanese species. The animals move back and forth, in a more or less orderly procession, along apparently established routes from shelter among upper boulders at high tide down to the lower beach to feed. The procession is led by older members of the tribe. The casual impression gained from watching our own *Ligia* is that of an aimless, disorganized rabble, dispersing in various directions.

Zones 2 and 3. High and Middle Intertidal

Here mussels and goose barnacles begin to occur in large beds, which afford shelter to a number of animals not themselves adapted to withstanding surf. Indeed, many of the animals more characteristically found on the protected coast will also be found here occasionally, taking advantage of shelter provided either by the dense, multi-layered beds or by other protective nooks, crannies, and crevices.

§163. Probably the most obvious of these is the predaceous worm *Nereis vexillosa* (Fig. 183), known variously as the mussel worm, the clam worm, and the pile worm. Certainly it is the most important from the standpoint of sportsmen, who seek it along thousands of kilometers of coastline to use for bait. Possibly because of its proclivity for letting other animals provide its shelter from surf, *Nereis* is one of the few animals that may be found in nearly all types of regions, from violently surf-swept shores to the shores of bays and completely protected inlets, and in such varying types of environment as rocky shores, gravel beaches, and wharf piling. Not only does it occur from Alaska to San Diego, but it seems to be abundant throughout its range, whereas other animals occupying so great a stretch of coast usually have optimum areas. Thus, *Nereis* might be considered in almost any section of this handbook. It is treated in this particu-

Fig. 183. A nereid, probably the mussel worm *Nereis vexillosa* (§163).

lar place because the largest specimens are commonly found in the mussel beds of the open coast, where they vary from 5 to 30 cm in length. They are usually colored an iridescent green-brown.

The animals are very active, and they squirm quite violently when captured, protruding and withdrawing their chitinous jaws, which terminate a wicked-looking, protrusible pharynx and make carnivorous *Nereis* a formidable antagonist. These powerful jaws can deliver a businesslike bite to tender wrists and arms, but in collecting many hundreds of the worms bare-handed we have rarely been bitten, always taking the precaution of not holding them too long.

In its breeding habits *Nereis* is one of the most spectacular of all shore animals. In common with other segmented worms, its sexual maturity is accompanied by such changes in appearance that early naturalists considered *Nereis* in this condition to be a distinct animal, *Heteronereis*—a name that has been retained to denote this phase of the animal's life cycle. The posterior segments (red in the male) swell up with eggs or sperm, and the appendages normally used for creeping become modified into paddles for swimming. When moon and tide are favorable, the male heteronereis, smaller than the female, leaves his protective shelter, seemingly flinging caution to the winds, and swims rapidly and violently through the water, shedding sperm as he goes. Probably a large number of the males become food for fishes during the process, and no doubt a similar fate awaits many of the females, who follow the males within a short time, releasing their eggs.

It has been determined experimentally in connection with related species that females liberate their eggs only in the presence of the male, or in water in which sperm has been introduced. If a specimen of each sex can be obtained before the ripe sexual products have been shed, the experiment can be performed easily. The female, in a glass dish of clean seawater, will writhe and contort herself for hours with frenzied and seemingly inexhaustible energy. Introduce the male—or some of the water from his dish, if he has shed his sperm in the meantime—and almost immediately the female sheds her eggs. When the process is over, the worms, in their natural environment, become collapsed, empty husks and die, or are devoured by fishes, birds, or other predaceous animals.

We have many times found the large heteronereis of *N. vexillosa* in the mussel clusters on surf-swept points—abundantly in late January, less frequently in February or early March. M. W. Johnson found spawning heteronereid adults at Friday Harbor in the summer of 1941 and identified gelatinous egg masses of the species. According to Johnson (1943), *N. vexillosa* spawns only at night, usually an hour or two before midnight. Shedding specimens of *Nereis grubei*, a smaller nereid of mussel beds and piling, are common in Monterey Harbor (§320).

§164. A second nereid, not as common as *Nereis vexillosa* but sufficiently startling to put the observer of its heteronereis stage in a pledge-signing state of mind, is the very large *Neanthes brandti* (formerly *Nereis*), which differs from the similar *Neanthes virens* (§311) in having many, instead of few, paragnaths on the proboscis. Specimens may be nearly a meter long, and are broad in proportion—a likely source for sea-serpent yarns. To the night collector, already a bit jumpy because of weird noises, phosphorescent animals, and the ominous swish of surf, the appearance of one of these heteronereids swimming vigorously at the surface of the water must seem like the final attack of delirium tremens. We have found this great worm, not in the free-swimming stage, under the mussel beds at Moss Beach, San Mateo County. It has been reported from Puget Sound to San Pedro and has been collected from southeastern Alaska.

§165. Associated with *Nereis* in the mussel beds are the scale worm *Halosydna brevisetosa* (§49), the porcelain crab *Petrolisthes* (§18), and, at times, the pill bug *Cirolana harfordi* (§21). In some regions the nemertean worm *Emplectonema gracile* is so common that it forms tangled skeins among the mussels. This rubber-band-like animal is usually dark green on its upper surface (sometimes a yellowish green) and white on its lower. It is slightly flattened, but is rounder than most ribbon worms, and is from 2 to 10 cm long when contracted. We have found great masses of these at Mora on the outer coast of Washington and at Santa Cruz, California; the recorded range is from the Aleutian Islands to Ensenada. In California *Em-*

plectonema seems to be a rather unselective carnivore, and Dayton (1971) reported that on the coast of Washington it eats barnacles and small limpets. Many nemerteans have the unpleasant habit of breaking up into bits when disturbed; but this one—a highly respectable animal from the collector's point of view—has that trait poorly developed and may therefore be taken with comparative ease.

Two other nemerteans should be mentioned, since they may be common here: *Paranemertes peregrina* (§312), which eats nereids (§§163, 164); and the pale-pink or whitish *Amphiporus imparispinosus*, which eats crustaceans. A fuller account of nemerteans is given in §312.

Another haunter of this zone is a snail superficially resembling the rock snail *Nucella*, but it often occurs higher up than *Nucella*. This is *Acanthina punctulata*, also known as the spotted unicorn or thorn snail; the popular name in both cases refers to the short spine carried on the outer margin of its lip. A predator, this snail makes a diet mainly of littorines, *Tegula*, and barnacles. Gastropod prey are drilled, and it takes about 28 hours for the snails to drill and consume an average-sized littorine. They attack barnacles in two ways, either by drilling or by ramming the shell spine into the barnacle's opercular plates, often repeatedly, and prying the plates apart. Some observations (Sleder 1981) suggest that the spine may be used to apply, or to assist in the application of, a toxin that paralyzes the barnacle. Apparently, use of the spine is the more efficient method, since snails that have had their spines removed and are thus forced to drill feed at about half the rate of intact snails. In the Monterey area, individuals spawn in May and June. Clusters of flask-shaped capsules each containing 400 to 500 eggs are deposited, but only about 10 percent of the eggs in a capsule ever complete development; the remainder (nurse eggs) serve as nourishment for the select few. The snail ranges from Monterey south into Baja California.

A second species of *Acanthina*, *A. spirata* (Fig. 184), is often common in high and middle zones on somewhat more protected rocks and on piling and breakwaters at the entrance to bays. *A. spirata* is not always easily differentiated from *A. punctulata*, and the two were long united under the name *A. spirata* (as in previous editions of this book). Usually, however, *A. spirata* has a relatively longer spire, a prominent keel at the shoulders, and spiral rows of prominent black spots; *A. punctulata* is rounder and has spiral rows of less-prominent black spots. Where the two occur together, such as in the Monterey Bay area, *A. spirata* occurs lower on the shore. Its range is Tomales Bay to northern Baja California.

Most characteristic of all is the short-spired purple, or rock, snail *Nucella emarginata* (Fig. 185; formerly *Thais*), which is not only abundant on rocky shores where the surf is fairly heavy but is almost entirely restricted to these regions. Purple snails are not so named for the color of their shells, which are usually gray-brown, greenish, or white, with darker bands (young specimens are occasionally a bright orange); rather, they are reputed to be the snails from which the Tyrian dye of ancient times was made. Irreverent

Fig. 184. A thorn snail, *Acanthina spirata* (§165), of open-coast mussel beds; about natural size.

modern investigators find the dye to be a rather dull purple that fades badly. It is suspected that the color sense of ancient peoples was based on standards that would seem strange to us. No doubt the colors that thrilled the ancient Romans would arouse contempt in a schoolboy artist of today, since the civilization of Tiberius knew nothing of our lacquers and brilliant coal-tar dyes.*

Nucella emarginata ranges from the Bering Sea to northern Baja California. It usually feeds on mussels and barnacles, and in some areas on limpets as well. The basic method of attack is the same in each case: a hole is drilled in the shell by the radula (Fig. 185), aided by a secretion from the foot, which softens the shell; then the snail's long proboscis extends through the hole to devour the soft parts. Individual rock snails apparently differ somewhat in their food preferences. Lani West followed individual *N. emarginata* at Pacific Grove and found that different snails selected different species of prey and that an individual tended to attack the same prey species several times in a row.

Spawning is not highly seasonal in this species, and each mature female deposits several clusters of yellow egg capsules during the year. The vase-shaped capsules (Fig. 185), sometimes called "sea oats," are attached by a short stalk to the sides of rocks or to shells in mussel beds. An individual female in California deposits about 8 or 9 such capsules in a cluster, each capsule requiring about 30 minutes to form and attach. Other females may add their eggs to the cluster so that as many as 300 capsules may be found in a mass. There are usually 500 or so eggs in each capsule, but a large portion of these are sterile nurse eggs. These generally degenerate and are consumed by the larvae that develop from the relatively few fertile eggs. Although this has been referred to as "cannibalism," it is not clear whether any of these viable embryos actually kill each other within the capsule, or if they merely consume their already dead kin. In any event, a relatively small number of eggs, only about 16 on the average, usually hatch and crawl away from the capsule.

*Unfortunately, Ed Ricketts did not see Pompeii; he would have revised his opinion of Roman talent for pigments as well as enjoyed the subject matter on the walls.—J.W.H.

Fig. 185. The rock snail *Nucella emarginata* (§165), the most common carnivorous snail of the open coast: *right*, the animal, slightly reduced; *center*, egg capsules, greatly enlarged; *below*, the radula (this scanning electron micrograph is 420 times natural size).

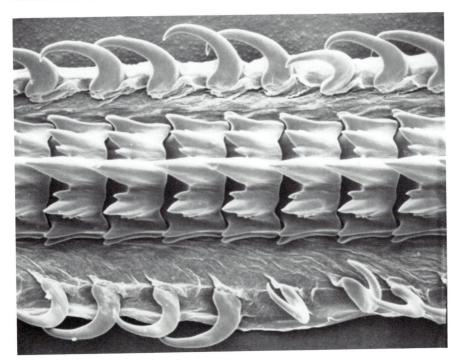

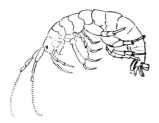

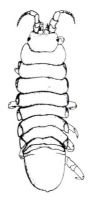

Fig. 186. *Elasmopus rapax* (§165), a tiny amphipod abundant in mussel beds; 6 times natural size.

Fig. 187. *Idotea wosnesenskii* (§165), a large dark isopod of the middle intertidal region.

Two small crustaceans will be found at this level. The minute amphipod *Elasmopus rapax** (Fig. 186) is exceedingly abundant in the middle-zone mussel beds. This is the commonest small crustacean at Santa Cruz, in numbers more than equivalent to the mussels that make up the clusters in which it characteristically occurs. The much larger isopod *Idotea wosnesenskii* (Fig. 187), up to 35 mm long and very dark in color, ranges from Alaska (and the USSR) to Estero Bay, San Luis Obispo County. This and other idoteid isopods are commonly found in the stomachs of fish.

§166. Great honeycombed colonies, often dome-shaped, of the tube worm *Phragmatopoma californica* (Fig. 188) may share available areas with the mussel beds. The colonies are formed not by asexual division but by the gregarious settlement of the worm's larvae; the larvae, according to Rebecca Jensen, require contact with adult tubes to metamorphose. Specimens of *Phragmatopoma* are very likely to be found taking advantage of the slightest bits of shelter, such as overhanging ledges and concave shorelines. Their thin-walled tubes, each sporting a flared rim, are made of sand cemented together. The worm itself, rarely seen unless one chops into a colony, is firm, chunky, and dark, with a black operculum that stoppers up the tube when the animal is retracted. When submerged at high tide, the worm extends a crown of ciliated, lavender tentacles from the tube, using these to gather food and sand grains from the surging water. The similar, generally more northern *Sabellaria cementarium* has a rough, amber-colored operculum. This worm sometimes occurs as a solitary tube, but more often in a colony. Like other tube worms, both of these are dependent for food on what chance brings their way, although they can assist the process by setting up currents with the cilia on their protruded tentacles.

Phragmatopoma occurs from Sonoma County to northern Baja California. *Sabellaria* occurs from Alaska to southern California.

---

*Two former subspecies of *Elasmopus rapax* are now commonly elevated to the species level: *E. mutatus* Barnard and *E. serricatus* Barnard.

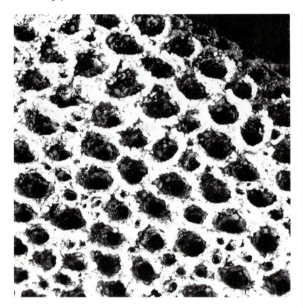

Fig. 188. *Above,* a colony of the tubed worm *Phrag-matopoma californica* (§166). *Right,* a worm within its tube, which has been secreted against a pane of glass; the expanded feeding tentacles and black operculum are evident. Both about twice natural size.

§167. In situations similar to the above, but where the shore formation is tipped more horizontally, beds of the small aggregated anemone *Anthopleura elegantissima* often occur. They are not so characteristic here, however, as in the protected outer-coast environment (§24).

§168. A fairly large barnacle, dull brick-red in color and conical in shape, with the semiporous consistency of volcanic slag, is *Tetraclita rubescens* (formerly *T. squamosa* ssp. *rubescens*). Average specimens are easily twice the size of the *Balanus* (§4) found in the same association, and they are more solitary, rarely growing bunched together or one on top of another as is the case with the white barnacle. The shell wall of this species consists of only four plates, instead of the six characteristic of other acorn barnacles on our coast. Broods of larvae are released during the summer in central California, and individuals that escape predation by various gastropods and seastars may live for up to 15 years. The range is from San Francisco Bay to Cabo San Lucas, Baja California.

§169. *Nuttallina californica* (Fig. 189) is a small sea cradle, or chiton, rarely more than 4 cm long, that is pretty well restricted to the middle intertidal of fairly exposed shores. It can be distinguished by its rough uncouth appearance, spiny girdle, and color—dull brown streaked with white. This chiton, which is often covered with an impressive growth of algae, may oc-

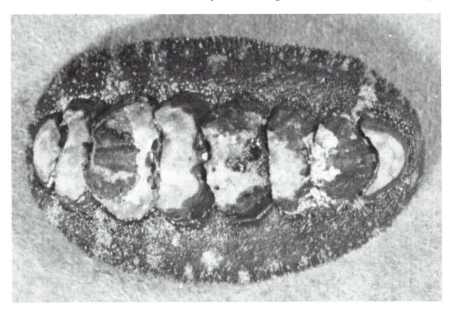

Fig. 189. The chiton *Nuttallina californica* (§169), characteristic of the middle intertidal of exposed shores.

casionally be found well up toward the high-tide line, but never much below the middle zone. At Laguna Beach and other places in the south this chiton lives in sculptured furrows in soft sandstone. If these furrows are of the chiton's own making, as they appear to be, they are comparable to the excavations made in rock by the owl limpet and the purple urchin. In all three cases the object is apparently to gain security of footing against the surf, although occupation of the furrows may also aid in resisting attacks by predators, such as the western gull in the case of *Nuttallina*. *Nuttallina's* recorded range is Puget Sound to Bahía Magdalena, Baja California.

Below the middle zone, *Nuttallina's* place is taken by the larger *Katharina tunicata* (§177). Both of these chitons are perfectly able to take care of themselves at times of high surf, for they are very tough and attach to the rocks so tightly that a flat, sharp instrument must be used to pry them away. Detached from their supports, chitons promptly curl up like pill bugs.

§170. Where there are rock pools at this level, congregations of purple urchins may be found, with occasional hermit crabs (*Pagurus samuelis*), rock crabs (§14), and purple shore crabs (§27); but these are all treated elsewhere, in the zones and habitats where they find optimum conditions. Theoretically, any animal common in the middle tide-pool region of protected outer shores may be found also in these more or less protected rock pools, but they are scarce or local, and not characteristic of the region in general as is the *Mytilus-Pollicipes-Pisaster* association, which finds its pinnacle in this zone.

§171. Plants are outside the province of this book, but in connection with methods of resisting wave shock it is worthwhile to mention the brown alga called the sea palm, *Postelsia palmaeformis* (Fig. 190). It is restricted to the temperate Pacific coast of North America, and occurs only where the surf is continuous and high. Along the central California coast, great forests of these beautiful plants are found on rocky beaches and flats. Instead of sustaining the shock of the towering breakers by rigid strength, as do most of the animals in the region, the sea palms give to it. Under the force of a powerful breaker a row of them will bend over until all fairly touch the rock, only to right themselves immediately and uniformly as the wave passes. This particular form of resistance is made possible by the sea palm's tremendously tough and flexible stalk, which is attached to the rock by a bunchy holdfast of the same material. It will be noted that specialized individuals of a limpet considered previously, *Collisella pelta* (§11), may be found on the swaying stalks of *Postelsia*, thus extending their range into an environment otherwise too rigorous on account of surf—were it not for the protection and amelioration offered by this plant's flexibility.

Fig. 190. The brown alga *Postelsia palmaeformis* (§171), the sea palm, a tough, resilient inhabitant of wave-swept rocks.

Zone 4. Low Intertidal

On violently surf-swept cliffs and tablelands where even the sea palm cannot gain a foothold, bare rock constitutes the typical tidal landscape. A few scrawny laminarians may occur, but never on the boldest headlands; and not until the observer descends (the surf permitting) to the extreme low-tide level will he find any common plant. Even then he is likely to overlook it unless he accidentally knocks a bit of inconspicuous red crust from a rock. This flat encrusting material, coralline algae of several species, provides a surface under which most of the small animals of the region are to be found. The animals so occurring, however, are on the completely exposed shore only by virtue of the protection that this encrusting plant gives them, for all are protected-coast forms (and hence treated elsewhere). Many of them also occur in kelp holdfasts.

§172. The great green anemone *Anthopleura xanthogrammica* is very much at home here, growing to a size exceeded by few other anemones in the world. This giant beauty will always be found, however, in situations where it is reasonably sheltered from the full force of the surf. For this reason, and because it is quite likely to have been first seen in the tidepool regions, it has been treated as a member of the protected-outer-coast fauna (§64).

A more characteristic animal, occurring also under flat rocks or clinging to the boa kelp *Egregia*, is *Idotea stenops* (Fig. 191). This bulky, olive-green to brown form is one of the largest of the isopods, recorded from

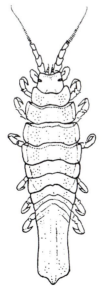

Fig. 191. One of the largest of the isopods, *Idotea stenops* (§172), shown about natural size; characteristically found in the low intertidal of exposed shores under rocks or clinging to the boa kelp *Egregia*.

Alaska to Punta Eugenia, Baja California. The smaller *I. schmitti* will be found under similar circumstances, and at San Remo, below Carmel, we have taken both at the same place. *I. schmitti* ranges from the Bering Sea to Punta Banda, Baja California.

§173. A great ostrich-plume hydroid, *Aglaophenia struthionides*, occurs in rock crevices, especially in large vertical crevices in the line of surf. This form, up to 8 cm tall, is considerably larger than members of the genus found on relatively protected shores. The color of plumes ranges from yellowish to light red, and the largest clusters may be so dark as to be nearly black. Otherwise there is little difference, and the reader is referred to Figure 88.

This is one of the few hydroids that occur in bona fide surf-swept environments, except on the open coast at Sitka. There tufted colonies of two great, coarse sertularians will be found. The ranker of the two, *Abietinaria turgida* (reported as common in the Bering Sea–Aleutian region), occurs in coarse growths up to 16 or 18 cm high, with short branches (which may be branched again) almost as coarse as the main stem, and with the cuplike hydrothecae closely crowded on stem and branch. The more delicate and elongate *Thuiaria dalli*, ranging from the Bering Sea to Puget Sound, has a straight, heavy main stem; but the branches, which take off at right angles, are slender and without secondary branches, and the cuplike receptacles into which the tentacles are withdrawn are small and comparatively few.

However, where the crevices are deep enough, even along the California coast, or where overhanging ledges simulate the semisheltered conditions of reef and tide pool, such other hydroids as *Eudendrium* (§79), *Abietinaria anguina* (§83), and even *Tubularia marina* (§80) may occur, with the sponge *Leucilla* (§94), the hydrocoral *Allopora porphyra* (§87), and the bryozoan *Bugula californica* (§113), all of which are treated in Part I.

Although it is mostly a subtidal species, especially in the southern parts of its range, the sea strawberry *Gersemia rubiformis* is occasionally found intertidally on almost barren rock in Alaska. This circumpolar, boreal-arctic species, a beautiful bright pink with eight-tentacled polyps, is now recorded as far south as Point Arena. It is a fairly common subtidal species in the Cape Arago region, where it may sometimes be found harboring an interesting pycnogonid, *Tanystylum anthomasti* (Fig. 192). This pycnogonid was described from soft corals in Japan, but Oregon, Point Barrow, and its type locality are the only places from which it has been recorded, although its potential hosts are widely distributed. Possibly it has been overlooked because it retreats into the crevices between the lobes of the *Gersemia* colony.

§174. Here the purple urchin, *Strongylocentrotus purpuratus* (Fig. 193), is distinctly at home, having made one of the most interesting adaptations of all to the pounding surf. The animal will commonly be found with at least half of its bristling bulk sunk into an excavation in the rock. For more than a hundred years the method of producing these excavations was a subject of controversy, but it is generally agreed that J. Fewkes stated the

Fig. 192. The pycnogonid *Tanystylum anthomasti* (§173); 5 times natural size.

Fig. 193. *Strongylocentrotus purpuratus* (§174), the purple urchin; individuals about 5 cm across, in pits. The many small holes evident in the upper part of the photograph are the openings to the calcareous tubes of the polychaete *Dodecaceria* (§63).

situation correctly before the turn of the century. He believed that the teeth and spines of the animals, aided by motions produced by waves and tides, were sufficient to account for the pits, however hard the rock and however frayable the urchins' spines, for the spines, of course, would be continually renewed by growth. Sometimes an urchin will be found that has imprisoned itself for life, having, as it grew, gouged out a cavity larger than the entrance hole made when it was young (or, more probably, made by another urchin, since it seems probable that these holes may be formed by successive generations of urchins). Holes abandoned by urchins may provide refuge for all sorts of nestlers, such as dunce-cap limpets, small chitons, and the like.

From their holes, urchins snag pieces of drift algae for food, using their tube feet and spines as snares and utensils. The animal's diet thus varies a good deal depending on what happens to drift by; but the purple urchin, like its giant red kin (§73), seems to prefer the kelp *Macrocystis*. Indeed, when free to exercise their preference (and many do not live in holes, especially where the rock is hard), urchins may dramatically affect the growth of kelp beds.

The main reproductive season of the purple urchin from Washington to Baja California is during January, February, and March, although some individuals with ripe gametes may be found a few months earlier or later. Sea urchin gametes have become important research material, and developmental biologists are now among the important predators of urchins, which also include sea otters, some fishes, and several species of seastars. The sunflower star, *Pycnopodia* (§66), is one of these predators, and urchins, both the purple and the red, respond defensively to its presence. The responses differ somewhat between the two, however, and often have different results. The purple urchin employs flight and the pinching pedicellariae (see §73) as its primary defensive measures. The red urchin, on the other hand, can call on both of these if necessary, but this form possesses an additional, and apparently more effective, weapon—its long spines. These can be used to pinch the soft rays of the seastar, often resulting in withdrawal of the would-be predator. If equal numbers of purple and red urchins are placed in an aquarium with a *Pycnopodia*, it is the purple urchins that are eaten first.

The purple urchin ranges from Alaska to Isla Cedros, almost entirely on the outer coast, unlike the green urchin of Puget Sound and northern inside waters. A good many red urchins occur here at a level below the pits of the purple urchins. In fact, they are possibly as common here as in the outer tide pools of the more protected areas.

All three of these urchins have a number of hangers-on, both inside and out. An intestinal flatworm has already been mentioned in connection with the giant red form. In addition to the flatworm, there may be practically a 100 percent infestation of the intestine by one or more of 12 distinct species of protozoans, some of them comparatively large (0.25 mm). There is also an isopod, *Colidotea rostrata*, that occurs nowhere except clinging to the spines of urchins. And under the host, the shrimp *Betaeus macginitieae* may be found, often in pairs.

§175. The gigantic horse mussel *Modiolus modiolus*, up to 23 cm in length, is no longer common, its depletion being the result, probably, of too many chowders, too many conchologists, and the animal's presumably slow rate of growth. Although it occurs a bit below the low-tide line, the

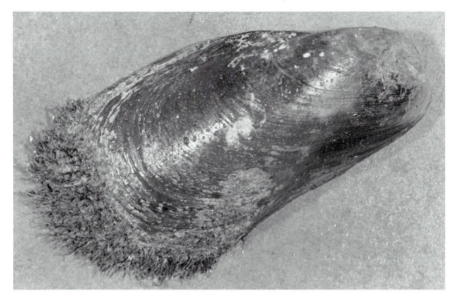

Fig. 194. The fat horse mussel, *Modiolus capax* (§175), with its tuft of brown hairs.

horse mussel's brown shell, naked except for a fine beard, is very notice-able, especially by contrast with the California mussels alongside it. To the south, the fat horse mussel, *Modiolus capax* (Fig. 194), up to 10 cm in length, may be found as a solitary individual in protected areas along the open coast, although it is more common on piling in bays. The shell of this species, which ranges from Santa Cruz to Peru, is usually tufted with dense, brown hairs. Such of the California mussels (§158) as occur this far down are giants of the tribe but are never clean-shelled as are the colonies higher up in the intertidal; many plants, from coralline algae to small laminarians, grow on their shells, providing almost perfect concealment. Some common seastars (§157) range down this far also, undeceived by the mussels' disguise, but their apparent base of distribution is higher up.

§176. In the Monterey Bay region, tide pools and beds of the foliose coralline *Cheilosporum* (the coralline being stiffened by the same encrusting red alga that grows on rocks in this zone) are almost certain to harbor two very tiny cucumbers, *Cucumaria curata* and *C. pseudocurata*, that look like bits of tar (Fig. 195). The two species are easily confused, and have been often. Critical separation apparently requires microscopic examination of the spicules. Sometimes, however, examination of the tentacles will distinguish the two: in *C. curata* all ten tentacles are approximately equal in size, whereas in *C. pseudocurata* the two ventral tentacles may be conspicuously smaller than the other eight. If one accepts that these are indeed separate species (and not everyone does), then *C. curata* is known with certainty only from Monterey Bay; *C. pseudocurata* has a much wider geographical

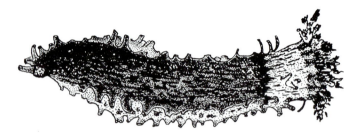

Fig. 195. A small black cucumber, *Cucumaria pseudocurata* (§176), found in Zones 3 and 4 on protected as well as exposed shores; 4 times natural size.

distribution, extending from Monterey Bay at least as far north as the Queen Charlotte Islands. *C. pseudocurata* also occurs in a broader range of habitats, being found on protected as well as on exposed shores and from the upper level of mussel beds to about zero tide level.* Both of these small cucumbers (maximum size under 4 cm) feed on phytoplankton and attached diatoms that are caught by the tentacles. Likewise, their reproductive habits are similar. *C. curata* spawns in mid-December, and *C. pseudocurata* (in Monterey) spawns about a month later. Both brood their eggs under the body for about a month before the young hatch, emerging as an odd crawling stage that possesses at the outset only five tentacles and two locomotory tube feet.

§177. *Katharina tunicata* (Fig. 196), whose dead-black tunic has almost overgrown the plates of its shell, is one of the few chitons that do not retreat before daylight, or even sunlight; and next to the gumboot (*Cryptochiton*), it is the largest of the family. In the low-tide area *Katharina* assumes the position held in the middle zone of the same surf-swept regions by *Nuttallina*. *Katharina*, however, shows more-definite zonation, occurring, in suitable locations, in well-defined belts a little above the zero of the tide tables. In Alaska, *Katharina* is the most abundant intertidal chiton; it ranges plentifully as far south as Point Conception, where there is a great colony, and less commonly below there to the Coronado Islands. Strangely enough, on at least one occasion *Katharina* was found to be not only the commonest chiton but one of the most prevalent of all littoral forms, in a British Columbia locality almost completely protected from wave shock (but directly fronting a channel to the open coast)—an anomalous situation recalling the remarks of the ecologist Allee (1923) to the effect that if the search is long enough, one can turn out very nearly any animal in any environment, however farfetched.

From British Columbia to Monterey, the main spawning period of *Katharina* is June and July, although in some years a partial spawning occurs in the spring as well. The chitons feed on diatoms and on brown

*The cucumber referred to as *Cucumaria lubrica* in §176 of earlier editions was probably *C. pseudocurata*; *C. lubrica* is rare in California, being more common in protected waters to the north.

Fig. 196. *Katharina tunicata* (§177), the black chiton. The effects of past foraging excursions by this herbivore are apparent as cleared paths through the algal mat.

and red algae, and they grow to about 5 cm in length by the end of their third year.

§178. Where laminarians or other algae provide the least bit of shelter, the sea lemon, *Anisodoris nobilis* (see also §109 and Fig. 197), one of the largest of all nudibranchs, may be found. Its average length is around 10 cm, but 26-cm specimens have been taken. Their usually vivid yellow color sometimes tends toward orange; all have background splotches of dark brown or black, and the knoblike tubercles that cover the back are yellow. In common with some other dorid nudibranchs, especially the yellow ones, *Anisodoris* has a fruity, penetrating, and persistent odor. This species ranges from Alaska to Islas Coronados, Baja California. It is a consumer of various sponges and is often common on piling, such as in Monterey Harbor. We have seen these hermaphroditic animals depositing strings of eggs during the winter months.

§179. The black abalone, *Haliotis cracherodii* (Fig. 198), may be found under large rocks and in crevices. This abalone, with 5 to 7 open holes in its shell, is most abundant intertidally and occurs higher on the shore than any other, often well up on the rocks. Its recorded range is from Coos Bay to Cabo San Lucas, although it is most abundant from central California south. Like the red abalone, adults feed on pieces of drift algae trapped in their crevices or captured by the foot. They may also graze algae from the backs of their neighbors, perhaps explaining in part why the shell of the

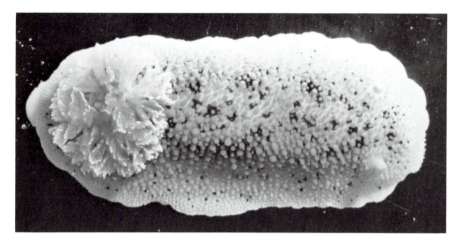

Fig. 197. The sea lemon, *Anisodoris nobilis* (§178), a large sponge-eating dorid nudibranch.

Fig. 198. The black abalone, *Haliotis cracherodii* (§179), which occurs higher on the shore than any other local member of the genus; this specimen, shown life-size, is under the legal limit of 12.7 cm (5 inches).

black abalone is usually clean and shining, unlike that of the red abalone, which invariably carries a small forest. Growth is rapid during the first 2 years, with average specimens reaching nearly 30 mm in length by the end of their first year and 55 mm by the end of their second year (in southern California). This rate may continue for another year or two, depending on food availability, but then it slows considerably, to 4 mm per year or less, after individuals reach 90 mm or so in length. Spawning occurs sometime during the late summer in the Monterey area. Within a population, the release of gametes is relatively synchronous, but for unknown reasons, populations separated by as little as 11 km may spawn several weeks apart.

Despite its small size, the black abalone has supplied an increasing share of the commercial catch as stocks of other abalones have become depleted (legal size is 5 inches, or 12.7 cm). In 1973–74, the harvest of blacks exceeded that of reds.

On the outer coast of British Columbia and Alaska is another abalone, whose thin, pink, and wavy shell is the prettiest of the lot; this is *Haliotis kamtschatkana* (§74), commoner in semiprotected waters.

§180. The largest of the keyhole limpets, which are related to the abalones and no more than a family or so removed from them in the phylogenetic scale, is *Megathura crenulata* (Fig. 199). A good-sized specimen is 18 cm long and massive in proportion. The flesh of the underside of the foot is yellow, and the oval shell, which has many fine radiating ridges, is nearly covered by the black mantle. Many animals lose their color after death, but this giant keyhole limpet is unique in that its black will come off in life if the mantle is rubbed. The range is from Monterey to Baja California, and we have found specimens particularly common in Bahía de Todos Santos.

§181. *Diodora aspera* (Fig. 200), another of the keyhole limpets, is small by comparison with *Megathura*, but still large for a limpet. Large specimens will measure 7 cm long. The diet includes encrusting bryozoans

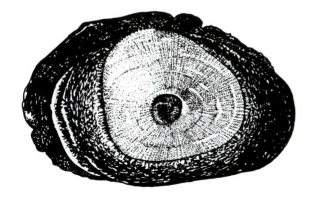

Fig. 199. The giant keyhole limpet *Megathura crenulata* (§180), whose mantle almost covers its shell; ½ natural size.

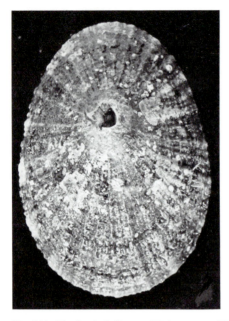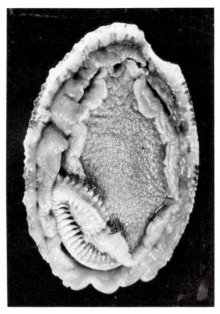

Fig. 200. The keyhole limpet *Diodora aspera* (§181): *left*, dorsal view, showing the "keyhole" at the top of the shell; *right*, ventral view, showing the commensal scale worm *Arctonoe vittata* (§76). Both pictures are about natural size.

as well as algae, and the recorded range is Alaska to Baja California. Any specimens found should be examined for the commensal scale worm *Arctonoe vittata*, which has been mentioned as occurring with *Cryptochiton* (§76). The length of the worm, considering the size of the host, is remarkable, and the worm often has to curl around the mantle so completely that its ends almost touch. Even more remarkable, perhaps, is the response this worm gives when its limpet host is attacked by the predatory seastar *Pisaster ochraceus*. The scale worm, as if understanding that the host's demise would also be its own, seeks out the seastar's tube feet, bites them, and usually causes the seastar to loosen its grip and withdraw. As might be expected, the host also has a response of its own to predatory seastars, and this is performed whether or not the commensal is present. Instead of fleeing, *Diodora* elevates its shell off the rock until it begins to resemble a mushroom. Then the limpet greatly expands two flaps of the mantle, sending one up to cover the shell and a second down to cover the foot. The result is a limpet surrounded by a soft mantle, and no place for the seastar's tube feet to get a grip.

A third member of the group, the volcano-shell limpet *Fissurella volcano*, is the smallest of the lot, but handsome nevertheless. It has been treated (§15) as a protected-outer-coast form, but it is practically as common in this environment. Readily distinguished from *Diodora*, *Fissurella* has a pinkish shell with an elongate "keyhole"; the apical opening of the gray-shelled *Diodora* is circular.

§182. The leafy hornmouth snail, *Ceratostoma foliatum* (§117), is occasionally found in surf-swept areas where there is a bit of shelter just at the low-tide line.

Occasional also, but sufficiently spectacular to cause comment, is the prettily marked purple-ringed top snail, *Calliostoma annulatum* (see Fig. 22), up to 25 mm in diameter and ranging from Alaska to northern Baja California. The foot is bright orange or yellow, and the yellow shell has a purple band along the beaded spiral. This handsome species may be found on the giant kelps just offshore.

Two other species of *Calliostoma* may be found in this zone, and readily distinguished. *C. canaliculatum*, whose range extends from Alaska to Baja California, is characteristically an inhabitant of the rocky shore, from the lowest zone far down into the sublittoral. Its light-brown or white shell with darker spiral bands is sharply conical, and the foot is tan. *C. ligatum* (Fig. 201), ranging from Alaska to San Diego, occurs sparsely in Zone 4 in California, being more common offshore on kelp; in the quiet waters of Puget Sound, however, it is quite abundant in the low zone. *C. ligatum* has a dark-brown shell with contrasting spiral lines of light tan. Its shell is more rounded than that of *C. canaliculatum*, and the foot has dark-brown sides with an orange sole. When freshly collected, the shells of both species may feel slippery, the result of a mucus layer deposited regularly by the snail's foot. Apparently, the mucus-covered shell makes the animal much more difficult for predators to grasp.

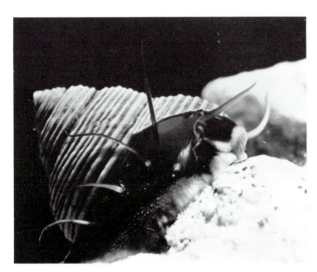

Fig. 201. *Calliostoma ligatum* (§182), a top snail having a dark-brown shell with contrasting lines of tan; enlarged.

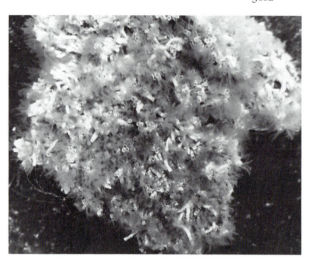

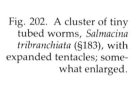

Fig. 202. A cluster of tiny
tubed worms, *Salmacina
tribranchiata* (§183), with
expanded tentacles; some-
what enlarged.

Although archaeogastropods such as *Calliostoma* have long been characterized as grazers on algae, all three species appear to be omnivorous.
When residing on kelp, both the plant and the attached animals are consumed. *C. annulatum* in particular seems to prefer animals, especially
hydroids.

§183. Two encrusting forms may puzzle the amateur collector. A
roughly circular growth that looks like a tightly adherent mass of corallines
is composed of tiny-tubed annelid worms, the serpulid *Salmacina tribranchiata* (Fig. 202). They are not unlike intertwining tubes of *Serpula* or
Sabellaria, but on a pinpoint scale. At first sight, the mass, produced by repeated asexual reproduction, may appear dingy, but on close examination
it will seem to have an encrusted, dainty, filigree pattern. En masse the gills
are a rusty red, individually a brighter red and rather attractive. The worm
occurs from British Columbia to southern California and may be locally
abundant, particularly in somewhat sheltered habitats.

The other form, occurring in similar situations on vertical cliff faces
almost at the lowest-water mark, is the sponge *Hymeniacidon sinapium*,
which may thinly encrust areas of half a square meter. This is the yellow-
green, slightly slimy form, with nipple-like papillae, recognizable by the
European observer as similar to *Axinella*, and originally so called in this
area. M. W. de Laubenfels noted that the open-coast habitat is rare, specimens occurring more characteristically in quiet water (§225), on oyster
shells, and the like.

4 & Open-Coast Sandy Beaches

Whereas the animals living on surf-swept rocky shores have solved the problem of wave shock by developing powerful attachment devices, the inhabitants of surf-swept sandy beaches (Fig. 203) achieve the same end by burying themselves in the sand. These sand dwellers must remain vigilant, however, because the surf regularly shifts loads of sediment from one area to another. Although nearly a continuous process, this shifting is most evident to the casual observer on a seasonal basis; the beach is cut to a steep slope by winter storm waves only to become shallow-sloped again during the summer as sand is redeposited. Some animals, like the mole crab and the razor clam, survive by being able to burrow with extraordinary rapidity. Others, like the Pismo clam, burrow more slowly, depending on the pressure-distributing strength of their hard, rounded shells. These animals have achieved the necessary great strength and resistance to crushing not by the development of such obvious structural reinforcements as ribbing, with the consequent economy of material, but by means of shells that are thick and heavy throughout. Ribbing would provide footholds for surf-created currents that could whisk the animal out of its securely buried position in a hurry. Natural selection has presumably produced a race of clams with thick, smooth shells, over which the streaming and crushing surf can pour without effect.

That the obstacles faced by sand dwellers on an exposed coast are all but insuperable is indicated by the fact that few animals are able to hold their own there. These beaches are sparsely populated in comparison with similar rocky shores. Actually, we know of only six or seven common forms that occur in any abundance on heavily surf-swept sand beaches. This reflects a situation quite different from that assumed by most amateur collectors, who would have one but turn over a spadeful of beach sand to reveal a wealth of hidden life.

Most members of the sandy beach community are uncertain in their occurrence: they have good years and bad years, and a beach that was teem-

Fig. 203. A sandy beach on the open coast.

ing with mole crabs or hoppers one year may be unoccupied a year later. The population may have been wiped out by a rapid change in the shape of the beach—the waves may have shifted direction, thus cutting the beach back to the dunes, or building it up faster than it can be populated. Because most of the animals living on sandy beaches have pelagic larval stages, the young must be set adrift and may therefore settle in another part of the world than their parents. Food is also uncertain—little is produced in the sand itself except minute algae among the sand grains, for the most part available only to animals adapted to ingesting sand grains and eating whatever may adhere to the surface. The major source of food in this region is either the plankton washed ashore by the waves or the seaweeds and occasional corpses of fishes, birds, and sea mammals cast ashore.

Nevertheless, the species that have adapted to living in the sand are, as species, successful in terms of numbers. In good years they may be there by the millions. Some have learned to shift up and down the beach with the tide and the waves; all have learned to subsist on an uncertain food supply. Most of them, except for the clams, are short-lived, living perhaps a year or two. But those with comparatively long life spans, the Pismo clam and the razor clam being two, are more vulnerable to the depredations of man, for it is more difficult for these populations to become reestablished once the breeding stock has been reduced.

As for the beach itself, it is a phenomenon of the meeting forces of sea and land, and an experienced observer can tell much about the condition of the sea from the slope of the beach, the sizes of the sand grains, and the kinds of minerals the sand is made of. Since this is beyond our scope, the reader is referred to the excellent general reference by Bascom (1964), an ardent and lifelong student of beaches.

§184. We have already cited several examples (notably the periwinkles) of animals with marked landward tendencies. On the sandy beaches of the open coast there is another, the little pill bug *Alloniscus perconvexus*, about 16 mm long. This isopod is an air-breathing form that will drown in seawater. It will be found, therefore, in the highest zone, above the high-tide line; and because of the obvious nature of its burrows, it is often one of the first animals to be noticed in this environment. The mole-like burrows are just beneath the surface, and in making them the animal humps up the surface sand into ridges. Buried during the day, *Alloniscus* emerges at night to feed on beach wrack.

Another air-breather, the isopod *Tylos punctatus*, a 6- to 12-mm oval form resembling *Gnorimosphaeroma* (§198), is restricted to the beaches of southern California and Baja. In many locations from about Santa Barbara south, this isopod dominates the high zone, occurring in densities of up to 23,000 per square meter. It is strictly nocturnal, remaining during the day under beach debris or buried down to 20 cm or more in the sand. At night, the animals emerge from their burrows and move down the beach, as far as 30 meters, to forage among the piles of wrack stranded by the receding tide. After their meal, consisting mostly of the kelp *Macrocystis* (but likely including some animal matter as well), the isopods return up the beach to rebury at the high-tide line. During the winter months this nightly migratory activity is suspended, and the animals remain buried in the sand, apparently undergoing a hibernation of sorts. The breeding season lasts from the summer to early fall, when females can be found carrying broods of 14 young on the average. Offspring become sexually mature in about 2 years and live for another year or two.

§185. During the night, or most noticeably at dusk or at dawn, the foreshore seems to be alive with jumping hordes of the great beach hopper *Megalorchestia californiana* (§154). They are pleasant and handsome animals, with white or old-ivory-colored bodies and head region, and long antennae of bright orange. The bodies of large specimens are more than 25 mm long; adding the antennae, an overall length of 60 mm is not uncommon. Like the other beach hoppers, this form avoids being wetted by the waves, always retreating up the beach a little ahead of the tide. These hoppers seem always to keep their bodies damp, however, and to that end spend their daylight hours buried deep in the moist sand, where they are very difficult to find. Night is the time to see them. Observers with a trace of sympathy for bohemian life should walk with a flashlight along a familiar surfy beach at half-tide on a quiet evening. The huge hoppers will be holding high carnival—leaping about with vast enthusiasm and pausing to wiggle their antennae over likely-looking bits of flotsam seaweed. They will rise up before the intruder in great windrows, for all the world like grasshoppers in a

Fig. 204. The sand crab *Emerita analoga* (§186) feeding: *left*, the feathery antennae are extended into the current; *right*, the accumulated food particles are scraped from one antenna by specialized appendages.

summer meadow. Too closely pursued, they dig rapidly into the sand, head first, and disappear very quickly. Breeding occurs from June (occasionally as early as February) through November in central California. Courtship and mating take place in the burrow, and the eggs are brooded in a thoracic marsupium as in the isopods.

§186. *Emerita analoga* (Fig. 204), the mole crab or sand crab, is the "sand bug" of the beach-frequenting small boy. Its shell is almost egg-shaped—a contour that is efficient for dwellers in shifting sands where the surf is high, since the pressure is distributed too evenly to throw the animal out of balance. Most of the crabs characteristically move in any direction—forward, backward, or sideways—but whether the mole crab is swimming, crawling, or burrowing, it always moves backward. Crawling is apparently its least efficient mode of locomotion, and swift burrowing its most developed, but it is also a fairly good swimmer—an action achieved by beating its hindmost, paddle-like appendages above the posterior margin of the shell. The same appendages assist in burrowing, but most of the work of burrowing is done by the other legs.

When in the sand, the mole crab always stands on end, head end up and facing down the beach toward the surf. Characteristically, the entire body is buried, while the eyes (tiny knobs on the end of long stalks) and the first pair of antennae (which form a short tube for respiration) project above the sand. When a wave starts to recede down the beach, the sand crab uncoils its large second pair of antennae (like small feathers) and projects them in a V against the flowing water to gather minute organisms, mostly dinoflagellates but also small plant particles and some sand grains. The feeding antennae work rapidly and are pulled through an elaborate arrangement of bottlebrush-like appendages to be scraped clean of food, perhaps several times during a receding wave. *Emerita* may gather food particles as small as 4 to 5 microns in this manner (a micron is 0.001 mm), and can handle objects up to 2 mm in diameter. Efford, who has studied the sand crab for several years (1965, 1966, 1967, 1970), has not been able to confirm opinions that the sand crab may feed on bacteria or on food adhering to sand grains, or that the animal can sift the sand out of its food.

Sand crabs often occur in dense patches or aggregations on the beach, the largest individuals at the lowest part of the beach and the smallest higher up. Since the males are smaller than the females, they will also be found somewhat higher up the beach than the females (large females are about 3 cm long). The reason for this aggregation is not understood, but the crabs may be packed so tightly that there is virtually no room between them. Perhaps such a dense stand facilitates mating and has some value in reducing the amount of sand taken in during feeding. Whatever the reasons, aggregation patterns can change considerably from beach to beach, and are likely a combined function of many variables, both physical and biological.

Sand crabs move up and down the beach with the tide, the bulk of the population staying more or less in the zone of breaking waves, where feeding is evidently best (although animals will live and feed submerged in an aquarium if the water is flowing). This movement requires some adjustment to tidal and seasonal rhythms, but it may also be explained to a large extent by the crab's behavioral reaction to the changing physical properties of sand in the wash zone. According to Cubit (1969), the crabs emerge from sand that has become "fluid" by water movement. On a rising tide, individuals at the lower end of the wash zone emerge from the sand when waves breaking on the beach suspend the sand; wave wash then carries the crabs up the shore where they reburrow. On a falling tide, animals emerge from firmly settled sand when waves run up the shore, temporarily wetting and liquefying the sand in the upper end of the wash zone; backwash then carries the crabs down the shore. Other researchers have found that aggregations of *Emerita* are also carried along the beach by currents in the longshore direction.

Emerita analoga has a larval life of 4 months or more, during which time the zoea and megalops stages may be carried for long distances. In some years, when the northward set of currents is favorable, larval swarms may populate beaches in Oregon and even Vancouver Island in British Columbia. The entire recorded range is from Alaska to Patagonia, but *Emerita* is most abundant from central California southward to Ensenada. In south-

ern California the megalops larvae arrive on the beach in the greatest numbers from March to May, but ovigerous females carrying bright-orange
eggs beneath the reflected tail may be found at all seasons. *Emerita* reproduces during its first year of life, and may not live more than 2 to 3
years. Farther south there is a tropical species, *Emerita rathbunae*, in which
the males are minute and attached to the legs of the female, thus carrying
the tendency for small males in this genus almost to the verge of parasitism
by the males.

The primary predators of *Emerita* are various shorebirds and fishes. It
does not appear that predation by birds is very serious, since birds tend to
space out when feeding and thus pass over aggregations of sand crabs
somewhat lightly. Fish predation may be more severe, especially by the
surfperch (*Amphistichus argenteus*). The crabs tend to aggregate to a greater
degree in the upper third of the wash zone, perhaps because fish predation
(but also wave shock) might be expected to be less intense there.

The spiny sand crab, *Blepharipoda occidentalis* (Fig. 205), is an inhabitant of subtidal regions of sandy beaches, and may occasionally be found at
very low tide. It is larger, with a flattened carapace 6 cm or so long. It occurs as far north as Stinson Beach (Marin County) and ranges south to
Bahía Santa Rosalía, Baja California. Unlike its nicely adapted relative
Emerita, *Blepharipoda* is a generalized feeder, a scavenger and predator like
most of the more ordinary crabs. Indeed, dead specimens of *Emerita* are an
important food source for *Blepharipoda*.

§187. At about the sand crab's level are small, shrimplike crustaceans,
Archaeomysis maculata, called opossum shrimps because, like the other my-

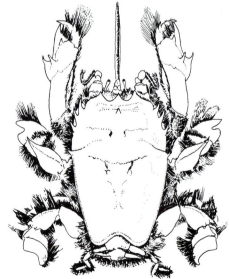

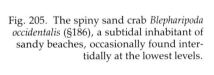

Fig. 205. The spiny sand crab *Blepharipoda
occidentalis* (§186), a subtidal inhabitant of
sandy beaches, occasionally found intertidally at the lowest levels.

Fig. 206. *Donax gouldii* (§188), the bean clam, a shallowly buried bivalve of open-coast sandy beaches.

sids, they retain the young in a marsupial pouch under the thorax. Often these animals hide just below the surface of the moist sand, but they may be brought into the open and collected when water filling a hole dug in the sand causes them to swim about. They may also be netted from the surf with a fine net. This form is related to the more visible mysids of the tide pools, although it lacks gills.

§188. The bean clam *Donax gouldii* (Fig. 206) is common from Santa Barbara to southern Baja California, and is occasionally found north to Santa Cruz County. This small, wedge-shaped clam, averaging 2 to 3 cm in length, is said to have been so common at one time that it was canned commercially at Long Beach. For many years it has not been available in commercial quantities, but the individual collector can still find enough for a delicious chowder by combing the sand just beneath the surface. The bean clam's hiding place is commonly revealed by tufts of a hydroid that grows on the shell and protrudes above the surface of the sand. This elongate hydroid, *Clytia bakeri* (Fig. 207), related to *Obelia*, is the only hydroid found on exposed sandy shores. It also occurs on the Pismo clam and the gastropods *Olivella biplicata* and *Nassarius fossatus*.

Like the sand crab, *Donax* forms dense aggregations and is sporadic from year to year in some places. When it is abundant, it may give the beach a pebbled appearance, since it is not completely buried in the sand. Our *Donax* is sedentary, digging deeper when waves strike the beach; however, species of *Donax* in some other parts of the world, like Japan and Texas, do just the opposite. These emerge from the sand and let the waves carry them up or down the beach. They thus "migrate" up and down the

Fig. 207. The hydroid *Clytia bakeri* (§188) growing on the Pismo clam, *Tivela stultorum*.

Fig. 208. The Pacific razor clam, *Siliqua patula* (§189), a fast-digging, thin-shelled clam that prefers surf-swept beaches; foot and siphons extended.

shore, with the wash zone, somewhat as *Emerita* does on Pacific shores, and dig rapidly into the sand between waves.

§189. A razor clam, *Siliqua patula* (Fig. 208), corresponds ecologically, on the open sandy beaches of Washington and Oregon, to the different-looking Pismo clam of similar stretches in California and Mexico. *Siliqua* is long (shell up to 17 cm) and thin, with fragile, shiny valves—just the opposite of what one would expect in a surf-loving animal. Apparently it depends on speed in digging for protection from wave shock. A clam that has been displaced by a particularly vicious wave can certainly be reburied under several centimeters of sand before the next comber strikes, for specimens laid on top of the sand have buried themselves completely in less than 7 seconds. The foot, projected half the length of the shell and pointed, is thrust into the sand. Below the surface, the tip expands greatly to form an anchor, and the muscle, contracting, pulls the clam downward. The movements are repeated in rapid succession, and with only a few quick thrusts the clam disappears. A digger must work quickly to capture the animal before it attains depths impossible to reach.

Spawning generally occurs during the late spring and summer, and interesting differences exist between northern and southern populations of the clams. In the south (Washington, etc.) spawning is simultaneous, all the clams along several kilometers of beach spawning on the same day when there is a sudden rise in water temperature (to about 13°C). This usually occurs at the end of May or early in June. Sometimes the set of young is enormous, but in other years the animals almost fail to spawn at all. In Alaska they are likely to spawn without fail, but not suddenly and not simultaneously, and the set of young is more uniform; spawning there is usually in July and early August. The larval stage lasts about 8 weeks for both regions; but after settlement, growth rates, maximum size, and longevity of the clams again vary between northern and southern populations. Clams in southern beds grow more rapidly at first, but then decline, reaching a lower final size and having a shorter life span than clams in northern beds. The average maximum age in California is about 9 years and in Washington 12 years; in Alaska, specimens up to 18 and 19 years old are known.

The razor clam is highly regarded as food, and from Alaska to Oregon, *Siliqua* is taken by commercial diggers as well as by carloads of weekend gourmets. As is usual for such a slow-growing animal, restraint, and on some beaches more than has been exercised, is required to avoid depleting this animal. Would-be diners should note too that specimens during the summer may be lethal as food, owing to mussel poisoning (§158). The recorded range of *Siliqua* is from the Bering Sea to as far south as San Luis Obispo, but it is rarely seen along the central California coast.

Up to 80 percent of these clams, according to canners, carry within the mantle cavity a commensal, the nemertean worm *Malacobdella grossa*. The percentage of clams so accompanied increases toward the north. This curious worm has been reported from British Columbia to central Califor-

Fig. 209. *Tivela stultorum* (§190), the Pismo clam, requires a sandy beach with heavy surf.

nia along the Pacific coast, where it occurs also in *Macoma secta*, *M. nasuta*, *Tresus nuttallii*, and doubtless other clams as well; the same species of worm has been found in over 20 species of Atlantic bivalves. These flat nemerteans, up to 38 mm in length, attach themselves to the clams' gills by a sucker and feed on minute plankton in the water passing through the gills. Apparently they have no harmful effect on the host.

§190. The Pismo clam, *Tivela stultorum* (Fig. 209), does not merely tolerate surf; it requires it. Clams removed from their surf-swept habitat to lagoons and sheltered bays to await shipment live but a few days, even though tidal exposure, temperature, and salinity are the same. Apparently *Tivela* has accustomed itself, through generations of living in an environment so rigorous that it would be fatal to most animals, to the high oxygen content of constantly agitated waters and has carried the adaptation to such an extreme that it can no longer survive under less violent conditions. An interesting feature of the clam's adaptation is its inhalant siphon. Most clams live in relatively clear and undisturbed waters, but the Pismo clam, living in waters that are commonly filled with swirling sand, must provide against taking too much sand into its body with the water that it must inhale in order to get food and oxygen. It does this by means of a very fine net of delicately branched papillae across the opening of the siphon, forming a screen that excludes grains of sand and at the same time permits the passage of water and the microscopic food that it contains.

Many years ago, when the Pismo clam was as common on exposed beaches in southern California as are sand dollars on bay and estuary beaches, teams of horses drew plows through the sand, turning up the clams by the wagonload. From 1916 to 1947 nearly 50,000 clams per year were collected by commercial diggers. Now, adults of the species are almost unobtainable in the intertidal zone. In 1973, fewer than 40 legal-sized clams were taken from the once-fine collecting areas of Pismo Beach and Morro Bay. Fortunately, subtidal populations have not been extensively exploited, and replenishment of the shore is still possible.

Because of this clam's commercial importance, a good deal is known of its natural history, particularly through the work of Weymouth (1923) and Herrington (1930). During stormy weather the clams go down to considerable depths to avoid being washed out, but normally they lie shallowly buried, often with the posterior tip of the shell just below the surface. They show a strikingly constant orientation in the direction of wave action, always being found with the hinged side toward the ocean. Though they are not as active diggers as the razor clams, they can, especially when young, move rather rapidly; and given a reasonable chance, they can dig themselves back in from the surface. Weymouth observed that "the ordinary action of the foot in burrowing appears to be supplemented in the Pismo clam by the ejection of the water within the mantle cavity . . . recalling the method of 'jetting' a pile, . . . an important factor in moving such a bulky shell through the sand."

The Pismo clam is a very slow-growing species, requiring from 4 to 5 years to reach its present legal size of 4½ to 5 inches (11.4 to 12.7 cm, the legal size depending on location; ten-clam limit). It may continue to grow, but not so rapidly, until it is at least 15 years old. Many legal-sized clams are at least 24 years old; the oldest clam reported was estimated to be 53 years of age! The rate of growth varies according to a definite seasonal rhythm— rapid in summer and slow in winter. The winter growth of the shell is darker and is deposited in narrow bands. The sexes are separate, and after the animal becomes sexually mature at the end of its second or third year, it spawns in the late summer and fall. Females produce eggs in quantities directly proportionate to their size, a large clam producing an estimated 75 million in one season.*

The Pismo clam's natural enemies are several, but with the exception of man, and in some locations the sea otter, the toll on adults is not heavy. Seastars do not frequent sandy beaches, and boring snails such as *Polinices* are unable to penetrate the hard, thick shells of adults, although clams less than 2 years old may frequently fall prey to the snail's drill. Gulls will de-

---

*For many years it was the general belief (recorded in previous editions of this book) that the Pismo clam was hermaphroditic. However, the sexes are separate, although a few cases of hermaphroditism have been reported in very young Pismos.

vour small specimens that are left lying on the surface by diggers or otherwise exposed, but they cannot get at the clams in their normal buried habitat.

The nearest relatives of this clam are tropical forms, and it can be considered subtropical. It ranges from Half Moon Bay to Bahía Magdalena, Baja California, but is not common in the northern part of its range. In Monterey Bay it is well known, but it is not abundant by comparison with its occurrence on the southern beaches. Like the bean clam, *Tivela* will be found occasionally harboring a cluster, several centimeters long, of the hydroid *Clytia bakeri* (§188), which, protruding above the surface, reveals the hidden presence of the clam.

§191. The annelid worm *Euzonus mucronata*, perhaps more abundant on protected beaches (§155), is not a stranger to some exposed beaches. On the exposed beach at Long Bay on the west coast of Vancouver Island the Berkeleys reported it (1932) "in vast numbers . . . , whole stretches of sand being tunnelled by countless millions. Judging by the complex system of furrows on the sand beds they inhabit, they seem to emerge from their burrows and crawl on the surface of the sand, but none were found exposed. Large flocks of sandpipers are frequently seen at low tide, extracting these worms and feeding on them."

In certain exposed sandy beaches at and below the lowest-tide level, bait gatherers frequently dig out the polychaete worm *Naineris dendritica*, which may be had sometimes by the shovelful. Except for its smaller size, this so nearly resembles *Nereis vexillosa* (§163) that bait collectors confuse them, swearing up and down that it is the height of folly to dig under mussels when the desired worms are so available in easily dug sand. *N. dendritica* has been reported from Alaska to southern California and occurs commonly in algal holdfasts and masses of bryozoans attached to rocks, as well as on sandy beaches.

Another polychaete, *Nephtys californiensis* (§155), may also be locally abundant.

§192. A good many popular food fishes, notably the striped bass, are found along sandy shores just outside, or even within, the line of breakers. The live-bearing surf perch occurs similarly. Although these can scarcely be considered intertidal forms, one interesting southern fish (which has been reported as far north as San Francisco) actually comes high into the intertidal zone for egg deposition. This is the famous grunion, *Leuresthes tenuis*, a smeltlike fish about 15 cm long (Fig. 210).

The egg-laying time of the grunion is holiday time for tremendous numbers of southern Californians. Along the coast highways cars are parked bumper to bumper for many kilometers, and the moon and thousands of beach fires light up the scene. The fish are caught with anything available, from hats to bare hands, and are roasted over the fires, making fine fare indeed. It is illegal to use nets, however.

The grunion's extraordinary spawning habits are as perfectly timed as

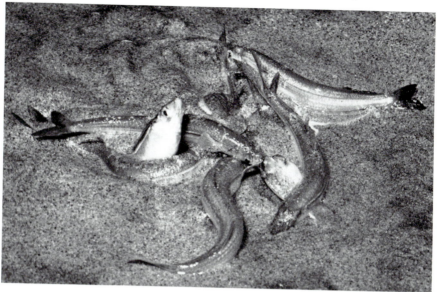

Fig. 210. The grunion, *Leuresthes tenuis* (§192), spawning at La Jolla; the female is half-buried in the sand, depositing the eggs.

those of the palolo worm of the South Seas, and the timing force is as mysterious. On the second, third, and fourth nights after the full moon—in other words, on the highest spring tides—in the months of March, April, May, and June, and just after the tide has turned, the fish swim up the beach with the breaking waves to the highest point they can reach. They come in pairs, male and female. The female digs into the sand, tail foremost, and deposits her eggs some 7 cm below the surface. During the brief process the male lies arched around her and fertilizes the eggs. With the wash of the next wave the fish slip back into the sea. Normally the eggs remain there, high and dry, until the next high spring tides, some 10 days later, come to wash them out of the sand. Immediately on being immersed, the eggs hatch, and the larvae swim down to the sea.

It is an astounding performance. If the eggs were laid on any other tide, or even an hour earlier on the same tide, they would probably be washed out and destroyed. If they were laid at the dark of the moon, they would have to wait a month to be hatched, for the full-moon tides are never as high as those of the dark of the moon. But the interval between the two sets of spring tides is the proper period for the gestation of the eggs, and the grunion contrives to utilize it. Incidentally, the succeeding lower high tides actually bury the eggs deeper by piling up sand.

The fish mature and spawn at the end of their first year, and they spawn on each set of tides during the season. During the spawning season their growth ceases, to be resumed afterward at a slower rate. Only 25 percent of the fish spawn the second year, however, 7 percent the third, and none the fourth.

§193. The storm wrack and flotsam cast up on the sandy beaches is sure to contain the usually incomplete remains of animals from other kingdoms—representatives of floating and drifting life and of bottom life below the range of the tides.

Shells of deep-water scallops, snails, piddocks, and other clams are very commonly washed up. Although perfect specimens of this sort are adequate for the conchologist, to the biologist they are merely evidence that the living animals probably occur offshore.

Some years in early spring vast swarms of the by-the-wind sailor, *Velella velella*, often mistakenly called "Portuguese man-of-war," are blown toward our coast, and great numbers of the little cellophane-like floats, with their erect triangular sails, may be cast ashore in windrows (Fig. 211). Often the fresh specimen is intact enough to place in an aquarium or jar of water to observe its details. The animal, beneath its transparent float, is bluish to purple; contrary to older zoological opinions, it is not a colony of specialized individuals, like the Portuguese man-of-war, but a highly modified individual hydroid polyp that has taken up life on the high seas (Fig. 212). This can be visualized as the upside-down hydranth of a hydroid like *Tubularia*, which has been developed from a larva that did not descend to the bottom to settle and grow a stalk but has instead settled at the surface and grown a float, so to speak. *Velella* and two or three lesser-known relatives are properly called chondrophores, and the term siphonophore is used for such animals as the Portuguese man-of-war and a host of lovely, delicate colonial creatures of the high seas. *Velella* is one of the few examples of high-seas life that a beachcomber may expect to find.

Examination of windrows of cast-up *Velella* will soon reveal something of interest to the casual beachcomber: the sail is situated on a diagonal to the long axis of the animal, and this diagonal is in the direction of northwest to southeast on the specimens cast up on northeast Pacific beaches (we should get used to the circumstance that the northeastern Pacific is our side of the ocean; the northwestern Pacific is the side of Japan and Kamchatka). Another form of *Velella*, with the sail running from northeast to southwest, occurs on the western side of the Pacific. In the Southern Hemisphere the distribution of the two forms of *Velella* is reversed. It is suspected, although no one yet has enough data to prove it, that the two kinds of *Velella* may be mixed together out in the middle of the Pacific Ocean (perhaps in about equal numbers) and that they are sorted by the action of the wind, so that "right-handed" and "left-handed" kinds are concentrated on different sides of the ocean.

When blown before a moderate wind, *Velella* tacks at about 45° away from the following wind, when viewed with its long axis at right angles to the wind direction (Fig. 212). The *Velella* that occurs on our beaches tacks to the left, so that light southerly winds tend to blow it away from the shore.

Fig. 211. Drifts of the chondrophore *Velella velella* (§193) cast up at Dillon Beach in spring 1962.

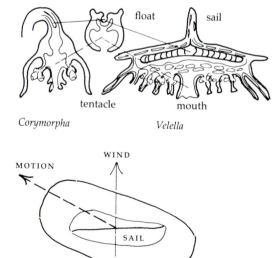

float   sail

tentacle   mouth

Corymorpha   *Velella*

WIND

MOTION

SAIL

Fig. 212. *Above, Velella* compared with *Corymorpha*, a hydroid polyp, to show the corresponding structures. *Below,* the relation of *Velella*'s movement to wind direction.

However, when the wind is strong, *Velella* tends to spin around rapidly and follow the wind at a much closer angle. This explains why the animal may appear on our shores after the first strong southerly or westerly winds of the year.

The sailing ability of *Velella* may be demonstrated in a broad shallow pool on the beach. George Mackie, one summer at Dillon Beach, designed an ingeniously simple experiment, using plastic bottle tops for controls (they sailed straight across the pool as good controls should). Computing the angle of arrival of *Velella* on the opposite side, he found (1962) that the "best sailors" could tack as much as 63° to the left of the wind. Thus, it would seem that the name "by-the-wind sailor" is not strictly accurate.

After a few days on the beach the animals die and disintegrate, leaving their skeletons to drift across the beach and into the dunes like bits of candy or cigar wrappers. Often the arrival of *Velella* on the shore is the omen that summer is not far behind.

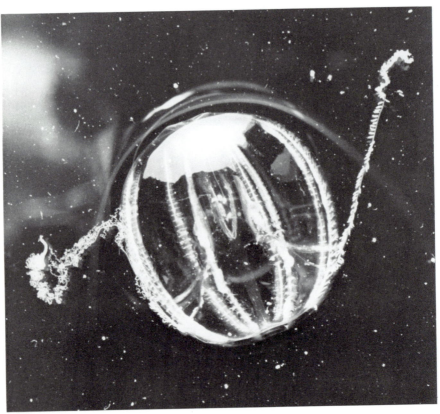

Fig. 213. The sea gooseberry, or cat's eye, *Pleurobrachia bachei* (§193); this is one of the ctenophores, or comb jellies.

"Gooseberries," the "cat's eyes" of the fishermen, are occasionally cast up on the beach, where succeeding waves roll them around until they are broken. These are comb jellies, or ctenophores, usually *Pleurobrachia bachei* (Fig. 213). The nearly transparent spheres, usually 1 to 2 cm in diameter, carry two long tentacles. Of these also we have picked up fresh and living specimens so perfect that the iridescent paddles of the plate rows started to vibrate the moment the animals were placed in seawater.

Various kinds of jellyfish, or scyphozoans, are stranded on the beach from time to time. *Aurelia* (§330) is commonly seen, as a flattened transparent blob; if not too broken up, it can be identified by its four horseshoe-shaped gonads, which are yellowish in the female and lavender in the male. A smaller and firmer transparent jellyfish may be *Polyorchis* (§330), sometimes in good enough condition to be revived in a bucket of water. Two large jellyfish that are occasionally stranded in tide pools as well as on the sandy beach are potentially dangerous to anyone who may have allergic reactions. One of these is the brown-striped *Chrysaora melanaster*. The other may appear as a shapeless mass of yellowish or pale-orange jelly with a great tangle of tentacles; this is *Cyanea capillata*, the lion's mane. Either of these may cause dangerous reactions when handled—from a severe rash to acute allergy shock, depending on individual susceptibility and past history of exposure. Sherlock Holmes once investigated, unsuccessfully, a mortality caused by the lion's mane.*

One of the charms of strolling on a beach is the possibility that almost anything may turn up, sooner or later—a case of scotch, perhaps, or some relic from the wreckage of a ship offshore. But usually it will turn out to be a block of wood or piece of timber bored by *Teredo* and festooned with a dense growth of goose barnacle, *Lepas anatifera*, and perhaps with other exotic creatures crawling among the crevices.

*See Joel W. Hedgpeth. 1948. Re-examination of "The adventure of the Lion's Mane." Baker Street Journal 3: 285–94.

PART III

Bays and Estuaries

The chief environmental factor that distinguishes the animals of bay shores, sounds, sloughs, and estuaries from all the others so far considered is their complete, or almost complete, protection from surf. This factor alone, it is true, will segregate only a few animals, for of the many already mentioned a scant half dozen actually require surf, however many may tolerate it. The absence of surf, however, will alter the habits of many animals, and some, encountering no rigorous wave shock, will venture up from much deeper water.

Such previously encountered environments as rocky shores and sandy beaches will be found in these sheltered waters; in addition, there are gravel beaches, eelgrass flats, and mud flats. Each type of shore has an assemblage of forms not found elsewhere, mingled with animals that occur almost anywhere and a few strays from neighboring environments.

A coherent treatment of the animals living on completely sheltered shores is desirable for a number of reasons. For one thing, the tides in these locations are invariably later—often, as in parts of Puget Sound and the inside waters of British Columbia and southeastern Alaska, many hours later. The result is that the intertidal areas are bared during the heat of the day or the chill of the night, and the shore animals must withstand temperature changes far greater than animals in corresponding positions on the outer coast ever have to face. Moreover, organisms of quiet waters are probably exposed to more desiccation stress, for periodic wetting by splash, common on the outer coast, is absent here. The "completely protected" shore animals must also tolerate more variable salinity, which is decreased by freshwater streams and sometimes increased to a considerable degree by evaporation from quiet, shallow areas. Altogether, the animal communities in this environment are characteristic and very different, however much of a potpourri they may be from a geographical and ecological standpoint.

A bay is simply a quiet arm of the sea. An estuary, by contrast, is a mixing zone between the saline waters of the sea and the fresh waters of the land, and the term "estuarine" connotes conditions between the two extremes. The mixing is effected primarily by the tidal action from the sea, and most of the inhabitants of the estuary, both plant and animal, are derived from the sea.

Estuaries are transient phenomena in the geological sense, and the greater abundance of organisms adapted to the changing salinity conditions of eastern and Gulf coast estuaries may be a reflection of the greater period of time that estuaries have had to develop there. On the geologically younger Pacific coast there are no old estuaries, and only a few young ones: San Francisco and Humboldt bays in California, Coos Bay and the smaller bays of Oregon, Willapa Bay in Washington, and the branches of Puget Sound.

Most of the bays of the Pacific coast are essentially marine bays, not estuaries. A few, especially in southern California, may be closed off from the sea, forming hypersaline lagoons. However, it is now rare to find a bay left in southern California. The bays are being chopped up into marinas, and the marshlands have become parking lots. From the naturalist's point of view, the transformation of Mission Bay, once an "unimproved" environment of sloughs and marshlands, into a complex of artificial islands, swank restaurants, and speedboat courses is hardly an improvement. Bays and estuaries, alas, are particularly vulnerable to progress, since so many of them lie near cities and are treated either as sewers or as wastelands to be filled up and converted into subdivisions. A true estuary, receiving a steady supply of nutrient materials washed down by the rivers and flushed out of the bordering marshlands, is a rich and thriving place. Unfortunately, only a few of the west-coast isolated estuarine areas are still rich in life.

5 🐟 Rocky Shores of Bays and Estuaries

The inhabitants of estuaries are characteristically euryhaline, that is, they can adapt themselves to changes in the salinity of the water; but most of them can survive, if not thrive, in oceanic water as well. This may account for the wide distribution of some estuarine species, and for the tendency of some of them to appear in scattered parts of the world, including the Pacific coast. Interestingly enough, some organisms from the sea are more capable of adjusting to reduced salinities than freshwater ones are of accepting increased salinity; in fact, very few representatives from fresh water occur in salinities higher than 2 to 3 parts per thousand, which is less than 10 percent the salinity of ocean waters.

Because Pacific coast estuaries are so young, there are few estuarine fish, and no native estuarine oysters or large crabs; the large prawns of warm-temperature waters are also missing. These ecological niches thus left unoccupied are easily taken over by exotic species. One of the most astounding examples of fish transplantation is that of the striped bass from the Atlantic seaboard to the estuarine regions of the Sacramento–San Joaquin river system. This was accomplished in 1879 by Livingston Stone, who released 150 striped bass at Martinez in upper San Francisco Bay. Within 3 years the fish had established themselves. Obviously, this could not have happened if the environmental niche needed by the striped bass were already occupied by some other enterprising species; but California waters were poor in corresponding types of fish. Quite a few estuarine animals have needed no encouragement (indeed, they have invited themselves in to settle); but others, especially the desirable oysters, seem most reluctant to oblige man and become naturalized in Pacific coast estuaries. This may be due in part to different temperature requirements and in part to different food requirements at early stages.

The fauna of this environment is particularly well developed on the reefs and cliffs at the south end of Newport Bay, along the railway embankment in Elkhorn Slough, on the rocky shores of Tomales Bay, in the San Juan Islands and other parts of the Puget Sound region, and in hundreds of places in British Columbia and southeastern Alaska.

Zones 1 and 2. Uppermost Horizon and High Intertidal

Although, as on the outer coast, the highest fixed or lethargic animals are barnacles, limpets, and periwinkles, it follows from the lack of surf that their absolute level (allowing for tidal differences) is lower. That is, since spray and waves will not ordinarily come up to them, they must go down to the water, and instead of ranging a meter or more above high-tide line they will be found at or a little below it (see Fig. 4).

§194. An extremely common small barnacle of fully protected waters throughout our area is *Balanus glandula*, which has already been discussed as the dominant barnacle of the protected outer coast. As was pointed out in §4, this great variability of habitat bespeaks a tolerance and generalization that are uncommon among marine invertebrates. In the outer-coast environment there are wave shock, low temperature, high salinity, and plenty of oxygen. In bays and estuaries the animals must put up with variable and often high temperatures, variable and often low salinities (because of the influx of fresh water), and relatively low oxygen content in the water, gaining no apparent advantage except escape from wave shock. To make the situation even more puzzling, *B. glandula* actually thrives best, according to Shelford et al. (1930, see 1935), in such regions as Puget Sound, living in enclosed bays where these apparent disadvantages are intensified and avoiding the more "oceanic" conditions of rocky channels that have swift currents and more direct communication with the sea. The term "thrive," translated into figures, means that although *B. glandula* is never compressed by crowding in Puget Sound, there may be as many as 70,000 individuals per square meter.

Some of the great success of *Balanus glandula* in such quiet-water situations is undoubtedly due to the relative scarcity in these environments of important competitors and predators more common on the open coast (see §4). In this connection, perhaps the ecological catholicity of *B. glandula* will enable it to resist invasion by *Elminius modestus* (the New Zealand barnacle that now appears to be taking over the barnacle lebensraum on many European shores), should *Elminius* arrive on our shores. It should be noted that colonization of open-ocean shores by invaders seems to be rare.

Another barnacle, the much larger *Semibalanus cariosus* (Fig. 214; formerly *Balanus*), comes very near to reversing *B. glandula*'s strange predilections. In Puget Sound this form prefers steep shores with strong currents and considerable wave action, that is, the nearest approach to oceanic conditions that these waters afford. When this same barnacle moves to the protected outer coast, however (and it is fairly common as far south as Morro

Fig. 214. *Semibalanus cariosus* (§194), a strongly ridged conical barnacle often with a thatched-roof appearance.

Bay), it avoids oceanic conditions most assiduously, occurring only in deep crevices and under overhanging ledges in the low zone, where it finds the maximum protection available. The net result, of course, is that it maintains itself under nearly identical conditions everywhere—a much more logical procedure than that of *B. glandula.*

The young of *Semibalanus cariosus* have strong, radiating ridges in a starry pattern and are beautifully symmetrical. An adult specimen that has not been distorted by crowding is conical, sometimes has shingle-like thatches of downward-pointing spines, and is up to 5 cm in basal diameter and slightly less in height. Under uncrowded conditions in Puget Sound, *S. cariosus* may attain a diameter of 13 to 15 mm within a year of settling in the spring. At this size it becomes nearly invulnerable to attack by a major predator, the gastropod *Nucella* (§§165, 200). The largest specimens are found to the north of Puget Sound; but under ideal Sound conditions the animals sometimes grow in such profusion and are so closely packed together that "lead-pencil" specimens develop, which reach a height of 10 cm with a diameter of less than 12 mm. Test counts have shown 15,000 of these per square meter. *S. cariosus* occurs, as a rule, at a lower level than *B. glandula,* and where the two are intermingled *B. glandula* will often be found attached to the larger form rather than to the rock.

Still a third barnacle, *Chthamalus dalli*, occurs abundantly from Alaska to San Francisco, filling in that region the position held by *C. fissus* in the south and even extending, though not in great numbers, as far south as San Diego. It is the smallest of the lot, being about 6 mm in basal diameter and half as high. It is a definite and clean-cut form, never crowded and piled together like *Semibalanus cariosus*, although one test count in the San Juan Islands showed 72,000 to the square meter. In the Puget Sound region it will be found chiefly interspersed with *B. glandula*, a potential competitor for space. If space is indeed limited, the contest almost invariably goes to *B. glandula*, whose faster-growing shell crushes, smothers, or undercuts the smaller *Chthamalus*; apparently competition rarely or never occurs between *Chthamalus* and *S. cariosus*, which generally occurs lower on the shore. The compensating advantage of *Chthamalus* is its hardiness. It is capable of residing higher on the shore than any other acorn barnacle. *C. fissus* can tolerate both high salinity and small enclosed bays where there is so much fresh water or decaying vegetation that no other barnacles can survive. Only in the latter environment will it ever be found in pure stands. This small form shows particularly well the six equal plates or compartments that form the shell. The various species of *Balanus* (and *Semibalanus*) have six also, but they are unequal in size and overlap irregularly, so that it is often difficult to make them out.

Many species of barnacles resemble each other so closely that it is difficult for the beginner to try to identify them surely. However, the collector who is sufficiently interested to dry the specimens and separate the valves can probably determine their species by consulting the chapter by Newman in Smith and Carlton (1975).

§195. Several familiar limpets occur in these quiet waters. *Collisella pelta* is foremost (§11); it is by far the commonest limpet in southern Puget Sound and often the only one. Great aggregations of small specimens occur high up with the small barnacles, on smooth rocks, and on the shells of mussels. Farther down—and these animals occur at least to the zero of tide tables—they are solitary and larger. The dingy little *C. digitalis* and/or *C. paradigitalis* (§6) also occurs, but only in the more exposed parts of the Sound, and both the plate limpet (§25) and the low-tide dunce-cap limpet (§117) exceed it in abundance, according to the published lists.

A common Puget Sound periwinkle, especially in the quiet waters where the barnacle *Balanus glandula* and the quiet-water mussel thrive, is *Littorina sitkana* (Fig. 10), usually under 1 cm in diameter and having a gaping aperture. In the less-protected parts of the Sound, with the *Semibalanus cariosus*–California mussel association, the small, cross-stitched snail *L. scutulata* (§10) is the most common form. Farther north a few of the variety of *L. sitkana* formerly called *L. rudis* will be seen. *Littorina sitkana* is also the larger periwinkle of somewhat protected ocean shores in Oregon, where it extends as far south as Charleston. Apparently the scarcity of suitable, sheltered habitats, and possibly increased predation by the crab *Pachygrap-*

sus crassipes, prevents its becoming established farther south. This littorine prefers damp, sheltered crevices, perhaps because its eggs are laid in gelatinous masses susceptible to drying. There is no planktonic dispersal stage, and this characteristic may contribute to the snail's failure to extend its range farther south. Tagging experiments indicate that the littorines are great stay-at-homes, never migrating from the immediate neighborhood of the pool in which they have been born.

The purple shore crab *Hemigrapsus nudus* (§27) cavorts around rocky beaches in Puget Sound, and on gravel shores it mixes with its pale and hairy brother *H. oregonensis* (§285). The latter, however, distinctly prefers regions where there is some mud, whereas the former is equally partial to rocks. In Oregon bays and in San Francisco Bay *H. oregonensis* is the common form.

§196. A small hermit crab with granulated hands and bright-red antennae, *Pagurus granosimanus*, appropriates the shells of the northern periwinkles. These hermits, which we have previously noted as occurring in Monterey Bay with the dominant *P. samuelis* (also with red antennae), differ from their very similar relatives in a few slight anatomical characters. Their external characteristics, however, will identify them in a reasonable percentage of cases. In the first place *P. granosimanus* lacks the more-southern form's slight hairiness; second, its legs are unbanded; and third, it is typically the smallest intertidal hermit to be found in Puget Sound. Beyond that lies color—and confusion. Not only is there likely to be a great color variation in a large number of the animals herein listed, but it is apparently impossible to find any two persons who will use the same color terms in describing an animal, even if they are looking at the same specimen. As an experiment, we submitted one specimen to three people and got three sets of color terms with almost no overlapping. One writer describes this particular hermit as "buffy olive to olive" with "porcelain blue" granules on the claws. The harassed amateur can do no better than find an undoubted specimen and make his own color chart—a chart that will likely have to be altered to fit the next specimen he finds.

Pagurus granosimanus ranges from Unalaska to Bahía de Todos Santos. Contrary to previous reports, it is common throughout California, where hermits of breeding size typically occupy shells of *Tegula funebralis*.

§197. The animals mentioned so far differ only in species from comparable forms familiar to the reader. The same thing is true of the mussel *Mytilus edulis*, but it is a predominantly quiet-water animal and notable for its wide distribution—literally around the world in northern temperate regions, as well as along South American shores. The rich beds along the coasts of England and France provide great quantities for food and bait for commercial fisheries, and on the west coast of France the mussels are cultivated very much as oysters are cultivated in other places. When "farmed," they grow larger than in the typically overcrowded natural beds.

M. edulis is wedge-shaped and rarely more than 5 cm long, although specimens twice that size have been collected; its shell is smooth and thin, whereas that of *M. californianus* (§158) is ridged and thick. Sometimes the battle for living space drives it to establish itself on gravel beaches. We found a particularly fine example of this in Hood Canal, a long arm of Puget Sound, where the mussel beds formed a continuous belt along the beach. The byssal threads were so closely intertwined that one could tear the bed away from the gravel in a solid blanketlike mass, exposing the hosts of small animals that lived beneath it.

In Puget Sound and north to Alaska this mussel occurs in tremendous quantities. It may also be expected in nearly every quiet-water bay south of Puget Sound as far as Baja California, but never as commonly as in the north. A few anemic-looking specimens can be found on gravel banks and rocky points in San Francisco Bay, but most of those that have survived the Bay's contamination occur on wharf piling and the undersides of floating docks.

Under certain circumstances, the bay mussel may be found on shores exposed to heavy surf, although the thinner byssal threads of *Mytilus edulis* would seem less well suited to these conditions than the more strongly attached *M. californianus*. Nevertheless, along the open coast of Washington, *M. edulis* regularly occupies a distinct band above the level of *M. californianus* and often occurs within the latter's vertical range where storms have ripped loose and cleared large patches from beds of California mussels. Bands of *M. edulis* may also be found occasionally below the level of *M. californianus*, but these tend to be mostly temporary, presumably disappearing in large part through the activities of predators, several of which prefer the thinner-shelled bay mussel to the California mussel when presented with both.

M. edulis, however, is best suited for life in quiet water, and in this connection, one other trait deserves mention. The bay mussel is considerably more mobile than *M. californianus*, and it can move rapidly to avoid being buried by the sediments continually being deposited in harbors. In mixed clumps with *M. californianus*, small individuals of *M. edulis* crawl to the surface and survive; by contrast, the California mussel remains firmly attached, as would be appropriate if it were in its open-coast habitat. Thus, in the bay environment, *M. californianus* may find itself attached beneath a layer of bay mussels where it can be suffocated by accumulated sediment.

Spawning generally occurs in the late fall or winter, and recruitment success varies considerably from one year to the next. After settlement, the mussel's growth rate depends on the amount of time submerged, and thus feeding. This, in turn, varies with the position on the shore. Bay mussels exposed to greater wave action grow more slowly and have thicker shells than their quiet-water counterparts.

§198. Under the crust formed by mussels of gravelly shores there is an important and characteristic assemblage of animals. There are few species but tremendous numbers of individuals; and the commonest of the

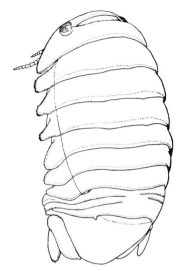

Fig. 215. The small isopod *Gnorimosphaeroma luteum* (§198), a common inhabitant of protected shores, especially in estuarine waters; 25 times natural size.

larger forms, to the eternal joy of bait gatherers, is a small, quiet-water phase of the segmented worm *Nereis vexillosa* (§163), rarely exceeding 10 to 12 cm.

Two closely related little pill bugs, *Gnorimosphaeroma oregonensis* and *G. luteum* (Fig. 215), each about 6 mm long, flaunt their belief in large and frequent families. The average under-crust population of these squat isopods with widely separated eyes will run many dozen in a tenth of a square meter in Puget Sound, and they are the commonest pill bugs in San Francisco Bay.

Formerly considered varieties of the same species, the two occupy different habitats. *G. oregonensis* (= *G. o. oregonensis*) occurs on the open coast and frequents the more saline parts of bays from Alaska to San Francisco Bay, whereas *G. luteum* (= *G. o. lutea*) is found in dilute estuarine water and in fresh water near the sea from Alaska to the Salinas River in Monterey County. A third species, *G. noblei*, described from Tomales Bay, occurs from Humboldt Bay to the Palos Verdes Peninsula.

There are several ribbon worms here, notably our acquaintance *Emplectonema gracile*, the dark-green rubber-band nemertean (§165) from under open-coast mussel clusters, and *Paranemertes peregrina* (§139), the purple or brown form that occurs under rocks and on mud flats. Small individuals of a fish, the high cockscomb (*Anoplarchus purpurescens*, §240), will also be seen.

The infrequent bare spaces on the sides of mussel-covered stones, especially in crevices, may be occupied by an encrusting sponge, *Haliclona* sp. (similar to the outer-coast form treated in §31), described originally as a *Reneira* and listed as *H. rufescens* in previous editions of this book.

Zone 3. Middle Intertidal

As in the higher zones, many of the species found here are the same as, or close relatives of, those encountered on the open coast. The middle zone, however, which is largely uncovered by most low tides and covered by most high tides, is richer in both plant and animal life than the rocks higher up.

§199. The quiet-water form of the seastar *Pisaster ochraceus* (§157), *P. o. confertus*, haunts the mussel beds in completely sheltered waters as does its rough-water relative the beds of larger mussels on the open coast. *P. o. confertus* may be found in the lower zone as well as in the middle zone, but we have never seen it below the low-tide line—a fact entirely consistent with its unfriendly relationship with mussels. On certain quiet-water gravelly and rocky shores in Puget Sound and British Columbia, to which region the animal is restricted, most of the specimens are colored a vivid violet.

§200. The common quiet-water snails are the purples (or dogwinkles) —various species of *Nucella* (Fig. 216), especially *N. canaliculata*, the channeled purple, and *N. lamellosa*, the highly variable wrinkled purple. The former is generally found within the mussel beds, or at least at about that level, whereas the latter is generally lower, below the beds. Both range at least from Alaska to Monterey Bay and are common from Puget Sound to San Francisco Bay; they also occur on sheltered outer-coast rocky shores and on jetties. Both are predators on barnacles and mussels, although barnacles seem to be preferred by *N. lamellosa* and mussels by *N. canaliculata*. In all cases, a hole is drilled in the shell of the prey by the rasping action of the radula aided by a chemical secretion that softens the shell. Following this process, which may take a day or two, the soft tissues of the prey are removed.

Once a year in the winter or spring, mature individuals of *Nucella lamellosa* (those 4 years old or more) aggregate in small groups to breed. Despite the fact that the snails are dispersed for much of the year, these annual breeding groups apparently represent persistent, natural units, since individuals are likely to return and breed with the same group the following year. After mating, females deposit their eggs in vase-shaped, yellow capsules, each about 1 cm long, which are attached in crevices and under overhanging surfaces. The area covered by the clusters varies, being proportional to the size of the breeding aggregation, but on the average about 300 cm^2 of surface is covered, making the communal mass of clusters quite conspicuous. Each female produces about 1,000 eggs each year, and unlike the situation in *N. emarginata* (§165), where some of the "eggs" serve only as food for other developing larvae, nearly all eggs of *N. lamellosa* have the potential to develop into young snails. Even so, mortality after hatching is very high, and rarely do 1 percent of the fertilized eggs survive to a year of age.

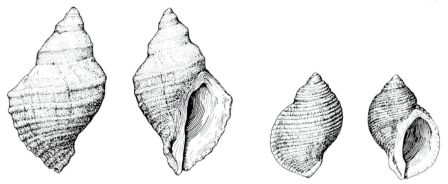

Fig. 216. Snails of the genus *Nucella: left, N. lamellosa* (§200), the wrinkled purple, of somewhat lower levels; *right, N. canaliculata* (§200), the channeled purple, of rocky shores at about
the level of mussel beds.

Fig. 217. *Ceratostoma nuttalli* (§200),
a murex from southern waters;
natural size for a large specimen.

Locally, *Searlesia dira*, the elongate "dire whelk," may be abundant, especially to the north, where it becomes the commonest littoral snail in the quiet pools and channels of northern British Columbia. The dire whelk is attracted by, and seems to prefer for its food, dead and injured animals, which are consumed, without drilling, roughly in proportion to their abundance in the habitat. Limpets, especially *Notoacmea scutum* (§25), are eaten, as are barnacles, chitons, and snails. The sexes are separate, and from September to May, females may be found depositing low, convex egg capsules in clusters on the rock walls of shady crevices.

A murex, *Ceratostoma foliatum* (§117), also occurs in these northern waters—surprisingly for a relative of these tropical snails—as one of the three very common snails in certain locally rich pockets, like Fisherman's Cove, south of Prince Rupert. In southern waters from Point Conception into Baja California, a second species, *Ceratostoma nuttalli* (Fig. 217), occurs on rocks and piling; it is especially common on breakwaters where barnacles are abundant.

§201. The shell of *Nucella* provides a portable home for the hairy hermit, *Pagurus hirsutiusculus*, which, in quiet northern waters, attains great size. At Friday Harbor, specimens have been taken with 5-cm bodies and an overall length of more than 7 cm. The animal's extreme range is Alaska to San Diego, but south of Puget Sound the specimens are smaller and less hairy. They are fairly plentiful at Elkhorn Slough, however, and are the common hermits of San Francisco Bay. The legs of this otherwise generally brown hermit are banded with white or bluish white.

To one whose acquaintance with hermits has begun with the outer-coast forms, those found in Puget Sound and northern inside waters will exhibit a strange trait—a trait that we deduce to be a direct result of the great difference in environment. The outer-coast hermits will never desert their shells except when changing to another or when dying. Usually they cannot be removed by force unless they are caught unawares (a difficult thing to do) or are literally pulled apart. In Puget Sound, however, the animals will abandon their shells very readily—so readily, indeed, that on more than one occasion an observer has found himself holding an empty shell, the hermit having deserted in midair. Our unproven inference is that the presence or absence of surf is the factor determining how stubbornly a soft-bodied hermit will cling to its protecting shell.

§202. The ubiquitous warty anemone *Anthopleura elegantissima* (§24) is found in the quiet waters of Puget Sound as an attractive pink-and-green form. This anemone, with its many variants, is certainly an efficient animal, for it is universally characteristic of the northern Pacific. A visitor from Japan reported it as common there, but he considered the center of distribution to be on the American coast, since it is more variable and more highly developed here.

On erect rocks of the inner waters at Sitka, an anemone that may possibly be *Charisea saxicola* is very abundant. It grows fairly high up, colonizing rock crevices and vertical and sharply sloping surfaces; the average specimen is longer (to 5 cm) than it is thick, and is generally dingy white to buff in color. We found it nowhere else, and Torrey (1902) recorded it only from "the shore rocks at Sitka," where it was taken abundantly by the Harriman Expedition.

§203. Everywhere on the Puget Sound rocky shores the purple shore crab, *Hemigrapsus nudus* (§27), replaces *Pachygrapsus*, the rock crab of the California coast. Conditions in Puget Sound would lead one to expect *Pachygrapsus* there also; but for some reason, possibly the somewhat lower winter temperatures, it does not extend that far north. The purple crab modifies its habitat somewhat with the change in regions, occurring on the California coast chiefly in the middle zone, below *Pachygrapsus*, but reaching considerably higher in Puget Sound. Northern specimens at their best, however, are smaller than those on the California coast, and the highest specimens are very small. Hart (1935) recorded females in "berry" (ovigerous) in April and May at Departure Bay, and has studied the development, finding five zoeal stages and one megalops.

Numerous other species already treated are frequent in Puget Sound, such as the lined chiton (§36) and the keyhole limpet *Diodora* (§181). Small and undetermined polyclad worms similar to *Notoplana* (§17) are very common. The estuary fauna in the south derives largely from open-coast species. In Newport Bay great twisted masses of the tubed snail *Serpulorbis* (§47) may almost cover the rocks, and the rock oyster *Chama* (§126) grows so thickly that it appears to form reefs.

§204. Along the shores of Hood Canal and generally in the Puget Sound region we have found great hordes of the small crustacean *Nebalia pugettensis* (formerly *Epinebalia*) mostly in pockets of silt among the rocks and in the organic mud under half-decayed masses of seaweed (*Fucus* and *Enteromorpha*). It also occurs in abundance among the beds of green algae and eelgrass in upper Tomales Bay, and has been found in similar sheltered situations in Monterey Bay, Morro Bay, Los Angeles Harbor, San Diego, and southward to San Quintín. Though it thrives on organic detritus, it seems favored by situations where the tidal action is enough to prevent completely foul conditions. Cannon (1927), who studied the feeding habits of the North Atlantic *Nebalia bipes*, a closely related species (Fig. 218), describes the manner in which this animal feeds. Like many other small filter-feeding arthropods, the nebaliid sets up a food current by oscillating its abdominal appendages. The food is then strained by the bristles, or setae, on the anterior appendages. *Nebalia* can also feed on larger particles.

These curious little crustaceans are of particular interest to students of classification because they are representatives of the superorder Phyllocarida (order Leptostraca), which exhibits characters transitional between the smaller, less-specialized crustaceans (the Entomostraca of the older textbooks) and the larger, more-specialized crustaceans (the Malacostraca). They have a small bivalve shell, but their loss of the nauplius larva characteristic of the entomostracans and their arrangement of appendages and sex openings have earned them a place as the lowest division of the Malacostraca. The adults differ from those in most other groups of the Malacostraca by having seven segments in the abdomen (six being usual) and by having caudal rami (two long terminal bristles on the abdomen) in the adult stage.

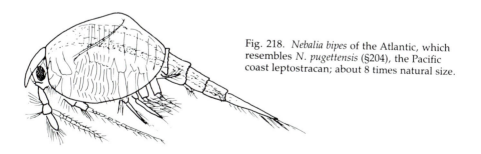

Fig. 218. *Nebalia bipes* of the Atlantic, which resembles *N. pugettensis* (§204), the Pacific coast leptostracan; about 8 times natural size.

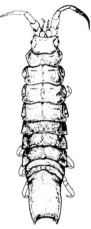

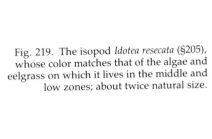

Fig. 219. The isopod *Idotea resecata* (§205), whose color matches that of the algae and eelgrass on which it lives in the middle and low zones; about twice natural size.

§205. Under the rocks there are the usual pill bugs and worms, occurring in such variety and abundance as to distract the specialist. As in other environments, the algae provide homes for numerous isopods, the predominant species being *Idotea resecata* (Fig. 219), also common on eelgrass (§278), and *Idotea urotoma* (§46). The first is usually a yellow-brown form, just the color of the kelp that it eats and to which it may often be found clinging with great tenacity. On eelgrass this form is green; thus, it appears to camouflage itself by matching the color of its host plant. *I. resecata*'s body is long (nearly 4 cm) and narrow, and terminates in two jawlike points. We found a particularly rich culture near Port Townsend, and it occurs from Alaska to the southern tip of Baja California. *I. urotoma* is less than 2 cm long, and without the two terminal points. It ranges from Puget Sound to Baja California.

Another abundant isopod of this genus is *Idotea montereyensis*, found from British Columbia to Point Conception. This isopod really seems to prefer the surf zone where *Phyllospadix* flourishes, but the young that are hatched there cannot hang on, and they are swept inshore. As they settle on red algae in more-protected areas, their color changes from green to red. A molt is required for full expression of the color change because the animal's overall color is primarily due to pigments deposited in the cuticle. After the isopods grow older in the more sheltered situations, they move out toward the *Phyllospadix*, where they can now hang on without being washed away. In the process their color changes from red to green; there are even transitional brown stages, and an occasional piebald one. Thus it appears that we have a neat adaptation to circumstances by the populations of old and young animals; Lee (1966a,b, 1972), who studied this animal at Pacific Grove and Dillon Beach, refers to the process as "remarkably precise and beautiful." As a result, the species can use a large variety of plants for food and habitat, avoid competition between young and old individuals, and remain camouflaged from the many species of fish that commonly take this and other species of *Idotea* as food.

Fig. 220. Shell of the rock cockle *Protothaca staminea* (§206).

§206. Occasional geyserlike spurtings that will be noticed wherever a bit of substratum has no covering rock can usually be traced to clams, often to the chalky-shelled *Protothaca staminea*, the rock cockle (Fig. 220). This well-rounded clam (called a cockle because its strong, radiating ridges suggest the true cockle) is also known popularly as the littleneck clam, the hardshell clam, the Tomales Bay cockle, and the rock clam. As the number of popular names would suggest, it is well known and widely used for food. In suitable localities it occurs so thickly that the valves often touch, and on the level gravel beaches of Hood Canal and such southern Puget Sound places as Whollochet Bay we have found it in such superabundance that two or three shovelfuls of substratum would contain enough clams to provide a meal for several hungry people.

The rock cockle is a poor digger, and for that reason it never lives in shifting sand, where rapid digging is essential. It will be found in packed mud or in gravel mixed with sand but seems to prefer clayey gravel, where it usually lies less than 8 cm below the surface. Where these favorable conditions exist, it grows almost equally well in isolated bits of the protected outer coast. At several points, especially Tomales Bay, the clam assumes some commercial importance, and at many other places it is an attraction to tourists. Open-coast specimens should be used warily during the summer, on suspicion of mussel poisoning (§158). Typical specimens seldom exceed 7 cm in length, although larger ones may be taken from below the low-tide level.

In the waters of British Columbia and Alaska, spawning takes place during the summer. Growth is relatively slow. At the end of the second year in British Columbia, when about half of the animals first spawn, the average length is only 25 mm; when the balance spawn, a year later, it is only 35 mm. Growth is slower still in the colder waters of southern Alaska, and slowest in south-central Alaska, where the clam requires 8 years to reach a shell length of 30 mm. But water temperature is not the only important variable, since rock cockles in Mugu Lagoon, southern California, grow considerably more slowly than those in British Columbia. The rate of growth, being determined largely by the extent and constancy of the food supply, seems to depend on the animal's position in relation to the tidal current and its degree of protection from storms, in addition to water temperature. Whatever the locality, the animal conveniently records its growth rate by producing distinct annual growth lines in the shell. The range of *P. staminea* is the Aleutian Islands to the southern tip of Baja California.

The Japanese littleneck clam, *Tapes japonica* (*Protothaca semidecussata* of some authors), was accidentally introduced to our coast in the 1930's (perhaps earlier), presumably along with the Pacific oyster, *Crassostrea gigas*. The clam is now abundant in San Francisco and Tomales bays and in Puget Sound, where it has become important commercially, with approximately a million pounds harvested annually in the late 1970's. Somewhat smaller than its relative *Protothaca staminea*, it is considered to have better flavor than the native species. In appearance it is more elongate, and although its shell color is highly variable, brownish or bluish banding is often found on one end of the yellow or buff shell. The clam's success and commercial desirability suggest the possibility that it will be introduced into other bays and estuaries—this time purposefully—along its eastern Pacific range of British Columbia to southern California.

Psephidia lordi is a small clam of some importance in Puget Sound and northward, having a total range of Unalaska to Baja California. It has a polished and gleaming white or light-olive shell that is somewhat triangular in shape, flatter than that of the rock cockle, and usually only a fraction of its size.

Zone 4. Low Intertidal

Here, as on the protected outer coast, the low intertidal region is the most prolific. Hence it seems desirable to divide the animals further, according to their most characteristic habitats.

Rock and Rockweed Habitat

In the Puget Sound area the most obvious and abundant animals of the low-tide rocky channels are a large seastar, a green urchin, a tubed serpulid worm of previous acquaintance, a great snail, and a vividly red tunicate.

§207. The seastar *Evasterias troschellii* can, it is suspected, be considered a mainly subtidal animal, but in quiet waters that are too stagnant for *Pisaster* it is very much a feature of the low intertidal zone and therefore deserves consideration here even though its proper habitat is below our scope. Elsewhere, it takes up the seastar's ecological niche where the common *Pisaster* drops it. Although the average *Evasterias* has a smaller disk, more-tapering arms, and a slimmer and more symmetrical appearance than the common seastar, the two are much alike, and they are sure to be confused by the amateur. Since both occur so frequently in the same region, the interested observer will want to differentiate them. This can be done readily with a pocket lens, or even with a keen unaided eye. On the underside of the rays, and among the spines just bordering the groove through which the tube feet are protruded, *Evasterias* has clusters of pedicellariae (§157). These are lacking in the common *Pisaster*.

At Departure Bay, and elsewhere on the inside coast of Vancouver Island, this seastar occurs in great numbers; but although it is recorded from Unalaska to Carmel Bay, the shore observer will seldom see it south of Puget Sound. Fisher (1930) distinguished several forms, but considered the animal to be one of the most variable of seastars. Specimens with a diameter of 60 cm or more are not rare. Sexual maturity comes in June, July, and August.

Evasterias, like *Pisaster*, feeds on a variety of animals, its diet varying with the abundances of various prey species in its habitat. However, *Evasterias* apparently has somewhat different table manners than its cousin. Christensen (1957) found that *Evasterias* prefers force to stealth when feeding on a bivalve, because digestion is slowed down unless all lobes of the stomach can encompass the meal. To maximize the rate of digestion, the clam must be opened as wide as possible; if necessary, however, the seastar can also extend its stomach through any narrow opening between the shells and digest a closed clam.

A good many specimens of *Evasterias* have a polynoid worm, *Arctonoe fragilis*, living commensally in the grooves under the rays. Several of these scale worms, which also occur as commensals on at least seven other seastars, may be found on a single seastar.

§208. The very fragile seastar *Orthasterias koehleri*, also chiefly subtidal and occurring probably less than 1 percent as frequently as *Evasterias*, may occasionally be found in the lowest intertidal zone at Departure Bay and Friday Harbor. It is a striking and brilliant animal. The noticeable spines, often surrounded by wreaths of pedicellariae, are white or light purple and are set against a background of red marked with yellow. Strangely, this apparently quiet-water animal is fairly common in the subtidal zone in Monterey Bay, straying, on rare occasions, just above the lowest of low tides. Even more strangely, it has been found there in surf-swept regions, but probably creeps up into the influence of waves only when the surf is light. Occasionally it is found on the seaward side of Tomales Point at very low tide. The entire recorded range is Yakutat Bay, Alaska, to Santa Rosa Island, off southern California.

The diet of *O. koehleri* is varied. On cobbled bottoms, adults dig clams, and use the pull of their tube feet to break a small opening in the

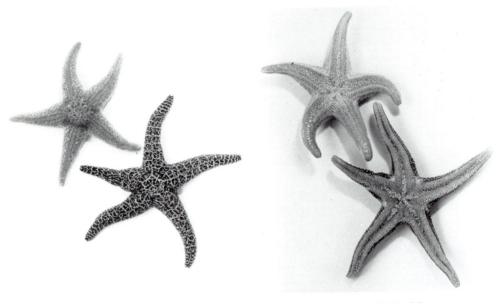

Fig. 221. *Pisaster brevispinus* (§209) and the common seastar *P. ochraceus* (§157): *left*, upper (aboral) surfaces; *right*, lower (oral) surfaces. *P. brevispinus*, pink-colored in life, is the smaller, paler specimen.

outer layer of shell; the stomach is then inserted through the opening. On rock bottoms, the diet includes small snails, chitons, barnacles, and tunicates.

§209. In Puget Sound, *Pisaster brevispinus* (Fig. 221), a pink-skinned, short-spined edition of the common seastar, attains gigantic proportions. We have seen many specimens, notably in Hood Canal, that were more than 60 cm in diameter and massive in proportion. They occurred there on soft bottoms below rocky or gravelly foreshores, usually just below the line of low spring tides. *P. brevispinus*, normally a deep-water form (drag-boat hauls from the 60-fathom mud bottom of Monterey Bay bring them up by the thousand), ranges into the intertidal on sand, mud, or wharf piling only in the most quiet waters. Its softness and collapsibility mark it as an animal that is obviously not built for long exposure to the air.

The diet of *Pisaster brevispinus* varies with its habitat, and this seastar is apparently equally at home subtidally on rock, sand, or mud. On piling and rock bottoms, it feeds mainly on barnacles, mussels, and tube-dwelling polychaetes. On soft bottoms, it feeds on several species of clams and snails and on the sand dollar *Dendraster excentricus*. Two distinct methods are used to capture clams. They may be slowly unearthed by digging (this process takes over a day). Alternatively, the seastar may extend the tube feet of its central disk down into the substratum for a distance roughly equal to the radius (arm plus disk) of the seastar. Once the tube feet attach to the buried shell, the prey can be hauled to the surface. If too much resistance is encountered, the seastar's stomach can be extended down the burrow, at least as much as 8 cm for a specimen with a radius of 11 cm; the prey is then digested in place. Apparently, bivalves boring into rock are captured and consumed by *P. brevispinus* in a similar manner. Spawning occurs in the spring in Monterey Bay and Bodega Harbor. The recorded range of this pink seastar is from Sitka to Mission Bay.

§210. The small, common, six-rayed seastar *Leptasterias hexactis* (§68) occurs in these quiet waters. Its biology is well known, but we should note that the systematics of *Leptasterias* has proved a perplexing problem, one still not completely solved. As Fisher (1930) said, "among the numerous small 6-rayed sea stars of the northwest coast of America specific lines are exceedingly difficult to draw." More recently, Chia (1966b) has decided that in most cases the lines should not be drawn; he suspects that "all the six-rayed sea stars of the genus *Leptasterias* of the North Pacific may belong to one or only a few polytypic species." From his work, based on studies in the San Juan Islands, it seems that several of the earlier-named forms (e.g., *L. aequalis*) are simply variants of *L. hexactis*. The situation in California, however, is not necessarily the same as to the north and needs careful reconsideration.

§211. A number of other seastars may be seen. The slim-armed red seastar *Henricia leviuscula* (Fig. 74) is common in these quiet waters of the north as well as on the protected outer coast. We have taken one specimen that harbored a commensal scale worm, *Arctonoe vittata*, much as *Evasterias* entertains its similar guest.

Dermasterias imbricata, the leather star (Fig. 222), is a "web-footed" form resembling *Patiria*. This is another of the very beautiful animals for which the tide pools are justly famous. The smooth, slippery covering of skin is commonly a delicate purple with red markings. The tips of the rays are often turned up, and a strong smell of garlic or sulfur often accompanies the animal. Ranging from Sitka to Sacramento Reef, Baja California, the leather star, nowhere very common, is most numerous in the northern half of its range, and the specimens found in Puget Sound are gigantic (diameter up to 25 cm) by comparison with those in the tide pools of the protected outer coast. *Dermasterias* feeds largely on sea anemones, but where anemones are not abundant, the diet includes sea urchins, sea cucumbers, and a variety of other invertebrates.

Specimens of *Dermasterias* may harbor *Arctonoe vittata*, an occasional resident also on *Cryptochiton*, the limpet *Diodora aspera*, and *Henricia*. *A. vittata*, like its cousin *A. pulchra*, is attracted to its host by chemicals released into the water. In an unusual twist, however, experiments have shown that the host is also attracted to the commensal; perhaps *Dermasterias* derives some benefit from its association with *A. vittata*.

The sunflower star, *Pycnopodia* (§66), also occurs in rock pools of the

Fig. 222. The leather star, *Dermasterias imbricata* (§211), a slippery-skinned purple seastar with red markings.

lower intertidal, as well as the sun star, *Solaster* (§67). For some reason, *S. dawsoni*, which makes its living eating other seastars, does not attack *Dermasterias*. Perhaps a chemical defense is involved.

§212. The common urchin in the Sound and northward into southeastern Alaska is the mildly green *Strongylocentrotus droebachiensis*, of circumpolar range. Except for its color, its pointed spines, and its usually smaller size, it resembles the southern purple urchin (Fig. 193). Furthermore, juvenile specimens of *S. purpuratus* are often greenish white, and the beginner may confuse these with *S. droebachiensis*. Many specimens of *S. droebachiensis* occur in the rocky San Juan Islands, and they appear to be crowding out the purple urchins that were once common there. The latter still retain unquestioned dominance on surf-swept shores along this coast, although on the east coast and in Europe the green urchin populates the open shore as well as protected regions. Adults on both the eastern and western American coasts spawn in the spring. Then, in common with other urchins, the green urchin passes through several free-swimming stages not even remotely resembling the adult. The green urchin is an omnivorous feeder but lives mainly on fixed algae, exhibiting feeding preferences similar to those of the giant red urchin (§73).

The red urchin, so common along the California coast, occurs also in

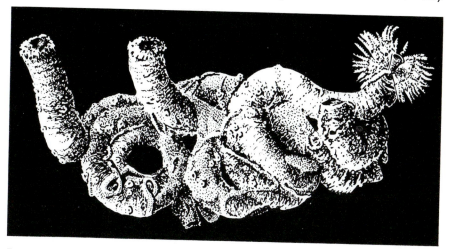

Fig. 223. The limy-tubed worm *Serpula vermicularis* (§213), one specimen with gills exposed. Many of these clusters are much larger than the one shown; 1½ times natural size.

northern inside waters but is found mainly below the low-tide line. It does range into the intertidal, however, at least as far north as Juneau.

§213. Great twisted masses of the limy-tubed worm *Serpula vermicularis* (Fig. 223) cover the rocky reefs in the Puget Sound–Strait of Georgia region. Common also in central and northern California, this cosmopolitan worm attaches to hard substrata in exposed (§129) and sheltered habitats from Alaska to San Diego. As far south as Elkhorn Slough, *S. vermicularis* may be found on the sides of intertidal rocks in quiet circumstances, but in southern California it generally occurs subtidally. Except that these worms have red "stoppers" to their tubes and exhibit vividly red gills that snap back into shelter on slight provocation, they occur just as do the tubed snails on the rocks in Newport Bay.

Serpula, like other tube worms, is dependent for its food on such microscopic particles of organic matter as chance may bring it. It assists the process, however, by setting up currents with the hairlike cilia on its delicate gill filaments. When removed from its tube, this worm often obligingly extrudes red eggs or yellowish-white sperm. The sexes are separate, but there is no apparent external difference between the male and the female, although probably, as with Roland Young's fleas, "she can tell, and so can he." Spawning has been observed at Pacific Grove during the spring; Puget Sound specimens are sexually mature in July and early August.

§214. The hairy Oregon triton, *Fusitriton oregonensis* (Fig. 224), is the largest of the rocky-intertidal snails in our entire territory, for large specimens reach a length of almost 15 cm. The animal has a handsome, brown, coiled shell that is decidedly hairy, a character that is thought to inhibit the

Fig. 225. The checkered hairy
snail *Trichotropis cancellata* (§215).

Fig. 226. The northern slipper
snail *Crepidula nummaria* (§215);
about twice maximum size.

Fig. 224. *Fusitriton oregonensis* (§214), the hairy Oregon
triton; this is the largest intertidal rock snail of the
Pacific coast.

Fig. 227. *Cnemidocarpa finmarkiensis* (§216), a bright-red
tunicate with a smooth outer
tunic; twice natural size.

settlement of boring and encrusting organisms. *Fusitriton* is a feature of the extreme low-tide rocks and reefs near Friday Harbor and has a recorded range of Alaska to La Jolla. During the summer, egg capsules are found on the same rocks. A murex, *Ceratostoma foliatum* (§117), may be found this far down also, associated with the Oregon triton, but its center of distribution is thought to be higher in the intertidal.

Only the largest intertidal hermit crab on the coast would have any use for so huge a shell as the Oregon triton's, and we find this hermit in *Pagurus beringanus*. It has granules and short, scarlet spines on the hands, but no true hair. The general color is brown with scarlet and light-green markings. The only other large hermit common on shore in the Puget Sound region and lower British Columbia is a smaller, slimmer, hairy-handed hermit, *P. hirsutiusculus* (§201).

§215. Two other common shelled snails, *Trichotropis cancellata*, the checkered hairy snail, and *Crepidula nummaria*, the northern slipper snail,* are illustrated in Figures 225 and 226. The blue top snail, *Calliostoma ligatum* (Fig. 201), which ranges from Alaska to San Diego on the protected outer coast, occurs in these quiet waters also and is particularly abundant in Puget Sound, under overhangs and in rocky crevices fairly well up, and from there clear into the subtidal.

§216. The tunicate *Cnemidocarpa finmarkiensis* (Fig. 227) has an extraordinarily bright red, smooth but leathery outer tunic, or test, which makes it one of the most conspicuous animals in the Puget Sound region. The craterlike projections—the excurrent and incurrent siphons that provide the animal with water for respiration and feeding—are rather prominent. In collecting and preserving specimens we have noted that these orifices have considerable power of contraction, closing almost completely under unfavorable conditions. We have found eggs in August.

The elongate cylindrical tunicate *Styela gibbsii*, as found under rocks at Pysht (near the entrance to the Strait of Juan de Fuca) and probably in many other places, is a small and short-stalked relative of *S. montereyensis*, so well developed on wharf piling. *S. gibbsii* is interesting in that it plays host to several other animals: a commensal pea crab, a parasitic and degenerate rhizocephalan cirripede, and a vermiform copepod, *Scolecodes huntsmani*, which more nearly resembles a degenerate isopod.

Pyura haustor, another large Puget Sound tunicate (often several centimeters long), has a bumpy test covered with all sorts of foreign material—a veritable decorator crab among tunicates. This tunicate, with its roughly

---

*Crepidula nummaria is extremely difficult to distinguish from another white slipper snail, *C. perforans*. Both range at least from British Columbia to southern California, and apparently both are found in similar habitats: inside gastropod shells occupied by hermit crabs, in holes made by boring clams, and under rocks (the last especially for *C. nummaria*). Further complicating matters, two Atlantic species with light-colored shells, *C. fornicata* (§29) and *C. plana*, have been introduced to the Pacific coast, bringing to at least four the number of our "white slipper snails."

Fig. 228. *Pododesmus cepio*
(§217), the rock oyster, or jingle;
about natural size.

globular body and relatively long siphons, ranges from Alaska to San Diego; a subspecies or variety, *P. haustor johnsoni*, is common under and alongside rocks in Newport Bay, and more rarely in Elkhorn Slough.

Several other tunicates occur, some obviously and many commonly, but for their sometimes difficult determination the reader must be referred to the additional works cited at the back of this book.

§217. The rock oyster or jingle *Pododesmus cepio* (Fig. 228) occurs fastened to Puget Sound reefs, just as a similar but smaller jingle, *Anomia peruviana*, occurs in the quiet bays of southern California. Curiously, both of these are associated with similar-looking but widely differing tube masses, *Serpula* in Puget Sound and *Serpulorbis* in southern California. *P. cepio* is one of various animals that are called jingles, presumably because of the sounds made by dead shells on shingly beaches. The valves are thin, and the adherent valve has a notched foramen, or opening, for the calcified byssus that attaches the animal to its support. The inner surface of the upper valve is highly polished and an iridescent green, the color being due in part to minute algae living within the shell. The flesh of *Pododesmus* is bright orange, and it has an excellent flavor. Large specimens may be more than 7 cm in diameter. The range is from British Columbia to Baja California.* Below Puget Sound, specimens will be found most commonly on breakwaters and on wharf piling. Spawning occurs during July and August in Tomales Bay.

Old shells of this and other mollusks, north of Puget Sound and clear into Alaska, are likely to be honeycombed to the point of disintegration by small, yellow-lined pores, evidence of the work of *Cliona celata* var. *californiana*, the boring sponge. Break apart one of the shells so eaten—they are fragile and crumbly if the process is well advanced—and note how the inte-

*In earlier editions *Pododesmus cepio* was referred to as *P. macroschisma*, which is an Alaska species.

rior is composed almost entirely of the yellow, spongy mass. In the Monterey Bay region *Cliona* attacks the shells of abalones, but usually only those growing in deep water, and the sponge is not in any case the important littoral feature that it becomes up north. The mechanism by which *Cliona* excavates its galleries and channels in shells has been the subject of recent investigation. Apparently, the sponge employs a two-step attack, the localized chemical dissolution of the calcium carbonate substratum, followed by a mechanical dislodging of small chips, which are carried to the surface in a system analogous to a mining operation. The burrowing life-style of *Cliona* is likely, at least in part, to be related to protection against predators, and to this end, the sponge's inhalant and exhalant orifices are borne on contractile papillae that respond to various chemical and mechanical stimuli. Despite the burrowing life-style and the contractile papillae, however, *Cliona* is reportedly eaten by the nudibranch *Doriopsilla albopunctata*.

Oyster shells in southern California estuary regions may be overgrown with a sponge, *Hymeniacidon* (§183), also encountered in wave-swept crevices.

§218. Enclosed in Newport Bay is a stretch of fully protected reef that is the counterpart of the Puget Sound reefs. In this small area at least two groups of animals, neither of which commonly occurs as an intertidal animal elsewhere along the California coast, maintain an extreme northerly outpost. The first is a rather beautiful gorgonian, *Muricea*, which is related to the "sea fans" and "sea whips" and hangs in graceful, tree-shaped, brown clusters from the vertical cliff face. The minute zooids that extend from the branches when the animal is undisturbed are a pleasantly contrasting yellow.* Clusters may be more than 15 cm in length and as large around as one's fist. This gorgonian, common in the kelp beds and offshore pilings, can occasionally be found in the lowest intertidal zone of outer Los Angeles Harbor.

The second group consists of the pale urchins *Lytechinus anamesus* and *L. pictus*, the latter often considered an ecological variant of the former. *L. anamesus* is about 4 cm in diameter and light gray, and it occurs abundantly in the Gulf of California (see also §269). The Newport Bay region appears to be more plentifully supplied with urchins (in number of species) than any other similar area on the Pacific coast of the United States. Both the purple urchin and the giant red urchin may be taken with this pale-gray form on the same rocky reef. On the neighboring sand flats the ubiquitous sand dollar occurs, with an occasional heart urchin or sea porcupine (§247).

§219. The small native oyster, with its irregular white shell, is familiar to most coast dwellers from Sitka to Cabo San Lucas, Baja California. *Ostrea lurida* (Fig. 229) occurs characteristically in masses that nearly hide the rock

---

*Two rather similar gorgonians are common in southern California: *Muricea californica*, with yellow polyps, and *M. fructicosa*, with white polyps. Previous editions here referred to *M. californica*, describing it as having white polyps.

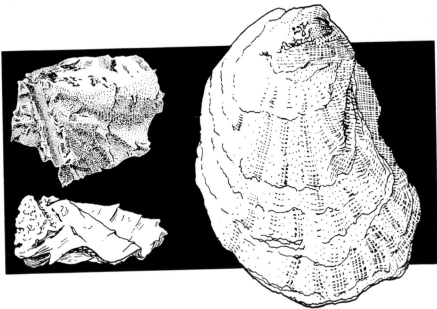

Fig. 229. *Left*, the native oyster of the Pacific coast, *Ostrea lurida* (§219), generally known as the Olympia oyster. *Right*, the larger eastern oyster, *Crassostrea virginica* (§219). All slightly reduced.

to which the animals are fastened. For many years it was scorned as food because of its small size, and even now this is probably the chief factor in preventing the animal from being used for food as extensively as it deserves. People acquainted with its delectable flavor, however, often pronounce it more delicious than the larger Eastern oyster, and probably its virtues will come to be more generally appreciated. The cultivation of *O. lurida* is a considerable industry in Puget Sound, where it is known as the Olympia oyster.

In Oyster Bay, near Olympia, the method of cultivation is as follows: Low concrete dikes are built to partition off many acres of mud flats, which are carefully leveled; the object is to retain water when the tide goes out, so that the growing oysters are never exposed. Since oysters would smother in mud, quantities of broken shell are scattered over the surface, and to these bits of shell the "spat," or free-swimming young, attach. Nelson (1924) studied the method of attachment of the Eastern oyster and found that a larva moves about over the selected area, testing it with its foot (it will not settle on a shell that is badly riddled). After choosing an attachment site it circles about, wiping its large and turgid foot over the area and extruding enough cement to affix the left valve. The cement, which hardens in 10

Fig. 230. A cultivated bed of *Crassostrea virginica* exposed by low tide in Tomales Bay; the stake fence keeps out stingrays and angel sharks.

minutes, is secreted by a gland similar to the one that produces a mussel's byssal hairs.* After 2 years the spat are transferred to more extensive beds, and after 5 or 6 years they reach maturity and a diameter of about 5 cm; they are then ready to market. It takes from 1,600 to 2,000 "shucked" oysters to make a gallon, solid pack.

Efforts to establish the Eastern or Virginia oyster, *Crassostrea virginica* (Fig. 229), have for the most part failed on this coast, although there are some small colonies in Willapa Bay. In addition, considerable quantities are grown in Tomales Bay and Puget Sound from imported spat, usually in subtidal beds (Fig. 230).

A third oyster is being grown commercially from imported spat of *Crassostrea gigas.* This animal, the Pacific or Japanese oyster, does reproduce naturally to some extent in Puget Sound and British Columbia, spawning in July or August and settling 15 to 30 days later; in British Columbia waters, *C. gigas* is abundant, forming massive growths in Departure Bay

---

*More recently, settlement of the European oyster, *Ostrea edulis*, has been studied in great behavioral and morphological detail by Cranfield (1973, 1974). In this species, the foot lays down the first cement, followed soon after by additional cement produced by the mantle folds; these folds continue to secrete cement into the adult phase. Settlement of the Olympia oyster has also been investigated (Hopkins 1935, 1937).

and the Georgia Strait area. In California, spawning also occurs in July or August, but larvae rarely survive. A few young oysters have settled naturally in several localities south of Puget Sound, such as Willapa, Humboldt, and Tomales bays, but the industry still depends to a considerable extent on spat imported annually from Japan. Once planted, spat of the Pacific oyster grow more rapidly than those of the native oyster and reach market size in 2 to 3 years. This rapid growth is due in part to the Pacific oyster's ability to filter even the finest plankton (nannoplankton) from the water; the native oyster, unable to trap nannoplankton, grows more slowly on its restricted diet of coarser plankton and detritus.

Oyster culture in Pacific coast waters seems to be at the crossroads: there are opportunities for a growing market that offset the uncertainties of raising bivalves in the presence of various enemies, foreign and domestic. In northern waters these include great hordes of the burrowing shrimps *Upogebia* and *Callianassa*, which dig up the beds and cover the oysters with mud, and a noteworthy predator, the bat stingray *Myliobatis californicus*. At least seastars are not the oyster pests in our waters that they are in the east, perhaps because they do not care for the reduced salinity in some regions. The oysterman's chief enemy is man himself, who builds roads across choice grounds, dumps his wastes into streams and bays, and digs up bottoms for boat basins. Along with these troubles is the one of ever-increasing expense: oyster-growing takes a lot of tedious labor, and labor is becoming more and more expensive, even in Japan.

What is hoped for is a tasty oyster that grows rapidly and without much care. We have tried cultivating the Eastern oyster, which neither grows nor reproduces well, and the Pacific oyster, which grows well but does not reproduce satisfactorily. Now we are going to try the European flat oyster *Ostrea edulis*. Plantings have been made in Tomales Bay and Drake's Estero, but without successful propagation.

The sexual vagaries of the oyster are worth detailed mention. It had been known for some time that the European oyster changed its sex, and it was later discovered that our Eastern oyster is also what is known as a protandrous hermaphrodite. Coe, working at La Jolla (1932), showed that *Ostrea lurida* belongs in the same category. It is never completely male or completely female, but goes through alternate male and female phases. The gonads first appear, showing no sexual differentiation, when the animal is about 8 weeks old. At 12 to 16 weeks the gonads show primitive characteristics of both sexes, but thereafter the male aspect develops more rapidly until, at the age of about 5 months, the first spermatozoa are ready to be discharged. Before the clusters of sperm are discharged, however, the gonads have begun to change to female, and immediately after the release of the sperm the first female phase begins—a phase that reaches its climax when the animal is about 6 months old. Coe reported: "Ovulation then occurs, the eggs being retained in the mantle cavity of the parent during fertilization and cleavage and through development until the embryos have become provided with a bivalved, straight-hinged shell—a period of approximately ten to twelve days, perhaps. It is not improbable that ovulation takes place only when the animal is stimulated by the presence in the water of spermatozoa of other individuals." As before, the phases overlap, so that the second male phase has begun while the embryos are still developing within the mantle cavity. The second male phase produces vastly greater numbers of sperm than the first. When they are discharged (except a few clusters that are retained, as always) the body of the oyster is "soft, flabby and translucent, presenting a marked contrast to its plump whitish condition preceding the discharge of the gametes." A period of recuperation follows before the assumption of the next female role. Apparently this cycle of male, female, and recuperative phases is continued throughout the animal's life. A lower temperature will arrest or prolong a particular phase. Coe remarked: "It is not unlikely that in the colder portion of the range of the species a single annual rhythm or even a biennial rhythm may be found to occur, as is the case with the European oyster in some localities." In Puget Sound, spawning occurs over a 6-week period in the early summer, when the water reaches 13°C; but in southern California, where Coe worked, spawning takes place during at least 7 months of the year (April to November) and requires a higher temperature, at least 16°C.

For *Ostrea lurida* the period from the fertilization of the egg to the time when the shelled larva leaves the parent is 16 to 17 days. The free-swimming period lasts about 30 to 40 days, after which the larvae settle down as previously described. They reach marketable size in 3 to 5 years in California.

Recent studies of the sexuality of oysters suggest a complicated interplay of genetic and environmental circumstances. In *Crassostrea virginica*, for example, a small percentage of the individuals in a population are permanently males or females with no capacity to shift to the other sex. The rest are potentially hermaphroditic, with varying degrees of expression. Some of these function only as males, others only as females, and about 30 percent may have equal male and female phases. Usually the male phase is first (protandry); examples of complete hermaphroditism, in which both sexes are functional simultaneously, are rare. Larviparous oysters such as the Olympia oyster are alternately male and female on a more regular basis. Though the different kinds of sexuality may possibly be determined by genetic factors, the actual expression of sexuality may be a response to environmental circumstances. Somehow the matter is arranged in nature so that there is an adequate release of the gametes of both sexes simultaneously—or nearly so—to ensure continuation of the species.

Importations of Pacific and Virginia oysters, made in the days of ecological innocence, have had interesting sidelights, in that unwanted settlers came with them. With the Pacific oysters has come the Japanese oyster drill *Ceratostoma inornatum* (listed as *Ocenebra japonica* in the Fourth Edition), now established in Puget Sound as a serious pest. It has also become naturalized in Tomales Bay. This unwelcome guest was preceded by the related Atlantic drill *Urosalpinx cinerea* and by the dog whelk *Ilyanassa obsoleta* (also known as *Nassarius*). *Urosalpinx* (see also §286) is said to be common at times in San Francisco and Tomales bays and at Elkhorn Slough, and its recorded range on the Pacific coast is now British Columbia to Newport Bay in southern California. It feeds on barnacles, mussels, and oysters, and tolerates brackish conditions well. *Ilyanassa* is abundant in San Francisco Bay and has been recorded on this coast from British Columbia to central California. An omnivore and deposit feeder, the dog whelk, or eastern mud snail, will also scavenge whatever opportunity brings it (see also §286). *Urosalpinx* and *Ilyanassa* have both been extensively studied on the east coast.

Another immigrant from Japan is the parasitic copepod *Mytilicola orientalis*, found in the intestine of oysters and even more often in mussels (especially *Mytilus edulis*). It is reported to be a minor pest of oysters in Puget Sound, and copepod-infested oysters have now been found in British Columbia, Oregon (Yaquina Bay), and California (Humboldt, Tomales, and San Francisco bays).

Somehow, perhaps with the oysters and their various associated mollusks, a small crab also has made its way from Atlantic to Pacific waters. This is *Rhithropanopeus harrisii*, a crab smaller than the native estuarine *Hemigrapsus oregonensis* and recognizable by its three spurs on each side of the carapace and by its whitish claws. It was first noticed in San Francisco Bay about 1940, and it is now abundant in the sloughs of northern San Francisco Bay and up the river as far as Stockton. In 1950 James A. Macnab found this crab in the sloughs of southern Coos Bay in Oregon.

Perhaps the most unwelcome invader to come to notice is the Atlantic green crab, *Carcinus maenas*, which turned up in Willapa Bay in 1961. This crab attacks young oysters and could become a serious pest. Its progress is being watched with apprehension (Fig. 231).

§220. In the Puget Sound area the scallops, *Chlamys hericius* (formerly *Pecten*) and other species, are the basis of an important industry. Trawlers sweep the bottom with nets, scooping up hundreds in a single haul. Although scallops in commercially valuable quantities occur only in deep water, the observer along rocky shores will be certain to see them in the intertidal zone and below the low-tide line. All scallops are characterized by projecting "ears" on either side of the hinge and by prominent, radiating ridges on their handsome shells. The species mentioned, which ranges from the Bering Sea to San Diego, is essentially similar in outline to *Argopecten aequisulcatus* (§255; Fig. 232) but has finer and more numerous radiating ridges; also, the left valve is a darker pink than the right. *C. hericius*, like most other species, remains symmetrical throughout its life. Scallops have only one large muscle for closing the shell, instead of two muscles (often unequal in size) as in the mussels and clams. It is usually this single muscle that is marketed, the rest of the body being thrown away, although the entire animal is edible—delicious, in fact.

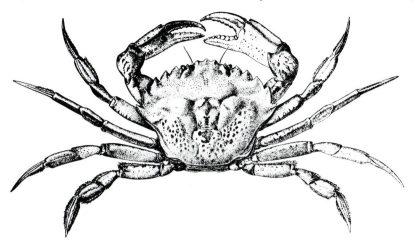

Fig. 231. The green crab *Carcinus maenas* (§219), a sinister new arrival from the Atlantic.

Fig. 232. *Argopecten aequisulcatus* (§220), the thick scallop of southern waters, similar in outline to *Chlamys hericius*, one of the commercially important scallops of Puget Sound; about natural size.

Next to the octopus and the squid, the scallop is the "cleverest" of all the mollusks. It is quick to take alarm, being warned of impending danger by a row of tentacles and shining functional eyes along the mantle fringe, and darts away by clapping the two valves of its shell together and emitting powerful jets of water. The animals may make a single "backward" leap, moving hinge first; or they may rise above the bottom and swim "forward," repeatedly opening and clapping shut the valves. When the valves open, water enters the mantle cavity on a broad front, but when they close, the water is ejected in two jets, one on either side of the hinge; in swimming, the animal thus appears to advance by taking bites out of the water ahead.

Scallops may be stimulated into galloping away in the aquarium by placing a seastar in the tank. It may not be necessary for the scallop to see or touch the seastar, since some diffusible essence of seastars will often suffice to stimulate the reaction, and the animal will flap abruptly away from the source of danger.

M. W. de Laubenfels remarked that very few *Chlamys* in the Puget Sound area lack the sponge *Myxilla incrustans* (formerly *Ectyodoryx parasitica*) or the less-frequent *Mycale adhaerens*. Both of these occur primarily on scallops, and no other sponges occur thus. The *Mycale* "may be distinguished by comparatively coarser structure, and best by this: when torn, it reveals prominent fibers thicker than thread, absent in *Myxilla*."

The close association of these sponges and scallops has recently been shown by Bloom (1975) to be a mutualism, benefiting both associates in their dealings with predators. The sponges are protected from predatory dorid nudibranchs by the scallop's galloping defensive response. In turn, the covering of sponge helps to protect the scallop from certain predatory seastars by decreasing the ability of the predator's tube feet to hold on and also by providing tactile camouflage. A similar mutualism has been reported between the sponge *Halichondria panicea* and the scallop *Chlamys varia* off the Atlantic coast of Ireland.

§221. Of chitons, the black *Katharina* (§177) can be expected on relatively exposed shores. The giant, brick-red *Cryptochiton* (§76) occurs also. And one or more species of *Mopalia* (§128) will be found: *M. hindsii* and *M. muscosa* (Fig. 134) are both known to occur in sheltered localities such as San Francisco Bay.

§222. Not many crabs will be found walking about on top of the rocks in this low zone—a situation quite the opposite of that existing higher up, where the purple shore crabs scamper over every square meter of rock surface. The small Oregon *Cancer* may be taken occasionally on the rocks, but is commoner in the under-rock habitat (§231). A lanky, "all legs" spider crab (*Oregonia gracilis*, §334), which occurs more characteristically on wharf piling, also strays into this rocky-shore region. Another small spider crab, *Pugettia gracilis*, the graceful kelp crab, seems to be characteristic here, but it occurs also in eelgrass and in kelp. It looks like a small and slim *P. producta* (Fig. 106), which is also occasional in this association. Like that larger kelp crab, *P. gracilis* keeps its brown, yellow, or red carapace naked and smooth, not deigning to disguise itself with bits of sponge, bryozoans, or algae as the more sluggish spider crabs do.* These kelp crabs, very active and having hooked and clawed legs that can reach almost backward, are

Pugettia gracilis may occasionally and sparingly decorate the rostrum and sides of the carapace.

amply able to defend themselves without recourse to such measures as masking. Most crabs can be grasped safely by the middle of the back, but the safe belt in the kelp crabs is narrow, and one who misses it on the first try will pay the penalty. *P. gracilis* ranges south from Alaska apparently as far as southern California, but it is uncommon south of Monterey Bay.

§223. The only other noticeable exposed crustaceans are barnacles. Except in Puget Sound and comparable waters to the north, barnacles are rare in the very low intertidal zone, although they are everywhere tremendously common in the higher zones. One might expect that these low-zone barnacles would be of species different from those previously treated, since the relative periods of exposure and immersion are so different. This was once assumed to be the case; but it is well established now that the commonest barnacle of this low zone is *Semibalanus cariosus* (§194), the same form that occurs in such tremendous numbers in the upper zone of quiet-water regions. However, some of these may be *Balanus nubilus* (§317), the second-commonest form, which seems to achieve its greatest development on fairly exposed wharf piling. On Puget Sound rocks the largest examples reach the considerable size of 7.6 cm in basal diameter by 6.4 cm tall.

§224. Anemones, abundant on all the quiet-water rocky shores from Alaska south, are especially noticeable in Puget Sound. Elsewhere in the Pacific the white-plumed *Metridium senile* (§321) occurs only on piling and in deep water, but in the Sound it is often found on the lowest intertidal rocks and ledges and in the pools, just as it occurs on the coast of Maine. North of Nanaimo, British Columbia, we have seen large numbers of *M. senile* hanging from the underside of strongly overhanging ledges.

The large, red-tuberculate *Tealia crassicornis* (§65) is occasional, and the small, semitransparent *Epiactis prolifera* (§37), with its eggs and young in brood pits around the base, certainly attains its maximum in size, though probably not in abundance, in these quiet waters.

One of the common Japanese and cosmopolitan anemones, *Haliplanella luciae*, has apparently been introduced on this coast—probably with oysters, for it has appeared at or near places where Pacific oysters are known to have been introduced. We have seen it above the Strait of Georgia, on Puget Sound floats with *Metridium*, and in Tomales Bay and Elkhorn Slough, in central California. It is a small form, usually under 12 mm in diameter, but it is attractively colored—often dark green, with 4 to 48 orange stripes on most individuals. A circumboreal form, it appeared suddenly at Woods Hole, Massachusetts; and owing to its competitive ability and rapid asexual reproduction, it colonized very quickly in areas where its colors and small size conceal it effectively. It tolerates a wide variety of environmental conditions, occurring high on rocks in Puget Sound and low on mud-flat rocks or algae in San Francisco Bay. Because of its small size, durability under a variety of natural and experimental conditions, and primarily asexual mode of reproduction, *Haliplanella* has become a frequent object of physiological studies.

Fig. 233. An erect sponge similar to *Isodictya rigida* (§225); the height of this specimen was 20 cm.

§225. Reaching its apparent southern limit at Pender Harbor, British Columbia, a large erect sponge occurs (Fig. 233) that is similar in appearance to the *Isodictya rigida* (formerly *Neoesperiopsis*) reported by L. M. Lambe from deep water north of Vancouver Island. Specimens from Refuge Cove are up to 20 cm long. They are brightly colored with reds or lavenders, which fade immediately in preservative. Forests of these upright sponges, growing at or below the lowest tide line, add something of a coral-reef atmosphere to the northern fauna.

A most interesting sponge, *Suberites ficus* (formerly *Choanites suberea* var. *lata*), is often the home of the hermit crab *Pagurus dalli* in Puget Sound. Clusters of this yellow cheesy or carrotlike sponge may be more than 7 cm in diameter and are almost invariably pierced with an opening that harbors this, or another, hermit. The mutual advantages are obvious: the crab gets a protective covering, and the sponge gets plenty of food and oxygen by being so much on the move. *S. ficus* is also to be found on shaded rocks in the very low intertidal down to depths of at least 36 meters; specimens have been reported from the Gulf of Alaska to off San Elijo Lagoon in San Diego County. The range of *P. dalli* is the Bering Sea to Oregon.

M. W. de Laubenfels remarked that the slimy, papillate sponge *Hymeniacidon* (§183) occurs abundantly in these quiet waters; we have noted it in connection with oyster shells.

§226. In this low zone of the quiet-water rocky environment, sea cucumbers are common in the south, entirely absent in Elkhorn Slough, and occasional in Puget Sound. Under rocks, however, they are common in the northern regions (§235). One of the strange, sluglike, creeping cucumbers, *Psolus chitonoides* (Fig. 234), is a fairly frequent migrant into the intertidal zone from deep water. If seen below the surface, and unmolested, *Psolus* will be noted and recognized at once. There will be a crown of gorgeously expanded tentacles at the head end, often in rosy reds or purple, white-tipped and contrasting with the pale yellow or orange of the scales that cover the back; on the soft belly are the tube feet that provide tenacious attachment but not overly rapid locomotion. As adults they are nearly sessile, moving only if living conditions become unsuitable, such as might occur if the boulder on which one was living turned over. The larvae of *Psolus*, however, are quite active, and these settle gregariously, either on or around an adult. After metamorphosis, juveniles migrate from the adult into nearby shaded habitats where they remain indefinitely, feeding on suspended particulate matter captured by means of adhesive papillae on the tentacles.

Fig. 234. The armored creeping cucumber *Psolus chitonoides* (§226), with tentacles expanded; natural size.

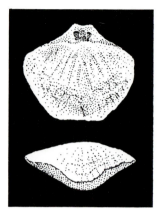

Fig. 235. Shell of the brachiopod *Terebratalia transversa* (§227); slightly enlarged. Primarily a subtidal species, *T. transversa* is restricted to the lowest part of the intertidal zone.

§227. *Terebratalia transversa* (Fig. 235) represents a group of animals not previously mentioned, which will be met only once again in this account. The brachiopods, sometimes called lampshells, are bivalved animals resembling clams but in no way related to them. They constitute a line tremendously important in past geological ages but now restricted to a few hundred species, most of which occur in deep water. The valves of clams lie on either side of the body, and in most (oysters and their kin excepted) the two valves are much alike. The two valves of brachiopods are dorsal and ventral to the body, and are usually dissimilar in size and shape. The lower is usually the larger, and bears the fleshy peduncle by which the animal is attached to the rock.

Although *Terebratalia* and allied brachiopods are sessile animals, we observed that specimens taken in British Columbia were able to rotate the shell considerably on the contractile peduncle. A cluster of the living animals will show all the shells slightly agape but ready to snap shut immediately at the least sign of danger. They feed on minute organic particles brought to them by the currents created by the ciliated tentacles on the lophophore—a "coil spring" organ that captures particulate food and carries it to the mouth. Thus brachiopods, like other plankton-feeding forms, must keep constantly on the job, opening their shells the instant they are unmolested.

Reproduction and growth of *Terebratalia* have been followed in the Puget Sound region. Breeding occurs from November to February, and recruits grow to a width of about 12 mm in a year; individuals reach maximum size (about 56 mm in width) at 9 to 10 years. *Terebratalia* is restricted to the lowest part of the low intertidal zone, where it is a feature of the rocky fauna of British Columbia. It is taken as far north as Alaska and occasionally as far south as Friday Harbor; in deep water it extends to Baja California.

In the summer of 1932 we found a great colony of *T. transversa* in a land-locked marine lake, Squirrel Cove, in south-central British Columbia. This unusual environmental condition was reflected in the great size of the individual brachiopods, in their large number, and in the dark color of their shells. Interestingly, the very abundant *Parastichopus californicus* (§77) taken in this rocky lagoon was also more blackly purple than usual, higher in the littoral (this species is more common subtidally), and small, climbing all about the tops and sides of the rocks. Specimens of *Dermasterias* (§211) taken in this assemblage were also darker than usual.

§228. At Elkhorn Slough, and probably elsewhere along the coast, the observing amateur will sooner or later find a group of ivory-white polyps, each about 1 cm long, attached to a rock on the side sheltered from the current. If disturbed, they immediately contract into stalked, translucent lumps less than half of their normal length. These tiny, unspectacular polyps are the parents of highly spectacular pelagic animals—the great jellyfish (the scyphomedusae) that are sometimes stranded on the beach. It will be remembered that there is an alternation in the life history between the relatively large and plantlike hydroids, and the small, free-swimming jellyfish that are the sexually reproducing adults. The giant jellyfish also show such an alternation, but the attached polyp is small and rarely seen. It is this polyp, the scyphistoma of *Aurelia aurita* or similar species, that will be found at or below the lowest tide line at Elkhorn Slough. We have transferred a good many examples to aquaria, where they live well, being extraordinarily tough and resistant, and have watched them with interest. They are accomplished contortionists. Fully relaxed specimens are pendulous, with dainty tentacles extending the equivalent of the length of the polyp; the mouth forms a raised crater in the center of the tentacles. When the animal contracts, the tentacles almost disappear, and the scyphistoma becomes nearly a ball at the end of a stalk. When producing jellyfish, the polyp elongates and becomes cut up into slices by transverse constrictions, so that it appears like a tall "stack of saucers." This is the strobila stage. The "saucers," which are budded off one by one, become the free-swimming young, the ephyrae, which grow into giant jellyfish. The jellyfish, normally to be seen only offshore (except in northern inside waters, §330), produces eggs that are liberated as ciliated larvae, the planulae. These larvae, frequently drifting inshore, settle down to become polyps, and the cycle is complete. Much of this cycle has been observed in Bodega Harbor, where young polyps of *Aurelia* are first observed in February on piling and floats. These strobilate and bud off ephyrae through March. The medusae grow rapidly, and by June they have reached reproductive size and have spawned. Most die shortly thereafter; a few, however, live another year, reproducing continuously.

§229. Along these sheltered rocky shores there are, finally, many hydroids and bryozoans, and occasional sponges, mostly of types already mentioned. Some of the ostrich-plume hydroids (*Aglaophenia*, §84), but more frequently the "sea firs" (*Abietinaria*, §83) and the delicate, almost invisible *Plumularia* (§78), will be seen. Collectors from the south will recognize a sponge cluster, *Leucosolenia nautilia*, similar to the California species considered in §93.

In the Puget Sound area we have found the erect, treelike bryozoan *Bugula pacifica* (§114), with some of its exposed clusters colored a vivid salmon-pink; but the more flabby European form *Dendrobeania murrayana*, restricted in the Pacific to this region, may be seen more commonly. The white, crinkly bryozoan *Tricellaria* (§112) occurs in these northern sounds on the roots and holdfasts of marine plants just as it does in less-sheltered areas along the California coast.

The "sea-lichen bryozoan" *Dendrobeania lichenoides* (Fig. 236) occurs on shells, rocks, and worm tubes as a leafy encrustation. It ranges from Alaska to central California and is common in Puget Sound. The white *Membranipora* (§110) encrusts on kelp, as do the small and dainty-tubed colonies of *Tubulipora flabellaris*. Colonies of the latter are 6 to 12 mm across, and the tubes are as hard as those of serpulid worms. This form ranges from the Sound to San Diego. A warm-water form, known from Los Angeles Harbor and south to Central America, is the soft *Zoobotryon verticillatum* (formerly *Z. pellucidum*), which forms great, flexible, tangled, treelike masses. The listing of bryozoans might go on indefinitely, for more than 150 species are known from Puget Sound alone. In the Strait of Georgia, C. H. O'Donoghue took some 76 species in a single dredge haul, and almost half of this total may be taken inshore. Most of us can scarcely hope to recognize more than a few.

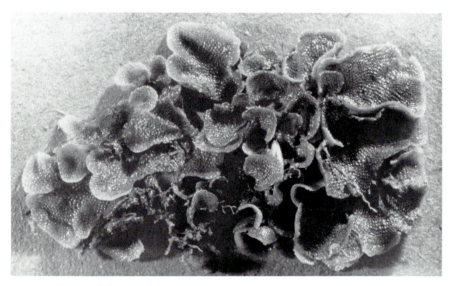

Fig. 236. The "sea lichen" *Dendrobeania laxa*, an encrusting bryozoan similar to *D. lichenoides* (§229); somewhat enlarged.

Exceedingly common in the Nanaimo region, and sporadically so on piling and protected rocks all the way to southern California, are great gelatinous masses of a rare and curious wormlike form, *Phoronis vancouverensis*, related to but not resembling the more solitary phoronids (§297). The clumps of intertwined individuals distinctly identify *P. vancouverensis*; the animal's body is pinkish to white.

Under-Rock Habitat

It would be expected that bay and estuary regions in the south would have their rocky-shore fauna pretty well concentrated under rocks as protection from sun and desiccation, and this is what we find. In Puget Sound, however, the under-rock habitat is also well populated—a fact probably attributable to the late-afternoon low tides, which bare the intertidal areas during the heat of summer days.

§230. The energetic upturner of rocks will find that in sheltered waters the most obvious under-rock animals are crabs, worms, and sea cucumbers, and, in the north, nemerteans. Other previously encountered animals are the eye-shaded shrimp *Betaeus* (§46) and the small and active broken-back shrimp *Heptacarpus* (§§62, 150), which occurs throughout our territory. Another shrimp, *Hippolyte californiensis*, is fairly well restricted to quiet waters and will be found, in the daytime, hiding under rocks and in crevices. At night it may be seen in great congregations swimming about among the blades of eelgrass. This graceful, green (quickly turning white after preservation), and excessively slender shrimp has an absurd rostrum, or forward projection of the carapace. It is known from Bodega Bay to the Gulf of California, but is common only as far south as Elkhorn Slough. The sometimes bulging carapace is a pathological condition due to infection by a parasitic isopod, *Bopyrina striata*.

§231. The Oregon cancer crab, *C. oregonensis* (Fig. 237), the roundest of the family, is the commonest *Cancer* of the Puget Sound region and is almost entirely restricted to the under-rock habitat. In Puget Sound, this crab is primarily carnivorous, crushing and eating barnacles whenever available. Egg-bearing females are found mainly from November to February, several months after having molted and mated (April to June). The extreme range of *C. oregonensis* is from the Aleutians to Palos Verdes, and possibly into Baja California, but it is rare south of Point Arena.

The black-clawed *Lophopanopeus bellus* (similar to *L. leucomanus heathii*, Fig. 137, but with slimmer big claws) cannot be mistaken, since it is the only member of its genus in the Puget Sound region and is uncommon below there, although ranging south to San Diego.* Eggs are carried in April and hatch from May to August, as determined for British Columbia speci-

*Two subspecies, differing slightly in appendage morphology and reproductive biology, occur: *Lophopanopeus bellus bellus* (Stimpson) ranges from Resurrection Bay (Alaska) to Cayucos (southern California), and *L. diegensis* Rathbun ranges from Monterey Bay to San Diego.

Fig. 237. *Cancer oregonensis* (§231), a rounded cancer crab, common under rocks in the Puget Sound region; natural size.

mens by Hart (1935); a second brood is produced by many females, with eggs of this brood being carried until August when hatching begins. Northern specimens of both of these crabs are often parasitized by the rhizocephalan barnacle *Loxothylacus panopaei*, similar to *Heterosaccus* (§327), which occurs on spider crabs; at Sitka possibly 25 percent of the tide-pool specimens will be so afflicted.

A quiet-water anomuran crab, *Petrolisthes eriomerus*, may be taken, among other places, in Puget Sound and in Newport Bay; the entire range is Alaska to La Jolla. This crab differs only slightly in appearance from the outer-coast form shown in Figure 30, most obviously in that the body is often mottled with blue. Egg-bearing females have been reported in Puget Sound from February through October; two broods of young are commonly produced by a female each year.

A reddish or brown hairy crab, *Hapalogaster mertensii* (resembling *H. cavicauda*, §132, except that it has tufts of bristles instead of fine hairs), also occurs in the under-rock environment to the north, ranging from the Aleutians to Puget Sound.

§232. Tube worms of several types will be found here, some with pebble-encrusted, black tubes up to 25 cm long. Dark-colored, almost purple, specimens may be turgid with eggs, or, if light-colored, with sperm. The terebellid *Neoamphitrite robusta*, occurring at least from Alaska to southern California, is one of the largest and commonest. Field directions for distinguishing several genera will be found in §53, but only a specialist interested also in ecology could untangle this situation with any degree of certainty. The varying environments have undoubtedly exerted a selective influence on the various species. We ourselves have noted differences in tube construction and appearance between the worms predominating in wave-swept, cold-water, California habitats and those in quiet-water sounds but have not been able so far to correlate the data.

Any of the several scale worms may take to living in the tubes of these larger and more self-reliant worms. Our friend *Halosydna* (§49) is frequent, and *Grubeopolynoe tuta* (§234) also occurs. Meticulous search will also reveal, in practically every tube, one or more of several pea crabs—most often *Pinnixa tubicola* (§53) or *P. franciscana* juveniles (Fig. 238; this species also occurs in *Urechis* burrows and with the ghost shrimps). Juveniles of *P. schmitti*, the echiurid commensal (§306), have also been taken, but *P. tubicola* is by far the commonest pea crab found thus, everywhere on the coast. The combination of tube builder, polynoid, and pea crab is particularly characteristic of quiet waters. The smaller commensals apparently have adapted themselves to gathering the proverbial crumbs. There is a nice economy about the situation, for what one animal rejects as too large or too small exactly suits one of its partners.

§233. For a number of years we have been taking a gray, sluggish "bristle worm" from the muddy interstices of rocks dredged up from below 40 fathoms in Monterey Bay. During the summer of 1930 we were surprised to find this *Pherusa papillata* (formerly *Stylarioides*), apparently perfectly at home, at several points along the shore in Puget Sound. Here is another case of an animal quite competent to live in the intertidal zone once

Fig. 238. The pea crab *Pinnixa franciscana* (§232). As juveniles, these crabs inhabit the tubes of various worms; as adults, the burrows of *Urechis* and ghost shrimps.

the menace of surf is done away with. It has often been said of Pacific coast invertebrates that northern-shore forms may be found in the south, but in deep water; and it has been assumed that the determining factor is temperature. In Puget Sound, however, the summer midday temperature that these animals must tolerate at low tide is greater than the maximum along the Monterey shore at any time, and the northern winter temperatures are correspondingly lower. It is easy to believe, therefore, that the absence of wave shock permits a normally deep-water animal to migrate upward until it has established itself successfully in the lowest intertidal zone.

§234. Many worms will be turned up here. *Neanthes virens* (§311) makes semipermanent burrows where the substratum is readily penetrable. *Phascolosoma* (§52) occurs; and under Puget Sound mussel beds on a gravel substratum one may find *Nereis vexillosa* (§163), a variety of the glycerid worms (§280), and the slaty-blue *Hemipodus borealis*, also a glycerid.*

But there could be almost no end to the enumerating of worms. Certainly more than a hundred could be taken, and the unfortunate aspect of the worm situation is that any of them might prove to be common at some of the good collecting places that we have necessarily missed. Furthermore, worms abundant one year may be rare or absent the next.

In any event, one more dainty little polynoid must be mentioned: this is *Grubeopolynoe tuta* (also as *Hololepidella*). It has a serpentine gait, and will literally wiggle itself to pieces when disturbed too persistently. When first molested, it will protestingly shed a few of its beautifully colored scales; lifted from its favorite under-rock surface, it immediately breaks in two, and before the transfer to jar of water or vial of alcohol can be completed the disappointed collector is left with a number of short fragments that continue the characteristic wiggle. We have never been able to keep a specimen intact long enough to photograph it. It occurs frequently as a commensal with various tube worms (terebellids, §§53, 232).

§235. The large, reddish cucumber seen so abundantly under rocks in Puget Sound, or between the layers of friable shale ledges, is *Cucumaria miniata* (Fig. 239), which reaches a length of 25 cm. The tube feet, which are arranged in rows paralleling the long axis of the body, hold fast to the rock very efficiently but rarely seem to be used for locomotion. Surrounding the mouth are long and much-branched feeding tentacles that are drawn in when the animal is disturbed or when the ebbing tide leaves it high and dry. In an aquarium they will flower out beautifully. The animal's color varies, in different specimens, from yellow and red to deep purple. *C. miniata* has been variously considered as synonymous with an Oriental cucumber and with a cucumber occurring on both sides of the Atlantic; in any case, it is the North Pacific representative of a circumpolar and almost cosmopolitan form. The pentamerous symmetry that marks cucumbers as

*Hartman (1968) listed the color of this worm in life as red; the reported range is British Columbia to San Diego.

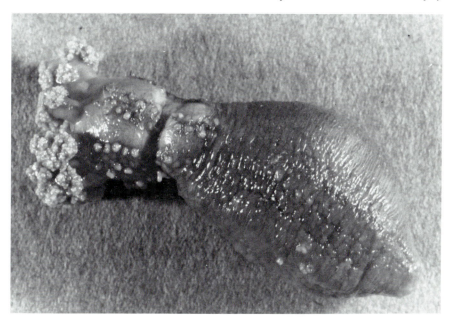

Fig. 239. The reddish *Cucumaria miniata* (§235), very common in Puget Sound; a circumpolar species. This specimen is contracted, and the characteristic rows of tube feet appear as slight bumps.

relatives of the seastars, brittle stars, and urchins is rather apparent in *C. miniata*.

C. *miniata* occurs commonly at least as far north as Sitka, south to Sonoma County, and sparsely at Carmel; it has been recorded as far south as Avila Beach. In British Columbian channels such as Canoe Pass it is one of the most spectacular of animals; great beds are zoned at 0.3 to 1.5 meters below lower low water. The delicate beauty of these expanded aggregations, with diagrammatically extended, translucent, coral-red tentacles literally paving the rocks, surprises the zoologist who has known them only from preserved specimens or has seen them only as contracted animals exposed at low tide.

A smaller white or cream-colored cucumber, *Eupentacta quinquesemita* (Fig. 240), may be considered as living primarily under rocks, since it is rarely found elsewhere in the intertidal, although it occurs in all sorts of situations below the low-tide line. The two double rows of long, fairly stiff tube feet give the animal a bristly appearance. This form and *Cucumaria miniata* differ from the *Parastichopus* we have considered earlier in having their tentacles around the mouth fairly long, branched, and yellow or orange. E. *quinquesemita* is very temperamental: specimens we took in the summer of 1930 at Pysht, in Juan de Fuca Strait, achieved a record of 100 percent evisceration in a few hours and before we could get them into anes-

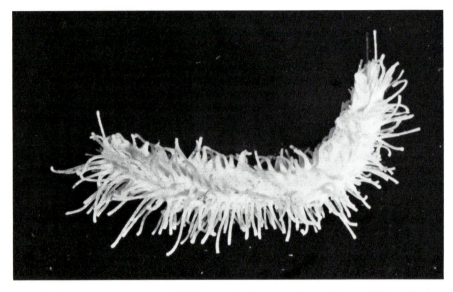

Fig. 240. *Eupentacta quinquesemita* (§235), a small white cucumber with long, fairly stiff tube feet; natural size for a large specimen.

thetizing trays. This is the more surprising when it is realized that the temperatures in the seaweed-filled buckets could not have been appreciably higher than those customarily tolerated on the tide flats during the late midsummer lows, when the sun and wind dry the algae to the point of crispness. This white cucumber ranges from Sitka to Baja California, but it occurs infrequently, and then so far down as to be practically a subtidal form. In Puget Sound it is often found associated with the reddish *C. miniata*. In southeastern Alaska a very similar form occurs, described by Elisabeth Deichmann as *Eupentacta pseudoquinquesemita*, which has heretofore been known from subtidal depths only. It is scarcely distinguishable from the more southerly-ranging white cucumber except that it is thinner-skinned, so that in preservation it partially constricts itself, producing many ringed wrinkles in the body wall.

Cucumaria vegae is a small, black holothurian taken from low-tide rock crevices at Auke Bay, north of Juneau. Another occasionally intertidal form in the Puget Sound area (but pulled up at Pacific Grove in subtidal holdfasts of kelp only) is the white, pepper-and-salt *Cucumaria piperata*. A large, often dark cucumber, with a straw-colored ventral surface, *C. lubrica*, also occurs here.

With the northern cucumbers, and elsewhere under rocks, a really gigantic flatworm, tough and firm, will sometimes be turned out, large enough to be the paternal ancestor of all the polyclads the southern observer will have seen. This is *Kaburakia excelsa*, which we have taken at

Sitka, in British Columbia at several points, and on the rocks near Port Orchard, Washington; it is reported to be common in the San Juan region and occurs uncommonly all the way to southern California. In addition to specimens taken intertidally, others may be found among mussels and other fouling organisms on boat bottoms brought to dry dock. Individuals are reported up to 10 by 7 cm. Most of the polyclads are photonegative—light-shunning—and *K. excelsa*, taken on the undersides of rocks or in deep crevices, is no exception, although a diver working in these rocks at high tide or at night would probably find them crawling about on the upper surfaces actively enough. A specimen taken during the summer of 1935 seems to have attempted to engulf the soft parts of a limpet when, en route from tide pool to laboratory, it and others were placed in a jar of miscellaneous animals.

Another smaller (to 4 by 1 cm), thinner, and more delicate polyclad, often secured with the above, is probably the *Freemania litoricola* reported as *Notoplana segnis* by Freeman (1933), and said by Freeman to be abundant around Friday Harbor.* This flatworm has been reported from Monterey Bay to British Columbia, and occurs abundantly on the protected coast of the Bodega Bay region. In the Bodega region, *Freemania* may be found gliding across the surface of moist boulders during low tides at night. It preys on *Collisella digitalis*, *Notoacmea scutum*, and probably other limpets, apparently ingesting only the soft parts while leaving behind the shell. Upon contact with *Freemania*, the limpet *N. scutum* usually reacts in a vigorous and characteristic manner: it lifts the shell off the substratum and departs with such great haste as to astound anyone accustomed to viewing limpets inactive at low tide or browsing slowly for food.

The brittle star *Ophiopholis aculeata* (§122) occurs frequently enough in beds of *Cucumaria miniata* (especially north of Puget Sound) that the presence of one can almost be considered an index of the other. South of Prince Rupert and at Sitka we have turned up colonies of specimens up to 13 cm in diameter, and associated with *Cucumaria* such that several individuals would be found clinging to each cucumber. Mixed up into the assemblage, characteristically, are dozens of the peanut worm *Phascolosoma* (§52); these are larger, and darker in color, than the possibly more numerous individuals available at Monterey.

§236. Along the California coast south of Los Angeles, octopuses (§135) have moved into estuaries where there are suitable pools containing rocks arched over the mud. Living in these mud-rock caverns in quiet waters, they attain, when unmolested (which they rarely can be in this heavily populated country), sizes that are large for their usually small species. Wherever they are numerous enough, they might be considered the dominant animals, for their size, strength, and cunning put them pretty well in control of their surroundings.

---

*Although more delicate than the gigantic *K. excelsa*, *Freemania* is thicker and firmer than *Notoplana acticola*, another common polyclad (§17).

§237. In the northern waters there are several under-rock chitons that we have met before, noticeably the small (to 4 cm long), red-marked *Lepidozona mertensii* (§128). In Puget Sound there is a still-smaller, greenish to tawny chiton, *Lepidochitona dentiens* (formerly *Cyanoplax*). This species, which occurs as far south as northern Baja California, extends into the middle intertidal in the Monterey area, often hiding under algae or sometimes (on steep surfaces) under another chiton, *Nuttallina californica*. In waters south of Monterey will be found the very similar but slightly larger *Lepidochitona keepiana*.*

§238. Under rocks in such varied quiet waters as those of Departure Bay and Pender Harbor, British Columbia, and Elkhorn Slough, we have found a flat encrusting sponge that was described originally from the central California estuary as *Mycale macginitiei*; it is yellow, smooth, and slightly slippery to the touch, although it has stiffening spicules and is marked by furrows. *Mycale* adheres closely to the rock, probably by slightly penetrating the rock pores. A species of *Halisarca*, *H. sacra*, is known to occur at Elkhorn Slough along with *Mycale*, but always below low water. Since there are various additional species which, on the basis of field characters alone, could be confused with the above and with previously treated species, the value of careful laboratory work should be kept in mind.

§239. The rubbery nemertean worms will often be seen under rocks in the north. Two especially common forms are sure to be seen sooner or later and will occasion excited comment. Both are a brilliant scarlet that is brighter than most reds even in the brilliant tide pools, both are large (to 3 meters, although the average will be less than a meter), and both occur under loosely embedded stones, especially on gravel substratum, from Dutch Harbor clear into Puget Sound. But the two may easily be differentiated. *Tubulanus polymorphus* is soft and rounded, very distensible, sluggish, and apt to be coiled in angular folds. The head is wide, rounded, and pinched off from the rest of the body. We have found this to be one of the characteristic under-rock inhabitants of quiet waters. In lower Saanich Inlet, about Deep Cove, for instance, in the summer of 1935, almost every suitable stone upturned revealed one of these brilliant worms in an apparently fabricated mild depression.[†] *Cerebratulus montgomeryi*, by contrast, is ribbonlike and flattened, with longitudinal slits on the sides of its white-tipped head, and with lateral thickenings that extend to thin edges (probably correlated with its swimming ability) and run the entire length of the body. *Cerebratulus* is famous for fragmenting at even the slightest provocation;

***Lepidochitona dentiens* can be distinguished from *L. hartwegii* (§36) by being smaller, smoother, and more elongate. *L. keepiana* (which occurs only south of Monterey) may often be distinguished from *L. dentiens* by the former's much-lighter color (often with orange-brown or cream) and straight side slopes (convex in *L. dentiens*); but unequivocal identification will most likely require a specialist.

†*Tubulanus polymorphus* (formerly *Carinella rubra*) may be an orange form as often as a red one. The recorded range is the Aleutian Islands to San Luis Obispo County.

but this species, Coe remarked (1901), lacks that trait to the extent that specimens may be killed whole by immersion in hot formalin.

§240. At least one vertebrate belongs with the under-rock fauna all the way from Alaska to the southern tip of Baja California, the grunting fish *Porichthys notatus*—possibly better known as the plainfin midshipman, since the pattern of its luminescent organs is reminiscent of the buttons on a naval uniform. The animals, normally inhabitants of deeper water, come into the shallows to spawn. The eggs, each the size of a small pea, are deposited in handful clusters, which may be seen under and around rocks during the summer, guarded by an attending parent. On one Puget Sound beach, in the summer of 1929, we found a small boy capturing the fish for food, using a method that must have seemed, to the fish, to violate Queensberry rules. He poked a slender stick under each likely rock and listened. If a midshipman grunted, he reached under with a heavier stick made into a gaff by the attachment of a large fishhook. A quick jerk, and another betrayed grunter was added to his string. One of them came off the string, however, after certain negotiations, and was added to our chowder, with satisfactory but not phenomenal results.

Blennies (or other blenny-like fish) will also be found commonly in this zone under rocks in Puget Sound—usually the high cockscomb, *Anoplarchus purpurescens*, and similar forms (see *Xiphister*, §51) that will have been seen higher up.

Burrowing Habitat

Nearly all of the animals that occur in the low-zone substratum in protected outer-coast areas may be taken here also—after all, the degree of protection in the substratum does not vary greatly between the two regions. We also treat here two common bivalves that bore into rock. In some areas, these bivalves may be more important eroders of rock than strictly physical processes are.

§241. The rocks in Newport Bay, for instance, are often completely honeycombed by the boring date mussel, *Lithophaga plumula* (Fig. 241), an elongate form covered with a glossy, brown periostracum. The method by which the date mussel bores into rock is still problematical. Older studies, noting that *Lithophaga* most commonly bores into limestone or mollusk shells, suggested that an acid is secreted that dissolves calcareous matter. More-recent studies have confirmed the involvement of a chemical agent, but have concluded that the chemical is a nonacidic substance that binds calcium. In either case, the thick layer of horny material that covers the date mussel's own calcareous shell likely provides some measure of protection against self-dissolution. (It is the color of this protective layer, as well as the shape of the shell, that has given the animal its popular name.) Precisely how *Lithophaga* bores into very hard, non-calcareous rock, which several investigators report that it does in some locations, remains unclear.

Fig. 241. The date mussel, *Lithophaga plumula* (§241), a rock-boring form; about natural size.

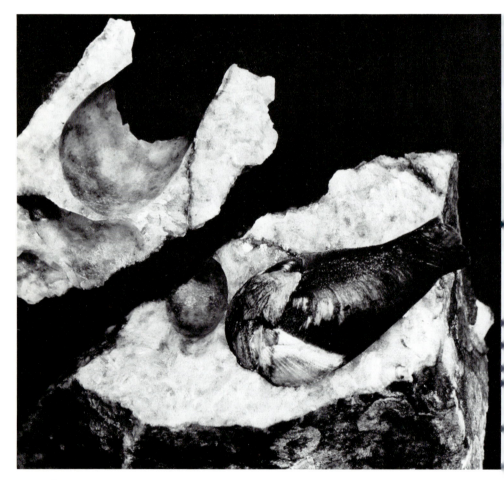

Fig. 242. *Penitella penita* (§242), the common piddock or rock clam; slightly enlarged. The rock has been split open to show the animal in place in its burrow.

From within its snugly fitting burrow, *Lithophaga* maintains communication with the outside world through its siphons, obtaining oxygen and microscopic food, like all of its relatives, from the stream of water that flows in and out. By attaching byssal threads to the burrow wall and by contracting the byssal retractor muscles, the date mussel is capable of moving forward and backward within its burrow, to a limited extent at least; rotation is also possible. The range of *Lithophaga* is Mendocino County to Peru.

§242. Another bivalve, *Penitella penita*, the common piddock or rock clam (Fig. 242), drills into rock so hard that nothing short of a sledge-hammer powerfully swung will break into the burrows; and it apparently drills without the aid of chemicals, using mechanical means only. The animal has a round flat foot, much like that of a limpet, which grips the rock at the forward end of the burrow and presses the rough valves against the rock. The valves are then repeatedly rasped against the walls of the burrow by alternate contractions of the anterior and posterior adductor muscles, while the clam slowly rotates in its burrow. This method of drilling is necessarily slow in direct ratio to the hardness of the rock, with boring rates varying from 4 to 50 mm per year. Eventually, the animal ceases boring, and when it does, a number of changes take place. A calcareous dome closes the shell anteriorly, the foot atrophies, growth ceases, and sexual reproduction begins. Curiously, the timing of sexual maturity varies with rock hardness (and, thus, with boring rate): in soft stone, an animal may reach maturity in 3 years, whereas in very hard rock it may take 20 years or more. Spawning seems to occur during July in Oregon.

This rock clam occurs, in deep water at least, throughout its range from Alaska to Baja California, but it seems most common intertidally in quieter water. In many localities, particularly in Oregon and Washington, it is clearly the species most responsible for the biological erosion of coastal shale. It has also caused considerable damage to concrete harbor works on this coast. One investigation revealed that more than half of the concrete-jacketed piles at four different places in Los Angeles Harbor were being attacked, sometimes with seven or eight borers in a tenth of a square meter. In some cases the animals had turned aside after penetrating the concrete jacket in order to avoid entering the wood.

The rock-boring clams are sought after as food, mostly in areas where they burrow into relatively soft sandstones or mudstones that are easily broken up with a pick. Unfortunately, when a size limit is set, the digger will reject the undersized clams and break away more rock. In places like Duxbury Reef and parts of the Oregon coast (where the rock is not eternal basalt) the clammer, as well as the clam, has become a potent geological force.

§243. The burrows of dead rock borers are likely to be appropriated by any of several home seekers. The most numerous of these is the porcelain crab of wharf piling (§318). But also occupying a borrowed home may be a large, suckered flatworm (probably *Notoplana inquieta*) that resembles *Alloioplana* (Fig. 14) in outline. This worm has the gray-brown color so common to retiring animals, and is very sluggish. It is firm in texture and outline, but when routed out of its quiet nest, it curls up very slowly. The worm has been reported from British Columbia, Puget Sound, Monterey, and Los Angeles, and also may be found under stones in the low intertidal. According to Hurley (1975), this polyclad is a major predator on the barnacle *Balanus pacificus*.

§244. In the substratum between rocks, and ranging in its circumpolar distribution at least as far south as Puget Sound, is the burrowing anemone *Edwardsiella sipunculoides*, which is similar to the *Edwardsiella* of Figure 285. It is a dull-brown form, with long transparent tentacles, and shows but faintly the eight longitudinal bands that are reputed to distinguish it from other anemones. A more readily apparent difference is that it has no attachment disk, but instead lives loosely buried in the soil.

§245. Because of their abundance, mention should be made of several species that are given more detailed treatment in other sections. Among these are the transparent, wormlike cucumber *Leptosynapta* (§55), the soft, blue burrowing shrimp *Upogebia* (§313), and the clam *Protothaca* (§206), which spouts from between tide-pool rocks in Puget Sound. In the south the mantis shrimp *Pseudosquillopsis* (§146) is taken infrequently, far under loosely embedded rocks. In the fully protected waters of Alaska (as at Jamestown Bay, Sitka, and Auke Bay, Juneau) there are such characteristic forms as the echiuroid worm *Echiuris echiuris alaskensis* (§306), the ubiquitous peanut worm (§52), and a slimy, apodous cucumber, *Chiridota albatrossi*. This last form, slim and sluglike and covered with peppery black dots, may be 12 to 15 cm long, the otherwise smooth skin bearing calcareous white particles—the "anchors" already mentioned in connection with *Leptosynapta*.

6 🦅 Sand Flats

In quiet-water beaches of fairly pure sand, where there is no eelgrass, we find a region that must seem utopian to many animals. The usual environmental problems are almost entirely lacking—which undoubtedly accounts, in part, for the extraordinary richness of the sandbars in Newport Bay in its former days, in Tomales Bay, and in parts of Puget Sound. There is no wave shock to be withstood; the substratum is soft enough that feeble burrowing powers will suffice, and yet not so soft that it requires special adaptations to avoid suffocation; there is no attachment problem, for a very little burrowing is enough to secure the animals against being washed out by the innocuous currents; and a little more burrowing protects them from sunlight and drying winds.

There is usually, however, something wrong with apparent utopias. In this case, the elimination of the struggle against natural conditions apparently results in proportionately keener competition among the animals, and overcrowding is the rule. This is accentuated by the fact that these pure sands are likely to contain little organic food by comparison with the rich stores of mud flats and eelgrass beds, although every tide brings in some quantity of plankton. Little zonation is obvious in areas of pure sand, but the region corresponding to the lowest zone of other shores is, for most of the animals, the only one that is habitable. The crowding is therefore accentuated, and the upper beach is sparsely inhabited.

Estuarine sand flats offer an added obstacle to prolific colonization by marine animals—the presence of fresh water. Where this occurs abundantly or sporadically, the faunal aspect is correspondingly changed. But a good many of the crabs and some of the shellfish tolerate greatly lowered salinities, and fresh water, even if actively flowing over the flats, has little effect on the seawater retained in sand below the surface; thus even minimal burrowing ability would protect a sessile form until the return of the tide.

A startling example of this retention of salinity is found in a small Alaskan stream that is used as a spawning ground by pink salmon. The lower reaches of spawning gravel are covered only by high tides; even so, the salinity around the salmon eggs, which normally develop in fresh water a few miles from the sea, is about 8 parts per thousand. Nevertheless, spawning here seems to be reasonably successful.

§246. The sand dollars definitely belong in this environment, and they occur in tremendous numbers. Their round tests are 7 to 10 cm in diameter, very flat, and colored so deep a purple as to be almost black. The common form from southeastern Alaska to northern Baja California is *Dendraster excentricus* (Fig. 243), which has the star design off center and consequently a bit lopsided. We have seen great beds of them in Puget Sound, Newport Bay, San Diego Bay, and along the Oregon and northern California coast. Another species, the northern sand dollar, *Echinarachnius parma*, is flatter and more symmetrical, with the pentamerous design centering at the apex of the shell. It is known from southeastern Alaska along the coast into the Bering Sea and the Arctic, and from there to the northeastern United States. Two other species of sand dollars occur on our coast. *Dendraster vizcainoensis*, which has coarser tubercles on the apical surface, replaces *D. excentricus* in Baja California. And *D. laevis*, a smaller, fragile sand dollar with shorter petals that are nearly central in position, occurs in deeper water (6 to 55 meters) in the California Bight and Channel Islands area and is occasionally washed up on the beach.

At first glance, sand dollars seem to have little resemblance to other urchins, but urchins they are. The spines, instead of being 7 to 10 cm long

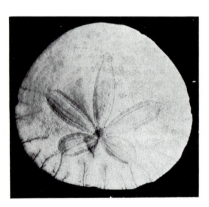

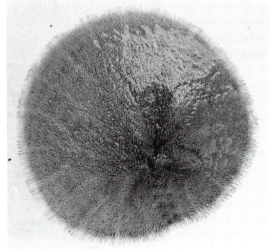

Fig. 243. *Dendraster excentricus* (§246), the common sand dollar: *above*, the dead test as found on the beach; *right*, a living animal.

as in the giant red urchin of the outer coast, are about 2 mm long and so closely packed that the animal looks and feels as though it were covered with velvet. The tube feet are there, too, and are visible to the unaided eye if one looks closely enough. Burrowing, and most locomotion, by sand dollars is accomplished by waves of coordinated spine movements, not by the tube feet as in the seastars. A temporary exception is juvenile sand dollars, which do indeed use their tube feet for locomotion. In motion, the spines and tube feet give an impression of changing light rather than actual motion. The five-pointed design on the back, another urchin characteristic, is visible in the living animal, but it is more obvious in the familiar white skeletons that are commonly cast up on exposed beaches. Many people believe that the living animals must occur near these beaches. They do—in deeper water. The sand dollar avoids surf at all times, and will seldom be found alive by the shore visitor except on completely sheltered flats.

When a low tide leaves the animals exposed, *Dendraster* will ordinarily be found lying flat, motionless, and partly or completely buried. Even when covered by the still waters of a sheltered lagoon, the majority lie flat on the bottom, but under these conditions, they may frequently move about. In areas of moderate or even slight water movement, however, most sand dollars stand vertically with only a third or more of the disk buried in the sand; their inclination to the substratum and their orientation to the current varies with the strength of current, but whatever it is, most individuals in the bed will be oriented in the same way. In areas of very rough water, the animals again lie flat and are at least partly buried.

The position and orientation of *D. excentricus* influences its feeding habits. When lying prone or buried, the sand dollar feeds on detritus, diatoms, and deposits swept by ciliary currents toward the mouth. When standing vertically, however, the animal becomes a suspension feeder, catching small mobile prey and large fragments of algae with its spines, pedicellariae, and tube feet, and gathering smaller particles by using its cilia. By either method, considerable sand is consumed with the food, and this sand has special significance to juvenile sand dollars. Juveniles, as it turns out, selectively ingest heavy sand grains and retain them in an intestinal diverticulum, as a weight belt, until a diameter of about 3 cm is reached. At this size, the adolescent sand dollar presumably has gained enough body weight to remain stable in the shifting sands of its environment.

Reproduction is similar to that of other urchins, eggs and sperm being extruded from different individuals for chance union and the development of free-swimming larvae that bear no resemblance to the adults. In Puget Sound sexual maturity comes in the late spring, and in central and southern California the major spawning period is from May through July. Sand dollars surviving predation by fish, seastars, and crabs may reach an age of 13 years in southern California, more commonly 6 to 10 years.

§247. A red to rose-lavender heart urchin, or sea porcupine, *Lovenia cordiformis* (Fig. 244), will be found occasionally on the southern sand flats,

Fig. 244. The heart urchin
Lovenia cordiformis (§247);
¾ natural size.

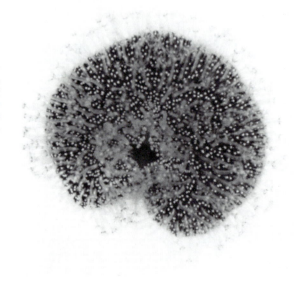

Fig. 245. *Renilla kollikeri*
(§248), the sea pansy, a
cnidarian common on southern mud flats. The polyps
covering the purple disk of
this expanded specimen are
nearly transparent; maximum
size is 8 cm across the disk.

where it lies half buried near the extreme low-tide line. This animal, by no means common in the intertidal, still occurs often enough from San Pedro to Panama to provide an excellent argument against barefoot collecting; and unless it is handled carefully, it can inflict painful wounds on bare hands. Ordinarily the long sharp spines point backward, and their thrusting effect helps the animal walk through the sand; but they bristle up in a formidable manner when their owner is disturbed.

§248. *Renilla kollikeri* (Fig. 245), the sea pansy, is one of the most obvious animals of the southern mud flats. When seen in their natural habitat, sea pansies seem misnamed, for they lie with their heart-shaped disks almost covered with sand and their stalks buried. A specimen should be transferred to a porcelain or glass tray containing clean seawater and left undisturbed for a time. The disk will there enlarge to three times its size when taken from the sand, and scores of tiny, hand-shaped polyps, beautifully transparent, will expand over the purple disk. Not the least of the sea pansy's charms is the blue phosphorescent light that it will almost invariably exhibit if mildly stimulated with a blunt instrument after being kept in the dark for an hour or so. In nature this response is triggered by the attack of one of its predators, the nudibranch *Armina californica*. Another important predator is the seastar *Astropecten armatus*.

According to work by Kastendiek (1982), the sand dollar *Dendraster* is a competitor of the sea pansy. In subtidal areas 6 to 9 meters deep the sand dollar effectively outcompetes *Renilla* for space and can largely exclude it from a band of shoreline. However, this apparently negative relationship is not without its benefits to the sea pansy. It seems that the sand dollar bed is also difficult for seastars to cross; in essence, it serves as a protective barrier, keeping the predator *Astropecten* seaward of the main *Renilla* population.

Reproduction occurs in the Santa Barbara area from May to early August, during which time a colony may spawn many times. In Newport Bay, sea pansies occur by the hundreds, in El Estero de Punta Banda by the thousands, and they are said to have been plentiful in San Diego and Mission bays before the days of commerce and industry. Southern California is the northernmost part of the range. The similar, broad, kidney-shaped *Renilla amethystina* extends at least to Panama.

The eight-tentacled polyps of *Renilla* are nearly always infested with minute parasitic copepods of the family Lamippidae.* Other copepods of the same family are known to infest various alcyonarians all over the world.

§249. The three true crabs that occur on sand flats are found nowhere else in the intertidal zone. *Heterocrypta occidentalis* (Fig. 246), since it lacks any other name, might be called the elbow crab, for its large forelegs, which terminate in absurdly small claws, suggest an energetic customer at a bargain counter. This weird creature is the only northern representative of a group of crabs well represented in tropical and equatorial regions. We have taken *Heterocrypta*, with the other two, at Newport Bay, and it has been reported as occurring at extreme low water in outer Los Angeles Harbor; it is known from deep water elsewhere along the coast.

Randallia ornata (Fig. 247) has the round, bulbous body and gangly legs of a spider, but it is not closely related to the "spider" crabs. The carapace is mottled purple, brown, or dull orange, and the legs are usually curled underneath it as the animal lies half buried in the sand.

---

*These copepods, listed as *Lamippe* sp. in previous editions, apparently still have not been described. Furthermore, since the old genus has been split, it is possible that even the generic placement is no longer accurate.

Fig. 246. *Heterocrypta occidentalis* (§249), the elbow crab; primarily a subtidal species, it is restricted intertidally to the lowest regions of sand flats; specimen shown is 2.5 cm across the carapace.

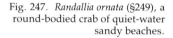

Fig. 247. *Randallia ornata* (§249), a round-bodied crab of quiet-water sandy beaches.

Portunus xantusii (Fig. 248) is a swimming crab, one of the liveliest of its kind. Its last two legs are paddle-shaped, and with them the crab swims sideways rather rapidly, with its claws folded. Two long spines, one at each side of the shell, are effective weapons of defense, and in addition the crab is unduly hasty about using its sharp and powerful pincers, so that a collec-

Fig. 248. *Portunus xantusii* (§249), a swimming crab with paddle-shaped hind legs and a talent for using its sharp spines and pincers defensively; ½ natural size.

tor who captures one without shedding any of his blood is either very lucky or very skillful—probably both. *Portunus* is a small relative of the prized blue crab of the Atlantic. It occurs as far north as Santa Barbara, but the other two crabs are not likely to be found in the intertidal above Balboa.

§250. A seastar, *Astropecten armatus* (Fig. 249), occurs on sand flats from San Pedro to Ecuador, usually below the waterline but occasionally exposed. Now and then a specimen will be found having an arm spread of more than 25 cm. They are beautifully symmetrical animals, sandy gray in color with a beaded margin and a fringe of bristling spines. Normally they lie half buried in the sand. *Astropecten* has a broad diet and often feeds on mobile mollusks, including the snail *Olivella biplicata*. Especially swift for a seastar, *Astropecten* quickly approaches an *Olivella* buried nearby in an aquarium, apparently using chemoreception to locate its prey. Capture is equally rapid: the seastar forces its arms into the sand around the snail, shifting positions if necessary, and then swallows the prey whole.

§251. Several snails and at least one nudibranch are members of the

Fig. 249. *Astropecten armatus* (§250), a swift carnivorous seastar of sand flats.

sand-flat assemblages. In El Estero de Punta Banda we once saw tremendous hosts of the nudibranch *Armina californica* (Fig. 250), which were living, peculiarly for nudibranchs, half buried in the sand—another of the interesting adaptations to this fruitful environment. Thin, white, longitudinal stripes, alternating with broader stripes of dark brown or black, make *Armina* a conspicuous animal. It occurs frequently with the sea pansy *Renilla*, on which it feeds; a similar association involving other species of *Armina* and *Renilla* occurs in the Gulf of Mexico.

The range of this opisthobranch, which is characteristic of clean sand bottoms, is from Vancouver Island to Panama. In the northern part of its range it must feed on something else besides sea pansies, and in Puget Sound it has been reported eating the sea pen *Ptilosarcus gurneyi*. Occasional specimens of *Armina* turn up in Tomales Bay, but sea pansies have not been found there—yet.

§252. The moon snails, or sand-collar snails, are common inhabitants of the sand flats throughout the whole stretch of coast from British Columbia to Mexico. There are two similar species. The larger (10-cm shell length), heavier, somewhat high-spired species is *Polinices lewisii* (Fig. 251), which may occur in somewhat muddy sand; *P. draconis* is somewhat smaller (8 to 10 cm) and shorter in height. *P. lewisii* and *P. draconis* both occur as far

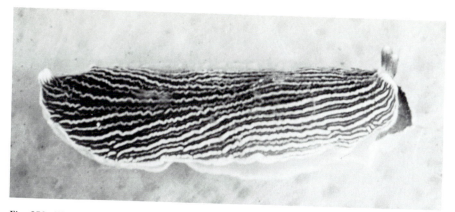

Fig. 250. The nudibranch *Armina californica* (§251), most commonly found burrowing just below the sand or mud surface; body is 2–5 cm long.

Fig. 251. The giant moon snail *Polinices lewisii* (§252): *above*, an adult, about ½ natural size; *left*, the collar-like egg case, greatly reduced.

south as northern Baja California, with the former extending to British Columbia and the latter to Trinidad, Humboldt County. When expanded, the body of the moon snail is much larger than its shell.

A third species, the strictly southern moon snail *Polinices reclusianus*, is much smaller—usually less than 5 cm in shell length—and, like *P. lewisii*, shows a preference for muddy sand. It can be distinguished from the other moon snails by its occluded umbilicus; that is, the "navel" is grown over and that part of the shell is smooth. Frequently the shells of *P. reclusianus*, which ranges from Mugu Lagoon (Ventura County) to Mexico, are occupied by a giant hermit crab (§262).

Any of the three will commonly be found partially buried in the substratum, through which they plow with apparent ease. At low tide a lump in the sand may indicate the presence of a moon snail a centimeter or so below the surface. The snail's slow, gliding movement across the sand is accomplished by means of millions of cilia beating on the surface of the broad foot. *Polinices* is also capable of a somewhat faster rate of locomotion, produced by waves of muscular contractions passing down the foot. All of our moon snails are capable, also, of contracting the immense fleshy foot enough to stow it away completely within the shell, which is then closed with the horny door common to other snails. This process involves the ejection of considerable quantities of water from perforations around the edge of the foot, and the animal cannot live long when contracted, since it cannot breathe.

The moon snail's method of food-getting is varied, but generally it clamps its foot around a clam or other mollusk and drills a neat, countersunk hole through the shell with its radula; the drilling is aided by chemical secretions produced by a specialized pad of tissue on the lower lip of the snail's proboscis. Shells perforated by the moon snail or some of its relatives are often found along the beach. In British Columbia, the clams most commonly drilled are *Protothaca staminea* and *Saxidomus nuttalli*, and in Bodega Harbor, *Macoma nasuta*. A less-spectacular method of food-getting consists of suffocating the victim by holding it inside the foot until dead. In this case, the proboscis is merely inserted between the shells of the dead animal and the contents are cut out. Apparently the great snail will also feed on any dead flesh that comes its way.

Cannibalistic members of its own tribe and the multi-rayed sunflower star, *Pycnopodia*, seem to be the moon snail's only natural enemies. In an aquarium a sunflower star has devoured two moon snails in 3 days. Humans also take enough moon snails to require the snails' coverage under California law: the daily limit is five, and none may be taken north of the Golden Gate Bridge (1984 regulations).

The egg cases of these giant snails are, to most laymen, one of the puzzles of the intertidal world, for they look like nothing so much as discarded rubber plungers of the type plumbers use to open clogged drains (Fig. 251). Many years ago an author of this book (J.C.) squandered a great deal of nervous energy on a fruitless search through unfamiliar literature in an attempt to find out what manner of animal they might be. It was some consolation to him to learn eventually that for a long while naturalists had made the same mistake. Certainly there is no obvious reason for connecting the rubbery, collar-shaped egg cases with the snails that make them. The eggs are extruded from the mantle cavity in a continuous gelatinous sheet, which, as fast as it emerges, is covered with sand cemented together with a mucous secretion. The growing case travels around the snail, taking its shape from the snail's foot as it is formed. In time the egg case crumbles, generally in the late summer, releasing a half million or so free-swimming larvae. There is some poetic justice in the assumption that many of the minute larvae will provide food for bivalves, which will later on be eaten by adult moon snails.

§253. The beautifully marked and polished *Olivella biplicata* (Fig. 252), the purple olive snail, often leaves a betraying trail on the sand by plowing along just under the surface. During the day, and when exposed at low tide, most of these snails are buried in the sand and inactive. But at dusk, at least when under water, many emerge from the sand and are active on the surface. It is apparently the onset of darkness that stimulates the snails to become active because snails placed artificially in darkness during the day soon emerge from the sand. Light will cause them to rebury in the laboratory, but the natural stimulus to rebury is less obvious; the snails start the process well before the first light of dawn.

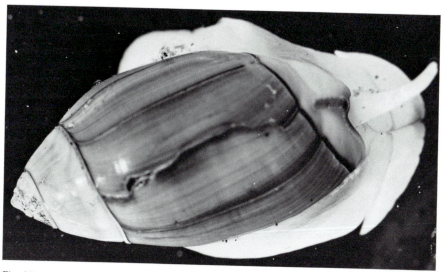

Fig. 252. *Olivella biplicata* (§253), the purple olive snail, about 5 times normal size. The snail's snowplow-like anterior end assists it in moving through the sand while the long fingerlike siphon maintains contact with the water overhead.

Juvenile *Olivella* occur lower on the shore than adults. The young snails grow rapidly, and reach sexual maturity in about a year (at 16 mm in length). Thereafter, growth slows to about 3 mm per year as energies are diverted to reproduction. Spawning occurs year-round in California, and courting pairs will often be seen. In these pairs, the female is in front, trailed by the usually larger male, with his foot firmly attached to the apex of her shell by a mucous thread. *Olivella* is apparently omnivorous, although one recent report (Hickman & Lipps 1983) indicated that foraminiferans are selectively ingested. Other reports described it eating algae and a variety of living and dead animals, and the regular presence of sand in its gut suggests that the snail is not averse to feeding on deposits. Its predators are many, and include octopuses, moon snails, shorebirds, and seastars. Defensive responses are triggered by some (but not all) seastars, the most common response being rapid movement away from the predator followed by burrowing. On occasion, however, the snail responds more violently and with spectacular results. On these occasions, the large lateral lobes of the foot are flapped up and down so vigorously that the snail may flip over backwards or may even appear to "swim" off the bottom.

Olivella biplicata ranges from Vancouver to Baja California; *O. baetica*, a smaller and slimmer species, will be found most commonly south of Point Conception, although its total range is about the same as the plumper olives.

The fragile bubble-shell snail may be first noticed in pools on bare sand flats, but it is much more at home in beds of eelgrass (§287).

§254. Two edible animals, a shrimp and a scallop, are found here, usually in pools in the sand or buried in the sand, but occasionally exposed. *Crangon nigricauda* (Fig. 253), the black-tailed shrimp, positively cannot be captured without a dip net unless it is accidentally stranded on the sand. It has, even on its legs and antennae, pepper-and-salt markings that make it extremely difficult to see against a sand background. This is one of the common market shrimps in California, along with several others found only in deep water. Shrimp-dredging boats net these animals from depths of from a few fathoms in San Francisco Bay to more than a hundred in Puget Sound, and it is in such depths that *Crangon* presumably finds its optimum conditions, since only occasional specimens are found in the intertidal. The black-tailed shrimp ranges from British Columbia to Baja California. Market specimens will occasionally be found with the carapace swollen out on one side into a blister, the result of infection with an asymmetrical isopod (bopyrid), *Argeia pugettensis*; this parasite is especially common on commercially netted shrimps in Puget Sound, where one out of possibly every 20 or 30 specimens of *Neocrangon communis* will be noticeably afflicted.

§255. The thick scallop *Argopecten aequisulcatus* (Fig. 232), frequently found buried in the sand, is fairly common. The peculiar swimming habits of the relatively "intelligent" scallops have been mentioned before (§220),

Fig. 253. *Above, Crangon nigricauda* (§254), the black-tailed shrimp, half-buried and well camouflaged. *Below,* the closely related *C. nigromaculata* against a more revealing background; about 7 cm long.

as have their toothsome qualities. In addition to this sand-flat form, which ranges from Santa Barbara to southern Baja California, and the rocky-shore forms (§220), another scallop will be found in eelgrass (§276).

§256. Although all the sand-flat animals so far considered burrow somewhat, the remainder are strictly burrowing forms. Perhaps the most obvious of these, when it occurs at all (and it does not occur intertidally north of southern California), is a great burrowing anemone (similar to *Pachycerianthus aestuari*) that lives in a black, papery tube covered with muck and lined with slime. More energetic diggers than the writers have taken specimens with tubes 2 meters long. The muscular lower end of the animal is pointed and adapted to digging, and at the upper end there are two concentric sets of tentacles—a short, stubby series about the mouth and a long, waving series around the border of the lividly chocolate-colored disk. Unfortunately, specialists have not had an opportunity to study this animal and ascertain its identity. We have found a much smaller *Pachycerianthus*, probably a different species, in the sand flats at Ensenada.* A similar form is shown in Figure 254.

§257. Another burrowing anemone, common in southern California (e.g., Los Angeles Harbor and San Diego Bay), is *Harenactis attenuata* (Fig. 255), which may occur north to Humboldt Bay. It is a tubeless form, sandy gray in color and with short tentacles numbering exactly, and distinctively, 24. The long, wrinkled, wormlike body is largest in diameter at the disk, tapering downward until it swells into an anchoring bulb. When annoyed, *Harenactis* pulls in its tentacles in short order, retracts into the sand, and apparently sends all its reserve body fluids into this bulb, thus swelling it so that it is impossible for a hungry bird or an avid human collector to pull the animal out. The bulb of a large specimen may be 30 cm below the surface, and digging out a specimen with a spade is very difficult, since the shifting sand fills up the hole almost as rapidly as it is made. While the excavation is in progress, the anemone retracts still more and makes its bulb even more turgid. Having done that, it can do nothing but sit tight and rely on Providence—a fatalistic philosophy that is usually justified in this case. Only if the sand can be shoveled away from the side of the bulb can a specimen be secured uninjured.

§258. A third anemone, found from Alaska to southern California, is the burrowing *Anthopleura artemisia* (Fig. 256), which is related to the great green anemone of the lower rocky tide pools. This species occurs in flats of

*Several species of the Order Ceriantharia have been described from the Pacific coast, but it would be only a guess to name the particular one that Ricketts saw. One possibility would be *Pachycerianthus aestuari* (mentioned in §256), which occurs in southern California. Another one, a close relative that is fairly common in the muddy bottoms of bays from southern California to British Columbia, is *P. fimbriatus*, which inhabits a tough, slimy, black tube that projects above the sediment; large individuals are said to occupy tubes a meter or more in length. This form extends into the low intertidal in southern California, but it is primarily subtidal in northern California, though some specimens can be found at extreme low water in Bodega Harbor and southern Tomales Bay.

Fig. 254. *Pachycerianthus fimbriatus* (§256), a burrowing, tube-dwelling anemone.

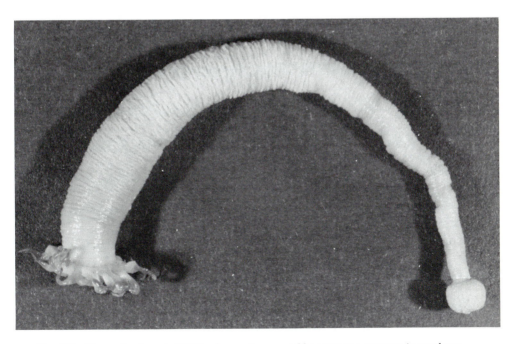

Fig. 255. *Harenactis attenuata* (§257), a burrowing, wormlike anemone common in southern California. The 24 short tentacles at one end and the anchoring bulb at the other are distinctive.

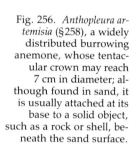

Fig. 256. *Anthopleura artemisia* (§258), a widely distributed burrowing anemone, whose tentacular crown may reach 7 cm in diameter; although found in sand, it is usually attached at its base to a solid object, such as a rock or shell, beneath the sand surface.

rather pure sand, but only where small cobblestones or cockleshells are available underneath the surface for basal attachment. At low tide *A. artemisia* retracts completely beneath the surface and is to be found only by raking or shoveling. It also frequently occurs on the open coast in pockets of sand caught in clefts or holes abandoned by rock borers. It is a much more abundant anemone than usually appreciated, undoubtedly because of its cryptic and retiring habits at low tide. The contracted specimens are white or gray-green, and might be mistaken for *Metridium* (§321) at first glance. Placed in a jar of seawater, they soon flower out beautifully, their long, wavy tentacles colored with lovely greens and pinks and their disks semitransparent.

§259. Considering a range of the Bering Strait to Cape Horn, Bernard (1983) listed over 1,000 species of bivalves on the eastern Pacific coast. Many, of course, are too small or too rare to be considered in this account; and so that the volume will not grow out of all proportion, we must further limit ourselves to the large, the obvious, or the interesting forms. After the scallop, with its peculiar swimming proclivities, the basket cockle *Clinocardium nuttallii* (Fig. 257) is the bivalve most likely to be noticed by the casual observer. Reaching its maximum development in British Columbia and Puget Sound, the basket cockle extends southward in decreasing numbers as far as Baja California and northward to the Bering Sea. In the north, specimens are found very frequently in flats of "corn meal" sand that would be very shifty if surf could ever get to it. Basket cockles are occasionally found entirely exposed, but they are active animals and can dig in quickly with their large feet. They burrow shallowly, however, for their siphons are little more than openings in the margin of the mantle. When

Fig. 257. *Clinocardium nuttallii* (§259), the basket cockle, with foot extended; about maximum size. This is an active bivalve that burrows shallowly.

touched on the soft parts by a predatory seastar, such as *Pycnopodia he-lianthoides* or *Pisaster brevispinus*, the powerful foot of *Clinocardium* is acti-vated in a remarkable, leaping escape response. The valves of the cockle open, the long foot is extended like a stout pole, and the foot is thrust vig-orously against the substratum, sending the cockle flying into the water (see Fig. 350). The foot may withdraw momentarily at this time, but usually it soon emerges for another leaping performance, and another, like a pole vaulter gone mad on the sand surface. We, like the seastars, can attest that this cockle makes excellent (if rather tough) food; however, it is too scarce to be used extensively, although it appears in markets to the north.

Fraser (1931) has reported on the natural history of *Clinocardium* in British Columbia. Most of the 760 specimens examined were 3 to 4 years old, and none was more than 7 years. Annual growth rings form in north-ern waters when growth slows during winter months (the lines are not ob-vious in most California specimens). Bands correlating with tidal cycles may also be seen in some specimens, reflecting periods of exposure at low tide. This genus is hermaphroditic, with ova and sperm usually shed si-multaneously, during a long spring spawning period that affects all ani-mals over 2 years old.*

*Clinocardium adults in Garrison Bay, San Juan Island, have recently been reported to spawn from April to November, usually for the first time during their second year of life.

Specimens will frequently be captured with tiny commensal crabs, juvenile *Pinnixa littoralis* (§300), in the mantle cavity.

§260. One would scarcely expect to find a clam throwing off a part of itself as a crab throws off a leg, but *Solen rosaceus* (Fig. 258), one of the jackknife clams, actually autotomizes rather readily. *Solen's* siphon, too large to be entirely retracted within its pink-tinged, nearly cylindrical shell, has prominent annulations. If the animal is roughly handled, or if its water gets somewhat stale, it drops off its siphon bit by bit, the divisions taking place along the lines of annulation. Presumably this autotomy was developed as a defense against the early bird (or fish) in search of a worm and not averse to substituting a juicy clam siphon. This clam, which averages around 5 cm in shell length, ranges from Humboldt Bay to Mazatlán, Mexico, and a similar but somewhat larger species, *Solen sicarius* (§279), occurs occasionally in mud flats and commonly in eelgrass roots as far north as Vancouver Island. *S. sicarius* has a slightly bent, glossy, yellow shell. Both species occupy permanent burrows 30 cm into the substratum and can move rapidly down the burrow when disturbed. Unearthed, both dig rapidly and well, although perhaps not as efficiently as their cousin *Tagelus* of the mud flats.

Other sand-flat clams are also common: *Chione undatella* (Fig. 259) and two other species of *Chione*, all known as the hardshell cockle and all ranging at least from Ventura County to southern Baja California; the purple clam *Nuttallia nuttallii* (Fig. 260; formerly *Sanguinolaria*), another southerner ranging from Bodega Bay to Baja California; and any of several small tellens, *Tellina bodegensis*, etc., most of which extend throughout the range considered in this book. A tellen, *T. tenuis*, is one of the characteristic components of the Scottish fjord fauna, and its natural history is well known. Stephens (1928, 1929), investigating the Scottish species' distribution, found small individuals (but in great numbers, up to 4,000 per square meter) at the low level of spring tides. The optimum zone for larger specimens, however, was higher up, toward high-water mark, where there were fewer individuals—an interesting zonation that reverses the usual situation. He also found that growth was more rapid higher up (probably a function of the lessened competition), that it decreased in regular progression with the age of the animal, and that it almost stopped in winter. There were four age groups, representing four annual spawnings; the males became sexually mature in May, the females in June. Specimens were found to feed on plant detritus, and in the spring on diatoms. These data on growth rates coincide with what is known of razor clams (§189).

Still another sand-flat clam is the white sand clam *Macoma secta*, which reaches a length of 9 cm and buries to a depth of 45 cm. It rather closely resembles its mud-flat relative, the bent-nosed clam (§303; Fig. 297), but it is typically larger and lacks the "bent-nose" twist in the shell at the siphon end. The shell is thin, and the left valve is flatter than the right. This species makes very good eating; but the intestine is invariably full of sand from its deposit-feeding habit, and the clams should be kept in a pan of

Fig. 258. The jackknife clam *Solen rosaceus* (§260), a fast mover within its permanent burrow.

Fig. 259. *Chione undatella* (§260), a hardshell cockle of sand flats in southern California; about natural size.

Fig. 260. *Nuttallia nuttallii* (§260), a common deeply buried clam of sand and gravel substrata in bays and on the open coast. The shell exterior is grayish but largely covered with a shiny dark-brown layer, the periostracum; the shell interior is purple; slightly enlarged.

clean seawater for a day or so before eating. The range is British Columbia to southern Baja California.

§261. A white mole crab occurs in this environment—*Lepidopa californica* (Fig. 261)—that never under any circumstances subjects itself to the pounding waves that delight its surf-loving cousin *Emerita* (§186). Visitors to the sheltered sand beaches at Newport Bay and southward* will find it submerged in the soft substratum, with its long, hairy antennae extending to the surface to form a breathing tube. This form burrows rapidly in the shifting sand, but not with the skill of *Emerita*, whose very life depends on its ability to dig in quickly enough to escape an oncoming breaker.

§262. One December afternoon we found, in the unpacked sand along the channel in El Estero de Punta Banda, thousands of hermit crabs, *Isocheles pilosus* (Fig. 262; formerly *Holopagurus*), all living in the shells of moon snails. The largest were giants of their tribe—larger than any intertidal hermits we have seen south of Puget Sound—but the really remarkable thing about them was their habitat. Many of them, probably most of them, were completely buried in the sand. Above the waterline they could be detected only by their breathing holes, below the water by occasional bubbles. Although the center of distribution is undoubtedly subtidal, here hosts of them could be turned out by merely raking one's fingers through the sand. Wicksten has also noted them to be abundant intertidally at Cabrillo Beach, San Pedro. She has further reported (1979) that when these crabs are unearthed and placed on top of fine sand, most can bury themselves completely in 15 to 60 seconds. These very hairy and rather beautifully colored hermits range as far north as Bodega Bay.

§263. At the same time and place, and in the same manner, a considerable number of brittle stars were taken. Occasionally a waving arm showed above the surface, but the vast majority were completely buried. These were *Amphiodia urtica* (formerly *A. barbarae*), once thought to be exclusively deep-water animals, having been reported as ranging from San Pedro to San Diego in depths up to 120 fathoms. Except that its disk is thicker, this animal resembles *A. occidentalis* (Fig. 54).

§264. A large cucumber with a most un-holothurian appearance, *Caudina arenicola* (Fig. 263; formerly *Molpadia*), is taken now and then at Newport Bay, in El Estero de Punta Banda, and probably at other places. Habitués of the Newport intertidal regions call it the "sweet potato," and the name is rather appropriate. A sweet potato as large and well polished as one of these animals, however, would be a sure prizewinner at a county fair. The first specimens we saw were dug near Balboa and put in an aquarium for the edification of a collecting party. We took them to be giant echiurid worms, and certainly there is little about them to suggest their actual identity: the mottled, yellowish-brown skin is tough, smooth, and slippery; there are no tube feet, and no obvious tentacles. *Caudina* feeds by

---

*Listed in previous editions as *Lepidopa myops*, *L. californica* is taken occasionally as far north as Monterey Bay; the main population is southern.

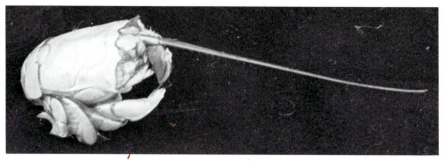

Fig. 261. The quiet-water mole crab *Lepidopa californica* (§261).

Fig. 262. *Isocheles pilosus* (§262), a very large and hairy hermit crab sometimes abundant on sand and mud flats; it lives in moon-snail shells.

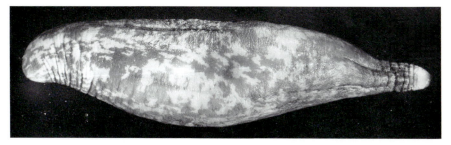

Fig. 263. The "sweet potato," *Caudina arenicola* (§264), an unusual cucumber with a tough, smooth skin (mottled yellowish-brown) and no tube feet; ½ maximum size.

passing masses of sand through its digestive tract for the sake of the contained detritus. Since it lives in sand that appears to be fairly clean and free from organic matter, it must be compelled to eat enormous quantities of inert matter to get a little food. We have no notes on the speed with which the sand mass moves through the animal, but the better part of the weight of a living specimen, and much of its bulk, is in the contained sand. Remove the sand, and the rotund "sweet potato" collapses.

Caudina has one cucumber trait, however, in that it always has guests in its cloaca. The pea crab *Pinnixa barnharti* occurs so commonly as to be almost diagnostic, although about 25 percent of the *Caudina* in the San Diego area have been reported to contain *Opisthopus transversus* (§76) instead of *Pinnixa*. When specimens are being narcotized with Epsom salts for relaxed preservation, the pea crabs are likely to come out, just as the *Opisthopus* escapes from *Parastichopus* (§77) under the same circumstances.

§265. Sand flats support the usual host of segmented worms. The tubed *Mesochaetopterus taylori* is a striking and usually solitary form (in some localities around Puget Sound it occurs in colonies). The tough, parchmentlike tubes, coated with sand grains, are oriented vertically in the substratum and may be very long, exceeding a meter in length. Unfortunately, the worm breaks apart so readily that it can rarely be taken whole (see *Chaetopterus*, §309); for the investigator willing to wait, however, partial worms will eventually regenerate in the laboratory to form whole animals (MacGinitie & MacGinitie 1968). The range of *M. taylori* is from British Columbia to Dillon Beach; a similar or identical form may be found at Elkhorn Slough, Morro Bay, Newport Bay (California), and El Estero de Punta Banda, Baja California—it seems to be rare intertidally in the south, however. *Mesochaetopterus* is somewhat similar in appearance to *Chaetopterus* (Fig. 302) and probably shares many aspects of its biology with this more thoroughly studied form.

The tubed and sociable "jointworm," or bamboo worm, *Axiothella* occurs in this environment, but it is more characteristic of areas of hard-packed muddy sand (§293).

Every shovelful of sand turned up in search of lancelets (§266) will rout out one or two specimens of a third segmented worm, a small and nereid-like species of *Nephtys*. The species turned up in the sandy muds of sheltered bays and lagoons is likely to be *N. caecoides*; another species, *N. californiensis*, lives in cleaner, sandy beaches of the protected outer coast (§155). The collector intent on snatching the swift lancelets the instant they are exposed will often grab a specimen of one of these smaller and slower animals before he can check the motion of his hand. A similar, shorter form taken in similar environments of loosely packed sand is *Armandia brevis* (formerly *A. bioculata*).

MacGinitie (1935) said of *Nephtys*: "It can burrow very rapidly, and in its activities reminds one somewhat of *Amphioxus* . . . can swim through loose sand as rapidly as some worms are able to swim in the water."

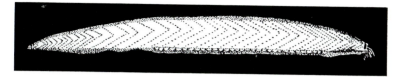

Fig. 264. The lancelet *Branchiostoma californiense* (§266), a chordate and a lightning-fast burrower; natural size.

§266. The lancelet, a famous animal that is seldom seen, comes into its own on this coast in southern California and (at least) northern Baja California; it rarely occurs intertidally in the north. Where there is a low-tide sandbar opposite the mouth of a sheltered but pure-water bay and far enough in to be protected from wave shock, these primitive vertebrates are likely to be found. Our species, *Branchiostoma californiense* (Fig. 264),* is reputed to be the largest lancelet in the world, adult or sexually mature individuals being up to 7 cm long. Specimens of that size will very rarely be taken, however, and the average size is less than half of the maximum. Violent stamping on the packed sand in just the right area will cause some of the animals to pop out. They will writhe about for a moment and, unless captured quickly, dive back into the sand. The usual method of capture is to turn up a spadeful of sand, trusting to one's quickness and skill to snatch some of the animals before all of them disappear. Until one has actually seen them in the act, it is hard to believe that any animal can burrow as rapidly as amphioxus, for the tiny, eel-like creatures are as quick as the proverbial greased lightning. Seemingly, they can burrow through packed sand as rapidly as most fish can swim. They will lie quietly on the bottom of a dish of seawater, but if disturbed, they will swim rapidly with a spiral motion.

Though the drawing shows the shape of the animal, it is utterly impossible for any illustration, with the possible exception of a motion picture, to convey a clear notion of the appearance of a living lancelet in its natural habitat.

Amphioxus has achieved unique fame in the last half century through being interpreted as a form ancestral to the other vertebrates. An important theory of phylogeny makes this lancelet a vital step in the evolution of all the higher animals. Off the coast of China a similar species is dredged by the ton for food, and the Chinese, with their epicurean palates, consider it a great delicacy. On this side of the Pacific, we have taken amphioxus at New-

---

Amphioxus was the original name of the genus *Branchiostoma*. By the time the name was changed, biologists (the only people who discuss the animal very much) had got into the habit of referring to the lancelet as "*Amphioxus*," and the name had become almost colloquial. It is now customary to print it as "amphioxus," without italics or capital letter. The word, however, is still used grammatically as if it were a proper generic name, to the occasional bewilderment of editors and proofreaders.

port Bay, at San Diego Bay, and in El Estero de Punta Banda, and Hubbs (1922) reported it from Monterey Bay to Bahía San Luis Gonzaga in the Gulf of California. Although it has occasionally been found in very low intertidal sand in Elkhorn Slough, it is no doubt common in deeper water below the effect of waves, for it has been taken subtidally in Tomales and Monterey bays and dredged off San Diego and in the Gulf. Probably amphioxus is not as rare as has been supposed, even on our warmer shores, for its small size, retiring habits, and great speed in escape make it a form easily overlooked.

Living nestled in the sediment, sometimes with amphioxus but more often on mud flats, is a small fish, *Gillichthys mirabilis*, the longjaw mudsucker.

§267. The acorn-tongue worms, or Enteropneusta, are also considered to be closely related to the primitive stock that gave rise to the ancestral vertebrate, and are usually treated as a separate phylum, the Hemichordata. A number of genera are present on our Pacific coast, although they are infrequently encountered intertidally.*

Balanoglossus occidentalis, up to 50 cm or more in length, is the most robust enteropneust found on Pacific shores to date. It has been common in Puget Sound and Mission Bay, and has also been taken at Palos Verdes and San Pedro. The worm's proboscis and collar are a pale yellow, and the hepatic caeca are a dark greenish brown; the genital regions are orange in males, gray in females.

A much smaller, proportionately more elongate worm, with a long, slender, pale-orange proboscis and deep-orange collar, is *Saccoglossus pusillus*. The remainder of this form's body is predominantly yellow to cream, although the mid-trunk region may have a tinge of green; the genital region in mature females is bright orange, somewhat duller in males.

Acorn worms usually construct burrows in the sediment. Those of *Saccoglossus pusillus* and others of this genus tend to be irregularly coiled, U-shaped tunnels with one or more entrances. All the intertidal species are substratum engulfers, and their bodies, almost always gorged with sediment, are very easily broken. When disturbed, these worms secrete large quantities of mucus and have a not unpleasant but pervasive and persistent odor that has been likened to iodoform. This will scent the hands and clothing of the digger for several days and is sometimes a useful clue to the presence of enteropneusts. Often the worms' presence can be detected beforehand by the casts deposited on the surface; indeed, on densely populated shores, these spiral deposits of processed sediment may dominate the scene. Some species of *Balanoglossus* can also be detected by the feeding funnels they form, each a few centimeters in diameter. The distance between the feeding funnel and the large, coiled cast varies with the length of the worm. In adult specimens, the two may be 15 to 30 cm apart.

---

*A forthcoming monograph, based on material studied by W. E. Ritter and several subsequent workers, describes some 24 species in 7 genera of 3 families.

7 🐚 Eelgrass Flats

Since the eelgrass *Zostera* (Fig. 265) occurs on flats of many types, from almost pure sand to almost pure muck, it seems desirable to list the animals associated with this seed plant before going on to the mud-flat forms.

§268. The start of a bed of eelgrass is an important step in the conversion of a former ocean region into wet meadowland and, ultimately, into dry land. The matted roots prevent the sand from being readily carried away by wind and tide, provide permanent homes for less-nomadic groups of animals than inhabit the flats of shifting sand, and enrich the substratum with decaying organic matter until it can be classed as mud. Eelgrass supports a rather characteristic group of animals, which live on its blades, about its bases, and among its roots in the substratum.

§269. In El Estero de Punta Banda we have found small and delicate sea urchins crawling about on the blades of eelgrass, possibly scraping off a precarious sustenance of minute encrusting animals but more likely feeding on the grass itself. They are so pale as to be almost white, but with a faint coloring of brown or pink. This urchin, *Lytechinus anamesus*, is known from shallow water south of Santa Rosa Island, along the west coast of Baja California, and in the Gulf of California. The largest recorded specimens are just over 25 mm in diameter. Ours are less than 12 mm; they resemble bleached, miniature purple urchins. In subtidal habitats, this urchin is known to feed on kelps and other algae. Most spawning occurs from June through September (see also §218).

§270. In an adjoining eelgrass pond in the same region were many thousands of small snails, the high-spired *Nassarius tegula* (formerly *Nassa* and *Alectrion*). Apparently these dainty little snails lead a precarious life, for about half of their shells were no longer the property of the original tenant, having been preempted by a small hermit crab.

These snails and hermits, and the urchins mentioned in the previous section, were all seen in the winter. The summer sunlight and temperature conditions in this region being decidedly unfavorable, it is likely that every

Fig. 265. The eelgrass *Zostera*, which supports a characteristic group of associates on its blades and among its roots.

animal able to do so would move downward beyond the chance of exposure. It is quite possible that El Estero, which we have not visited in hot weather, might then be a relatively barren place.

Many of the little *Nassarius* shells, whatever their tenants, are covered with white, lacy encrustations of the coralline bryozoan *Diaperoecia californica* (formerly *Idmonea*). On superficial examination, the erect, calcareous branches of this form, which may also be found on rocks and other hard substrata, cannot be distinguished from tiny staghorn coral. It occurs from British Columbia to Costa Rica, primarily below the level of lowest tides, and is fairly common in southern California. It will be quickly noticed wherever found, and almost certainly mistaken for a coral. When encrusted on *Nassarius*, it makes the shells at least half again as heavy.

§271. Any of several hydroids—*Aglaophenia*, *Plumularia*, or *Sertularia*—will be found attached to eelgrass, as previously mentioned, or on other forms specific to this environment. The cosmopolitan *Obelia dichotoma* is common in situations of this sort all up and down the coast and, in fact, in temperate regions throughout the world. Compared with other delicate and temperamental species of *Obelia*, *O. dichotoma* is very hardy and obliging. It withstands transportation well even when its water becomes somewhat stale, lives for several days under adverse conditions, and expands beautifully when its proper conditions are approximated. The delicate white clusters of this and *O. longissima* (§316) are conspicuous on the tips of eelgrass in such places as Elkhorn Slough. Judging from specimens determined by C. M. Fraser, it is this form and *O. dubia* (§332) that were frequently called *O. gracilis* in the past; in any case, the differences are slight. *O. dichotoma* ranges on this coast from Alaska to San Diego. We have not so far taken this form on the outer coast, but it occurs on Puget Sound wharf piling with *O. longissima*, etc. At least some of the eelgrass hydroids taken at Jamestown Bay, Sitka, turn out to be the ubiquitous *O. longissima*, and it may be that the two are mingled there in the same habitat as they seem to be at Elkhorn Slough.*

A minute nudibranch, *Eubranchus olivaceus* (formerly *Galvina*), has been reported among these hydroids, presumably feeding on them, and tangling them up at certain times of the year with its white, jelly-covered egg strings. Some of the *Obelia* nudibranchs are among the smallest occurring within our geographical range—a few millimeters in length and identifiable only with a binocular microscope.

§272. Associated with the eelgrass of Elkhorn Slough, Tomales Bay, Los Angeles Harbor, and similar places in central and southern California where the bottom is compact, sandy mud, there is a skeleton shrimp, *Caprella californica* (Fig. 266). It is large, even by comparison with *Metacaprella kennerlyi* (§89) of the rock-pool hydroid colonies. Sometimes great numbers of *Caprella* may be seen at night in the beam of a flashlight, under the surface of the water. Occasionally they will even be found on shores of pure mud. The animal is often brown, but may be bright green, red, or dull-colored, depending on whether it occurs on eelgrass, red algae, or muddy bottoms. It is an omnivorous feeder, scraping small organisms from the substratum, feeding on detritus, or capturing smaller amphipods as various opportunities present themselves.

§273. Some of the many eelgrass beds in Puget Sound support large populations of the fixed jellyfish *Haliclystus auricula* (Fig. 267), which also has been reported on eelgrass in northern California. This species is actually one of the scyphozoan jellyfish, like those seen offshore in the summer pulsating lazily at the surface of the sea; yet it is small and has lost almost all power of motion. The small, stalked attachment disk has some power of contraction, and the tentacle-studded mouth disk is able to fold somewhat, but the animal swims not at all. We have never seen any of them so much as change their position on the grass blades, but it is presumed that they can glide about slowly on a smooth surface, like an anemone. Specimens de-

---

*Cornelius (1975) revised the genus *Obelia*, and according to his revision, all of the species of *Obelia* cited in this book—*O. commissuralis*, *O. dichotoma*, *O. dubia*, *O. gracilis*, and *O. longissima*—are one and the same species, referable to *O. dichotoma*! Because these changes are rather drastic, far from universally accepted, and involve several forms whose biology is very poorly known, the species designations of the Fourth Edition are retained here until more work is done and the dust settles a bit.

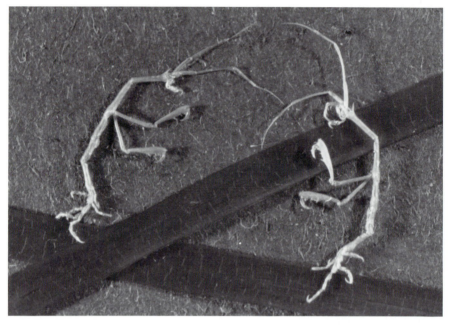

Fig. 266. *Caprella californica* (§272), a large skeleton shrimp, on eelgrass; body to 3.5 cm long.

Fig. 267. The fixed jellyfish
Haliclystus auricula (§273), about
2 cm in diameter.

tached from their grass will get a new hold with their tentacles and bend the stalk down until it can be reattached. An interesting feature of the digestive system is the presence of filaments that mix with the food in the body cavity, thus greatly increasing the area that the enzymes can act upon. The animal's colors are subtle rather than striking, but it is large enough (2 cm or so in diameter) to be seen without difficulty.*

*Another related but undescribed form, *Manania* sp., can sometimes be found on coralline and other algae in low intertidal regions of central and southern California. Both of these species were probably erroneously identified as *H. stejnegeri* in previous editions; *H. stejnegeri* is known only from Alaska to Japan.

§274. Necessarily small snails of several kinds frequent the slender leaves of eelgrass. *Lacuna porrecta*, shaped like *Littorina* but only one-fourth to one-half its size and having an aperture that takes up most of the front of the shell, may be found either on the grass or by screening the soil about the roots. George MacGinitie remarked that this form waddles like a duck. It ranges from the Bering Sea to San Diego. The related *L. variegata*, of zigzag markings, may be found similarly on *Zostera* in the Puget Sound–Georgia Strait region.

Haminoea vesicula is a small, white, bubble-shell snail, up to 2 cm in shell length by usually about half of that, ranging from Alaska to the Gulf of California. Like the great *Bulla* (§287), with which it sometimes occurs, it cannot completely stow away its comparatively large body in its fragile shell. In some specimens the beating heart can be seen through the shell. The eggs of this species are laid in coiled, deep-yellow ribbons about 1 cm wide and up to nearly 20 cm long.

A small, dark slug, related to the large and shell-less sea hares, will be found sporadically in great numbers at such places as Elkhorn Slough, where it burrows in the surface layer of mud, moving through a slime envelope of its own secreting. Actually, there are two such small slugs, often difficult to distinguish and both ranging from Alaska to southern California.* *Aglaja ocelligera* is brown to black with distinct yellow and white spots. The somewhat smaller *Melanochlamys diomedia* (formerly *Aglaja*) is cream-colored, but with brown or black mottling that is so dense in some specimens the slug often appears dark brown or black. Neither is likely to be confused with the much-larger *Navanax inermis*, which is a feature of the southern eelgrass beds and adjacent mud flats, where it feeds on *Haminoea*. All these forms—*Bulla, Haminoea, Aglaja, Melanochlamys,* and *Navanax,* as well as the sea hare *Aplysia*—were once lumped together as "tectibranchs," opisthobranchs with covered gills, to distinguish them from the nudibranchs, with naked gills. But the term tectibranch is now rarely used. Instead, the current practice is to refer directly to the various orders of opisthobranchs, such as the Cephalaspidea (e.g., *Aglaja, Bulla, Haminoea*), Anaspidea (e.g., *Aplysia, Phyllaplysia*), and Nudibranchia.

Another such opisthobranch with covered gills, but one that looks much more like a nudibranch than like the forms just mentioned, is a pretty green animal with India-ink lines of stippling. It is *Phyllaplysia taylori* (Fig. 268), reported from Nanaimo to San Diego and very common in Tomales Bay. This well-camouflaged animal, usually 25–45 mm long, spends its entire, but short, life (probably less than a year) on the blades of eelgrass, from which it grazes diatoms and small encrusting organisms. Ovipositing individuals of *Nassarius fossatus* (§286) will also be found among blades of eelgrass during the late summer, depositing encapsulated eggs. Even an annelid worm, the small *Nereis procera*, known more commonly

*Information about the two species is mixed in MacFarland's monograph (1966; plate 2, Fig. 4, is *A. ocelligera*, not *M. diomedia*).

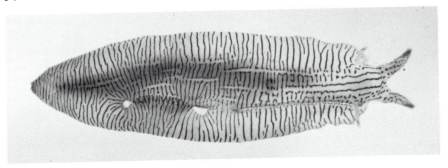

Fig. 268. The opisthobranch *Phyllaplysia taylori* (§274), green with dark lines, matches the background of its host plant, the eelgrass *Zostera*.

from dredging in Monterey Bay, will be found at home here on the *Zostera* (and on *Enteromorpha*), where it constructs permanent membranous tubes on the blades and about the roots. Another annelid, *Platynereis bicanaliculata* (§§153, 330), will also be found here.

A very narrow limpet, the painted *Notoacmea depicta*, is restricted to this habitat from near San Pedro to southern Baja California. Though not at all obvious, it is readily discoverable if one runs a quantity of eelgrass through one's hands; where common, it may occur on every tenth blade. A similar limpet, but without the well-marked pattern of brown lines on a white surface (and occurring in a different habitat), is *Notoacmea paleacea* (§119) of the open-coast surfgrass (*Phyllospadix*) beds.

§275. *Melibe leonina* (formerly *Chioraera*) is a large nudibranch of eelgrass areas, readily identified because its head end is expanded into a broad, elliptical hood bordered by tentacle-like processes (Fig. 269). The cerata along the back are flattened, leaflike structures, and the body is pale yellow, with the grayish branches of the liver showing clearly in the cerata. *Melibe* feeds in an unusual manner. While remaining firmly attached to the substratum by its narrow foot, the animal extends the broad hood into the water and then sweeps it down to the right or left. If the hood contacts something in its downward sweep, the two sides are brought together, the tentacles fringing the hood interlock, and the prey is trapped inside. Further contractions of the hood force the prey, primarily copepods and other small crustaceans, into the mouth, which, curiously, lacks the rasping radula of most gastropods.

Although *Melibe* remains attached to the substratum most of the time, it can swim. If harassed by a hungry kelp crab (§107) or if accidentally dislodged from its perch, the animal takes to the water, swimming actively (and usually upside down) by flexing its body from side to side. This remarkable animal is common in some years in the Friday Harbor region, particularly on Brown Island. In 1928 approximately 1,000 appeared at one time, and of these an estimated 150 pairs were copulating, mostly in mu-

Fig. 269. *Melibe leonina*
(§275), a large hooded
nudibranch with flattened
cerata; natural size,
although large specimens
may approach 10 cm long.

tual coitus. Guberlet wrote (1928): "During the act of copulation the animals lie side by side, or with the bottom of the foot of each together, with their heads in opposite directions, and the penis of each inserted in the vaginal pore of the other . . . and the animals so firmly attached that they can be handled rather roughly without being separated." After they have spawned the animals take on a shrunken appearance, and many, if not all of them, die, the indications being that they live only one season and spawn but once. The eggs are laid during the summer in broad spiral ribbons 8 to 10 cm long, which are attached to eelgrass or kelp.

The recorded range is from Alaska to the Gulf of California, but south of Puget Sound it usually occurs offshore on kelp (sometimes in harbors). In southern Puget Sound we have found occasional solitary forms only, but in central British Columbia we once came upon a great host of them about one of the rare eelgrass beds in this ordinarily precipitous region.

§276. The southern scallop *Leptopecten latiauratus* (formerly *Pecten*), is commonly seen at Newport Bay and southward (less commonly northward to Point Reyes) attached to eelgrass clusters or swimming about their bases in the peculiar manner of the scallop tribe. It differs from *Argopecten aequisulcatus* in being flatter and having wider "ears" at the hinge line.

§277. A massive and most amazing sponge, *Tetilla* (§298), is sometimes attached to eelgrass clusters, but it is more common on the mud flats, where it leads an apparently normal life under conditions abnormal for a sponge.

§278. In and about the roots of eelgrass, and elsewhere in quiet muddy bays, lives a pistol (snapping) shrimp, *Alpheus californiensis* (formerly *Crangon*), that differs from its relative of the outside pools only in minor characters, such as having a slightly smaller snapping claw of somewhat different shape. It ranges from San Pedro to central Baja California.

Another quite different shrimp to be found here, and also in mud-flat pools, is the lovely, transparent *Heptacarpus paludicola* (Fig. 66; formerly *Spirontocaris*). It is scarcely distinguishable from the *H. pictus* described in §62. In Puget Sound a grass shrimp, *Hippolyte clarki*, occurs commonly in this association even in the daytime, its green color matching that of the eelgrass to which it clings so well that the observer may be working in a group of them without realizing it until one of the graceful animals swims away. This species has a recorded range of Alaska to the Palos Verdes Peninsula; and from Bodega Bay south it is joined by a similar form, *H. californiensis* (§230), which continues into the Gulf of California. In California, *H. clarki* occurs most commonly among blades of the giant kelp *Macrocystis*, whereas *H. californiensis* occupies the eelgrass beds. The isopod *Idotea resecata* (§205) will also be found clinging lengthwise to the blades, often abundantly, without its presence being suspected; specimens found here will likely be green, whereas those found on kelps will be brown. Farther north other crustacea may be found; at Sitka *Heptacarpus stimpsoni* (formerly *Spirontocaris cristata*) and *H. camtschatica* have been taken, among others.

§279. The jackknife clam *Solen sicarius* is, for its kind, an extremely active animal. It can move rapidly up and down its permanent vertical burrow, and if unearthed, digs so rapidly that it can bury itself in 30 seconds with only a few thrusts and draws of its foot. But this clam has even more striking accomplishments: according to George MacGinitie, it can jump several centimeters vertically and can swim by either of two methods. Both involve the familiar "rocket" principle. In the first method, water is forcibly expelled from the siphons, thus propelling the clam forward; in the second, water is expelled from the mantle cavity through an opening around the foot, and the resulting darting motion is accelerated by a flip of the extended foot, the combined forces sending the animal some 60 cm through the water in a reverse direction. Similar gymnastics are stimulated in an Atlantic razor clam (*Ensis directus*) by contact with predatory snails of the genus *Polinices*; whether our species respond in this manner is not known but seems likely. Under more normal circumstances, when the animal is buried, the expulsion of water anteriorly around the foot serves to "soften" the sediment ahead of the digging animal, thus increasing burrowing speed.

Solen sicarius has a glossy-yellow, bent shell, often reaches a length of more than 7 cm (large specimens may be twice this), and ranges from Vancouver Island to Baja California. Like its somewhat smaller, straight-shelled relative of the sand flats, *S. rosaceus* (§260), it autotomizes its siphons at the annular constrictions. The two may at times be found together in the sand flats, but *S. sicarius*, although it is occasional in mud flats also, is far more common among the roots of eelgrass.

§280. Of many of the worms that writhe about in the roots we freely confess our ignorance. Of others, which we can recognize, so little is known that a full list would "little profite the Reader" and of them all only three will be named. The first of these is the slender and iridescent *Glycera americana*, which looks not unlike a *Nereis* with a pointed head. The difference becomes obvious when *Glycera* unrolls a fearful introvert almost a third as long as itself—an instrument armed with four black terminal jaws that are obviously made for biting; when the introvert (proboscis) is extended, the head loses its characteristic pointedness. The worm's proboscis is used for burrowing as well as for prey capture. To burrow, the retracted introvert is first unrolled (not simply pushed, which would be very difficult) into the sand or mud ahead; then, the tip swells to form an anchor, and the rest of the worm's body is pulled forward and envelops the proboscis again. The efficiency of this method can be easily seen by placing a worm on the surface of even firmly packed sand. In a series of jerky movements, the worm quickly disappears below the surface. Specimens of *G. americana* occur singly from British Columbia to Baja California and may be 20 cm or more in length. Reproduction occurs in summer months when mature worms leave their burrows as swimming epitokes.

§281. The second worm to be mentioned is *Cistenides brevicoma* (formerly *Pectinaria*), whose body is tapered to fit the slender, cone-shaped tube that it builds of sand grains cemented together. This worm digs by shoveling with its mouth bristles, which may be of brilliant gold, as in the case of one specimen that we dredged from the mud bottom of Pender Harbor, British Columbia. A model woven from cloth of gold could not be more striking. We have specimens from the vicinity of Juneau also, and they occur in Tomales Bay and Elkhorn Slough, as well as off southern California. The minute and highly specialized pea crab *Pinnixa longipes* (§293), widest of the lot, is a common commensal. A similar worm, *Pectinaria californiensis* (Fig. 270), tends to inhabit sands of finer grain than *Cistenides*; the latter form may be found in coarse sand or even gravel. The two may also be distinguished by the shape of the tube—straight in *Pectinaria* and curved in *Cistenides*. *Pectinaria* occurs primarily from central California to Baja, although its complete range extends north to Puget Sound and possibly Alaska.

Watson worked with Fabre-like patience on these animals, and his observations, hidden away from popular reach in an English biological periodical (1928), seem to us as popularly fascinating as those of that French entomologist. "Each tube is the life work of the tenant," he wrote, "and is most beautifully built with grains of sand, each grain placed in position with all the skill and accuracy of a human builder." Each grain is fitted as a mason fits stones, so that projecting angles will fill hollows, and one appreciates the nicety of the work when one perceives that the finished wall is only one grain thick. Furthermore, no superfluous cement is used. Suitable grains for building are selected by the mouth and applied to the two lobes of the cement organ, "after which the worm applies its ventral shields to

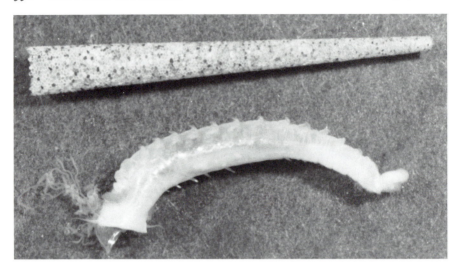

Fig. 270. The worm *Pectinaria californiensis* (§281), with its tube of sand grains, a masonry masterpiece in which each grain is fitted carefully against its neighbor; about 1½ times natural size.

the newly formed wallings, and rubs them up and down four or five times, apparently to make all smooth inside the tube. . . . The moment when an exact fit has been obtained is evidently ascertained by an exquisite sense of touch. On one occasion I saw the worm slightly alter (before cementing) the position of a sand grain which it had just deposited." Only very fine grains are used, but one specimen in captivity built 2 cm of tube in 2 months.

Cistenides lives head downward, usually with the tube buried vertically in the substratum, although sometimes, and especially at night, 2 cm or so of the small end of the tube may project above the surface. Digging is accomplished by means of the golden bristle-combs that project from the head, "tossing up the sand by left and right strokes alternately." The worm eats selected parts of the sand that it digs: that is, it passes sand through its body for the contained food. The digging therefore creates a hollow about the digger's head, and frequent cave-ins result. Thus the worm is able not only to dig its way to new feeding grounds, but to remain in the same place for a time and make its food fall within reach of its mouth. There are strong hooks at the hind end of the body that enable the animal to keep a firm grip on its tube while digging. For respiration and the elimination of waste matter, two currents of water are pumped through the tube in opposite directions. Cistenides is able to repair breaks in its tube, appears to be light-sensitive, and probably has a tasting organ. Like some other sand-burrowing forms, it may be here one year and somewhere else the next; a large aggregation may simply disappear over the winter.

§282. The third worm, perhaps the most striking of all but the least common, is a magnificent species, *Themiste perimeces*, taken at Elkhorn Slough and Bodega Bay. Large specimens of this slim sipunculid are almost 25 cm long when expanded. Except for its slimmer body and usually larger size, this worm does not differ greatly from *T. pyroides* (the species illustrated in Fig. 163); but closer examination will reveal that the tentacles of *T. perimeces* branch closer to the base, whereas the lower part of the smaller form's tentacles is bare. The delicately branched and beautifully flowering tendrils presumably gather food particles, which are transferred to the intestine by rolling in the introvert.

§283. In a good many of the quiet-water channels of Puget Sound the subtidal sandy or gravel shores support a variety of eelgrass that is always at least partially submerged on even the lowest tides. The characteristic fauna that belongs to these deep plants can scarcely be considered intertidal, but the shore visitor will fringe this zone and will surely see, among other things, the hydromedusa *Gonionemus vertens* (Fig. 271) during a summertime visit. This graceful, little, bell-shaped medusa (up to 25 mm in diameter), almost invisible except for an apparent cross inside the bell formed by its reddish-brown gonads, has some 60 to 100 long, slender tentacles, with which it stings and captures small fish and crustaceans. It seems to be restricted to these particular growths of deep eelgrass or, less frequently, to shallow bays blanketed with layers of sheetlike algae (e.g., *Ulva* or *Laminaria*). Many such locations when examined in summer turn out a few specimens. When the medusae are not visible, thrusting an oar into the eelgrass or algae, or merely walking slowly through it, will send some of them pulsing toward the surface, often in considerable numbers.

C. Mills (1983 and unpublished) has found that *Gonionemus* medusae are released from their polyps in May and June and live for 2 to 3 months. It takes young medusae about 1½ months to attain sexual maturity, at which point they are nearly full-size. They apparently live about 30 days more, spawning every evening, and the entire population of medusae dies by mid-September. Although mostly benthic by day, the medusae swim off the bottom at dusk and tend to remain near the water surface all night. They spawn about 45 minutes after dark, and females produce up to 10,000 eggs per day. After fertilization, the egg may develop into a swimming planula, or in some cases it may simply remain a sessile, relatively undifferentiated mass of cells until metamorphosing into the hydroid stage. The hydroid of *Gonionemus*, less than 1 mm tall with 2 to 6 tentacles, is a solitary polyp that lives on rocks and bits of shell and wood. In the spring, each polyp can produce a single medusa, which buds off the side of the column. Additionally, *Gonionemus* polyps can produce asexual buds year-round. These are pinched off the parent's column, from which they creep away to become new polyps.

Gonionemus is a species that has suddenly appeared in a number of locations, including Santa Barbara, far outside its normal range, which is

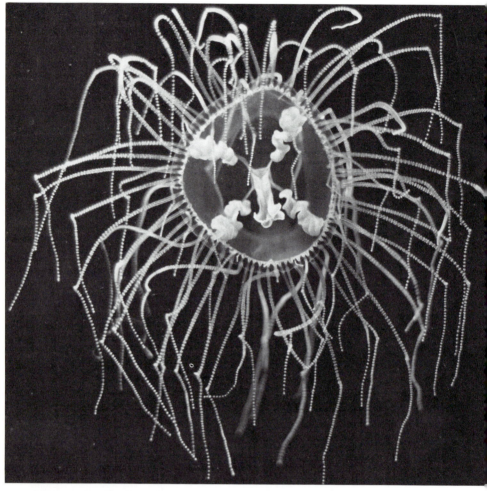

Fig. 271. A hydrozoan jellyfish, *Gonionemus vertens* (§283), with many long slender tentacles and four reddish-brown gonads; up to 25 mm in diameter.

from Puget Sound north. Medusae have even appeared sporadically in various locations in Europe, their presence there probably a result of importing oysters from our coast during the late nineteenth century (Edwards 1976).

Other occasional visitants from deep water may sometimes be seen in this environment. The cucumber *Parastichopus californicus* (§77) is common underfoot. Shrimps occur, especially at night, including such forms as the "coon stripe," *Pandalus danae*, a colorful shrimp marked with red, white, and blue spots and transparent areas; but they are rare and solitary. Although considerations of this nature are beyond the scope of this work, it may be interesting to note the answer (worked out in Berkeley 1930) to a question frequently asked by shrimp fishermen, who take the pandalids commercially (*P. danae* is usually incidental in catches of other, more common pandalids). They have wondered why it is that the only males taken of a given species are small, while all the females are large. The rather spectacular answer is that the five species investigated by Berkeley invariably undergo a sexual metamorphosis. Young specimens are always male. At the age of from 18 to 40 months, depending on the species, they become females and presumably spend the rest of their days as functioning members of that sex. This type of sexual alternation is known as protandry.

8 🪵 Mud Flats

A prominent feature of mud flats is the extent to which the inhabitants modify their own surroundings (Fig. 272). By continually reworking the sediments, burrowers and deposit feeders create channels in the sediment and alter its chemical and physical properties. They form mounds and pools, and their fecal pellets settle on the surface as a rich flocculent layer that is used by many animals, large and small. Signs of this "biogenic" activity, many characteristic of a particular species, are everywhere on the flats, and much can be learned from a close examination of the sediment before, as well as after, the shovel is used. The problems of respiration and food-getting in mud flats and the lack of attachment sites make for a rather specialized fauna. Seastars and urchins are poorly suited to this environment, and, with the exception of an occasional *Pisaster brevispinus*, none occur. There are only one sponge and one hydroid, and there are no chitons or bryozoans; but the many worms, clams, and snails make up for the scarcity of other animals.

Even in the south the animals extend farther up into the intertidal on the mud flats proper than they do on the sand flats anywhere on the coast. Some of them, like the fiddler crab, seem to be changing into land dwellers— a tendency that reaches an extreme in Japan, where one of the land crabs has worked up to almost 600 meters (2,000 feet) above sea level.

Zones 1 and 2. Uppermost Horizon and High Intertidal

§284. The little fiddler crab *Uca crenulata* (Fig. 273) is characterized, in the males, by the possession of one relatively enormous claw, which is normally carried close to the body. It looks like a formidable weapon, but is used to signal for females and to fend off other males and as a secondary aid to digging. In battle the two contestants lock claws, each apparently endeavoring to tear the other's claw from its body, although injury seldom results. Sometimes, according to one capable observer, a crab loses its hold

Fig. 272. A mud flat (Bodega Harbor) showing mounds, depressions, and holes created by the biogenic activities of the infaunal inhabitants.

Fig. 273. The fiddler crab *Uca crenulata* (§284), of southern beaches: *left*, a male about twice natural size; *above*, fiddlers on the sand in upper Newport Bay (July 1962).

on the substratum and is "thrown back over his opponent for a distance of a foot or more"—a performance that suggests a judo wrestling match. At breeding time, and particularly when a female is nearby, the males make a peculiar gesture, extending the big claw to its full length and then whipping it suddenly back toward the body. One might conclude, with some logic, that the big claw is a distinctly sexual attribute comparable to the horns of a stag.

The Japanese call the fiddler crab *siho maneki*, which translates as "beckoning for the return of the tide." It is too picturesque a name to quibble over, but one might reasonably ask why Mahomet does not go to the mountain, for the presumably free-willed fiddler digs its burrow as far away from the tide as it can get without abandoning the sea entirely. There it feeds, like so many other animals occurring on this type of shore, on whatever minute plants and animals are contained in the substratum. Instead of passing quantities of inert matter through its body, however, the fiddler crab daintily selects the morsels that appeal to it. The selecting process begins with the little claws (the female using both claws and the male his one small one) and is completed at the mouth, the rejected mud collecting below the mouth in little balls, which either drop off of their own weight or are removed by the crab.

Dembrowski (1926), working with the sand fiddler of the east coast, made careful observations of the animal's behavior and came to the interesting conclusion that its behavior is no more stereotyped than that of man. More recent work, however, suggests that their behavior patterns are, in fact, rather preset. Their courtship behavior in particular is so ritualistic, both in time between gestures and in the sequence of gestures, that different species can be recognized in the field by their characteristic courtship motions (see Crane 1957 for a more modern treatment). In any event, Dembrowski's observations, if not all of his conclusions, remain valid today. He found that in their natural habitat the fiddlers construct oblique burrows up to a meter long, which never branch and usually end in a horizontal chamber. He observed their burrowing behavior in sand-filled glass jars in the laboratory, and found that they dig by packing the wet sand between their legs and carapaces and pressing it into pellets, which they then carefully remove from the burrow. The smallest leg prevents loose grains from falling back while the pellet is being carried. Sometimes hours are spent in making the end chamber: Dembrowski thinks that its depth is determined by how far down the fiddler must go to reach sand that is very moist but not wet. In pure sand, water would filter into the burrow at high tide, but the crab always takes care to locate in substratum that has so high a mud content that it is practically impervious to water. Furthermore, it lines its burrow and plugs the entrance before each high tide. Thus, at high tide or during heavy rains the animal lives in an airtight compartment.

Under experimentally unchanging conditions the crabs showed no signs of the periodicity that might be expected in an animal that stoppers up its burrow against each high tide. Evidently the door-closing process is due not to memory of the tidal rhythm but to the stimulus of an actually rising tide, which would moisten the air chamber before the surface of the ground was covered. If a little water was poured into their jars, the crabs started carrying pellets to build a door. When the water was poured in rapidly, the crabs rushed to the entrance and pulled pellets in with great haste, not pausing to construct a door. If the jar was then filled up carefully, so as not to damage the burrows, the crabs would stay in their air chambers and not stir for a week. Dumping water in, however, filled the burrows and caved them in, burying the animals. Their response this time was to dig out and wait on the surface of the sandy mud for the "tide" to ebb.

Each "high" high tide destroys part of the burrow, and the animal must dig itself out by somewhat reversing its former operations. By digging, it "causes the chamber to rise slowly in an oblique direction, still keeping its volume unaltered, as the sand is always carried from the roof of it to the bottom," again by means of pellets. The procedure varies widely with the "individuality" of the animal.

Under natural conditions the fiddlers are probably never in water longer than is necessary to moisten their gills; nevertheless, several animals lived experimentally in seawater for 6 weeks with no apparent ill effects. They can also live in air for several weeks without changing the water in their gill chambers. Dembrowski concluded that they are true water-breathers but have interesting land-living adaptations.

Uca crenulata is the only fiddler crab north of southern Baja California, and it is known from only as far north as Playa del Rey (Los Angeles County). Our *Uca* is a creature of higher marshlands—regions peculiarly susceptible to "improvement" by man. Where any of the right kind of tidal marsh is left, a few *Uca* will still be found, as in upper Newport Bay, but fiddlers are becoming rarer every day.

It has been remarked that specimens are common only in summer. We got an inkling of why that should be, in January several years back, when we persistently dug out a burrow in the estuary south of Ensenada. The living animal was at the 102-cm level in a tunnel that extended to a measured 122 cm from the surface. The water in the pool at the bottom was bitterly salt.

§285. Another burrowing crab, superficially resembling a fiddler crab and with somewhat the same mannerisms, but living at all tide levels, was once abundant in Newport Bay. This animal, *Malacoplax californiensis* (Fig. 274; formerly *Speocarcinus*), is apparently now rare. This crab shares some mannerisms with the fiddler: both bristle up at the mouths of their burrows to intimidate an intruder, and both scurry down into the security of their burrows before danger comes near. The lively fiddler crab never goes much below the high-tide line, however, and the less-active *Malacoplax* roams through the whole intertidal region. *Malacoplax* has fringes of hair on its legs and carapace edges, and short eyestalks; the fiddler is hairless and has long and prominent eyestalks.

North of *Malacoplax*'s range, which is Mugu Lagoon to Bahía Mag-

Fig. 274. *Malacoplax californiensis* (§285), a hairy burrowing crab of high intertidal mud flats; maximum size is about 2 cm across the carapace.

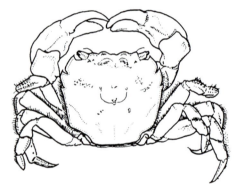

Fig. 275. *Hemigrapsus oregonensis* (§285), the shore crab of Puget Sound and San Francisco Bay; natural size.

dalena, Baja California, the common mud-flat crab of the higher tidal reaches is *Hemigrapsus oregonensis* (Fig. 275). This species looks like a small edition of the purple crab of rocky shores (*H. nudus*) that has been bleached to a uniform light gray or muddy yellow on top and white underneath, deprived of its underside spots, and given in compensation a slight hairiness about the walking legs. Night, half tide, and an estuary like Elkhorn Slough furnish the ideal combination for seeing an impressive panoply of crab armor. Hordes of these aggressive yellow shore crabs rear up in formidable attitudes, seeming to invite combat. The gesture is largely bluff: if hard pressed they resort to the comforting philosophy that he who runs away will live to fight some other time when he is more in the mood. At night, when not blustering in temporary defiance, the mud-flat crab feeds, mainly on diatoms and green algae; it is not averse, however, to scavenging a meal of meat should the opportunity arise. *H. oregonensis* ranges southward from Alaska all the way into Baja California, and is the common shore crab of San Francisco Bay. In Puget Sound and British Columbia it is extraordinarily abundant. Hart (1935), investigating the development, found that the young are hatched from May until August at Departure Bay; a second brood may be produced by some females in August.

The purple shore crab occurs to a limited extent along the clay banks in the upper parts of sloughs, but will generally be found pretty closely associated with rocks.

Two immigrant bivalves will be found in great numbers on the mud flats of certain bays. *Ischadium demissum*, the Atlantic ribbed mussel, has found a new home to its liking in the southern part of San Francisco Bay, as well as in Los Angeles Harbor and upper Newport Bay. Colored brownish to black and sporting prominent radial ribs on the shell, this mussel continues to be the object of considerable taxonomic discussion and uncertainty, being variously assigned to the genera *Volsella*, *Modiolus*, *Geukensia*, and others. It is well adapted to long periods of exposure, and occurs on the flats in such dense stands, attached to marsh plants or rocks, that consideration has been given to harvesting these mussels for pet food, chicken meal, or even some extract for space travelers. Shorebirds, particularly clapper rails, are apparently fond of dining on this mussel at low tide, and occasionally the tables will be turned. A bird can be trapped when the shells close on its foot, and drowned by the advancing tide.

The other immigrant, this time from Japan, is *Musculus senhousia* (formerly *Volsella*), a thin-shelled, greenish mussel with zigzag patterns. Reported now from several localities, including Samish Bay (Washington), Puget Sound, Tomales Bay, San Francisco Bay, and Mission Bay, it is sometimes abundant, forming extensive mats on the mud.

Zone 3. Middle Intertidal

Here, again, it is convenient to divide this prolific zone into different subhabitats—surface and burrowing.

Surface Habitat

§286. The channeled basket snail, *Nassarius fossatus* (Fig. 276; also formerly known as *Nassa* or *Alectrion*), is the commonest of the large carnivorous snails on suitable mud flats between British Columbia and Baja California, although it is predominantly a northern species. Incidentally, it is one of the largest known species of this genus, large specimens being

Fig. 276. The channeled basket snail,
Nassarius fossatus (§286), a common large
gastropod of mud flats along much of the
Pacific coast.

nearly 5 cm long, and the clean-cut beauty of its brown shell makes it a very
noticeable animal. Ordinarily these snails will be seen plowing their way
through flats of mud and muddy sand, probably in search of food. Primarily scavengers, they have well-developed chemical senses. MacGinitie
(1935) found that they are attracted by either fresh or decaying meat, and
on one occasion a dead fish in a narrow channel was seen to attract hungry
snails from as far away as 30 meters downstream. When feeding, *N. fossatus* wraps its foot completely around the food, hiding it until it is consumed. Other investigators have found that the snail's sensory abilities include the ability to detect predators, and this species of *Nassarius*, like
others, possesses defensive responses that are at times spectacular. Contacted by a predatory seastar such as *Pisaster brevispinus*, the snail may as-
tonish the observer by catapulting into the water, using the foot to execute
one or a whole series of violent, twisting somersaults. On other occasions
the snail's response to a predator may be less vigorous. It may simply turn
abruptly away from the source of danger and flee while rocking its shell
back and forth.

    In late winter and spring in central California, many individuals may
be found depositing eggs in "shingled" capsules on the blades of eelgrass
or other firm objects on the mud flat. MacGinitie also reported (1931) on
this process. The snail first explores the surface with its sensory siphon and

then cleans a spot with its radula. Next it forms a fold in its foot to connect the genital pore with the pedal gland, and through this fold the naked egg mass is passed. The lip of the pedal gland is pressed against the cleaned surface for 9½ minutes during the process. When the eggs have entered the pedal gland, the infolded tube disappears, and the foot is used to form a water chamber, presumably for aerating the egg capsules already laid. During this 9½ minutes both the body and the shell oscillate, for some unknown reason, and after this period "the anterior part of the foot is lifted upward and backward, and, as the capsule is now cemented to the object to which it is being attached, it pulls out of the pedal gland as a completed egg capsule." The entire process has occupied 12½ minutes, and the animal is ready to move on and repeat the operation on a newly cleaned area just ahead. One typical string a little more than 6 cm long contained 45 capsules and was 9 hours in the making. The flattened capsules, averaging just over 3 mm long, are laid with the base of one overlapping the base of the next, producing the "shingled" effect mentioned.

Another considerably smaller carnivorous snail is the oyster drill *Urosalpinx cinerea* (§219; Fig. 277), a predator that was accidentally introduced from the east coast along with the Virginia oyster. It is an inoffensive-looking animal, like a small dingy-gray *Nucella*, but once it gets established it can work considerable havoc with the oyster beds. Its method of attack—the same as that employed by *Nucella* and the great moon snail—is to drill a hole through the bivalve's shell and rasp out the soft body. It also eats barnacles.

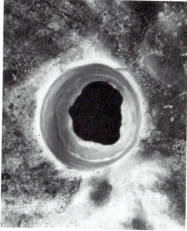

Fig. 277. *Left*, oyster drills, *Urosalpinx cinerea* (§286), feeding on an oyster. *Right*, bore hole of *Urosalpinx*; greatly enlarged.

A characteristic snail of mud flats in central and southern California is the tall-spired horn snail *Cerithidea californica* (Fig. 278), which extends to the upper reaches of such sloughs as Newport Bay, even where there is some admixture of fresh water. As might be expected, *Cerithidea* is a hardy animal, tolerant of the extreme conditions of salinity, temperature, and exposure regularly encountered in the high reaches of estuarine shores. It can, for example, readily withstand 12 days' immersion in water half the salinity of seawater. Activity and reproduction are distinctly seasonal. During the winter the snails hibernate, either on the surface or buried in the mud. From about March and extending through November, the snails are active and abundant on the surface, where they feed on diatoms and detritus coating the mud; during the summer in San Francisco Bay, as many as 1,000 snails per square meter are not unusual. From May through September, elongate egg capsules are deposited on the mud surface. The young develop directly, hatching as juvenile snails beginning in June. Reproductive maturity, at a length of about 22 to 24 mm., is reached after 2 or more years.

In San Francisco Bay *Cerithidea* has become restricted to only part of its normal habitat range as a result of the accidental introduction of an ecologically similar snail, *Ilyanassa obsoleta* (§219), from the east coast. According to work by Race (1981, 1982), *Cerithidea* normally inhabits several habitats within an estuary: marsh pans (small tide pools in the mud), tidal creeks, and mud flats. But in San Francisco Bay, *Ilyanassa* now claims the creeks and mud flats, and has restricted *Cerithidea*, which actively avoids the invader, to the marsh pans for most of the year. *Ilyanassa* reinforces its dominant position by eating the eggs and juveniles of the native snail. The range of *Cerithidea* is Tomales Bay to central Baja California.

From Elkhorn Slough northward to British Columbia, probably wherever Pacific oysters have been imported, will be found a tall-spired snail superficially similar in appearance to *Cerithidea* and even more similar in its ecological requirements. This is *Batillaria attramentaria* (Fig. 279; sometimes misidentified as *B. zonalis*), a Japanese import first noticed in the 1930's. Thus, in central California, our native *Cerithidea* must contend with a second introduced competitor. *Batillaria*, which is easily distinguished because it lacks the weltlike ribs of *Cerithidea* and because its aperture tends to form a short canal at the base, is doing quite well. It now occurs in dense aggregations at the mouth of Walker Creek near Millerton in Tomales Bay, and in Elkhorn Slough.

§287. In the south, a large opisthobranch, the bubble-shell snail *Bulla gouldiana* (Fig. 280), with a shell some 4 to 5 cm long, is one of the most common of all gastropods in the Newport Bay and Mission Bay region. The paper-thin shell is mottled brown, and the body, which is too large for the shell, is yellow. When the animal is completely extended and crawling about in the mud, the mantle covers most of the shell. The diet of *B. gouldiana* is uncertain; the snail is probably herbivorous, although one observer

Fig. 278. The tall-spired horn snail *Cerithidea californica* (§286); twice natural size.

Fig. 279. *Batillaria attramentaria* (§286), an introduced snail similar in ecological requirements to the native *Cerithidia californica.*

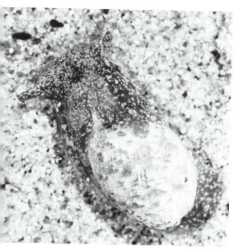

Fig. 280. The bubble snail *Bulla gouldiana* (§287), a shelled opisthobranch common on southern mud flats.

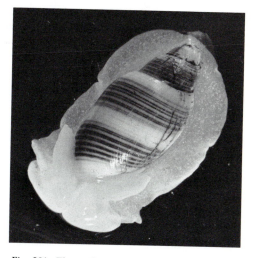

Fig. 281. The strikingly banded opisthobranch *Rictaxis punctocaelatus* (§287), a barrel-shell snail.

says that some bubble-shell snails eat bivalves and smaller snails, swallowing them whole. The long, yellow strings of eggs, lying about on the mud or tangled in eelgrass, are familiar sights in the summer.

The barrel-shell snail *Rictaxis punctocaelatus* (Fig. 281; formerly *Acteon*) is sporadically common in central California bays. The animal is less than 2 cm long but is so striking that it is sure to be picked out of its surroundings, for narrow black bands follow the whorls around its white shell. Sometimes its eggs, deposited in a gelatinous bag attached to the sediment by a slender thread, indicate the animal's presence when no adults can be found; the bulk of the population is subtidal. The recorded range is Alaska to Baja California.

Fig. 282. The large mud-flat hydroid *Cory-morpha palma* (§288); natural size.

§288. Mud flats would seem to be the most unlikely localities of all to harbor hydroids, but the tubularian *Corymorpha palma* (Fig. 282) is famous in the Newport–San Diego Bay area. Specimens attaining the very great size (for a shore hydroid) of 14 cm tall have been taken, and in some areas whole forests of these delicate fairy palms grow on the flats. The solitary individuals fasten themselves to the substratum by ramifying roots that come away as a chunk of mud. The animals are watery and transparent, and therefore will often be overlooked if the light is poor. When the ebbing tide leaves them without support, they collapse on the mud, there to lie exposed to sun and drying winds until the tide floods again. It is remarkable that so delicate an animal can survive this exposure, but it apparently suffers no harm.

Around the single flowerlike "head" of the animal (the hydranth) develop the tiny jellyfish, which, although they pulsate like free jellyfish in the act of swimming, never succeed in forcing themselves from the parent hydroid. Upon reaching maturity, sperm are shed into the water, but eggs are retained. Eventually the progeny emerge, and may be transported some distance by currents, but wherever they touch they settle and begin to grow as hydroids; there is no free-swimming stage whatever.

Corymorpha is one of that minority of Pacific coast intertidal animals that have been closely observed and studied. In aquaria it remains erect with spread tentacles so long as a current of water is maintained, but begins a slow and regular bowing movement as soon as the current is stopped. During each bow it sweeps the mud with its tentacles; then, as it straightens up again, it rolls up its long tentacles until its short tentacles can scrape the food from them and pass it to the mouth. The animal is well known for its remarkable powers of regeneration. If the entire hydranth is cut off, a new one will grow within a few days, and sections cut out of the "stem" will grow complete new individuals.

Like other hydroids, *Corymorpha* has its associated sea spiders. Near Balboa we once found, in August, a bank of muddy sand that supported a *Corymorpha* bed about which were strewn large numbers of a straw-colored and quite visible gangly pycnogonid, *Anoplodactylus erectus*, which occurs also with southern *Tubularia* (§333). Presumably it was a seasonal occurrence.* The sea spiders had probably just grown to an independent stage and released themselves from their hydroid nurses—quite likely by eating their way out.

Burrowing Habitat

§289. One of the noticeable substratum animals in the Newport Bay flats of sandy mud is the sea pen, or pennatulid, *Stylatula elongata* (Fig. 283), but whether or not it should be considered a burrowing animal will depend on the observer's reaction—which, in turn, will depend on the state of the tide. The visitor who appears on the scene between half tide and low tide will be willing to swear before all the courts in the land that he is digging out a thoroughly buried animal. But let him row over the spot at flood tide, and he will see in the shallow water beneath his boat a pleasant meadow of waving green sea pens, like a field of young wheat. The explanation comes when he reaches down with an oar in an attempt to unearth some of the "plants." They snap down into the ground instantly, leaving nothing visible but the short, spiky tips of their stalks. Like the anemone *Harenactis* (§257), they have a bulbous anchor that is permanently buried deep in the bottom. When molested, or when the tide leaves them exposed, they retract the polyps that cover the stalk and pull themselves completely beneath the surface. When cool water again flows over the region, they expand slowly until each pennatulid looks like a narrow, green feather waving in the current.

Pennatulids as a group are notably phosphorescent, and our local species are no exception. Both *Stylatula elongata* and *Acanthoptilum gracile* (Fig. 284), the second common in Tomales Bay, exhibit startling flashes of light when adapted for about half an hour in a dark room and touched with a needle or stimulated with an electric current.

§290. Even in this middle intertidal zone there are many burrowing animals besides the spectacular *Stylatula*. Another relative of the hydroids, jellyfish, and corals is the burrowing anemone *Edwardsiella californica* (Fig. 285), which also occurs in great profusion. Early in March 1931 we found a sandy mudbank east of Corona del Mar that must have averaged more than 50 of these animals in a tenth of a square meter. It goes without saying that they are not large, a good-sized extended specimen being possibly 5 cm long. The wormlike body, which normally protrudes slightly above the surface, is almost covered with a brown and wrinkled tube. When first dug up, the animal looks very much like a small peanut worm (sipunculid). In an aquarium the disk flowers out, during the afternoon, with symmetrically arranged tentacles, and the attachment bulb expands and becomes transparent, so that the partitions are visible as eight longitudinal bands. This species of *Edwardsiella* is known from southern California only, but a similar or identical form occurs (or occurred) rarely at Elkhorn Slough.

*By 1967 these pycnogonids were rare in the area, presumably because of the changes associated with "progress."

Fig. 283. The sea pen
Stylatula elongata (§289),
with its bulbous anchor
buried in the substratum
and its polyp-covered
stalk extended into the
water; up to 60 cm tall.

Fig. 284. The sea pen *Acanthoptilum gracile* (§289).

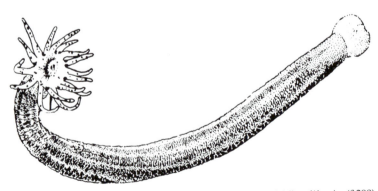

Fig. 285. A burrowing anemone of the mud flats, *Edwardsiella californica* (§290);
twice natural size.

§291. During some years the southern California muddy sand contains a considerable population of the stalked brachiopod *Glottidia albida* (Fig. 286), the second and last of its group to concern us. This small bivalved animal (which, although it resembles a thin, stalked clam, is not even distantly related to the mollusks) is almost a facsimile of the large tropical *Lingula*, the oldest living genus in the world. Remains of *Lingula* date back to the Ordovician period, about 500 million years ago (although living species of *Lingula* originated much more recently). It certainly speaks much for the tolerance and adaptability of an animal that it has lived from such a remote geological age to the present time and still persists in such numbers that races bordering the Indian Ocean use it for food, selling it in public markets as we sell clams.

Our Pacific coast specimens of *Glottidia* live in burrows, retracting at low tide so that only a slitted opening can be seen at the surface. The white, horny shells are rarely more than 2 cm long. The fleshy peduncle with which the animal burrows and anchors itself may be two to three times the length of the shell. The animal is not helpless when deprived of its peduncle, however; one specimen so handicapped was able to bury itself headfirst, presumably with the aid of the bristles that protrude from the mantle edge. When covered with water and undisturbed, the animal projects its body halfway out of its burrow, and at such times the bristles are used to direct the currents of water that carry food and oxygen into the body.

Glottidia is more abundant as a subtidal species in southern California waters, although there was once an intertidal bed of them near the laboratory at Corona del Mar. One bright morning years ago a group of conchologists raked them all up, every one, and they did not return; the memory of this infamous excursion used to turn the laboratory director George MacGinitie purple for years afterward. An occasional stray *Glottidia* has been collected in Tomales Bay, the northernmost extent of its range.

Fig. 286. The stalked brachiopod *Glottidia albida* (§291), a burrow-dwelling and primarily subtidal species.

Fig. 287. *Callianassa californiensis* (§292), the red ghost shrimp, inhabiting the middle zone of mud flats from Alaska to Baja California; about maximum size.

§292. One of the burrowing ghost shrimps, *Callianassa californiensis* (Fig. 287), will be found in this zone, often in sediments of mixed sand and mud. Next to the fiddler crab and the somewhat similar *Malacoplax*, this pink-and-white, soft-bodied ghost shrimp seems to be the most shoreward of the burrowing forms, with an optimum tide zone of 0 to +1 foot (30 cm). It occurs, too, over a wide geographical range, reaching southeastern Alaska to the north. We have seen it in the Strait of Georgia, Puget Sound, Tomales Bay, Elkhorn Slough, Newport Bay, and El Estero de Punta Banda.

Callianassa is neatly adapted to its life in burrows, but it will survive in aquaria without the reassurance of walls around it. Earlier reports that it would not survive outside the burrow may have been based on injured specimens. According to MacGinitie (1934), the animal digs its burrow with the claws of the first and second legs, which draw the sandy mud backward, collecting it in a receptacle formed by another pair of legs. When enough material has been collected to make a load, the shrimp backs out of its burrow and deposits the load outside. All of the legs are specialized, some being used for walking, others for bracing the animal against the sides of the burrow, and still others for personal cleansing operations.

The impermanent burrows are much branched and complicated, with many enlarged places for turning around. The work of adding new tunnels or extending and maintaining the old ones is unending. When the shrimps do pause in their digging operations, they devote the time to cleaning themselves; seemingly they are obsessed with the Puritan philosophy of work. This incessant digging overturns the sediment of the flats in a manner suggesting the labors of earthworms; MacGinitie estimated that a bed of *Callianassa* turns over the sediment to a depth of 50 cm in 6 months. This digging also makes the ghost shrimps undesirable pests on oyster beds in Washington and Oregon.

Callianassa feeds by ingesting organic detritus sifted from the mud, as well as by digesting microorganisms from the continuous stream of mud that passes through its digestive tract. Recent evidence suggests that they also gain nutrients in the form of plankton and detritus brought in with the respiratory current of water moved through the burrow by the pleopods. They are hardy animals, capable of surviving without oxygen for 6 days if necessary and tolerating salinities ranging from 30 to 125 percent that of normal seawater. Judging from their age grouping, these shrimps are surprisingly long-lived, perhaps reaching an age of 15 or 16 years. In central California, breeding is continuous throughout the year, with an optimum season during June and July.

As happens with many other animals that make permanent or semi-permanent burrows, the ghost shrimp takes in boarders, which are likely in this case to be a scale worm, *Hesperonoe complanata*, and the same pea crab, *Scleroplax granulata*, that occurs with the remarkable innkeeper *Urechis* (§306). Five other commensals are known to occur, including the copepod *Hemicyclops thysanotus* and, in Baja California, *Betaeus ensenadensis*, both on the gills. But the most common commensal is the red copepod *Clausidium vancouverense*.

§293. It is almost axiomatic to remark that there are the usual hosts of worms here. The slim and very long *Notomastus tenuis*, looking like a piece of frayed red wrapping twine, elastic, and up to 30 cm long, may be present in nearly every spadeful of mud, sometimes occurring so thickly as to bind the soil together; but it is so common and so ready to break that one soon loses interest in it. In the south will be found a long, white, wrinkled sipunculid worm with a threadlike "neck." This species, *Golfingia hespera*, sometimes occurs with the burrowing anemone *Edwardsiella* or the brachiopod *Glottidia*. Exceedingly common at Elkhorn Slough, as well as in Tomales Bay and Bodega Harbor, is *Lumbrineris zonata*—highly iridescent, slim, and long (MacGinitie 1935 recorded a specimen with 604 segments). Its full range extends at least from Alaska to Mexico.

Sandy tubes projecting a little above the surface in areas of hard-packed muddy sand are likely to be occupied by the fragile "jointworm," or bamboo worm, *Axiothella rubrocincta* (Fig. 288), whose segmented body is banded in dull ruby red. The U-shaped tubes may extend 30 cm below the surface and are filled with seawater during periods of intertidal exposure. The anal end of this deposit-feeding worm bears a funnel-shaped rosette, and the head end, which is slightly enlarged, embodies a plug for stopping up the tube when the animal is retracted. *Axiothella* occurs from British Columbia to Mexico, and has been taken in Puget Sound, in Tomales Bay, and at San Pedro; it occurs intertidally in central California.

Occurring in the *Axiothella* tubes is the most specialized of all commensal crabs, the pea crab *Pinnixa longipes* (Fig. 288), which, in adapting itself to moving sideways in the narrow tube, has become several times as wide as it is long and has achieved tremendous development of the third pair of walking legs. Seeing a specimen outside a tube, one can scarcely

Fig. 288. The jointworm, or bamboo worm, *Axiothella rubrocincta* (§293), about 4 times natural size, with its tube and commensal pea crab *Pinnixa longipes*. The worm's funnel-shaped anal rosette is at the upper left in the photograph.

believe that it could possibly insinuate itself through the small opening. The passage takes some pains, but the tiny crab manages it, inserting one huge leg first and then carefully edging itself in sideways. The range of *P. longipes*, which may also be found occasionally with the polychaetes *Cistenides* (§281) and *Pista elongata*, is from Bodega Harbor to Ensenada; it has been erroneously reported as occurring on the east coast of North America.

Two other burrowing worms will be taken frequently in this habitat. The first is *Cirriformia spirabrancha*, with its coiled tentacles writhing on the mud of such places as Elkhorn Slough and Tomales Bay (recorded range, central California to Bahía de San Quintín). This species at first will likely be taken for a cluster of round worms; but a little serious work with the shovel will turn out the pale, yellowish or greenish cirratulid, which lies hidden safely beneath the surface while its active tentacles keep contact with the outside world of aerated water. Similar in appearance to *C. luxuriosa* (§16), *C. spirabrancha* may be distinguished by its habitat and by its dark-green tentacles (as opposed to orange-red in *C. luxuriosa*) and numerous yellow setae (rather than a few black ones). Related to its occurrence in bays and estuaries, *C. spirabrancha* tolerates both brackish and hypersaline conditions. The reproductive season is from March to May in southern California and from June to July in Tomales Bay.

The second burrowing worm taken frequently in this habitat is the cosmopolitan tubed worm *Capitella capitata*, which forms great beds, acres in extent, in such areas as the muddy eastern shores of San Francisco Bay. The worm, several centimeters long, red and earthworm-like, lies head up in vertical, dirt-encrusted, black membranous tubes. The worms make their living ingesting the substratum, rich in organic debris, and thrive under conditions of pollution that few other animals can tolerate. As a consequence, they have been used in many pollution studies.

Zone 4. Low Intertidal

Here there are many channels and holes, which form pools when the tide is out. Therefore, besides surface and burrowing animals we will meet several that are characteristically free-swimming.

Surface Habitat

§294. Many snails crawl about on the mud flats in this low zone. The largest are slugs, not popularly recognized as true snails but actually belonging with this group of animals. *Navanax inermis* (Fig. 289), an animal closely related to the sea hares, is large (up to 12 cm) and strikingly colored with many yellow dots or lines and a few blue ones on a brown back-

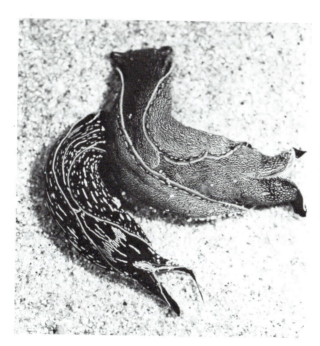

Fig. 289. *Navanax inermis* (§294), a large opistho-branch of mud flats in the low zone; these two show the striking and variable markings, which may be yellow and blue on a tan to nearly black background.

ground. It is common in southern California and northern Mexico, and has been taken frequently at Morro Bay and occasionally at Elkhorn Slough. We find it commonly on bare mud, but almost as often in association with eelgrass.

Navanax is a voracious feeder on other opisthobranchs, which it tracks down by following their mucous trails. Overtaken prey are engulfed whole by a sucking action of the pharynx. In bays it feeds mostly on the shelled opisthobranchs *Bulla* and *Haminoea*, and on species of *Aglaja*. When on rocky shores, it will eat all sorts of nudibranchs at the rate of three to five a day, including *Hermissenda* and others presumably loaded with nematocysts from a coelenterate diet. *Navanax* will even eat small fish, and often consumes the young of its own species.

Like all opisthobranchs, *Navanax* has never heard of birth control. In the La Jolla area it produces eggs the year around; large egg masses may contain 800,000 eggs. These are laid in light-yellow, stringy coils, woven together in pleasing designs.

The sea hare *Aplysia* (§104) is a frequent visitor to the mud flats. At Elkhorn Slough huge specimens weighing 5 kilograms (10 pounds) or more (most of this is water) may frequently be seen when none are to be found on the rocky shore outside.

§295. *Hermissenda crassicornis* (see also §109), a yellow-green nudi-branch about 25 cm long with brown or orange processes along the back and a well-defined pattern of blue lines along the sides, is often common in bays, but it is also found in rocky tide pools. It ranges from Alaska to Baja California, but like all opisthobranchs, *Hermissenda* is of uncertain occurrence in any given part of its range. This nudibranch's varied diet includes hydroids, phoronids (§297), annelids, other *Hermissenda*, and most any dead animal. Its egg masses, resembling a coil of tiny, pink sausage links, will be found attached to blades of eelgrass and algae.

§296. The heavy-shelled bivalve *Chione* (§260) may be taken on flats of sandy mud, often, strangely, at the surface. There are also many kinds of shelled snails: a tropical brown cowry (*Cypraea spadicea*, §136), which ranges as far north as Point Conception (rarely to Monterey), and a cone snail (*Conus californicus*, §136) are fairly common. The second packs about more than its allotment of the dark-brown slipper snail *Crepidula onyx* (Fig. 290)—a combination that extends into the upper reaches of the estuary at Newport; the slipper snail is also to be found on rocks and piling, often abundantly. Ovipositing specimens of the olive snails (§253) have been noted in August at Elkhorn Slough. The visitor at Newport may also run across the flared *Pteropurpura trialata* (Fig. 291; formerly *Pterynotus trialatus*), prize of the conchologists in that region. The largest of the moon snails (*Polinices lewisii*, §252) is almost as characteristic an inhabitant of the mud flats as it is of the sand flats, and in San Francisco Bay, another large preda-tory snail will be found. This is the channeled whelk, *Busycotypus cana-liculatus* (formerly *Busycon canaliculatum*), which was introduced from the

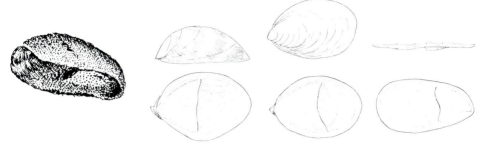

Fig. 290. Slipper shells. On the extreme left is *Crepidula onyx* (§296), the southern California slipper snail, shown about twice natural size. The others, left to right, are *C. fornicata*, *C. convexa*, and *C. plana*, all east-coast species that have been reported from northern bays on this coast.

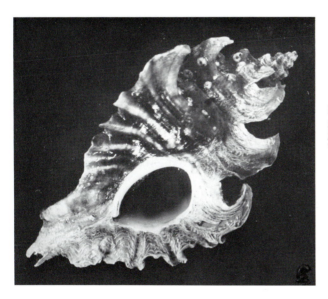

Fig. 291. An especially gaudy murex, *Pteropurpura trialata* (§296); somewhat enlarged.

Atlantic coast and is now common subtidally; specimens are sometimes found at low tide, and strings of egg capsules half a meter long or so may be washed ashore.

§297. The most noticeable animal of all, but one that is by no means obtrusively common, is the magnificently orange-plumed *Phoronopsis californica*, which has been seen at Newport, its gelatinous body protected by a tube that is buried in the mud. This anomalous "worm" extends down into the substratum, retracts immediately at the least sign of trouble, and is difficult to dig. Although occurring solitarily, it is a near relative of the green-plumed *Phoronopsis viridis*, which sometimes grows in great beds at Elk-

scales at 4½ kilograms. Growth to such size is slow, and the animal may live 15 years or more. The shells are relatively lightweight and not even large enough to contain the animal's portly body, let alone the siphons; even in living specimens the valves always gape open 2 cm or so.

The recorded range is Alaska to Scammon Lagoon, Baja California. Although common in the northern part of its range, particularly in Puget Sound, it is uncommon in California; specimens are occasionally taken, however, in Morro Bay, Elkhorn Slough, and other quiet-water localities in the south. In British Columbia, the geoduck spawns in April and May. Little else is known of its biology, and anyone who has ever tried to dig one will understand why the animal is known so slightly and is used so infrequently for food.

§300. Another large clam that resembles the geoduck in appearance, habits, and size and is often mistaken for it is variously known as the gaper, the summer clam, the rubberneck clam, the big-neck clam, the horse clam, the otter-shell clam, and the great Washington clam. After this appalling array of popular names it is almost restful to call the animal *Tresus nuttallii* (Fig. 294; once familiarly known as *Schizothaerus*!). The individual gaper-etcetera is readily located by its siphon hole in the mud—"squirt hole" in the vernacular—from which, at fairly regular intervals, it shoots jets of water a meter or so into the air. These jets are particularly powerful when the clam is disturbed, and sometimes hit with deadly accuracy; but at any time, a succession of the largest geysers produced by the clams betrays the presence of a bed. Digging them out, however, is no small job, for they lie from ½ to 1 meter below the surface. Their average size is smaller than the geoduck's, but the shell may be 20 cm long, and the clam may weigh 1.8 kilograms (4 pounds). As with the geoduck, the easily broken shell is incapable of closing tightly over the large body, and the huge siphon must shift for itself. The siphon is protected by a tough, brown skin and by two horny valves at the tip. Nevertheless, the siphon, or part of it, must often be sacrificed, for the tips are commonly found in the stomachs of halibut and bottom-feeding rays.

Spawning takes place in central California mainly from February to April, when water temperatures are lowest, but some spawning probably occurs throughout the year. The new recruits reach 50 mm in length by the end of their first year, and 70 mm after 2 years, achieving sexual maturity at about this time. Young clams burrow actively, a trait necessary for success in the unstable upper layers of sediment. But as *Tresus* ages and occupies deeper, more stable layers of substratum, there is a progressive loss of burrowing ability; clams over 60 mm long burrow very slowly. Specimens are often heavily infested with larval tapeworms, the adults of which are found in the bat stingray; they are harmless to man. The range of *T. nuttallii* is British Columbia to Scammon Lagoon, Baja California.

A second species of *Tresus*, *T. capax*, is common in northern waters, extending to Kodiak Island, Alaska, but uncommon in central California,

Fig. 294. The gaper, *Tresus nuttallii* (§300), another large deep-burrowing clam whose shell is incapable of completely enclosing the body and siphons.

the southern limit of its range. The two species are similar in many respects, and occur together in many protected locations, such as in Puget Sound. They are perhaps most readily distinguished by the siphonal valves, which are much heavier and harder in *T. nuttallii*, and by the shape of the shell, which is proportionally longer in *T. nuttallii*. Spawning by *T. nuttallii* seems to occur primarily during the winter throughout its geographic range. The large pea crab *Pinnixa faba* is quite likely to be present in the mantle cavity of this form, or occasionally, the similar *P. littoralis* will be taken;* only one species of crab, however, will be present on any one clam.

Of the big clams, the two species of *Tresus* are the least used for food (although they are popular locally), partly because their gaping shell cannot retain moisture and therefore makes them unfit for shipment. Appearances are against them, too; the mud-flaked brown neck is unappetizing to look at. It has been said, however, that the Indians used to dry the siphons for winter use, and that at Morro Bay the siphons of *T. nuttallii* are skinned, quartered, and fried, although the bodies are discarded. As a matter of fact, the bodies are perfectly edible and, after being run through a meat grinder, make a not unsatisfactory chowder. Both species are protected by California law, ten per person being the limit for one day.

Although *Tresus* decidedly prefers quiet bays, small, rather battered-looking specimens may also be found on the outside coast. Specimens taken "outside," or even from those inlets directly adjacent to oceanic

---

*Previous editions of this book listed *T. nuttallii* as the host of these pinnotherid crabs. According to Pearce (1965), however, the crabs occur almost exclusively on *T. capax* (not on *T. nuttallii*) when the two clams co-occur, and may occur on *T. nuttallii* only to the south, below the range of *T. capax*.

Fig. 295. *Saxidomus nuttalli* (§301), the Washington clam of bays in California; similar to the smaller *S. giganteus* of northern waters.

waters, ought never to be used for food during the summer because of the danger of mussel poisoning (§158). Predators on *Tresus* besides man include the moon snail (*Polinices lewisii*), crabs, and the seastar *Pisaster brevispinus*.

§301. *Saxidomus nuttalli* (Fig. 295), the Washington clam, butter clam, or money shell (this last because the California Indians used the shells for money), ranges from Humboldt Bay to northern Baja California. Clam diggers, however, do not distinguish between this clam and the very similar but smaller *S. giganteus*, which is the most abundant clam on suitable beaches in Alaska, British Columbia, and Puget Sound and occurs, but not commonly, as far south as Monterey. The same popular names are used for both. The shells of *S. nuttalli* sometimes rival those of the geoduck in size, but are characteristically only 8 to 12 cm long, whereas *S. giganteus*—again proving that there is not much in even scientific names—averages about 8 cm. Both appear in the markets, but not commonly except in the Puget Sound region, where the smaller form is used rather extensively. Two pea crabs, the small *Pinnixa littoralis* and *Fabia subquadrata*, may be found occasionally in the mantle cavity; but the Washington clam is, on the whole, remarkably free from parasites.

In a study of the distribution and breeding of *S. giganteus*, Fraser and Smith (1928) examined 2,600 British Columbia specimens. Tidal currents are an important factor in distribution because of the animals' pelagic larvae. With the exception of clams in their seventh year, the result of a particularly prolific season, the age distribution was fairly regular from the fourth year to the tenth. About half of the clams spawn at the end of the third year. At Departure Bay, spawning took place in August, but at other

beaches along the inner coast of Vancouver Island there was much variation. The larvae appeared as bivalved veligers in 2 weeks, and at the end of
another 4 weeks, when still less than 5 mm long, they settled down on the
gravel. Once established, the animal may live for 20 years or more. It is reported that mussel poisoning (§158) may be present during the summer to
a dangerous degree in Washington clams occurring (as they do rarely) on
the open coast or in inlets directly adjacent to oceanic waters.

§302. The eastern soft-shelled clam *Mya arenaria* (Fig. 296) was first
noticed on the Pacific coast in 1874, when Henry Hemphill collected it in
San Francisco Bay. There is no record of any deliberate introduction, but
apparently the first oysters from the East were planted in San Francisco Bay
in 1869, and the clam may have been introduced at that time. Whether introduced or not, *Mya* is now one of the common bay clams of the Pacific
coast from southern Alaska to Elkhorn Slough and is important commercially, especially in northern bays.

Mya is egg-shaped in outline, averages less than 13 cm in length, and
is characterized by a large, spoon-shaped internal projection on the left
valve at the hinge. Only one other species (*Platyodon*, §147) has this projection. The shell is light and brittle, and the siphons dark. The adults of the
species are incapable of maintaining themselves in shifting substratum,
having lost all power of digging, and hence require complete protection.
They thrive in brackish water, however, if it does not become stagnant, and
can stand temperatures below freezing. Their life span, at least in Massachusetts, is 10 to 12 years. Along our northern coast *Mya* is likely to carry
the same pea crab as the gapers.

In common with several other clams that can live in foul estuary mud,
Mya exhibits, but to a striking degree, a facility for anaerobic respiration—
the ability to live in a medium absolutely lacking in free oxygen. Experimentally, these clams have been known to live in an oxygen-free atmosphere for 8 days. They produced carbon dioxide continuously but showed
no subsequent effects beyond a decrease in stored glycogen (an animal car-
bohydrate) and a considerable increase in the metabolic rate after being replaced in a normal environment. During periods of low oxygen the shell
serves as an alkaline reserve to neutralize the lactic acid produced by anaerobic metabolism.

§303. The bent-nosed clam, *Macoma nasuta* (Fig. 297), is a small,
hardy species (shell usually to 6 cm long, occasionally to 11 cm) that may be
turned out of almost every possible mud flat between Kodiak and the
southern tip of Baja California. At Elkhorn Slough it is the commonest
clam. It can stand water so stale that all other species will be killed; hence, it
is often the only clam to be found in small lagoons that have only occasional
communication with the sea. Also, it can live in softer mud than any other
species. At rest the clam lies on its left side at a depth of 10 to 20 cm with
the bend in its shell turned upward, following the upward curve of the separate yellow siphons. When burrowing it goes in at an angle, sawing back

Fig. 300. The echiuroid worm *Urechis caupo* (§306), the fat innkeeper. *Above*, the living animal, fairly relaxed. *Below*, with guests, in its burrow: *a*, two drawings of the innkeeper, showing the variety of shapes it may assume; *b*, a permanent guest, the scale worm *Hesperonoe adventor*, ½ natural size; *c*, the pea crab *Scleroplax granulata*, almost 1½ times natural size; *d*, a transient, the goby *Clevelandia ios*, ⅝ natural size; *e*, slime net used for feeding.

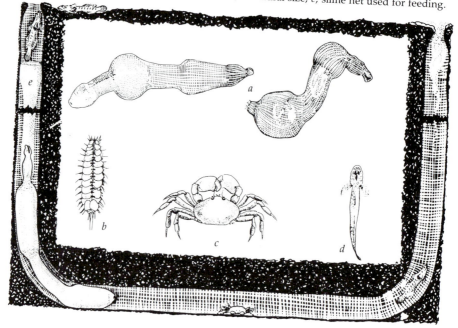

striction of the burrow entrance becomes useful, for it momentarily accelerates the velocity of the issuing water, thus enabling *Urechis* to expel rather large chunks of debris. Pieces too large to be so forced out are buried.

The animal's most remarkable trait is the spinning of a slime net, with which it captures its microscopic food. The net permits the passage of water and yet is so fine that particles less than one micron in diameter (one-millionth of a meter) are caught. The openings are invisible even under a light microscope. In preparing to feed, the innkeeper moves up the vertical part of the burrow at its head end and attaches the beginning of the net to the burrow walls close to the entrance. It then moves downward, spinning a net that may be from 5 to 20 cm long. At first the net is transparent, but as it collects detritus it turns gray and becomes visible. For about an hour *Urechis* lies at the bend in its burrow, pumping water through the net and increasing the force of its pumping activities as the net becomes clogged. When sufficient food has been collected, the animal moves up its burrow, swallowing net and all as it goes. The net is now digested along with its contained detritus. The innkeeper, a fastidious eater, discards all large particles as the net is swallowed.

It is scarcely to be expected that so ready a food supply would go to waste, and it does not. The innkeeper's three guests stand ready to grab all particles the instant they are discarded (Fig. 300). The most dependent of these is an annelid, the beautiful, reddish scale worm *Hesperonoe adventor*, from 1 to 5 cm long, which remains almost continuously in contact with the body of its host. To quote Fisher and MacGinitie: "It moves from place to place with its host, making little runs between peristaltic waves, and turns end for end when *Urechis* does. After *Urechis* spins its mucus-tube *Hesperonoe* may crawl forward and lie with its palps almost touching the proboscis. As soon as *Urechis* starts to devour the tube *Hesperonoe* also sets to, making absurd little attacks on the yielding material with its eversible pharynx." Only one *Hesperonoe* at a time is found in a burrow, for the polychaete confronts and forcibly drives others of the same species out of the inn.

The second guest, of which there are sometimes a pair, is the pea crab *Scleroplax granulata*, usually not more than 10 mm across the carapace.* The third guest is a goby, *Clevelandia ios*, and there may be from one to five specimens in a burrow. This little fish is a transient guest, foraging outside much of the time and using the innkeeper's burrow chiefly for shelter. *Urechis* derives no apparent benefit from any of these commensals. On the contrary, "both crab and annelid interfere with the regular activities of *Urechis*, especially its feeding and cleaning reactions. A particle of clam dropped into the slime-net is immediately sensed by both commensals. Their attempts to reach it cause *Urechis* prematurely to swallow the tube when the clam morsel is stripped out the open end. It is usually snapped up by *Hesperonoe* and swallowed if small enough; otherwise *Scleroplax* will snatch it away, when the annelid must be content with what remains after the crab's appetite is satisfied. . . . Enmity exists between crab and annelid in which the latter is the under-dog. This feud may account for the close association of *Hesperonoe* with *Urechis*." A more congenial relationship exists between the crab and the goby, for the goby has actually been observed to carry a piece of clam meat that was too large for it to swallow or tear

*Another pea crab, *Pinnixa franciscana* (§232), occurs rarely, and *P. schmitti*, which is the commonest *Echiurus* commensal and occurs rarely with *Upogebia* and *Callianassa* as well, has also been taken here.

apart to the crab and stand by to snatch bits as the crab tore the meat apart. Even the crab and the scale worm forget their differences, however, when danger threatens, for at such times both rush to their host and remain in contact with its body wall until the danger is past.

In *Urechis* the sexes are separate. MacGinitie (1935) has shown that in its breeding habits (the eggs and sperm are discharged into the water) *Urechis* exhibits a specialization as remarkable as that of its feeding. The eggs or sperm mature while floating freely in the coelomic fluid, where very young, immature, and fully ripe sexual products are mixed together indiscriminately. A remarkable set of spiral collecting organs—modified nephridia—pick up only the ripe cells and reject all other contents of the body cavity, and provide an instant means of disposal for sexual products as soon as ripe. After being removed from the body, these collecting organs function outside as autonomous organs. But even in cultures of blood or of seawater, the excised collectors refuse to pick up eggs other than those of *Urechis* (which have a characteristic indentation through which the organs operate), although the foreign eggs are identical in size.

Although embryologists go to great trouble to provide the proper conditions for fertilizing gametes and raising embryos, it would appear that nature is not so exacting. Some years ago, while preparing a motion picture, the late Alden Noble worked assiduously to fertilize *Urechis* eggs and have trochophore larvae ready for the photographer at the right moment, barely getting enough for the picture. After the photographer departed, Noble discovered that a 30-cm bowl left on the table in the sun was almost solid with trochophore larvae of *Urechis*. This had been the jar into which he had discarded all the slop and failures from his previous efforts!

At Newport Bay, *Urechis* breeds during the winter and has usually spawned out before the end of the summer. The same is true at Bodega Harbor, where individuals seem to spawn more than once during a season. At Elkhorn Slough, however, specimens are normally ripe throughout the year.

According to recent work by Suer (1982), *Urechis* larvae in the laboratory require about 60 days to develop to a stage at which they are "competent" to settle from the plankton and begin life in the sediment. She has also discovered that competent larvae can delay settlement, at least by 1 to 3 weeks, until they encounter a suitable stimulus, the most effective trigger being sediment from around the burrow of an established *Urechis*. The active ingredient is a chemical adsorbed to the grains of sediment.

There must be the usual tremendous mortality among *Urechis* larvae, but once the animals become established in burrows they are relatively safe. Growth is seasonal, at least in Bodega Harbor, and fastest during the summer. At the maximum rates observed by Suer in the field, *Urechis* could grow to a large size in 3 to 4 years. Interestingly, large *Urechis* may actually shrink, as is the case for some other invertebrates as well. Because of this and the highly variable growth rate of the animal, size in *Urechis* is not a good indicator of age.

Predators on subtidal populations of *Urechis* (and intertidal populations at high tide) include several bottom-feeding fish such as flounders, turbots, and leopard sharks. Also, certain rays (e.g., the bat ray) eat echiuroids and other burrowing animals. Apparently, they are able to pop potential prey out of a burrow by using their broad, flattened bodies as a sort of plumber's friend. The principal predator on intertidal populations of *Urechis*, however, is probably man, and in some areas the worm has been severely depleted. The worms are used by the hundreds for bait by fishermen, as objects for study by zoology students, and as sources of gametes by researchers investigating the processes of fertilization and development.

Urechis is known chiefly from Elkhorn Slough, but has been found also in Newport, Morro, San Francisco, Tomales, Bodega, and Humboldt bays, as well as in Tijuana Slough. A related, smaller form, *Echiuris echiuris alaskensis*, with its pea crab *Pinnixa schmitti*, is exceedingly abundant in southeastern Alaska, occurring embedded in gravelly substratum where there is a clay admixture, as well as in the mud flats proper. The Alaskan form lacks the food net of *Urechis*, but the scoop-shovel proboscis achieves results just as effectively, if the abundance of individuals is an index of successful food-getting.

§307. Two species of echiuroid worms besides *Urechis* may occur intertidally along the California coast: the small green *Listriolobus pelodes* (in at least Tomales Bay and Newport Bay) and the larger, also greenish *Ochetostoma octomyotum*. According to recent studies by Pilger (1980), *Listriolobus* forms a U-shaped burrow in mud and uses its proboscis to feed on the layer of sediment deposited around the burrow openings. In addition to the isolated intertidal populations, this echiuran often occurs abundantly at depths of 18 to 155 meters from northern California to Baja. Subtidally off Palos Verdes, *Listriolobus* spawns in the winter and spring, and an influx of juveniles can be detected in the late winter and spring. The newly settled juveniles apparently reach sexual maturity in this location within a year. *Ochetostoma* is similar in habit to *Urechis*, but it has a longer proboscis and feeds by sweeping detritus from the walls of its burrow onto the mucus-covered surface of this appendage.

§308. The striking worm *Sipunculus nudus* (Fig. 301), cosmopolitan on warm shores, is common at Newport Bay and has been taken in Mission Bay and at Ensenada. The white skin is shining and iridescent, and shows the muscles in small rectangular patches. Like the "sweet potato" (*Caudina*, §264), this animal passes great quantities of the substratum through its intestines and extracts the contained nourishment. More than half the weight of a test *Sipunculus* at Naples consisted of sand in the intestine. It has often been remarked that such animals play a large part in turning over and enriching the shallow bottoms of bays, just as earthworms function in

Fig. 301. *Sipunculus nudus* (§308), a large white sipunculid worm.

the production of rich vegetable mold on land. *Sipunculus* finds its optimum conditions, on our coast, in a downward extension of the flats peopled by the burrowing anemone *Edwardsiella*, the brachiopod *Glottidia*, and the sea pens.

Sipunculus nudus is known as far north as Monterey Bay. To the north it is replaced by a similar sipunculid, *Siphonosoma ingens*, which differs principally (as far as external anatomy goes) in the structure of its introvert, somewhat shorter and more flower-shaped than that of *Sipunculus nudus*.

§309. The true segmented worms have a large and obvious representative in the tubed *Chaetopterus variopedatus* (Fig. 302), which is known at least from Vancouver to Baja California, and is widely distributed elsewhere in warm and temperate seas. In its typically U-shaped burrow it constructs a fairly thick, parchmentlike, woody-brown tube, which may be a meter or so long; the curiously shaped worm, divided into three distinct regions, itself measures from 15 to 30 cm. Many generations of tube dwelling have softened the worm's body to the point where it is helpless outside its tube, and it almost invariably registers its protest at being removed by breaking in two just behind the head. This reaction no doubt has developed as a response to predators. The worms, which can move freely back and forth within their tubes, often extend somewhat out of the burrow. If the worm is grasped by a predator while so extended, it simply jettisons the head region; this is later easily regenerated by the remaining body.

Chaetopterus pumps water through its burrow to meet its respiratory needs and to bring in fine particles of food, which are removed from the water current by filtering through a mucous bag secreted by the middle body region. The animal secretes a great amount of mucus. This slime covers its body and lines the tube, and it seems to have some connection with the worm's brilliant phosphorescence, for when the worm is touched, some of the slime will come off on one's fingers and continue to glow there.

§310. The low-tide, deep sandy or muddy flats in the Puget Sound

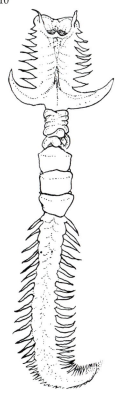

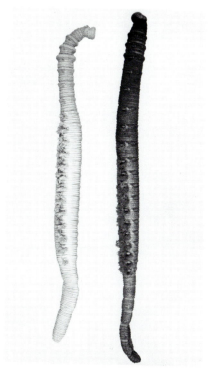

Fig. 302. A phosphorescent annelid tube worm, *Chaetopterus variopedatus* (§309); natural size.

Fig. 303. Two Pacific coast lugworms, genus *Abarenicola* (§310). The larger, darker animal is *A. vagabunda;* the other is *A. pacifica.*

region are the favored environment of a black, rough-skinned lugworm, *Abarenicola vagabunda* (Fig. 303), which is popular with fishermen as bait and with physiologists for experimentation. A somewhat similar species, lighter in color, is found in muddier bottoms at higher levels; this is *Abarenicola pacifica* (Fig. 303). *A. vagabunda* ranges from Japan through the Aleutian chain to Alaska and south at least to Humboldt Bay; the range of *A. pacifica* is similar.* The species found from southern California to Humboldt Bay is *Arenicola cristata.* Along with this species, especially from San Francisco Bay south, another species has been recorded, *Arenicola brasiliensis,* which is nearly worldwide in warm waters. Thus the lugworm situation

*The form of *Abarenicola vagabunda* (known also as *A. claparedii*) most common on the Pacific coast is *A. vagabunda oceanica.* A second form, *A. vagabunda vagabunda,* has been recorded only from the San Juan Islands and surrounding areas. In these quiet waters the latter replaces *A. vagabunda oceanica,* which, however, extends north and south of the San Juans along the outer coast.

in California seems to be similar to that in Puget Sound, but less is known of the ecological requirements of the California species. Lugworms are scarce in Tomales Bay, or at least rarely encountered; either the right place for them has not yet been found or they require a bit more freshwater drainage than is usually available in Tomales Bay.

Lugworms may attain a length of 30 cm, but 15 cm is about the average length for this coast. The animal's presence is indicated by little piles of mud castings, for, like so many other inhabitants of the mud flats, it solves the food-getting problem by eating the substratum. It usually lies in a U-shaped burrow of a size that will permit both its greatly swollen head end and its tail end to almost reach the surface. Sometimes, however, the arm of the U nearest the worm's head is partially filled with sand, and the head is buried deep in the substratum; the burrow thus more closely resembles a J. In either case the worm maintains a current of water through its burrow by dilating successive segments. This current, which can be made to flow in either direction, aerates the animal and assists its delicate bushy gills in respiration. At low water the current is up, that is, from head to tail; but when the animal is covered the current is reversed, thus being filtered through the sand in the tube.

Although most of the lugworm remains buried well below the surface of the sediment most of the time, the far tip of the tail is exposed at the surface several times each day when the worm defecates. Flatfish and other predators prey on these tips, usually removing only one or two segments per attack. At least in the case of the European species (*Arenicola marina*), such losses do not seem to be compensated for by the formation of new segments, and consequently the number of segments declines during the course of life of individual worms. The pea crab *Pinnixa eburna* occurs in *Abarenicola* tubes at Friday Harbor.

The whole mud-burrowing situation, as typified by the lugworm, offers an interesting example of adaptation to environment: by burrowing, the animals conceal themselves from enemies, escape wave action and drying, and place themselves in the midst of their food supply. The habitat requires, however, some provision against suffocation; and for the lugworm the protruded gills and the stream of water are the solution.

In the spring, great transparent milky masses, containing possibly half a million eggs, can be seen extending up from the burrows below the tide line; the egg masses resemble at first a stranded jellyfish, but are firmly rooted to the burrow sand by a tenacious stalk.

Arenicola is the only member of the bay and estuary fauna that can also live in the barren sand of the protected outer beaches, where very few animals seem able to exist. On those beaches the lugworm occurs, as previously noted, only where some silt is deposited around rocky outcrops.

§311. We have had biting worms and pinching crabs, stinging jellyfish and stinging sponges; in *Pareurythoe californica* we have a stinging worm. On mud flats in central and southern California (e.g., Elkhorn

Slough) the digger of *Urechis* or nemertean worms will sometimes see a long, pink, flabby-looking worm somewhat resembling *Glycera* but with white and glistening spicules about the "legs." If he picks it up, as he undoubtedly will the first time he sees it, he will regret the action instantly unless his hands are extraordinarily calloused. This and others of this worm's family are sometimes known as "fire worms" for the burning sensation felt by those who handle them. Once a specimen is preserved it may be handled with impunity; but the living spicules are quite capable of penetrating human skin and delivering an irritating toxin with mildly painful consequences.

The annelid proboscis worm *Glycera* is fairly common in mud flats; the relatively gigantic *G. robusta* (Fig. 304) occurs in beds of black mud (though more typically in sand or gravel, according to Hartman 1968). Of other worms, there are the forms mentioned in §293 as occurring higher up, and at Newport Bay and like places in central and southern California, there is a plumed tube worm, *Terebella californica* (similar in appearance to *Thelepus*, §53). *Terebella*, which can also be found among rocks and in the holdfasts of various algae, is interesting because it commonly harbors a commensal crab, *Parapinnixa affinis*, formerly known from a single specimen only. Here is another instance in which a seemingly rare animal can be taken almost at will, once its ecological station is known.

Neanthes virens occurs where there is an admixture of sand in the firm mud, building semipermanent tubes just as it does on the Atlantic coast. It appears that in *N. virens* we have a generalized and well-adjusted animal capable of coping with a variety of new situations. In its breeding habits it displays the sort of lunar rhythm for which polychaetes like *Nereis* and *Neanthes* are famous (see §151).

Fig. 304. The burrowing polychaete *Glycera robusta* (§311), with proboscis extended.

The giant *Neanthes brandti* (§164) may also be found in this zone. Frequently considered a subspecies of *N. virens*, the worms are quite similar and distinguished only with difficulty. Populations of the two may intergrade, and much of what is known about the biology of *N. virens* may apply to *N. brandti* as well. Specimens of *N. brandti* up to a meter long have been taken in the sandy mud of Elkhorn Slough.

Still another nereid, the ubiquitous *Nereis vexillosa* of mussel beds (§163), may be found here. On mud flats in the San Juan Islands (and presumably elsewhere), this worm forms semipermanent tubes to which it attaches pieces of drift algae. The attached algae, which may grow, serve as food for the worm, while providing some measure of protection from predators and from the physical stresses associated with desiccation and changes in temperature and salinity.

§312. As would be expected from the terrain, nemerteans (ribbon worms) are abundant, both in species and in number of individuals. A remarkable feature of these worms is a unique, eversible proboscis that can be shot out to capture prey. The proboscis may be nearly as long as the body itself, and when not in use, it lies tucked within a special body cavity. The proboscises of many nemerteans are armed with barbs or stylets, which, often accompanied by a venom, serve to capture prey. The stylets of local species are incapable of penetrating human skin, but at least one investigator has reported a sensation not unlike that of a bee sting when stabbed by a small Australian worm. Evidently, the stylet is lost at least fairly frequently, for in little pouches within the body the animal carries several spare stylets, much as a goodly yeoman carried spare arrows in his quiver.

Another conspicuous trait of certain nemertean worms is their habit of breaking themselves into pieces when persistently disturbed. It has long been known that in some species the separate fragments would grow into complete animals, and Coe demonstrated that a worm (*Lineus vegetus*, §139) may be cut and the resulting portions recut until eventually "miniature worms less than one one-hundred-thousandth the volume of the original are obtained" (1929). The limiting factor is that the wound will not heal, nor will regeneration take place, unless the fragment is nearly half as long as the diameter of the body. In other words, a worm 4 mm in diameter could be cut into 2 mm slices, or a little less, and all the fragments, barring accident, would become new worms. Decidedly, this is an item for "believe it or not" addicts. In a later paper (1930), Coe suggested that this extraordinary fragmentation constitutes the normal method of reproduction during warm weather, since the fragments (in an eastern *Lineus*) develop into normal adults, and since fragmentation can be induced experimentally at any season by raising the temperature. Coe also found that an adult worm, or a large regenerated fragment, can live for a year or more without food. The deficiency is compensated for by a continual decrease in body size—a trait that we have noticed in captive and improperly fed nudibranchs as well, although the nudibranchs cannot continue the process for more than a few weeks.

Probably the most abundant nemertean of mud flats between Tomales Bay and San Juan Island is *Paranemertes peregrina* (§139), reported as occurring from the Bering Sea to Ensenada. The natural history of this species has been studied extensively both in Washington and in central California by P. Roe (1970, 1976, 1979). At low tide, the worms emerge from their burrows about 15 minutes after the water leaves an area, and begin to search for food. Food in this case consists of various errant or shallowly buried polychaetes such as nereids, nephtyids, spionids, and syllids. The predator's search is haphazard until signs of prey are detected. Then, upon contact with suitable prey, the proboscis is everted, often rapidly and with great force. The stylet stabs, venom enters, and prey much stronger than the nemertean may be paralyzed within a few seconds. Feeding takes only a few minutes on the average; however, prey much longer than the nemertean are sometimes consumed, and in these cases, much more time (over an hour) is required. *Paranemertes* generally eats about one worm per day, and after feeding, it returns to the burrow it left, having spent a total of about 43 minutes on the surface of the flat. In its feeding, *Paranemertes* exhibits a distinct preference for nereid worms, and nereids (e.g., *Nereis vexillosa* and *Platynereis bicanaliculata*) exhibit in turn an escape response when contacted by *Paranemertes*. If water is present, the nereids' response is to swim rapidly away; if exposed at low tide, as is more often the situation when *Paranemertes* is foraging, they jerk away and crawl off rapidly. Other nemerteans that occur on the flats with *Paranemertes*, but that do not seem to eat nereids, do not elicit this escape behavior. *Paranemertes* spawns during June or July in Bodega Harbor, and somewhat earlier, during March, April, and June, in Washington. Most of the worms that survive to settle on the mud flat live about 1.5 to 1.8 years, generally spawning only once in their lifetime.

A wide, white *Micrura* is very common in this habitat, as is a dirty-pink, bandlike *Cerebratulus*. The latter has a firm consistency and may be 3 meters or more long, but when captured it immediately breaks into many fragments.

§313. Among the crustaceans native to this lowest zone of the mud flats is the long-handed ghost shrimp *Callianassa gigas*, which differs from the related *C. californiensis* (§292) chiefly in being very much larger and in having a longer and more slender large claw. It is usually white or cream-colored, but sometimes pink. Recorded from Digby Island, British Columbia, to Bahía de San Quintín, it probably parallels the range of *C. californiensis* and reaches into Alaska.

Burrowing in the lowest areas of mud bared by the tide, between southeastern Alaska and Bahía de San Quintín at least (but with small individuals rather high up in the north), occurs the most striking shrimp in this zone—the blue mud shrimp *Upogebia pugettensis* (Fig. 305), whose actual

Fig. 305. *Upogebia pugettensis* (§313), the blue mud shrimp, which is larger and found lower on the shore than the ghost shrimp *Callianassa*.

color is usually a dirty bluish-white. These animals are firmer, larger (up to 15 cm long), and more vigorous than the *Callianassa* found higher up. They are harder to dig out, also, because the ground in which they live is not only continually inundated by the tide but so honeycombed with burrows that the water pours in almost as fast as the mud can be spaded out. The blue mud shrimp's burrow is permanent and little branched, and has several enlarged places for turning around. It extends downward from the surface for about 45 cm, horizontally for 60 to 120 cm, and then to the surface again. It is inhabited, almost always, by one male and one female, and they are probably moderately long-lived. Egg-bearing females have been noted from December through April.

In burrowing, MacGinitie observed (1930), the animal carries its load of excavated material in a "mud basket" formed by the first two pairs of legs on the thorax. The third, fourth, and fifth pairs of legs are used for walking. On the abdomen, four pairs of swimmerets (pleopods) function as paddles to keep the water circulating through the burrow, while the mud basket doubles as a strainer to catch the minute food particles in the water. For this latter purpose the mud-basket legs are provided with hairs. The fifth pair of thoracic legs terminates in brushes, and is used not only for walking but also for cleaning the body. An egg-bearing female, especially, spends a great deal of time in brushing her eggs to keep them clean of diatoms and fungus growths that would otherwise kill the larval shrimps. She also keeps the eggs aerated using the swimmerets. The shrimp's recurved tail can be used to block a considerable head of water or, on occasion, can be flipped powerfully so as to cleanse the burrow by blowing out a swift current of water.

There are commensals, of course, one of them being distinctive in that it is a small clam, *Pseudopythina rugifera* (also as *Orobitella*), which attaches by byssal threads to the underside of the tail of its host. Another small clam, *Cryptomya californica*, will often be found as well, not attached to the host, but with its short siphons projecting into the shrimp's permanent burrow. This clam also taps into burrows of the innkeeper *Urechis* and the ghost shrimp *Callianassa*, with the benefit being the same—the clam can live a protected life, deep in the sediment, even though it is small. The little crab (*Scleroplax granulata*) that lives with *Urechis* and *Callianassa* is another guest of *Upogebia*. There is also the scale worm *Hesperonoe complanata*, found as well in the burrows of *Callianassa*, and the transient goby *Clevelandia ios* occasionally seeks protection in the burrows.

Upogebia as well as *Callianassa* are in ill repute with the oystermen of Puget Sound, for the mud that is dug up smothers many of the young oysters, and the burrows cause drainage from the areas that have been diked in to keep the oysters covered at low water.

Free-Swimming Animals

§314. Several familiar shrimps will be found in pools on the mud flats, and, more conspicuously, such fishes as flounders, skates, and rays. The very presence of these fish furnishes interesting sidelights on the food chains of estuary regions, since all were in search of food when the ebbing tide trapped them on the flats. The skates and rays were after clams, which they root out and eat after crushing the shells with their powerful "pavement" teeth. The stingray is to be avoided, and barefoot collectors should be particularly careful about stepping into pools. The wounds they inflict are never serious, but are said to be decidedly painful. Dr. Starks advises soaking the affected part in very hot water; and he adds, no doubt from experience, that the sufferer is perfectly willing to put his foot into water even hotter than he can painlessly tolerate.

Water brought into the laboratory from the Elkhorn Slough mud flats may contain free-swimming specimens of the active little amphipod *Anisogammarus confervicolus*, associated also with eelgrass, which is one of the most common amphipods of the northern Pacific. Ranging from Sitka to Elkhorn Slough, it is more common in brackish water than in oceanic salinity, especially in the Puget Sound region. Water taken from the mud flats at Mission Bay near San Diego will sometimes be teeming with a very common opossum shrimp, *Mysidopsis californica*, similar to the form mentioned in §61 but apparently adapted to the stagnant water and fairly high temperatures of this nearly landlocked shallow bay.

PART IV

Wharf Piling

Piling animals show variations apparently correlated with their degree of exposure to surf. Whether these differences are actually due to the presence or absence of wave shock, as is thought to be the case with rocky-shore animals, or to the different makeup of bay waters and open waters cannot be stated. It is known that the waters of bays and harbors differ from oceanic waters materially in oxygen content, salinity, degree of acidity or alkalinity, temperature, and especially nutrients. It may be that we must look to these factors for the explanation of the faunal differences; but until more-definite information is available the degree of protection from surf is a convenient method of classification.

For obvious reasons, no wharves or piers are built in fully exposed positions. Nevertheless, the outer piles of long piers, such as those at Santa Cruz and Santa Barbara, get pretty well pounded during storms, and it is noticeable that the animals most frequent on these piles are barnacles and California mussels, both adapted to surf-swept rocky headlands and perfectly able to take care of themselves in rough weather. The piling fauna, therefore, will be considered under two headings, exposed (using the term relatively) and protected.

Piling offers a conspicuous means of observing depth zonation, although for some reason this zonation is not as obvious and clear-cut on the Pacific as it is on the Atlantic. Additionally, fewer west-coast workers have been interested in the ecology of wharf piling, and much of the data reported below is more or less tentative.

9 🐾 Exposed Piles

Zones 1 and 2. Spray Area and High Intertidal

§315. Very few *Littorina* are to be found on piling, and the highest animals encountered are very small barnacles, often under 6 mm in diameter. They occur sparsely and grow almost flush with the wood. The statement in earlier editions that there were *Balanus balanoides* (now known as *Semibalanus balanoides*) at Monterey appears to have been some sort of *lapsus calami*; this species has not been recognized since from Monterey, and the specimens in question must have been *Balanus glandula* (§4).

Intermingled with these barnacles, and a little below them, are a few scattered clusters of California mussels. With the mussels, and a bit below them, some of the smallest examples of the rather prettily colored warty anemones (the aggregated form of *Anthopleura*) are likely to occur. A few of the little rock crabs (*Pachygrapsus*) may be seen walking about here, retreating, when danger threatens, into the larger mussel clusters just below.

Zone 3. Middle Intertidal

§316. This zone (and the upper part of the lower zone) is characterized by lush growths of an almost white hydroid, *Obelia*; by large clusters of big California mussels; and by the great barnacle *Balanus nubilus*. The hydroid *Obelia longissima* (Fig. 306) and similar smaller species form long trailers or furry growths, usually on barnacles or mussel shells but also on the bare wood where the area is not already occupied—a rare circumstance. *O. longissima* is the commonest hydroid of the piling environment in British Columbia, and is almost as common in Puget Sound (§332), ranging northward to Alaska and southward to San Pedro. The stalks of *Obelia* branch and rebranch in every plane. Each little "branchlet" is terminated by a cuplike hydrotheca that protects the retracted tentacles. When other, larger sacs, the gonangia, are swollen, they should be watched for the birth

Fig. 306. *Above*, the hydroid *Obelia longissima* (§316), a characteristic mid-zone animal of wharf piling; the inset shows a greatly magnified branch of *O. dubia* (§332), a similar species. *Below left*, a jellyfish (medusa) from the hydroid *Obelia*; greatly enlarged. *Below right*, a sea spider common on *Obelia* colonies, *Halosoma viridintestinale* (§316); enlarged.

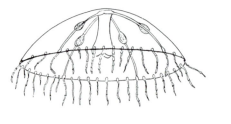

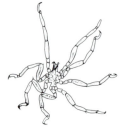

of minute jellyfish (Fig. 306). The young medusae, which usually have 26 tentacles, are given off most abundantly during the summer, but reproduction continues throughout the year. The newly born jellyfish swim away with all the assurance in the world, propelling themselves with little jerks. It is a pity that they are so small, for they are delicate, diagrammatic, and beautiful; they can barely be seen with the unaided eye if the containing jar is held against the light.

The growth of hydroids is very rapid—so rapid that many of them pass through several generations in a year. In one instance, 6 weeks sufficed for *Obelia* to cover a raft moored at sea, and a single month is sometimes long enough for a newly hatched larva to become a hydroid and release its own medusae.

Associated with these *Obelia* colonies there is almost invariably a minute sea spider, *Halosoma viridintestinale* (Fig. 306), not much larger than the hydranths on which it probably feeds. As its specific name implies, this pycnogonid has green guts, which show plainly through the legs; the significance of this color has not been investigated. A careful look at *Obelia*, especially in the late spring, may reveal the presence of immature pycnogonids developing in the hydranths. The species occurs from Monterey to Tomales Bay, where it is common. It also occurs on eelgrass (*Zostera*).

In this zone and lower, especially on piling and floats in San Francisco Bay, the cosmopolitan entoproct *Barentsia gracilis* occurs in brown, furry mats, often covering piling and barnacles alike. Anemones may attach to this matting; when they do so, they may be easily removed. Also common in San Francisco Bay is *B. benedeni*, a European estuarine species, which prefers waters with much suspended material. Easily confused with the cosmopolitan *B. gracilis*, which, however, prefers fairly clean water, *B. benedeni* can often be distinguished by the presence of muscular swellings on the stalks.

Two ectoprocts, or bryozoans, *Alcyonidium polyoum* (formerly *A. mytili*) and *Bowerbankia gracilis*, both forming colonies that are likewise brown and furry but close-cropped, very often encrust empty barnacle and mussel shells. The two may be distinguished by examining the individual zooecia that compose the colonies. Those of *A. polyoum* are contiguous, roughly hexagonal, and relatively squat, whereas those of *B. gracilis* are erect, tubular units that arise from a stolon.

§317. *Balanus nubilus* is the largest barnacle on our coast,* and probably one of the largest in the world. It ranges on this coast from southern Alaska to La Jolla. It is commonly 6 to 8 cm high, with a basal diameter

---

*Another large barnacle, *Balanus aquila* Pilsbry, may occasionally exceed the more common *B. nubilus* in size. Both species, but especially *B. nubilus*, have received the attention of physiologists who value their exceptionally large muscle fibers. Unfortunately, much of the work purporting to deal with *B. nubilus* was actually performed on *B. aquila*, through misidentification. *B. nubilus* can usually be distinguished by its large, flared aperture (small in *B. aquila*) and by the lack of striations on the scutal plates (present in *B. aquila*).

exceeding the height, and in some places specimens grow on top of each other until they form great clusters. We have such clusters from the Santa Cruz wharf that are more than 30 cm high. The animal's great size and accretionary habits are not entirely helpful to it, since clusters get so large and heavy that they often carry away the bark or part of the disintegrating wood and sink to the bottom, where it is likely that the barnacles cannot live. The shell of *B. nubilus* is frequently covered with *Barentsia* (§316), and may also provide an attachment base for anemones and tube worms. The mantle through which the cirri, or feeding legs, are protruded is gorgeously colored with rich reds and purples. *B. nubilus* is another of the long series of barnacles originally described by Darwin. His description still stands correct—a tribute to the carefulness of the work devoted to this difficult group by a thoroughly competent mind.

A smaller but even more striking barnacle that is characteristic of boat bottoms and is occasional on wharf piling (and even on rocks) is *Megabalanus californicus* (Fig. 307; formerly *Balanus tintinnabulum californicus*). The nearly cylindrical, ridged shell is a pinkish red with white lines, and the "lips" of the mantle, as with *B. nubilus*, are vividly colored. It is characteristically a southern California species, uncommon north of Monterey.

§318. The big-clawed porcelain crab *Pachycheles rudis* (Fig. 308), ranging from Kodiak, Alaska, to Baja California, seeks crevices, nooks, and interstices, and is often at home in the discarded and *Alcyonidium*-lined shells of barnacles. It is often the same size as the porcelain crab of rocky shores (*Petrolisthes*) but may be distinguished from the latter by the granulations on the upper surface of the big claws. Like *Petrolisthes*, it is a filter feeder, using the hairy maxillipeds to comb particles from the water. Ovigerous specimens have been noted in California throughout the year (except in September). The same is apparently true of populations in Puget Sound, where a study noted 200 to 2,000 eggs per brooding female.

§319. Crawling about on these same empty barnacle shells are specimens of *Pycnogonum stearnsi* (Fig. 309), the largest and most ungainly of all local sea spiders. A border design of these grotesque yet picturesque animals might surround the pen-and-ink representation of a nightmare. Most sea spiders spend part of their tender youth in close juxtaposition to a coelenterate—the larvae, in fact, usually feed on the juices of their hydroid or anemone host. This sea spider, white to pale salmon-pink in color, is found in a variety of habitats—on *Aglaophenia* fronds, among *Clavelina* clusters and the like, and around the bases of the large, green *Anthopleura*. It is especially common among the anemones of the caves and crevices of Tomales Bluff, Marin County; sometimes half a dozen occur on a single anemone.

Like the rest of the genus, *Pycnogonum stearnsi* feeds primarily on anemones and will often be found with its proboscis sunk into the column of its host. Although *P. stearnsi* is supposed to occur as far south as San Diego, it has not been found there. It is common among anemones in the

Fig. 307. A red and white barnacle, *Mega-balanus californicus* (§317); twice natural size.

Fig. 308. *Pachycheles rudis* (§318), a quiet-water porcelain crab with unequal tubercu-late claws and a thicker body than the more flattened *Petrolisthes* (§18) of the outer coast; enlarged.

Fig. 309. Two central California sea spiders (§319) associated with anemones: *left, Pyc-nogonum stearnsi; right, P. rickettsi.* Both enlarged about 3 times.

Monterey area, where it may also be found hidden among the accumulation around the bases of anemones set in pockets or crevices. It does not appear to occur north of Sonoma County in any abundance but has been recorded as far north as British Columbia. In the Friday Harbor area the common *Pycnogonum* is *P. rickettsi* (Fig. 309), a "reticulated" species that also occurs on Duxbury Reef in the same association as *P. stearnsi* (but not, as far as is known, on the same individual anemones); it was originally collected from anemones at Monterey by Ed Ricketts.

§320. There are a few clusters of goose barnacles on the outer piles, and some large solitary specimens of the great green anemone *Anthopleura*. Under the mussel clusters there are many acquaintances from other environments, among them the common scale worm (§49) of many loci and *Nereis grubei* (as *N. mediator* in previous editions), the pile worm of the bait gatherers.

Nereis grubei resembles *N. vexillosa* (§163) but is smaller, rarely reaching more than 5 to 10 cm. The worm's color is usually green, but females containing developing oocytes are a startling turquoise. Its diet, undoubtedly changing somewhat with habitat, includes a wide variety of plants and small animals. Two populations in California, one at Pescadero Point (San Mateo County) and one at Point Fermin (Los Angeles County), have been studied extensively, and differences with regard to size and reproductive cycles are noteworthy. In central California, mature individuals were large and reproduction seasonal; the population gave rise to swarms nearly monthly between mid-February and mid-June. In southern California, individuals were considerably smaller, and some epitokes were collected in all months of the year. *N. grubei* is found commonly in secreted mucus tubes among seaweed fronds and holdfasts, as well as on piling and in mussel beds. It has been reported from British Columbia to Mexico.

Zone 4. Low Intertidal

§321. The anemone *Metridium senile* (Fig. 310), usually white but occurring in several definite color patterns, prefers bare piling. Small specimens, however, may be found on dead shells (or even on the shells of living barnacles) and on the club-shaped tunicate *Styela*. Several times we have taken, from piling, kelp crabs that carried *Metridium* on their backs and claws. The largest specimens occur subtidally, extending clear to the bottom in water 6 to 8 meters deep, but small specimens extend into the intertidal, where, at low tide, they hang fully relaxed and pendulous; expanded, they are delicate and lovely.

Entire clusters of this anemone may be colored a rich brown, and reddish-yellow specimens are not uncommon. The fact that specimens of one color seem to segregate themselves is accounted for by the common method of reproduction via basal fragmentation (called pedal laceration by most specialists). This asexual method, actually a form of division, explains the presence of the many little specimens that seem to have been splattered about near larger individuals. *Metridium* also reproduces itself sexually, individuals discharging eggs or sperm from the central cavity through the mouth. As is often the case, the presence of sperm in the water stimulates females nearby to shed their eggs, and the unions of these products result in free-swimming larvae that carry the fair breed of *Metridium* far and

Fig. 310. The anemone *Metridium senile* (§321), usually white but sometimes tan, orange, or brown. Intertidal specimens generally measure 2 to 5 cm across the tentacular crown, but subtidal ones, such as those shown, may reach 25 cm in crown diameter.

wide—pretty well throughout the world in the Northern Hemisphere. On this coast we know them to extend from Sitka to Santa Barbara on wharf piling. Gigantic exhibition specimens are dredged from deep water; one specimen, from 110 meters, filled a 40-liter crock when expanded. Piling specimens will average around 5 cm in diameter, but 15-cm specimens have been taken.

Metridium generally appears to be a stolid and motionless animal, but time-lapse movies reveal that it is seldom actually motionless. Instead, its movements follow a well-defined rhythm of slow expansion and contraction, and when stimulated by a whiff of food, the movement is accelerated. Small and large anemones eat similar things, their diet in both cases consisting mainly of copepods and the larvae of various invertebrates, which are eaten in rough proportion to their abundance in the plankton.

Metridium, so delicate and tranquil in appearance, does have one behavior in its repertoire that is indeed quite violent. Individuals around the border of a dense clone, formed by asexual reproduction and thus genetically identical, often develop special tentacles called "catch" tentacles, which are used for aggression against non-clonemates and anemones

percent of the kelp crabs are infected. Infestation becomes more abundant with the increase in latitude on this coast. On the Atlantic coast, sacculinids are rare except on the Gulf coast, where infestation is common on mud crabs in Louisiana and on the blue crab in Texas waters.

§328. To many well-informed persons the biological connotations of wharf piling all center around the word "teredo." These wood borers have been the bane of shipping for at least 2,000 years, for they were known, unfavorably, by the ancient Greeks and Romans. It seems quite possible that the early Mediterranean custom of hauling boats ashore when they were not in use was motivated, in part at least, by the knowledge that frequent drying would protect them from attack by the shipworms. The species of *Teredo* and their near relatives are still, after many years of scientific research on preventives, the cause of tremendous destruction to marine timbers—destruction that often runs into millions of dollars a year in a single seaport.

The term "shipworm," which is popularly applied to the whole group, is a misnomer, for the animal is actually a clam, although its small calcareous shell—its boring tool—covers only the "head" end of its long, wormlike body. *Teredo* is a highly efficient woodworker, but countless generations of protected life in timber have made it helpless outside of that element, and once removed from its burrow it cannot begin a new one. In these days of steel hulls these invaders have had to confine their activities to wharf piling and other marine timbers, but that has affected neither their aggressiveness nor their numbers.

In open water the true *Teredo* rarely occurs, but the one shipworm that is native to this coast, the giant *Bankia setacea* (Fig. 313), known as the Northwest shipworm, operates in relatively exposed piles from the Bering Sea to as far south as southern Baja California, although below Monterey Bay it apparently does little damage. In San Francisco Bay proper it has been known since the earliest days of shipping; but since it is less resistant to lowered salinity than other shipworms, it does not extend its activities into the brackish waters of the adjoining San Pablo and Suisun bays, or into similar enclosed bays in other regions. Neither has it ever been as destructive, according to a report by the San Francisco Bay Marine Piling Committee, as the true *Teredo* (*Teredo navalis*, §337) that suddenly appeared there about 1910–13. However, if some of the plans discussed by engineers and politicians are carried out—shunting massive quantities of fresh water south to the San Joaquin Valley and Los Angeles—the salinity will likely increase up the river, and where there is any wood left, the shipworms will chew it up.

The work of the shipworm is begun by a larva, which drills a hole that is barely visible to the naked eye. For the next few months the animal grows rapidly, enlarging the burrow as it grows but leaving the entrance very small. A few centimeters inside the wood the burrow turns, usually downward, so as to follow to some extent the grain of the wood. The animal

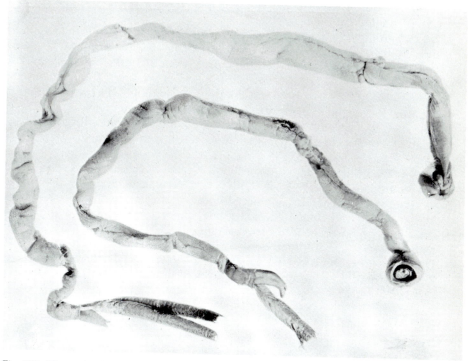

Fig. 313. The native Pacific coast shipworm, or Northwest shipworm, *Bankia setacea* (§328), a clam that burrows into relatively exposed piles; reduced.

turns aside, however, to avoid obstructions like nails, bolts, and knots, and to avoid penetrating the burrows of neighbors. Thus the actual course of the burrows is nearly always sinuous. *Bankia* concentrates its attack at the mud line, digging out burrows that are sometimes a meter long and 2 cm in diameter. The shipworm's method of boring is very similar to that employed by the rock-boring clams (§242). A heavy attack will reduce a new, untreated pile to the collapsing point in 6 months, and survival for more than a year is unlikely. The life of a pile may be prolonged to 3 or 4 years by chemical treatment, such as creosoting, but nothing keeps the animals out for long. Copper sheathing, which did good service in protecting ships for so many years, is impractical on piles because of its expense, the likelihood of its being stolen, and the ease with which it is damaged by contact with boats and driftwood. Also, the sheathing cannot extend below the vulnerable mud line, and the mud line is likely to be lowered by eroding currents. Numerous other jacketings have been tried, but not even concrete jackets are entirely satisfactory, for when the wood borers are thus thwarted, the concrete and rock borers (§242) come to their assistance. Steel and iron pilings have numerous disadvantages, aside from their great expense; so the battle between man and the shipworm goes on, with the shipworm still getting somewhat the best of it.

Bankia setacea likes low water temperatures, and extrudes its eggs, in San Francisco Bay at least, at that time of the year when the water temperature is lowest—from February or March to June. In southeastern Alaska it breeds about a month earlier, and in southern California some spawning occurs throughout most of the year, except in the late summer or early fall. Fertilization is external to the burrow and a matter of chance. The Bankia young develop first into functional males; later about half of these become females. Further considerations relating to the similar Teredo and Lyrodus will be found in §§337 and 338.

§329. William Dampier, the English buccaneer, who was something of a naturalist, is said to have written that his wormy ship was also being attacked by small white animals resembling sheep lice. Undoubtedly he was referring to a wood-boring isopod, the gribble, of which there are three species on this coast: Limnoria lignorum, L. quadripunctata (Fig. 314), and L. tripunctata. Shipworms perform their work on the inside of wood; Limnoria works, with almost equal efficiency, from the outside, attacking piles at all levels from the middle intertidal down to depths of 12 meters or more.

Fig. 314. The gribble Limnoria quadripunctata (§329), a wood-boring isopod; the photograph, enlarged about 10 times, shows the living animals in position in their burrows.

These animals are so tiny—scarcely 4 mm long—that hundreds of them may occur in a few square centimeters of heavily infested wood. Their legs are adapted to holding to wood in the face of severe currents. The digging is done by the mouthparts, and the wood is eaten as fast as it is gouged out, passing so rapidly though the intestine that a given particle remains in the digestive tract only an estimated 80 minutes. This speed of passage is what would be expected in an animal that derives all its nourishment from wood, as *Limnoria* apparently does, for the food content is low. There is no mechanism for filtering the contained plankton out of the water, and nothing but wood has ever been found in the digestive tract. This is a unique situation, for cellulose is highly indigestible and remains unchanged by the digestive juices of most animals unless, as with the termites, there are intestinal protozoa to aid in the conversion. In one experiment, however, specimens of *Limnoria* ingested pure cellulose in the form of filter paper and lived nearly twice as long as controls that were unfed. Some source of nitrogen, perhaps provided in nature by the fungi universally present in water-soaked wood, is required to balance the diet.

When digging, gribbles jerk their heads backward and forward, turning them slowly at the same time. They can completely bury themselves in from 4 to 6 days, starting their burrows at a rather flat angle. Presumably because of the difficulty of obtaining oxygen they seldom go more than 15 mm below the surface. Breathing in a solitary burrow would probably be difficult even at that depth, but the burrows of different individuals are usually connected by small openings that permit freer circulation of water than could be obtained in an isolated burrow.

Breeding is continuous throughout the year, and the number of eggs in each *Limnoria* brood depends on the species: 22 in *L. lignorum*, 9 or 10 in *L. quadripunctata*, and 4 to 10 in *L. tripunctata*. The newly hatched animal, unlike the larvae of shipworms, does not swim at all, but it gradually acquires some swimming ability as it grows. Experiments recorded in the *San Francisco Bay Marine Piling Report* show that at no stage of its life is the animal positively attracted to wood, and presumably gribbles do not fasten on wood unless they accidentally touch it. Migrations of swimming adults do occur (greatest in the spring and summer months for *L. tripunctata*), but at any stage, unlike shipworms, gribbles may be transferred to new locations without harm. Thus, driftwood floating among piling is assumed to be an important factor in their distribution, the spread being accomplished by adult animals instead of by free-swimming larvae. The taking in and discharging of ballast water by ships may aid *Limnoria* in spreading from port to port.

Limnoria lignorum is primarily a cold-water animal, occurring on the Scandinavian coast, within the Arctic Circle, and on this coast from Kodiak Island to as far south as Point Arena. *L. quadripunctata* occurs on this coast from Crescent City to La Jolla, and *L. tripunctata* from San Francisco to Mazatlán, Mexico. The various species are responsible for damage to piling in most of the great harbors of the world, and the warmer-water species, *L.*

tripunctata, is undeterred by the creosote with which most piling is treated to protect it from these depredations.

Over the last 35 years or so, the Navy has financed much research on the lives and times of borers, both bivalves and gribbles. It has been established that gribbles do have cellulase, the enzyme necessary to digest wood; and that they attack almost any kind of wood, although some exotic and expensive hardwoods are less susceptible. One investigator for the Navy has suggested that perhaps the Navy should consider breeding a resistant strain of trees to provide piling unappetizing or inedible to borers, since concrete pilings are cheerfully attacked by rock-boring piddocks and all other attempts to keep borers at bay have been unsuccessful.

§330. Whoever frequents wharves, piers, and floats, especially at not-quite-respectable hours, will be certain to see some of the more obvious free-swimming invertebrate animals that haunt the harbors. Though pelagic animals are not within the scope of this work, it would be a pity not to mention the bell-shaped jellyfish *Polyorchis penicillatus* (Fig. 315), whose transparent white against the dark of under-wharf waters makes it very conspicuous. It swims beautifully, and rather well, by rhythmically contracting and relaxing the bell, pausing frequently to drift with tentacles extended in a food-catching web. The average diameter is 2 cm or more. The gonads and some other internal organs are variably colored from yellow-brown to purple.

In quiet bays in Puget Sound and British Columbia the jellyfish *Aurelia aurita* (Fig. 316), nearly colorless except for its four horseshoe-shaped gonads (purple in males; yellow in females), sometimes occurs in such immense numbers that it is impossible to dip an oar without striking several of the beautiful, pulsating animals. A boat seems to glide through a sea of jellyfish rather than through water. These tremendous aggregations are but a brief phenomenon, usually lasting during one flood tide only. A few hours later it may be difficult to find a dozen specimens. Similar swarms occur in Tomales Bay during the summer months. The species has a world-wide distribution.

The ubiquitous *Nereis vexillosa* (§163) occurs in this environment also whenever clusters of mussels extend so far down. A smaller worm, probably *Platynereis bicanaliculata*, seems to be both abundant and characteristic, especially in its free-swimming (heteronereis) stage.* We discovered the interesting periodicity of this form quite by accident, in connection with a proclivity for roaming about at odd hours that is no doubt very annoying to conventional night watchmen. Swarming takes place in Monterey Harbor during the spring and summer months when, on moonless nights, the extreme high tides occur about midnight. At first only one or two of the viv-

*Although previous editions refer to the worm in this section as *Platynereis agassizi*, the correct name for the worm most likely seen is *P. bicanaliculata*. This common polychaete of mud flats, eelgrass, and piling swarms in bays and harbors from British Columbia to Mexico. It has a maximum length of 70 mm, but most specimens are far smaller than this.

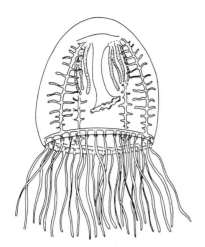

Fig. 315. The pelagic hydromedusa
Polyorchis penicillatus (§330), often seen
in harbors around piling; about natural
size.

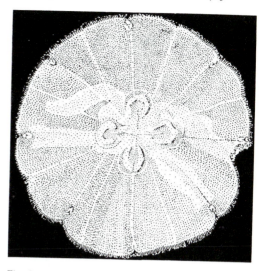

Fig. 316. The scyphozoan *Aurelia aurita*
(§330); about 1/10 natural size.

idly white and wriggly worms will be in sight; by the end of half an hour
there may be two or three score, apparently attracted to the area of observation by the rays of the flashlight. The male comes first—a tiny animal 1 cm
long or less—followed by the female, sometimes more than 2 cm long and
having red after-segments. When the two sexes are together, the heteronereis go completely crazy, swimming about in circles at furious speed and
shedding eggs and sperm. Only one who has tried to capture them with a
dip net knows how rapidly and evasively they swim. Specimens brought
into the laboratory have continued this activity all night, although in their
natural environment the swarming seems to last only a few hours at the
height of the tide. In the morning, captive specimens look withered and
worn out, no doubt an accurate reflection of their condition. In this connection, see also §163.

Finally there will be dozens, possibly hundreds, of smaller free-
swimming organisms of various types—microcrustaceans, minute jellyfish, larval stages of worms, mollusks, and echinoderms—all outside the
province of this book. It might be mentioned that one of the usually pelagic
euphausiids, *Thysanoessa gregaria* (or perhaps *T. longipes*), distantly related
to the opossum shrimps (§61), has been taken abundantly in Monterey
Harbor within a stone's throw of the shore.

10 ⫷ Protected Piles

As previously stated, we can draw only a vague line between relatively exposed and fully protected piling (Fig. 317). Between the obvious extremes there is great overlapping, and we have to remind the reader that this classification is offered tentatively. Further work may justify it as a working hypothesis or may show that other factors than exposure to waves are the primary ones.

Depth zonation is as apparent here as on more-exposed piling, but the distinctive animals are so few in number that they will be treated without formal zonal classification.

§331. The common small barnacles of protected piling in Berkeley and Newport Bay are the omnipresent *Balanus glandula* (§4), whose abundance here under conditions of fluctuating salinity and little water movement bears testimony to the remarkable adaptability of this form.

§332. The California mussels of the more-exposed piling are replaced in these quiet waters by the smaller, cosmopolitan *Mytilus edulis* (§197). In favorable environments these bay mussels often form great bunches that double the diameter of the piles on which they grow, and they may be found in probably every suitable port between the Bering Sea and northern Mexico. They attain their maximum development in the middle zone, so they are very obvious, even at half tide.

On some piling another mussel may be encountered and easily distinguished. This is *Modiolus capax* (§175), the fat horse mussel, which has a shell tufted with coarse, brown hairs.

The small mussel worm *Nereis grubei* (§320) will be found here no less frequently than in open-shore mussel clusters, only the mussels are different. The shield limpet, *Collisella pelta* (§§11, 195), occurs on these piles just as on rocks and gravel, minute specimens occupying the highest populated zone along with the small barnacles.

Interspersed among the mussel clusters are several bay and harbor hydroids. The cosmopolitan *Obelia commissuralis* was reported from San

Fig. 317. Fully protected piling sunk in a mud flat. Barnacles occupy the highest zones, with bay mussels (*Mytilus edulis*) and colonial tunicates lower down.

Francisco Bay in 1902, where it was adapted to the estuarine conditions of dirty water and lowered salinity. Clusters of *Obelia* may still be found in these situations about the wharves in western Oakland, in water so filthy that all animal life would seem to be precluded, and *O. commissuralis* is presumed to be the species involved (but see §271).

The sea spider *Achelia nudiuscula* may occur in hydroid clusters of this sort, and was once abundant on *Obelia* on pilings and junk along the eastern shore of San Francisco Bay—in very filthy water. It is known only from this, the type locality.

In cleaner stretches of quiet water, particularly up north, the tenuous *Obelia longissima* (§316) is known to be the predominant shore hydroid, avoiding pollution and direct sunlight but tolerating some fresh water. Fraser (1914) noted that *O. longissima* is "the commonest shallow water campanularian in the region. It grows throughout the year on the station float, Departure Bay, and medusae are freed at many times during the year." On continuously submerged floats, the situation seems to be that the very common and bristling *Clytia edwardsi* occurs in considerable sunlight, but that the lush clusters of *O. longissima* are either underneath the float or on portions protected from direct sunlight by some of the superstructure. In late July 1935, the colonies of *Clytia* in southeastern Hood Canal were producing round, four-tentacled medusae. *Obelia* and *Clytia* constitute up to 90 percent of the littoral hydroid population of this region (the rest is made up mostly of *Gonothyraea*, and of the smaller and more hidden *O. dichotoma*, §271).

On floats in the Puget Sound region (as at Port Madison, where medusae were released in July 1932) and on bell buoys, boat bottoms, etc., along the open coast at Monterey, the small and sparkling *Obelia dubia* (Fig. 306) occurs, so similar to *O. dichotoma* that some authorities consider them identical. We have never knowingly taken *O. dubia* on fixed piling, where *O. dichotoma* frequently occurs. *O. dubia* seems to be equally intolerant of stagnation, lowered salinity, pollution, and tidal exposure—a "touchy" form, difficult to preserve expanded unless narcotized within a few minutes after capture. This and *O. longissima* are the species of Monterey Bay, according to determinations by Fraser, who said (1914) that the juveniles of the two are sometimes impossible to distinguish (but see the footnote to §271, p. 343).

Four of the five commonest hydroids here release free-swimming medusae, thus increasing many times the opportunities for wide distribution of the race. Each hydroid colony produces thousands of minute, active jellyfish, which can be carried far from the parent hydroid by currents before reaching maturity. The union of the jellyfish's eggs and sperm produces planulae, free-swimming larvae that can drift still farther before they settle and develop into the polyp stage.

Gonothyraea clarki resembles *Obelia*, but it develops sessile medusae that are never released. Four or five of these little captives pop out of each vaselike gonangium and pulsate ineffectually at the ends of their stalks, releasing germ cells and eggs to be united by chance currents. In British Columbia and Puget Sound, *Gonothyraea* is plentiful in the spring and early summer, growing on rocks as well as on floats and piling, but it dies out later and is not to be found. It extends as far south as San Francisco.

§333. In the extreme low-tide zone enormous bushy clusters of the naked hydroid *Tubularia crocea* (Fig. 318) will be found banding the piles or floating docks with delicate pinks or reds. Its nearest relative is the small, solitary *T. marina* of rocky shores, but this clustered form, which may extend 15 cm out from the piles, is far more spectacular. It is the "heads" that provide the color; the supporting stems look like small flexible straws. The heads, or hydranths, autotomize readily; they seem to break off as a regular thing if the water gets warm or the conditions are otherwise unfavorable, after which, if conditions improve, the stem regenerates a new head within a few days. The sexual medusae are never liberated. During the summer, swimming sperm are released into the water, where they are attracted by some chemical substance released by the female reproductive structures; eggs are fertilized within the female medusoids. We have found *T. crocea* in Newport Bay, Elkhorn Slough, and San Francisco Bay; it is also known from many regions northward to the Gulf of Alaska and from both sides of the Atlantic.

Hilton (1916) found the *Tubularia* clusters at Balboa heavily populated with a small sea spider, *Anoplodactylus erectus* (which we have found rarely on compound tunicates at Pacific Grove). He worked out the life history.

Fig. 318. A cluster of the naked hydroid *Tubularia crocea* (§333), a pink form characteristic of the extreme low-tide zone on protected wharf piles; natural size.

The eggs, which are produced in the summer, develop into larvae that pierce the body wall of *Tubularia* and enter the digestive tract, there to live parasitically until further grown.

Flatter, tangled clusters of another naked hydroid are often interspersed with *Tubularia*, but the two will be differentiated easily. The tiny, bulbous heads of *Sarsia tubulosa* (Fig. 319; formerly *Syncoryne mirabilis*) bear clubbed tentacles and bud off free-swimming medusae from their lower sections. The medusae, after their release in the spring and summer, grow

Fig. 319. *Left,* a greatly enlarged drawing of the clublike polyps of *Sarsia tubulosa* (§333), a common hydroid of bays and harbors; *right,* a species of *Sarsia* similar to *S. tubulosa;* enlarged.

rapidly and produce the usual large number of gametes; the resulting larvae settle mainly during the winter. Louis Agassiz found this form in San Francisco Bay in 1865. There is at least a distinct possibility that it is a relic of the days of wooden ships, for the same species occurs on the east coast, and it seems unlikely that its natural distribution would account for its occurrence on this coast also. If it was so imported, it probably became established at several different places, since it now ranges along the entire coast as far south as Chile.

§334. Any of several crabs may be seen crawling about on the piling throughout the intertidal zone but keeping fairly well under water, whatever the height of the tide. Practically all of the rocky-shore forms have been reported, and two spider crabs seem to be characteristic. In Tomales Bay and southward to Colombia a thoroughly attenuated, gangly form, *Pyromaia tuberculata,* usually 15 mm or less in carapace width, is fairly common. Sponges and seaweeds mask this sluggish little crab, so that it will seldom be noticed until what appears to be a bit of fixed piling growth moves slowly away from the observer.*

Oregonia gracilis is very similar, but twice the size of *Pyromaia.* It may be seen occasionally on wharf piling in Puget Sound and in eelgrass beds, whence it ranges northward to the Bering Sea and southward, but in

---

*According to Mary Wicksten, Ricketts may have confused *Pyromaia* with *Podochela hemphilli* here. The former, although common in harbors, does little in the way of actively masking its body, whereas the equally gangly *Podochela* is a master of camouflage. Of course, as noted previously, even crabs that are not active maskers may be covered with growths of uninvited settlers.

deeper water, to Monterey. The masking habit of this group of crabs is discussed in §131.

Several cucumbers occur, in the Puget Sound area especially. Huge individuals of *Parastichopus californicus* (§77) crawl about on the piling frequently enough to be considered characteristic here, but they keep pretty well below the low-tide line.

An unexpected result of the Korean War (apparently) is the naturalization in San Francisco Bay of *Palaemon macrodactylus*, a brackish-water shrimp about 5 cm long. It was first noticed about 1954 in San Francisco Bay, and is now common in Marin County in small streams flowing into San Francisco Bay and in the main river as far as Collinsville and Antioch. Introductions can rarely be dated as closely as this one has been, and its progress into the Delta is being followed with interest. The shrimp appears to have been introduced from the seawater system of ships returning from the Korean peninsula, its native haunts. Populations have also been found now in Elkhorn Slough and Los Angeles Harbor. Common on wharf piling and among algae, *Palaemon* now seems more abundant than the native *Crangon*, from which it may be easily distinguished by its longer and apparently more numerous antennae and by its long, toothed rostrum. The obvious vigor and success of *Palaemon* may make it an important food resource for the fish populations of the region, and since this shrimp also thrives well in captivity, it may become an important experimental animal in our halls of learning.

§335. On these fully protected piles we find a good many of the animals already mentioned as occurring on more-exposed piling: the seastar *Pisaster*, both the common *P. ochraceus* and *P. brevispinus* (§209), and the usually white anemone *Metridium* (§321).

There are the usual tunicates; *Styela clava* (§322) is common in southern California harbors especially. The elongate simple tunicate *Ciona intestinalis* reaches a length of from 10 to 12 cm. Its translucent, green siphons are long and glassy, the basal portion being often covered with debris. This is one of the best-known ascidians, well distributed throughout the world in ports and bays but requiring relatively clean water. Some of the small rock-loving tunicates, notably the red *Cnemidocarpa* (§216), are found on Puget Sound piling, and the gelatinous clusters of the compound *Botrylloides diegensis*, in reds, yellows, and purples, are features of the piling in San Diego, Monterey, and Bodega bays.

Referring to another tunicate, *Pyura haustor johnsoni* (now regarded as merely a local variant of the more widespread *P. haustor*, §216), W. E. Ritter remarked: "A striking thing about this species is its great abundance in San Diego Bay, and the large size reached there by individuals, as compared with what one finds on the open shores. Its favorite habitat appears to be the piles of wharves where, at times, it makes almost a solid coating. Although it must be counted as a native of the whole littoral zone, we have found only occasional specimens at outside points."

In addition to the native tunicates, a small, dingy Atlantic tunicate, perhaps unnoticed in most places, appears to have successfully crossed the continent. *Molgula manhattensis* was found by J. E. Lynch on the piling at Marshall in Tomales Bay in 1949. It has spread fairly widely since then and now occurs regularly on floats and along the bottom in San Pablo and San Francisco bays.

§336. Boring forms are as common and destructive here as on less-sheltered piling. Boring isopods (*Limnoria*, §329) operate quite indiscriminately in these quiet waters, assisting the shipworms in making life miserable for harbor engineers. Another isopod of ill fame is the larger *Sphaeroma quoyana* (often formerly misidentified as *S. pentodon*), which grows up to about 11 mm long. Despite its size, however, this form cannot compete with *Limnoria* in destructiveness, at least in terms of piling. It rarely seems to attack until the wood has already been riddled by shipworms. Wood is for it apparently a secondary habitat, for the animal (which resembles *Cirolana*, Fig. 34) is found in great numbers in clay and friable rock. Indeed, the burrowing activities of *Sphaeroma*, combined with wave action, appear to have caused (or at least accelerated) the erosion and collapse of mud levees and banks in San Francisco Bay. The economic importance of this may be considerable. *Sphaeroma* bores for protection only, making oval openings up to nearly 10 mm in diameter that give the pile or rock a pitted appearance. Boring is accomplished with the mouthparts, as with *Limnoria*, but the loosened material (at least in the case of an observed specimen boring in chalk) is passed backward by the feet and washed out by the swimmerets. The fact that very little wood is found in the intestines of wood-boring specimens indicates that wood particles are handled in the same manner. Algae and other growths are the food supply.

Eggs are carried under the thorax of the female for a time, and young have been found with the adults in the spring, summer, and fall in San Francisco Bay; breeding probably lasts all year. The range is from Humboldt Bay to Bahía de San Quintín, Baja California; the same species also occurs in Australia and New Zealand.

§337. Seventy years ago the European shipworm, *Teredo navalis*, was unrecognized on this coast. Just when or how it gained a foothold is not known, but in 1914 it was discovered that piling at Mare Island, in the northern part of San Francisco Bay, had been extensively damaged. The attack was repeated in 1917, and within 3 years the damage reached unprecedented and catastrophic proportions. Ferry slips collapsed, and warehouses and loaded freight cars were pitched into the bay. All piling was attacked, and most of it destroyed. Since then *Teredo navalis* has spread to all parts of the bay. The animals can stand much lower salinity than other shipworms, which accounts for their devastating attacks in the northern parts of the bay, where piling was largely untreated because it had been supposed that the influx of fresh water from the Sacramento and San Joaquin rivers afforded immunity from borers. The fresh water does kill them off during seasons of heavy rainfall, but when conditions improve, they come back in force.

These animals are usually hoist by their own petard within 6 months, for their honeycombing of timber is so complete that by the end of that time it begins to disintegrate, the resulting exposure causing their deaths. They can reach a length of more than 10 cm in 4 months, however, and one 14-month-old specimen was nearly 30 cm long. The rate of boring is more rapid in comparatively young animals; a 3-month-old specimen bores about 2 cm a day. Thus the animals can ruin a pile during their short lives, making ample provisions in the meantime for another generation to take care of the next set of piles.

Unlike *Bankia*, *Teredo* breeds during the summer and early fall, the last lot of larvae being attached by December. The sexes are separate, but the same individual is first male, then finally female. Eggs are retained in the gill cavity until fertilized by spermatozoa contained in the incoming water. They are then stored in a brood pouch that forms in the gills until they are ready to be discharged through the parent's siphons as free-swimming bivalves known as veliger larvae. Apparently the breeding period is continuous for some weeks, for the brood pouch contains larvae in various stages of development, while at the same time the ovaries are enlarged with unfertilized eggs. Since the ovaries may contain an estimated million or more eggs at one time, the tremendously rapid spread of the animals is easy to understand.

In the first century, Pliny the Elder conjectured that teredos bored with their shells. His theory went unchallenged until 1733, when the Dutch naturalist Sellius announced that since the shells appeared to be inadequate for the work, the boring must be accomplished by the foot. Recent investigations have vindicated Pliny. R. C. Miller, working for the San Francisco Bay Marine Piling Committee, succeeded in cutting into a burrow and covering the opening with a piece of glass. Through this window observers watched *Teredo navalis* at work. The animal rasps the wood with its toothed shell by holding on with its foot and alternately contracting the anterior and posterior adductor muscles of the shell as do the rock-boring pholads. This rocks the valves in a motion that is repeated rhythmically 8 to 12 times per minute. Also as in the pholads, the animal rotates on its foot such that a perfectly cylindrical burrow is cut. Interestingly, there are two distinct grades of rasping surface on the shell, and the animal's intestines contain two correspondingly different sizes of wood particles.

All the wood removed from the head of the burrow passes rapidly through the animal's stomach into its intestine and is finally removed via the siphons. *Teredo* extracts some nourishment from this wood en route, but additional food comes, as with all bivalves, from plankton filtered from the respiratory stream by the ctenidia.

§338. In Los Angeles and San Diego harbors, and in Hawaii, the havoc-working shipworm is the small *Lyrodus pedicellatus* (Fig. 320; formerly *Teredo diegensis*). Reaching a length of 13 cm but averaging nearer 5 cm, it is the smallest shipworm occurring on the Pacific coast. Breeding apparently takes place throughout the year, but the heaviest infections by lar-

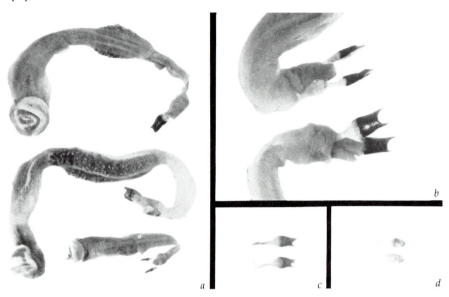

Fig. 320. *Lyrodus pedicellatus* (§338), a primarily southern shipworm: *a*, entire animals, showing enlarged brood sacs with larvae; *b*, posterior ends, showing the siphons and the flaplike pallets used to close the entrance to the burrow; *c*, normal pallets; *d*, pallets with horny tips removed.

vae take place when the water is warmest, from April to October. This is just the opposite of what takes place with *Bankia*, and differs somewhat from the breeding of *Teredo navalis*. As with *T. navalis*, the young are retained in a brood pouch until well along in their development, but they are larger and fewer in number. In the brood pouch of an average female 490 were counted. The brown-shelled, nearly globular larvae are plainly visible through the distended body wall of the parent.

Lyrodus has been found (1920) at one spot in San Francisco Bay—a small inlet near South San Francisco, where, apparently, it found survival conditions after being accidentally introduced from the south or from Hawaii. Although it has been collected in San Francisco Bay periodically since 1920, this warm-water species is unlikely to become an important destructive factor in San Francisco Bay's relatively cold waters. It has also been reported from Elkhorn Slough.

PART V

Between and Beyond the Tides

Any ecological arrangement will be inexact, suggestive rather than definitive. Inexactness and inconsistence, incidentally, might be regarded as prerogatives of the animals themselves. As though the flowing reality of natural things were conspiring against our attempt to catalog them neatly, as though nature refused to cooperate and disapproved our intellectual need for realizing all phenomena in discrete states: a given animal may occur in several environments, and even its primary assignment to one may not be certain until documented by quantitative methods. However, ecological arrangements need not be discarded because of these inadequacies. They ought rather to be used guardedly and for what they are worth, with due regard for their limitations. . . .

E. F. Ricketts
Zoological Preface to an unpublished
San Francisco guidebook (ca. 1940)

. . . both correlative pattern matching and seemingly rigorous experimentation can lead into a trap of perpetuating preconceived ideas. This trap can be seen in almost all scientific research where the emphasis is on the verification rather than falsification of hypotheses. The verification of ideas may be the most treacherous trap in science, as counter-examples are overlooked, alternative hypotheses brushed aside, and existing paradigms manicured. The successful advance of science and the proper use of experimentation depend upon rigorous attempts to falsify hypotheses.

P. K. Dayton and J. S. Oliver
(1980)

11 ⟨⟨ Intertidal Zonation

Studying the arrangement of the plants and animals of the shore and shallow sea in horizontal belts (Fig. 321) is probably the oldest and most "durable" concern of marine biology. Perhaps the phenomenon was discussed in some lost book of Aristotle's. In our time, at least, the scientific literature on the subject has been accumulating since the 1840's. By 1929, when Torsten Gislèn summarized the work on seashore ecology in a masterly review of an exhaustive bibliography (Gislèn 1930), there was already a vast literature; and by 1938, enough had been published on zonation to cause MacGinitie to express an opinion that the subject had "been decidedly overdone" (1939: 45). In view of what has been published since then, however, it would seem that discussion of the subject had hardly begun.

In a sense, *Between Pacific Tides* is an expanded paper on intertidal zonation. It was the first book for general readers to discuss the distribution of animals on the seashore according to their levels of occurrence and to attempt to use this zonal distribution as an aid to identification. Although the original edition of this book was essentially completed in 1936 (but not published until 3 years later), Ricketts, as early as 1930, was working on a paper on intertidal zonation. This paper, "The tide as an environmental factor, chiefly with reference to intertidal zonation on the Pacific coast," was never published. It was freely discussed with anyone available, and when Gislèn, from Lund, Sweden, spent the winter of 1930–31 at Hopkins Marine Station, the manuscript was, as he acknowledged, at his "free disposal." It was also made available to Willis Hewatt, who studied the zonation along a transect at Hopkins Marine Station from 1931 to 1934 for his doctoral dissertation (see 1937), and who was permitted to use one of the illustrations from the paper (Fig. 4 of this edition). It seems apparent that Ed Ricketts was developing his own ideas of zonation independently; and since he was blissfully oblivious of the academic need to publish or perish, he gave freely of his ideas.

Fig. 321. Zonation on an exposed rocky shore. The dark, middle band is a dense bed of the California mussel; below it, the surfgrass *Phyllospadix*; above the mussel bed, a light-colored zone of barnacles and tufts of the red alga *Endocladia*.

The Tides

Although we cannot afford a detailed discussion of the nature and mechanics of tides, we should say something about them, for tides, either directly or indirectly, greatly influence the distributions of plants and animals on the shore. Among the many interrelated factors that produce tides, the gravitational and centrifugal forces at work between the earth, moon, and sun are the most important.

The moon provides the dominant influence. As the earth and moon rotate about each other in space (like the knobs of a baton thrown in the air), they are held in position relative to one another by a balance of two critical forces. Gravitational forces tend to attract the two masses, and centrifugal forces keep them separate. On earth, the centrifugal force from this rotation is everywhere equal in magnitude and direction, the direction being essentially away from the moon. Directly opposing this centrifugal force is the moon's gravitational pull, which is directed toward the moon and is much stronger on the side of the earth closest to the moon, since gravitational force diminishes rapidly with distance. The net effect of these two opposing forces varies depending on one's location on the earth relative to the moon's position. Thus, on the side of the earth closest to the moon, strong gravitational forces override the centrifugal forces and result in a pull away from earth (toward the moon). On the opposite side of the earth, centrifugal forces override the weaker gravitational forces and again result in a net force directed away from the earth. To see what this has to do with tides, one might first imagine an earth completely covered with water. On such a globe, the combined forces would create two bulges of water on opposite sides, one closest to the moon and one farthest from it. The earth, rotating on its axis once every 24 hours, might then be seen to rotate under these bulges. A point on the earth's surface thus would experience alternately a high tide, a low tide, another high tide, and another low tide every 24 hours. Well, almost every 24 hours. During the time taken for the earth to complete one revolution on its axis, the moon has advanced somewhat in its orbit, requiring, as it turns out, an additional 50 minutes of the earth's rotation for a point to "catch up" to its previous position directly opposite the moon. Thus, a complete tidal cycle of two high tides and two low tides (a tidal cycle referred to as semidaily or semidiurnal) takes about 24 hours and 50 minutes, or one lunar day. This is why the low tides of tide tables occur approximately 50 minutes later each day.

The sun, in a similar manner, also generates tide-producing forces on the earth. However, the sun is so much farther away that, even though the sun is much larger, its influence is only about half that of the moon. Nevertheless, when the sun and moon are more nearly in a line with the earth (as during a new or full moon), their forces combine to produce tides of much greater range than average. These tides of greater range are known as spring tides (which has to do not with the spring of the year but with the Anglo-Saxon word "to jump," which we also use for certain mechanical devices). When the sun and moon are at right angles to each other, one week later, the tidal forces of the sun partially cancel those of the moon, and the result is tides of minimum range, known as neap tides. A neap tide therefore is not a low tide, but a tidal cycle of lesser range.

This fairly straightforward combination of lunar and solar influences seems sufficient to explain the more or less even progression of high and low tides, called equal semidaily tides, that are characteristic of much of the Atlantic coastline. However, the tides on the eastern side of the Pacific, California to Alaska, are characterized by pairs of high tides and pairs of low tides that differ greatly in magnitude. These are known as mixed semidaily tides (Fig. 322), and their varying magnitude can be related in large part to the declination of the moon, that is, to the moon's angle above or below the celestial equator. If the moon were always directly opposite the equator, the result would be two high tides and two low tides of equal magnitude. However, because of the tilt of the earth's axis (and to a lesser extent the tilt of the moon's orbital plane with respect to the orbital plane of the earth), the moon is rarely directly over the equator. Instead, as the moon moves in orbit around the earth, it moves higher and lower in the sky, somewhat north or south of the equator, taking the tidal bulges with it and

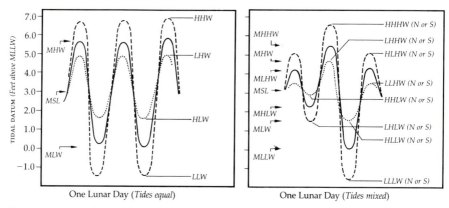

One Lunar Day (*Tides equal*) — One Lunar Day (*Tides mixed*)

Fig. 322. Types of tidal curves (after Doty): solid line, mean range of tides; broken line, maximum range; dotted line, minimum range; *M*, mean; *SL*, sea level; *N*, neap; *S*, spring; (*L*)*LLW*, (low) lower low water; (*L*)*HLW*, (low) higher low water; (*H*)*HLW*, (high) higher low water; etc.

Fig. 323. Largely because of the tilt of the earth's axis, the moon is seldom directly over the equator, and consequently the tidal bulges created by the moon are usually out of line with the equator. Thus a location on earth, *A*, would experience a moderately high tide, followed by a much higher tide about 12 hours later, when this location had rotated to position *A*₁.

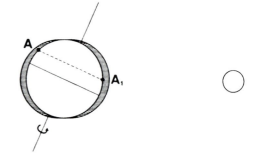

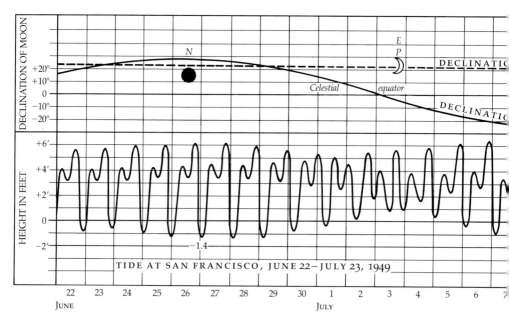

TIDE AT SAN FRANCISCO, JUNE 22–JULY 23, 1949

Fig. 324. The relation between the transits and phases of the moon and the ranges of the tides in a mixed tidal pattern.

daily changing the heights of the two sets of tides. For example, in Figure 323, location A on the earth's surface (e.g., Monterey) would experience a moderately high tide, followed by a much higher tide 12 hours and 25 minutes later when this location had rotated to position A_1.

Although each of the factors discussed plays an important part in determining the type and magnitude of the tidal changes characteristic of a particular location at a particular time, we would be remiss if we did not note that many more factors are involved. The position of land masses, for example, so conveniently ignored in the preceding discussions, has a significant influence on tidal regimes, and there are innumerable subtleties such as the fact that neither the moon nor the earth moves in a perfectly circular orbit. Rather, their orbits are more nearly elliptical, and as a consequence, the exact distance between the moon and the earth changes daily, thus affecting the magnitude of gravitational forces and the resulting tides. Indeed, not all the factors influencing tides are well understood.

As a practical matter, however, and despite all these difficulties in understanding precisely how tides work, tidal conditions for any given day on any selected stretch of coastline can be predicted years in advance, simply because the periodic nature of tides is so easily observed and recorded. Figure 324, for example, shows the predicted tidal cycle for a lunar month at San Francisco, together with the position of the respective heavenly bodies. One need not be a physicist to plan a trip to the shore.

Be aware, however, that predicted tide levels can be profoundly influenced by changing local conditions, such as weather, so that actual levels

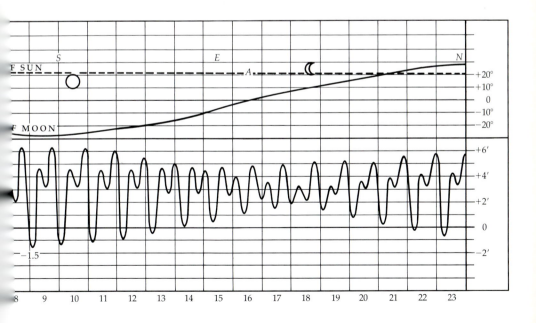

may differ considerably from those predicted. Visitors to the shore should keep this in mind. Low barometric pressure may raise the water level by a few centimeters to as much as several meters in rare cases. Onshore winds may pile water onto the coast half a meter higher than the predicted level. And incoming ocean waves two meters high are certain to make a low tide resemble high tide, even if, on the average, the level is as predicted.

Patterns of Zonation

With this introduction to the tides themselves, we can now look at patterns of intertidal zonation and how they relate to the tidal cycle. In this section, we will primarily identify and describe the zones; in Chapter 12, some of the physical factors and biotic interactions that create and influence zonation patterns will be discussed.

That zonation occurs is obvious to all who visit the shore. However, the number of zones that may be recognized depends not only on the complex variables of tide, climate, and the life subject to these variables, but also on the degree of refinement the observer seeks to attain. After all, we are considering the distribution of several hundred distinct species of animals and plants. Some of these individual distributions are wide; others are narrow. Some have sharp boundaries; others fade gradually. Several species may share precisely the same upper and lower limits of distribution; other distributions overlap in part, still others not at all. Standing back from the detail, however, some number of broad zones is apparent to everyone.

In Figure 325 some of the various ways in which zonation has been recognized on the Pacific coast are represented. At first glance, this appears hopelessly complex. Even Hewatt and Ricketts, working in the same locality, did not quite agree. The samples from Doty's study (1946), however, show that differences may occur in algal zonation in the same locality under different conditions of exposure. Rigg and Miller set up two schemes of zonation (1949: 342), one for algae and the other for invertebrates. These zones show an essential similarity to those of central California, despite the much greater tidal range at Neah Bay: "The major difference is one of nomenclature, the present writers having regarded mid-tide not as a zone, but as the boundary between the upper and lower intertidal regions. A very little vertical displacement would bring the two sets of concepts into harmony." To these various zonal arrangements we have added, principally for historical interest, the community terms of Shelford et al. (1935).

A few words of caution about these diagrams (Figs. 325, 326) are in order. They are, especially as presented in this interpretation, highly schematic, and it is, in a way, doing the authors an injustice to fix these zones on a scale of tidal heights. We are actually dealing with proportions related to tidal heights at specific localities, which are quite different from the fixed levels based on a standard established for a gauging station. Hence the

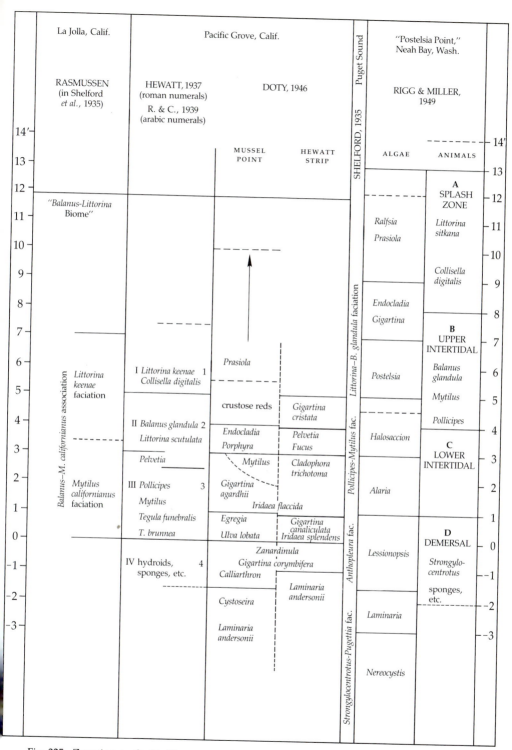

Fig. 325. Zonation on the Pacific coast, as observed by various authors. The numbers at either side represent the height in feet above or below mean lower low water (tidal datum).

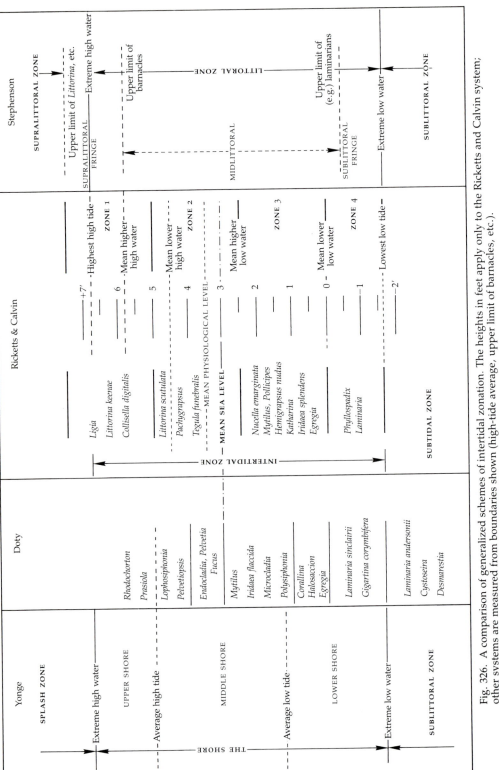

Fig. 326. A comparison of generalized schemes of intertidal zonation. The heights in feet apply only to the Ricketts and Calvin system; other systems are measured from boundaries shown (high-tide average, upper limit of barnacles, etc.).

measurements of Hewatt, Ricketts, and Doty, in referring to tide levels at San Francisco, are abstractions or idealizations of conditions that may not even occur on the rocks near the gauging station. The water levels of the schemes of Yonge (1949) and Stephenson and Stephenson (1972) in Figure 326 are not strictly comparable with those at San Francisco in the center of the figure, an indication of the difficulties inherent in reconciling the semi-daily and mixed tidal patterns. Zonation, like everything else in nature, is not a simple pattern, although we would sometimes like to have it so. Zones may overlap; they are quite different on a vertical face and a gently sloping reef, other things being equal; and in the case of some algae, they may change with the seasons. The Stephensons provided a detailed account of some of these variations in their Nanaimo papers (1961).

Although T. A. Stephenson insisted that zonation is not caused directly by the tides, the patterns of zonation are nevertheless associated with tidal patterns. A number of workers have noticed that in regions of mixed tides there are more-discrete zonal divisions than in regions of equal semidaily tides. The subzones recognized by the Stephensons at Nanaimo correspond to what Russian workers recognize as various "stories," or subzones, associated with complicated tidal patterns. Mokyevsky (1960), however, argued that the actual type of tide is not important in the "littoral bionomy," and that the differences merely cause "a certain shifting of some critical levels in the vertical distributions of the organisms." Nevertheless, it is this "certain shifting" that complicates zonation patterns.

Virtually all students of zonation agree that one break is well defined. This is the approximate 0.0 of tide datum, or mean lower low water, below which is the zone of laminarians, and of *Phyllospadix* on our coast (see Figs. 327, 328). There is also agreement on the zone above the average high tide (or mean higher high water of the Pacific coast), frequented by certain littorines and limpets. These two regions are (roughly) the sublittoral and supralittoral fringes of Stephenson's terminology, the *D* and *A* of Rigg and Miller, and, of course, Zone 4 and Zone 1 of this book (Fig. 4).

There remains the intervening zone or zones, the multilittoral zone of Stephenson's proposed universal system. Because of the dual application of "zone" to both the inclusive littoral (= intertidal) and the subsidiary midlittoral in this classification, we have reduced "midlittoral zone" to "midlittoral." Since it separates the fringes, and since one can hardly call the zone between two fringes a fringe, it is simpler to leave the term an unadorned adjective. The gap between Stephenson's supralittoral and littoral zones is, according to him, an inconsistency of nature rather than of his classification.

As can be seen from Figure 326, the broad midlittoral of Stephenson has at least two natural divisions on this coast, the break occurring at mean sea level (3.0 feet; 0.91 meters) according to Hewatt and at 2.5 feet (0.76 meters) according to Ricketts. Stephenson, during his trip to the Pacific coast in 1947, recognized this division and tentatively called the two parts

Fig. 327. The *Laminaria* Zone: laminarians ("kelp"), which grow in the low intertidal, exposed at low water.

Fig. 328. The surfgrass *Phyllospadix*, growing on rocks in the low intertidal.

the upper and lower "balanoid zones" in his privately printed *Report on Work Done in North America During 1947–48* (1948). The terminology is now abandoned, and he would call them upper and lower midlittoral. Doty's scheme can be reconciled with the four-zone system of this book by considering his *Endocladia-Pelvetia-Fucus* zone a lower subdivision of Zone 2, 2*b* perhaps; and similarly, his zone of *Corallina*, etc., becomes 3*b*.

Difficulty is sometimes encountered in distinguishing Zones 2 and 3 on the basis of animal distribution. Rigg and Miller, for example, found no animals on their Zones *C* and *D* abundant enough to serve as indicators (we have added *Strongylocentrotus* to the diagram); algal growth is so heavy that it obscures animal life at this locality. We must resort to the algae, and here Doty's work is especially useful. Although Figure 326 does not list all the algae that are restricted in their vertical distribution, it does name the most conspicuous species or genera. These should be recognizable with the aid of the drawings in the next section, which are arranged approximately according to their vertical distribution. Hopefully, no one will suppose that he can now get along without a handbook to the algae by using this handful of drawings; they are intended simply as a rough guide to the more pronounced zonation to be found on this coast, and nothing more.

In his exhaustive book on British shores, Lewis (1964) used the term Littoral Zone, which he divided into the Littoral Fringe, between the upper limit of *Littorina* and *Verrucaria* and the upper limit of barnacles (all of Zone 1 and perhaps a bit of Zone 2 in our terminology), and the Eulittoral Zone, between the upper limit of barnacles and the upper limit of laminarians (all of our Zones 2 and 3). The region below the *Laminaria* is termed the Sublittoral Zone.

The Stephensons were probably correct in postulating that there is a universal scheme of zonation applicable to all parts of the world, in the sense that there are generally three divisions that may be recognized on a seashore at low tide: the upper, middle, and lower parts. But our consideration of zonation is basically biological in approach, and when it comes to saying that a zone of algae in one part of the world is equivalent to a zone of ascidians in another because they are at about the same tidal level, we are not discussing biological equivalents. We cannot ignore the question: what is essentially different about a shore that supports a dense growth of filter-feeding animals, as opposed to one that is characterized by seaweeds at an equivalent tidal level? There is the danger, as Hodgkin (1960) observed, that preoccupation with universal schemes can lead one to fit observations into the scheme to the extent that what is really going on may be overlooked.

Although many more references to intertidal studies in various parts of the world will be found listed in the General Bibliography, we will not attempt to call the roll for all of them here. However, some important review papers or books since Gislèn (1930), Doty (1957), and Lewis (1964) are Stephenson and Stephenson (1972), Connell (1972), Newell (1979), and

Underwood (1979). A comparison between the later papers on this list, particularly those by Connell and Underwood, and the ones of an earlier vintage will also reveal a basic change that has occurred within the field of marine ecology. There has been a fundamental shift away from studies of a largely descriptive or correlative nature and toward studies that emphasize experimentation. The goal of these experimental studies is to understand the processes or variables that underlie or influence such phenomena as zonation. These processes and variables are examined in the next chapter.

Algal Zonation on the Pacific Coast

Following is a series of drawings showing the characteristic plants of each intertidal zone (Figs. 329–37). All the drawings have been taken from G. M. Smith's *Marine Algae of the Monterey Peninsula* (Stanford, 1944), except for *Pseudolithophyllum neofarlowii* (Fig. 329, right), which was taken from Abbott and Hollenberg (1976).

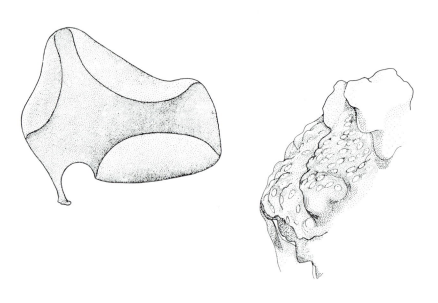

Fig. 329. Two upper-zone algae, from Zone 1: *left, Prasiola meridionalis*, a small, leafy green alga commonly found on the highest rocks (frequently among lichens) or in isolated high tide pools, where they often receive bird droppings (×18); *right, Pseudolithophyllum neofarlowii*, a crustose red alga (×3).

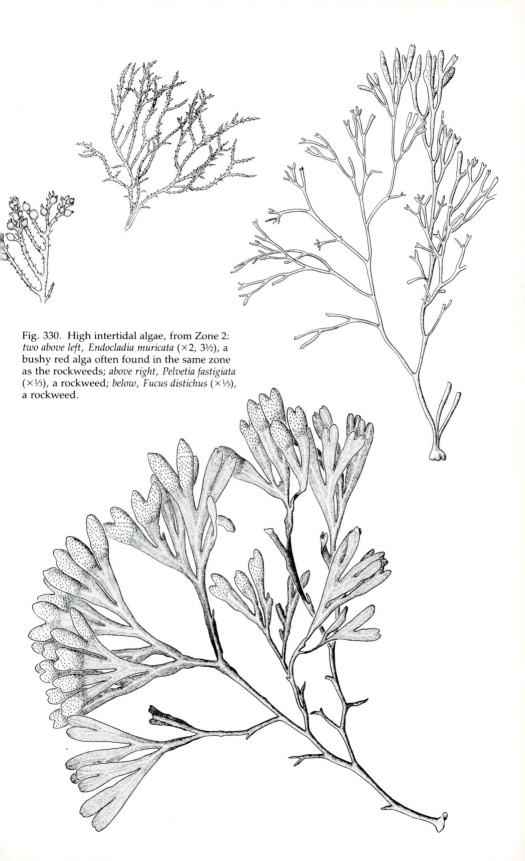

Fig. 330. High intertidal algae, from Zone 2: *two above left, Endocladia muricata* (×2, 3½), a bushy red alga often found in the same zone as the rockweeds; *above right, Pelvetia fastigiata* (×⅓), a rockweed; *below, Fucus distichus* (×⅓), a rockweed.

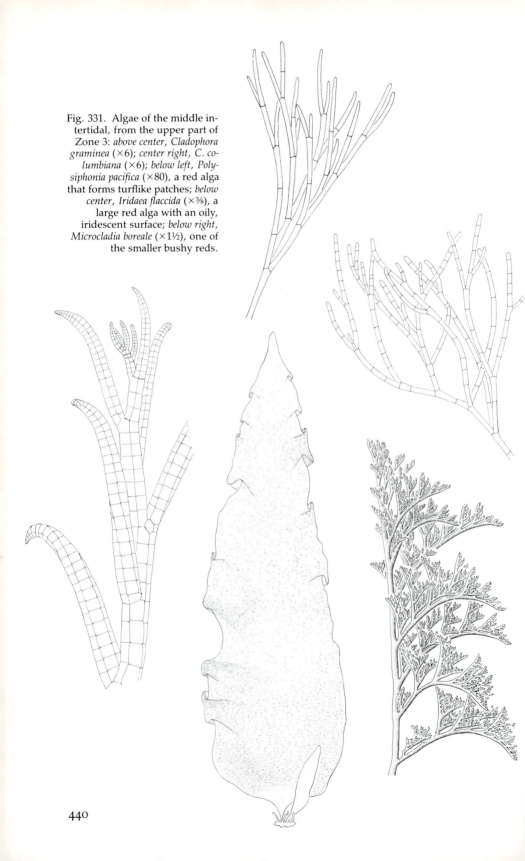

Fig. 331. Algae of the middle intertidal, from the upper part of Zone 3: *above center, Cladophora graminea* (×6); *center right, C. columbiana* (×6); *below left, Polysiphonia pacifica* (×80), a red alga that forms turflike patches; *below center, Iridaea flaccida* (×⅜), a large red alga with an oily, iridescent surface; *below right, Microcladia boreale* (×1½), one of the smaller bushy reds.

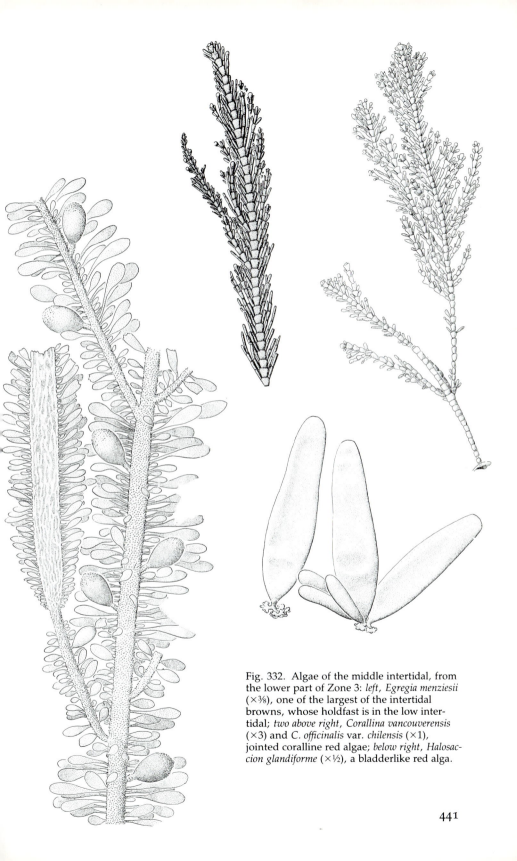

Fig. 332. Algae of the middle intertidal, from
the lower part of Zone 3: *left, Egregia menziesii*
(×⅜), one of the largest of the intertidal
browns, whose holdfast is in the low intertidal; *two above right, Corallina vancouverensis*
(×3) and C. *officinalis* var. *chilensis* (×1),
jointed coralline red algae; *below right, Halosaccion glandiforme* (×½), a bladderlike red alga.

441

Fig. 333. Algae of the *Laminaria* Zone, Zone 4, and below: *above left, Prionitis lanceolata* (×⅓), though common, is also extremely variable in size and branching pattern, and in its distribution from the intertidal to the subtidal; *center left, P. australis* (×½); *below left, P. lyallii*, frequently slippery-surfaced with elongate lateral blades; *right, Iridaea cordata* var. *splendens* (×¼), usually present in deep bands one or more meters deep and of a dark-purple to brown color.

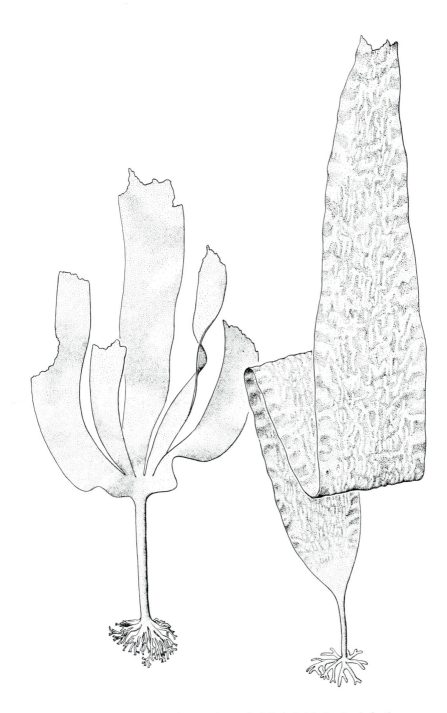

Fig. 334. Two of the laminarians, brown algae called "kelp": *left*, *Laminaria dentigera* (×¼), commonly marking the −2.0-ft tide level with their holdfasts, although subtidal plants are known with stipes up to 3 meters long; *right*, *L. farlowii* (×¼), low intertidal in deep pools in northern California but more commonly subtidal in southern California.

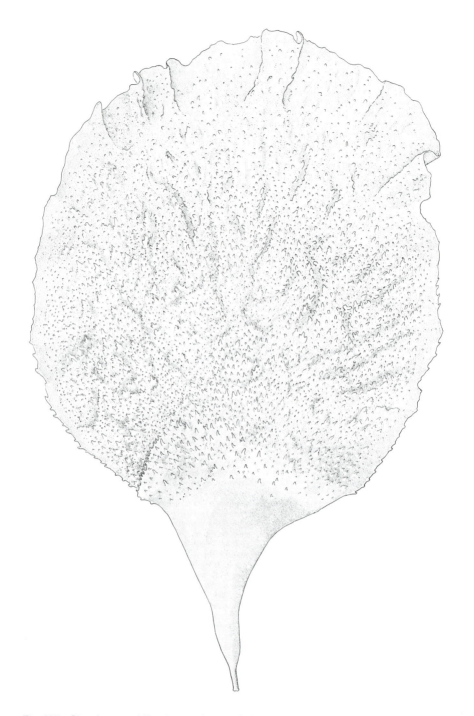

Fig. 335. *Gigartina corymbifera* (×½), a large red alga from the −1.0-ft tide level to below most kelp beds in central California (30–40-ft depth).

Fig. 336. *Cystoseira osmundacea* (×⅜), a large, brown, subtidal seaweed forming beds inshore from *Macrocystis* ("giant kelp").

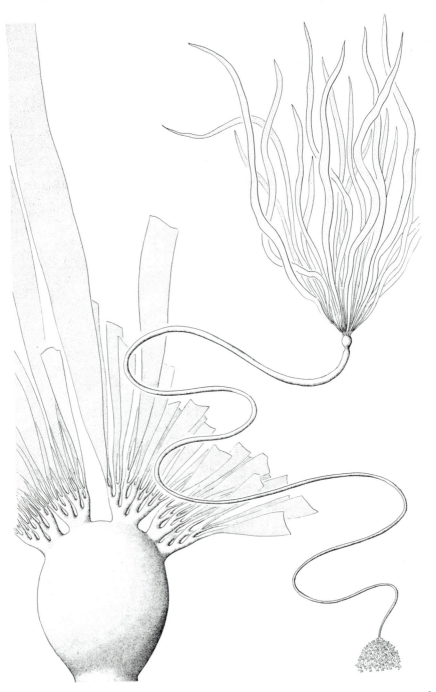

Fig. 337. The whiplike *Nereocystis luetkeana*, a subtidal species that is often cast up on sandy or rocky beaches (usually minus its holdfast): *right,* the entire plant, greatly reduced; *left,* the hollow float, or pneumocyst, and the bases of the blades (×½).

12 ⬧ Some Principles of Intertidal Ecology

Between Pacific Tides is basically a book about animals, what they look like, what they do, where they live, and how they are distributed on the shore. This is not a book about ecological "principles," at least as such, although many of the topics and principles that today form the basis of marine intertidal ecology are scattered among the various observations of organisms and habitats that occupy most of the preceding pages—Ed Ricketts was far ahead of his time. The purpose of the present chapter is to gather and organize these scattered topics, along with some new ones, into a general framework of variables that might be fruitfully considered when examining organisms of the shore. Three of these variables—wave shock, substratum type, and degree of tidal exposure—have such a great influence on distributions that they are used as the basis for the book's organization, first considered in the Introduction. We should also point out that these factors, and the others to be discussed, are important not only because they influence distributions, but because they have served, and continue to serve, as agents of natural selection. Studying them can help explain the myriad adaptations of form, function, and life history that are observed among the organisms of the shore.

Of the various factors that influence the distribution of organisms on the shore, we will focus on those influencing vertical, or tidal, zonation (Fig. 342). But we should keep in mind that the plants and animals that are considered characteristic of certain vertical zones do not always occur uniformly in those zones. Patches of organisms, or even bare space, frequently occur: one boulder in a boulder field may be covered with mussels, and next to it, another may be covered with algae. This mosaic in the horizontal dimension is a vital contributor to the richness of intertidal life. Variation along yet a third dimension, depth in the substratum, is an additional important consideration on some shores.

To be sure, many important factors influence intertidal distributions, and they are sometimes frustratingly interrelated. But such is ecology. It is the study of the interrelations between organisms and their environment, both physical and biotic. It is by definition about interrelations and about the nearly infinite number of different ways that life's variables, whether on a scale of molecules or community interactions, can be combined to produce successful organisms. Out of convenience, each factor will be examined separately, as a physician might examine the heart, lungs, and kidneys of a patient; but this separation is, of course, rather unnatural. Each factor is part of an interrelated whole, and the whole may be more than the sum of these parts.

Wave Shock

Anyone who has walked a stretch of coast extending from an exposed headland to a protected inlet already knows that the kinds of plants and animals change with the degree of exposure to wave shock (Fig. 338). Some of this change in species composition is directly due to the impact of waves. On rocky headlands, plants and animals must withstand the tremendous impact of tons of water breaking over them, and here, tough or hard bodies and strong attachment mechanisms are a specialty. Many of the plants and animals that are abundant in relatively protected locations are simply not equipped to take the battering and the shearing force of heavy surf, and as a direct result, fewer species are often found on the open coast than in more-protected regions. This relationship does not always hold true, because many species may be absent from protected areas owing to other factors (e.g., competitors, predators, or limited nutrient supplies). But severe wave shock precludes the success of many species on the open coast.

The degree of wave shock may also influence the distribution of organisms indirectly, through its effect on the activity of certain predators and herbivores. On sections of coast exposed to full wave impact, predators such as crabs or the snail *Nucella* may be reduced in numbers by the surf, or they may be largely restricted to protective crevices from which they can make only quick foraging raids on prey populations. Under these conditions, which restrict the abundance or foraging ability of predators, certain prey populations may flourish, while in quieter waters, predation intensity may be so high that all available members of a prey species are consumed. Similarly, herbivores such as littorines may be affected by wave shock, much to the benefit of some algae in exposed areas.

Another indirect effect of wave force, seen in many areas, is that the vertical distribution of certain organisms may become broader along a gradient of increasing exposure to waves (Fig. 339). This effect, most apparent with sessile, upper-intertidal forms such as barnacles, usually affects only upper limits of distribution, although lower limits may sometimes be affected as well. Presumably, more frequent wetting of the upper zones by splash and surf in exposed areas lessens the restrictive influence of desiccation and allows these organisms to live higher on the shore than they would under conditions of quieter water.

Fig. 338. A coastline experiencing various degrees of wave shock, from the heavy surf at Bodega Head to the quiet waters inside of Bodega Harbor. The substratum is equally variable: within a few kilometers, one encounters gently sloping beaches of sand and of mud as well as boulder fields and vertical cliffs of rock.

Fig. 339. The spreading of zones toward the more exposed, seaward side of a large rock. (Doran Beach, Sonoma County, May 1983.)

Substratum

The Pacific coast includes an enormous variety of shores, ranging from steep, vertical cliffs of rock, to gently sloping rock beaches and boulder fields, to flat expanses of mud (Fig. 338). Such variations in the essential nature or type of the substratum, and in the basic topography of the shore, are critical determinants of the distribution and abundance of many intertidal organisms.

So strong is the influence of substratum type, especially whether it is rocky or particulate, that entire assemblages of plants and animals change abruptly with shifts in the substratum. In general, animals that are adapted to cling to the surface of rocky substrata, the epifauna, characterize rocky shores, whereas these tend to give way on sandy or muddy beaches to animals that burrow into the substratum, the infauna. Of course, particulate substrata are not all alike; neither are rocky ones. Markedly different assemblages characterize sandy and muddy shores, for example, and here the size of particles is an important consideration. (Particle size in turn is related to, among other things, water turbulence, because the finest sediments of mud and clay can only accumulate where water is quiet enough to allow particles to settle to the bottom.) On rocky shores, the composition of the substratum, whether sandstone or basalt for example, can influence the kinds of inhabitants, and here too "particle" size can be an important consideration. Boulders, often quite large ones, can be turned over by waves, the rate of turnover depending on the size of the boulder (in addition to wave force). Since most organisms inhabiting a boulder would be killed if it were turned over, the size of a boulder can consequently influence the abundance and kinds of organisms present. Small boulders, turned over frequently, are likely to be inhabited by relatively few species. Although this effect can be recognized on beaches containing boulders of various size, it is perhaps most apparent on cobblestone beaches, which are among the most barren of all rocky shores.

The effects of topography, although usually more subtle than those associated with changes in the basic nature of the substratum, may still be quite influential (Fig. 340). On the open coast, for example, a gently sloping shore will dissipate the force of waves somewhat, providing a measure of protection for the inhabitants. It will also drain more slowly than a steep slope, allowing some of the organisms characteristic of lower zones to extend their ranges upward. Additionally, the slope of the shore will affect the foraging ranges of those many predators that move up in the intertidal to capture prey during high tide and then retreat to lower zones at low tide. On steep slopes, such predators can cover greater vertical distances in a given amount of time, and as a result, the vertical distribution of both predator and prey may be affected.

Another aspect of topography to be considered is its complexity, or

Fig. 340. The varied topography of a Pacific coast rocky shore: *above*, the coastline near Bodega Marine Laboratory; *below*, a closer view, showing the topographical complexity.

Fig. 341. The topography of a mud flat: *above*, ripples and pools on the mud flat in Bodega Harbor; *below*, a close-up of the mounds and depressions created by infaunal burrowers.

heterogeneity, since increased structural heterogeneity can increase the number of individuals occurring within an area, as well as the number of species. Irregularities such as cracks and crevices, the undersides of boulders, and tide pools can all serve as refuges from harsh environmental conditions or predators. Indeed, the abundance of some animals seems to depend absolutely on the availability of appropriate "irregularities." For example, on certain exposed rocky shores, the abundance and size of some littorines seem related to the abundance and width of the protective crevices available to them. To test this notion experimentally, Emson and Faller-Fritsch (1976) drilled large and small holes in boulders inhabited by *Littorina rudis*. Less than 2 years later, the density of littorines on boulders artificially supplied with a high density of holes had increased more than 800 percent over that on control (unbored) rocks. In addition, the mean shell lengths of the snails had increased on rocks with large holes relative to control areas. Of course, heterogeneity is not restricted to rocky shores, although it is most pronounced there. On sandy and muddy beaches, topographical relief is provided by mounds, depressions, and ripples of various sizes (Fig. 341), all of which potentially influence the distributions of organisms.

Tidal Position and Physiological Stresses

Whether intertidal plants and animals are influenced directly by the rise and fall of the tide, or only secondarily by other factors that are in turn related to tides, is a matter of some debate. At the very least, the tidal cycle sets up a predictably changing interface between water and air and establishes a crucial axis along which other factors may influence the distribution of organisms. In particular, an organism's height on the shore determines, first, the total percentage of time (averaged over many tidal cycles) it will spend submerged in water or exposed to air (emersed), and second, the maximum and minimum periods of time it will be submerged or emersed without interruption. These variables should be carefully distinguished because two positions on the shore that differ only slightly in the total amount of time they are submerged, on the average, may differ considerably in the maximum continuous exposure that occurs occasionally.

The basic circumstance is the same in all cases, however: most inhabitants of the intertidal are primarily marine organisms periodically exposed to air, rather than terrestrial organisms periodically exposed to oceanic conditions. As such, they encounter more and more stressful conditions (more and more terrestrial conditions) the higher they occur on the shore. We should probably remind ourselves occasionally that organisms now inhabiting a particular zone are likely to have become well adapted to that zone, whichever one it is. A high-zone barnacle such as *Chthamalus* would be unlikely to "feel" as stressed as a low-zone nudibranch if both were placed, for instance, at mean high water. In addition, we should note that organisms may often be found in optimum numbers in situations that are less than optimal physiologically. The mussel *Mytilus* may be able to grow faster and produce more gametes in the low zone, but because of predators it occurs in greater abundance under less-than-optimum physiological conditions higher on the shore.

In this context, several factors are seen as increasing in severity with increasing height on the shore (Fig. 342). One of these, desiccation potential, is generally conceded to exert a critical influence on the vertical distribution of many intertidal organisms. Intertidal position on the shore generally sets the length of exposure to desiccating conditions, and this length of exposure, together with the conditions of temperature, humidity, and wind speed, ultimately determines the severity of desiccation stress. Since the stress of desiccation increases with increasing height, it is specifically

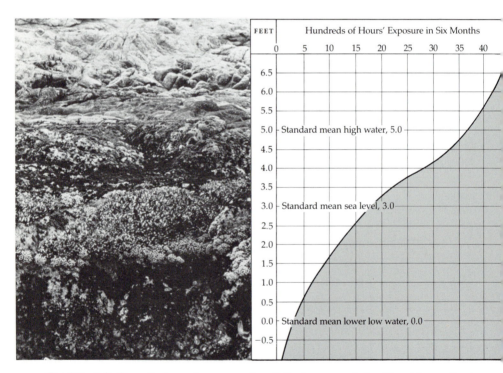

Fig. 342. *Left*, the vertical zonation on a rock wall: the low zone, dark with a rich growth of algae, contains a few seastars (*Pisaster*) and anemones (*Anthopleura*); in the middle zone are patches of mussels and goose barnacles; in the high zone the alga *Endocladia* makes a prominent dark band; and in the splash zone lichens produce a fourth dark band. *Right*, a graph showing intertidal height versus hours of exposure. The vertical levels in the photograph correspond approximately with the y-axis of the graph.

organisms near the upper limit of their vertical range that are most affected.

Although most biologists assume that desiccation is an important influence, relatively few studies have carefully tested this assumption. Notable among these are studies by Foster, Kensler, and Wolcott. Foster (1971) examined four species of barnacle in North Wales and determined that desiccation was indeed an important factor setting their upper limits of distribution. Cyprid larvae and young barnacles were killed in the laboratory by conditions that would be regularly encountered naturally in the upper intertidal zone. Furthermore, since barnacles are sessile as adults, the upper limits of the species' distributions were set by these more-susceptible young stages, even though adults were capable of surviving longer periods of desiccation and, consequently, could theoretically have lived higher on the shore. Foster also found that the success of the four species of barnacle in resisting desiccation was correlated with their relative positions on the shore; species characteristic of the high shore survived desiccating conditions longer than species of the lower shore. Similarly, Kensler (1967) tested the desiccation tolerance of several species of animals that inhabit intertidal crevices, and he too found a correlation between the desiccation tolerance of a species and its position on the shore. Kensler also found correlations within a species: individuals collected from sites higher on the shore had higher desiccation tolerances than individuals collected from lower sites.

The message of Wolcott's study (1973) was slightly different. He investigated five species of limpets common along the coast of California: three are characteristic of the high intertidal (*Collisella digitalis*, *C. scabra*, and *Notoacmea persona*), and two occur in lower zones (*C. pelta* and *N. scutum*). As in the previous studies, the animals higher on the shore survived desiccating conditions in the laboratory longer than those found lower on the shore. In fact, Wolcott calculated that the limpets of the lower zone could not tolerate the conditions of desiccation that were regularly encountered by the high-intertidal limpets. Significantly, however, he also noted that the low-zone limpets maintained a position on the shore well below their calculated upper lethal limit. Clearly then, desiccation tolerance is not the primary factor determining, at least in a proximate sense, the upper limit of these low-zone limpets. Rather, their upper limit is maintained behaviorally. Desiccation did seem to set the upper limit of the high-zone limpets, and many of these limpets were found dead in the field on several occasions, presumably from desiccation. On these occasions, the dead limpets were found at the upper limits of the species' intertidal ranges, and as with barnacles, smaller animals were more susceptible than larger ones.

Although probably secondary in effect to desiccation on temperate Pacific shores, temperature and salinity are two other physical factors that may influence the distribution of intertidal plants and animals. As with desiccation, their influence on vertical distributions occurs primarily during periods of exposure. When submerged, organisms experience relatively constant and benign levels of temperature and salinity, but when exposed to air or when submerged in shallow tide pools during low tides, they may be subjected to wide fluctuations in these variables. The probability that an organism will encounter such extremes of temperature and salinity, as well as the potential duration of exposure, increases with increasing height on the shore.

Temperatures below freezing have been shown to be fatal in some cases, and extremely high temperatures, in addition to increasing the rate of desiccation, may kill directly. As with desiccation, the temperature tolerances of some high-intertidal organisms (barnacles and limpets) tend to be higher than those of low-shore inhabitants, and in the case of some barnacles, periods of exposure to temperatures exceeding the observed tolerances have been at least occasionally encountered in the field. Beyond this, it is difficult to generalize about the effects of temperature extremes because the precise effect depends on not only the absolute temperature, but also the duration of exposure and the speed of temperature changes, as well as interactions with other variables.

Effects of salinity changes on intertidal distributions have been even less commonly observed, except of course in estuarine habitats where salinity obviously limits horizontal distributions. But occasionally, a rapid change in salinity, usually owing to rainfall, will have a striking impact on organisms of the open coast. Luckens (1970), for example, reported that 75 percent of the adult barnacles in one of her test plots in New Zealand were killed when a heavy rainfall coincided with a low tide.

The particular height on the shore also determines the time available for feeding, at least for some animals. Sessile filter feeders such as barnacles and mussels can feed only when submerged, and many mobile animals refrain from foraging when exposed. Thus, the upper limit of distribution of these animals may be closely related to the highest tide level at which sufficient food can be gathered to support growth and reproduction. Similarly, the distributional limits of some intertidal plants may be related to minimum levels of photosynthetic activity.

Finally, it seems worth noting that the various physiological stresses affecting organisms near their upper limit of distribution are likely to have cumulative or synergistic effects, and that stresses need not be immediately lethal to be important. An animal already under stress from inadequate food and prolonged desiccation is more likely to succumb to any additional stress such as that resulting from high temperature or low salinity. Furthermore, even if such an animal survives, it may have no, or greatly reduced, reproductive capacity. In the former case it might be considered "ecologically" dead, although not "biologically" dead. The most important variables relative to reproductive function seem to be temperature and food levels.

Microhabitats

Although a section of shoreline can be broadly described in terms of the general habitat features of wave shock, substratum type, and tidal exposure, scattered along that shore are likely to be numerous microhabitats (Fig. 343), which significantly alter the distributions of organisms. A rock in the middle of a mud flat may be covered with barnacles and limpets. A pocket of sediment in a rocky-beach tide pool may be teeming with burrowing polychaetes, and so on. To the resident of these microhabitats, of course, there is nothing "micro" about the place; it is simply home.

Microhabitats abound on rocky shores. Anywhere a large tide pool drains slowly through a narrow channel at low tide, the residents of the channel, protected from desiccation and provided with increased time for feeding, may extend their vertical range higher than they could just outside the channel. And the tide pool itself is likely to contain organisms that on bare rock could occur only much lower on the shore.

The sides of a large boulder are likely to yield differences in the resident organisms and their distributions, as are the tops and bottoms. In the Northern Hemisphere, the southern face of a boulder is exposed to more sun than the northern side, and the vertical range of organisms may be correspondingly higher on the shady, northern side. Likewise, the protected side of a boulder and the side receiving the full force of incoming waves may be inhabited by entirely different species of plants and animals. When the same organisms do occur on the two sides, their upper limits may be considerably lower on the protected shoreward side.

Furthermore, the simple presence of one organism may influence the distribution of another. Perhaps the most striking example is a mussel bed and its inhabitants: many of the organisms found in a mussel bed are characteristic inhabitants of shores protected from wave shock, but they may occur in great abundance on the exposed coast if provided with the protective microhabitat of the bed. Likewise, limpets and barnacles may find protection from desiccation high on the shore if a moist algal canopy is present. Holdfasts of algae trap sediment and provide protection from wave shock. And there are many other examples as well.

Microhabitats are not restricted to rocky shores, of course. What appears at first glance to be a simple muddy shore is more likely on closer examination to be a mosaic of small-scale patches, microhabitats differing in sediment size, organic content, and other physical and chemical features. In particular, the activities of large infaunal animals such as ghost shrimp, clams, and worms may markedly affect sediment characteristics for several centimeters around their burrows. The deposit-feeding holothurian *Molpadia oolitica*, for instance, produces stable fecal mounds about 2 to 3 cm high and 10 to 30 cm wide at the base. Loose sediment collects in depressions between the mounds and is so unstable that it confines sus-

Fig. 343. Microhabitats: *facing page, top,* water seeping from a high tide pool allows a wedge of algae to extend far up the shore; *bottom,* pools, cracks, and algal cover provide microhabitats for organisms; *above,* small rocks on a mud flat, with mussels, barnacles, and limpets attached.

pension feeders, such as the polychaete *Euchone incolor,* to the fecal mounds rising like mountains above it all (see Rhoades 1974). Other structural elements of soft bottoms are the various tubes and tunnels of burrowers and the roots of seagrasses. They also provide microhabitats and may reduce the mobility of other burrowing species (see, e.g., Brenchley 1982; Ronan 1975).

Competition

Wave shock, substratum, and the many factors influenced by tidal position (e.g., desiccation, temperature, and salinity) are all parts of the physical environment with which an organism interacts. Along with these "physical" factors, "biological" factors—the interactions between organisms—also significantly influence the distribution of many intertidal plants and animals. The importance of these biological influences has been appreciated for a long time, certainly since Darwin's time, but they have received increased attention during the last two decades as potentially critical influences on the distribution and abundance of intertidal organisms. Among the more important of these biological interactions are competition, predation, and herbivory.*

The notion of competition is deeply embedded in the fabric of ecology. The concept is at once intuitive and tautological, in that it flows logically from a small number of biological axioms. Darwin clearly viewed competition as an important component of the "struggle for existence," although just as clearly, he regarded it as only one of many such components.

Even though the concept of competition is on some level intuitive, a more formal statement of meaning is essential to any further treatment based on science. L. Birch's statement (1957) is generally useful in this regard: "competition occurs when a number of animals (of the same or of different species) utilize common resources the supply of which is short; or if the resources are not in short supply, competition occurs when the animals seeking that resource nevertheless harm one or other in the process." Competition, then, may occur for any resource in short supply. However, living space and food are probably the resources most often in short supply on the shore. Birch's meaning further includes two different components of competition that should be identified separately. One component, usually called interference competition, is the direct interference by one organism with another's attempts to utilize the limited resource. The second component, often called exploitation competition, involves an organism's ability to harvest or utilize the resource in short supply separate from any direct interactions between the competing organisms. Both of these components may be important in an interaction. Indeed, the "loser" in interference competition may compensate by being more adept at exploiting available resources.

Although the concept of competition is often held to be a cornerstone of intertidal ecology, its actual influence on the distributions of organisms in nature has been surprisingly difficult to demonstrate except in a relatively few cases. The clearest successes have involved sessile organisms, primarily barnacles, mussels, and algae; the impact on the majority of mobile organisms is far less clear and, in general, remains to be assessed carefully. However, the fact that competition does exist in the intertidal zone, at least sometimes, has been convincingly demonstrated by several investigators, and there is little doubt that it can be a significant factor influencing the distribution and abundance of some organisms.

The effects of interference competition have been easiest to see, and both intraspecific (within a species) and interspecific (between species)

---

*Parasitism and disease also are likely to have important biological influences on plants and animals of the shore. Parasites and agents of disease generally reduce the ecological potentials of their hosts by diminishing resistance to environmental factors and by reducing the host's ability to reproduce and compete for resources. However, too few studies have been done on intertidal organisms to comment with much certainty about the specific effects on abundance and distribution. A few references have been cited in the General Bibliography. Note, in particular, Johnson (1968), Kinne (1980), Sindermann and Rosenfield (1967), Sparks (1972), Tallmark and Norrgren (1976), Williams and Ellis (1975).

competition have been demonstrated. A striking example of intraspecific competition of the interference sort (described in detail on page 53) was provided by Francis (1973a,b). In essence, she demonstrated that individuals from two adjacent clones of the aggregated anemone *Anthopleura elegantissima* will battle each other, with nematocysts drawn, over the contested resource of space. The outcome of this competitive interaction in nature is often an obvious line delimiting the distributions of the two clones (Fig. 344). Connell's (1961b) study of barnacles on the coast of Scotland provided a clear example of interference competition between individuals of two species. In this case, larvae of the barnacles *Chthamalus stellatus* and *Semibalanus balanoides* settled in broadly overlapping zones, with *Chthamalus* somewhat higher on the shore. Adults of the two species, however, occurred mostly in separate zones; specifically, *Chthamalus* adults occurred in large numbers only above the zone inhabited by *Semibalanus*. Connell found that direct interference competition with *Semibalanus* was the primary factor reducing the density of *Chthamalus* on the lower shore. Few *Chthamalus* juveniles survived to adulthood if *Semibalanus* was present, even though *Chthamalus* young survived well at all tide levels when kept free from contact with *Semibalanus*. In the zone of overlap, *Semibalanus* generally grew faster, and smothered, undercut, or crushed the *Chthamalus*. Large numbers of *Chthamalus* survived to adulthood only at high shore levels unoccupied by *Semibalanus*. Dayton (1971) has reported that on the Pacific coast *Balanus glandula* employs similar methods to outcompete *Chthamalus dalli* (Fig. 345).

In contrast to these clear cases of interference competition, competition of the exploitation type is usually more difficult to demonstrate convincingly. Exploitation competition is by nature indirect, and it may be quite subtle. Generally, the existence of this type of competition must be inferred, often from a decrease in growth rate after a potentially competing species has been added. Despite the difficulties, results of several studies have inferred the importance of exploitation competition. For example, both intraspecific and interspecific competition seem to be important to the limpet *Collisella scabra*. Sutherland's study (1970) of this limpet's growth rate at several population densities suggested that individuals do indeed compete with each other for food, and Haven (1973) further concluded that *Collisella scabra* also competes with another species of limpet, *C. digitalis*, wherever these two co-occur.

Although competition has been demonstrated to be an important influence on the distribution of organisms in some situations (e.g., Peterson & Andre 1980; VanBlaricom 1982; Woodin 1974; in addition to the studies cited above), in other situations it seems to be without much impact, and in most, we simply have too little information to judge. The extent to which competition plays a role in structuring communities is controversial (see Schoener 1982 for a review). Yet, competition is so well entrenched as a concept in ecological thinking that it is frequently assumed to be an impor-

Fig. 344. Competition between two clones of the anemone *Anthopleura elegantissima* creates a prominent battle line.

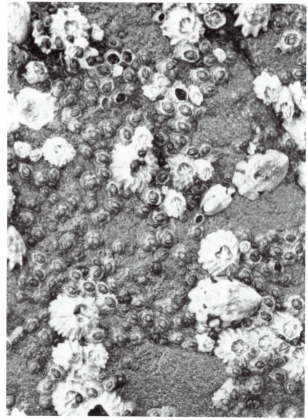

Fig. 345. Competition between the barnacles *Balanus glandula* (white) and *Chthamalus dalli* (dark); the larger, faster-growing *Balanus* overgrows, undercuts, or crushes the smaller *Chthamalus*.

tant mechanism of change even when mechanisms have not actually been examined. Unfortunately, these casual assumptions not only give us the appearance of knowing more than we do, they also tend to obscure reality and to stifle inquiry into other potentially important variables.

Consumers: Predation and Herbivory

Interaction between consumers, whether predators or herbivores, and the consumed, whether animals or plants, is another important factor influencing distributions on the shore. On the one hand, prey influence the distributions of consumers: predators and herbivores must be associated with their food items in some way and are often attracted to them. Several species of nudibranch, for example, are found almost exclusively with their particular sponge, hydroid, or bryozoan prey; the predatory snail *Nucella emarginata* is found near barnacles; and so on. Similarly, herbivores such as littorines and limpets are often attracted to specific plants, and may be essentially restricted in distribution to association with them. The limpets *Notoacmea insessa*, which occurs on the brown alga *Egregia*, and *N. paleacea*, on the angiosperm *Phyllospadix*, are two examples.

On the other hand, some consumers clearly influence the distribution of their food. On the outer coast, the distribution of the mussel *Mytilus californianus* is affected by its primary predator, the seastar *Pisaster ochraceus*. Since the seastar generally occurs lower on the shore than the mussels and moves up to feed during high tide, one would expect the lower portion of the mussel's distribution to be the one most affected. Paine (1974) tested this idea by removing all the *Pisaster* he could find from a stretch of shore in Washington. The result was clear. Over a 3-year period, the lower limit of the mussel bed in the "*Pisaster*-removal" area edged lower and lower on the shore, while in a control area, left undisturbed with its normal complement of seastars, the lower limit changed little. The lower limit of some barnacle populations is also determined by predators moving up in the intertidal. Connell (1961a, 1970) studied populations of *Semibalanus balanoides* in Scotland and *B. glandula* in Washington State and demonstrated that barnacles could survive at lower levels on the shore than usual if they were enclosed in cages to protect them from predatory snails. Without protection, nearly all the young barnacles within the foraging range of the predator were eaten, thus effectively limiting the natural distribution of adult prey to the tidal zone above the range of the predator.

Herbivores such as limpets and sea urchins may similarly influence the distribution and abundance of plants (Fig. 346). Their influence often seems not as strong as that of predators, however, perhaps in part because herbivores generally crop their food instead of killing it. Nevertheless, several studies have shown striking changes in the prevalence of certain algae depending on whether herbivores were present in an area or excluded by various experimental or natural barriers. In these studies (e.g., Dayton

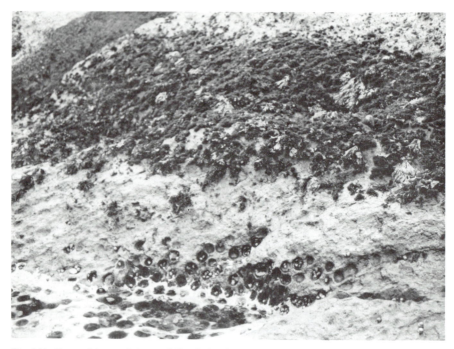

Fig. 346. Sea urchins and the snail *Tegula funebralis*, herbivores that appear to be setting the lower limit of this algal band. (Middle Cove, Cape Arago, Oregon, July 1978.)

Fig. 347. The effects of grazing by limpets (*Collisella digitalis*) in the high intertidal: a circle of rock from which limpets were excluded grew a dense mat of the alga *Urospora* while surrounding areas were grazed bare.

1971; Cubit 1974; Robles & Cubit 1981), barren-looking rocks quickly grew a luxuriant plant cover when protected from grazing (Fig. 347). As in the studies on predator influences, the herbivores tended to occupy lower positions on the shore than their food plant, and thus the lower portion of the plant's distribution was primarily affected.

Although being able to generalize about the influences of "predators" and "herbivores" would be comforting there are probably as many different "influences" of predators as there are different predators, prey, and situations. Certainly, predators are not all alike, and a variety of circumstances influences what impact a predator will have. Some predators are specialists, feeding on one or very few species of prey; others such as the seastars *Pisaster ochraceus* and *Leptasterias hexactis* are generalists. To be closer to the truth, however, even these simple generalizations need qualification. For instance, some predators seem to be specialists, but on different prey in different locations: the seastar *Dermasterias imbricata* may eat primarily sea urchins in one area but sea anemones, holothurians, or colonial tunicates in others. These seastars seem to be facultative specialists, concentrating on one kind of prey within a habitat but eating several kinds over a wide geographical area; a comprehensive analysis of their foraging behavior would make an interesting research project. As for generalist predators, most studies of their diets are in fact ambiguous, since they contain no information about the long-term diets of individuals. To illustrate, if 100 seastar predators are sampled only once, producing 60 individuals eating barnacles and 40 eating mussels, there is no way to tell whether all seastars in the population are generalists eating barnacles 60 percent of the time and mussels 40 percent of the time, or whether the sample contains 60 seastars that are specialists on barnacles and 40 that eat only mussels. The only way to distinguish these two possibilities is to sample marked individuals repeatedly through time, and this has been done infrequently.

Adding to the uncertainty, and the reality, of the situation is the fact that the diets of predators may also change over time. Certainly, individuals of the same species may consume different prey as they grow from juveniles to adults. But further, adults may change their diets and preferences over a relatively short time span. Indeed, some change in behavior with experience is probably a natural part of the lives of most invertebrates. Even though seastars are not known for their mental capacity, Landenberger (1966) demonstrated that when a meal was at stake, *Pisaster giganteus* could be trained quickly to associate a bright light with food being added to the aquarium. In other cases, results suggest that a predator's preference may change according to what it has eaten recently, a phenomenon known as "ingestive conditioning." Wood (1968), for example, found that individual *Urosalpinx cinerea* fed a diet of oysters (*Crassostrea virginica*) later chose oysters more often in laboratory tests than did a snail fed a diet of barnacles (*Balanus* spp.); conversely, snails maintained on barnacles chose barnacles more often. All of this is to point out that the influences of predators are changeable and may vary between situations. This variability certainly complicates any theory that hopes to predict accurately when predators will be effective and when they will not (optimal-foraging theory); however, the complications do not make the situation hopeless, just more interesting. In this connection, it seems worth noting, too, that predators have concerns in addition to catching prey. They, like their prey, must cope with wave shock, desiccation, and their own list of predators. Not everything a predator does relates directly to capturing prey.

Although predators obviously capture enough prey to maintain themselves and their populations, prey are not without defenses. Indeed, prey possess an impressive assortment of structural, chemical, and behavioral "tactics" that ensure their continued existence as well. These include passive defenses (those not triggered by the presence of a predator) and active defenses (those triggered by a nearby predator). Passive defenses include spines, thick shells, and a tough exoskeleton, all effective structural defenses against some predators (Fig. 348); but all, too, have been breached by some other predator. Noxious chemicals are another line of defense. Many dorid nudibranchs secrete acids and other noxious substances, and the tissues of most echinoderms are loaded with toxins (in this case, a group of steroid saponins). At least one animal may employ a form of chemical camouflage: the limpet *Notoacmea paleacea* (Fig. 349) incorporates into its shell a specific "plant chemical" taken from the host plant on which it is found, and one of the potential predators on this limpet, the seastar *Leptasterias hexactis*, seems to bypass the limpet in the field, as if the prey had not been detected.* There are literally hundreds of examples of structural and chemical defenses, and although the preceding examples all have involved animal prey, plants also employ similar kinds of defense. The calcified structure of coralline algae, for example, likely provides some discouragement to herbivores, as does the toxicity of *Desmarestia* and other algae.

Perhaps the most spectacular prey defenses, however, are behavioral (Fig. 350). The list of marine invertebrates that respond to predators is certainly long, so long in fact that prey failing to respond now seem of particular interest. Since many examples, from flight responses to biting back, can be found in the sections dealing with individual species, one set of examples should suffice here. The responses of limpets to seastars have been studied in some depth, and three general points have emerged. First, limpets may respond differently to approaching (but still distant) predators than to actual contact with a predator. In fact, the complete repertoire of an individual may contain several distinctly different responses: a limpet may move away from the predator or up shore, or twist the shell violently

---

*More work is clearly needed to demonstrate the validity of this suggestion (Fishlyn & Phillips 1980), but one might predict that since many invertebrate predators hunt using chemical cues, chemical crypsis could be an important defense.

Fig. 348. The sea urchin *Strongylocentrotus franciscanus* possesses an impressive armament of "passive" defenses: spines, hard skeleton, and noxious chemicals.

Fig. 349. The limpet *Notoacmea paleacea* (§119) incorporates into its shell a specific plant chemical taken from the surfgrass *Phyllospadix*, on which it feeds and lives. This may provide chemical camouflage against predators that hunt using chemical senses. The limpet's graze marks appear as a lighter area on the surfgrass blade.

from side to side, or loosen its hold on the rock and fall. Second, limpets do not respond equally to all species of seastars. Predators encountered in nature tend to elicit responses, whereas non-predators tend to be ignored. Indeed, non-predators are often used as experimental controls for behavioral tests. Third, not all species of limpets respond equally. Species of limpets that are highly responsive tend to co-occur with a predator, whereas species of less-responsive limpets tend either to occur high enough on the

Fig. 350. The European cockle *Cardium echinatum* reacts to the predatory seastar *Asterias rubens*. At first contact, the cockle-shell begins to gape (*facing page, top*); then, the long muscular foot is extended. Sometimes this powerful thrust of the foot pushes the predator away; at other times, it vaults the cockle into the water column in an attempt to escape.

shore that they rarely encounter a predator or to have developed an alternative defense such as noxious chemicals. A good deal of work on other marine invertebrates is consistent with these three points, and suggests a degree of generality.

Somewhat less spectacular than responses triggered by predators, but still important, are a number of behavioral patterns that might be considered "passive" defenses because they are not triggered by the presence of the predator. Whether passive or active, these patterns may effectively reduce predation. For example, the habit of foraging at night or only when awash by the tide may reduce predation by fish, birds, octopuses, and other fast-moving visual predators.

Camouflage is another defensive "tactic" employed by some inhabitants of the shore. One possible instance of chemical camouflage has already been mentioned; however, examples of visual crypsis are more common by far. Whether this skew in reported occurrence is real, or due to a perceptual bias of scientists who assume that the visual sense is as important for marine invertebrates as it is for them, remains to be seen (so to speak). In any event, examples of crypsis are numerous among intertidal invertebrates. The tiny red nudibranch *Rostanga* (§33) occurs almost exclusively on red sponges; the isopod *Idotea* (§205) changes color to match the red or green background provided by the plant on which it lives; and crabs of several species decorate their bodies with well-chosen bits and pieces of their biotic environment (Fig. 351).

Although there are clear examples of animals that match their backgrounds, the assumption that these matches always have defensive value is worth resisting or, better yet, testing. In a simplified experiment, specimens of *Idotea*, for example, might be distributed between two substrata—one that provides a match and one that does not—and then exposed to a predator, in this case several species of fish. If indeed there is defensive value to matching the surroundings, the unmatched isopods should be eaten more frequently. Of course, even with a simplified test of this sort, suitable controls are necessary. Most important, the survival of matched and unmatched isopods should be tested in the absence of the predator to

Fig. 351. A decorated and well-concealed crab, *Loxorhynchus crispatus* (§131).

see if other factors might have been responsible for the decline in numbers of unmatched individuals: perhaps the unmatched substratum did not provide adequate nutrition for the isopod or footholds against dislodgement by waves. Although the adaptive value of camouflage has been experimentally verified in a number of situations, its importance, like that of competition, has more often been assumed than tested, undoubtedly because in both cases the assumption seems so "reasonable."

This extensive battery of defensive tactics has provided prey with a considerable measure of success. This success is at least partly due to the fact that predators cannot deal simultaneously with all of the various types of defense that might be encountered. A predator that is successful against one type of defense is often ineffectual against another. Yet, many prey certainly are eaten. Why are prey defenses not more effective? One reason is that predators are continually evolving new countermeasures to deal with prey defenses. Another is that prey must be concerned with more than just the unwanted advances of one potential predator. Most marine invertebrates must deal with many different potential predators, and a successful defense against one may not be successful against another. The limpet *Notoacmea scutum*, for instance, is eaten by a formidable array of predators, including several seastars, predatory snails, flatworms, octopods, crabs, and shorebirds. Even so, predators are not the animal's only concern. The effects of desiccation, wave shock, food availability, and many other factors must also be accommodated. One should not be too surprised, then, when a prey species behaves in a way that seems to expose it to increased predation: predation is just one variable in the complete equation, and benefits from other factors may more than compensate for increased rates of predation.

Two more aspects of predation and predator-prey interactions must be discussed. The first is the indirect influence that predation may have on intertidal communities through its influence on competition. Predation can, and in some circumstances clearly does, reduce the intensity of competition. If the competitor eaten most often is also the most numerous or otherwise "dominant" species, then predation may prevent the dominant from monopolizing a limiting resource, thereby increasing the diversity of species in an area. Darwin again was among the first to express this concept, and several more-recent authors have contributed additional evidence. Paine (1974), for example, has shown that predation by *Pisaster ochraceus* on the mussel *Mytilus californianus* increases the number of species able to utilize the resource of primary space. A similar increase in diversity may result if the intensity of predation on a particular prey type increases in proportion to the relative abundance of that prey type. By contrast, we should note, as others have (e.g., Lubchenco 1978), that if a predator happens to select rare or less-able competitors, predation may actually hasten the monopolization of a resource and lead to lower species diversity. Lower diversity may also result if a predator has no particular preference but eats all prey voraciously.

A second point is that predator-prey interactions are just that: interactions. This simple but essential element is emphasized here because the organization of this section on predation largely reflects the dichotomous separation of research efforts in the field—namely, the influences of predators versus the defenses of prey. Those interested in predators see 25 percent of a prey population eaten annually; those interested in prey defenses see 75 percent of the prey population escaping. The "preferences" of predators and the defenses of prey are best seen as products of this interaction. Specifically, if natural selection favors a predator that maximizes its caloric intake per unit time (or maximizes some other important "currency"), as is often assumed, then the behavioral attributes of the predator associated with optimum choices of prey should be favored in the process. That is, the current preferences of a predator presumably reflect past selection for, among other things, ease of capture and caloric content of prey. Thus, prey characteristics such as defensive responses that reduce predator success have an influence on the behavioral predispositions of affected predators. Indeed, the apparent "ranking" of prey by a modern predator seems likely to be greatly affected by its prey, past and present. Predator preferences and prey defenses reflect the inseparable interaction of predator and prey now and over evolutionary time. So do the distributions and abundances of many of the current combatants.

Disturbance

Although ecologists have been known to argue at length over its proper definition, a disturbance is usually regarded to be some marked structural change, with fairly clear boundaries in space and time. A common connotation of disturbance, both in everyday usage and in ecology, is that it takes a system away from some normal pre-existent state or resets a developmental progression. Disturbances, however, seem to be a natural component of all, or certainly most, naturally occurring populations, so that if anything, disturbed conditions may be closer to being general on a local scale than they are exceptional. In any event, disturbances can be forceful agents of natural selection, and adaptations to resist and to exploit disturbance are not uncommon among plants and animals of the shore.

The sources of disturbance are many, and a partial list provides some indication of the scope and importance of this phenomenon. For convenience, disturbing agents are often classified as being physical or biological, although as always, some examples fit both places. On the physical side, storm waves may tear loose sections of mussel bed (Fig. 352) or overturn boulders. Floating drift logs, icebergs, cobbles, and even large boulders may batter the shore, and unusual weather conditions, freezes or hot spells, may decimate populations. Biological sources of disturbance are

Fig. 352. Sections of mussel bed torn loose by waves, creating various-sized patches of bare rock.

equally numerous. On rocky shores, limpets and other herbivores may scrape newly hatched larvae off the substratum as they forage, and fronds of algae sweeping back and forth with the waves may also remove larvae. In soft sediments, deposit feeders continually turn over the sediment through their feeding activities (Fig. 353); in fact, the entire surface of the muddy seafloor may pass through the benthos at least once, and in some cases several times, each year (Rhoades 1974). The activities of epifaunal macro-invertebrates such as crabs also disturb the upper layers of sediment and may smother the infauna. Brenchley (1981) provided one example of the effects of such "bioturbation" in False Bay, Washington. She set out a series of cages on the sediment, and to one set of cages she added the sand dollar *Dendraster excentricus*, a burrower and suspension feeder. Within about a

Fig. 353. Disturbance in a mud flat, here caused largely by the feeding and burrowing activities of the ghost shrimp *Callianassa californiensis* (§292).

week, the density of tube-dwelling polychaetes and crustaceans in the *Dendraster*-cage had decreased significantly relative to their levels in control cages. Laboratory tests showed that tube-dweller mortality was due to the physical effects of sediment deposition, not to predation by *Dendraster*.

Regardless of the specific kind of disturbing agent, the immediate and primary effect of a disturbance is usually to free a patch of resource. This "patch" may vary in size from a few centimeters in diameter to many kilometers of coastline, as in the case of the great Alaskan earthquake of 1964, during which great sections of coast were shifted up or down by many meters. Most commonly in the intertidal, space seems to be the critical resource being freed, although in some situations other resources (e.g., food, light) are important as well.

Since the effect of a disturbance is often to free a potentially limiting resource, one might expect disturbances to lessen the influence of competition when competition might otherwise be expected to be intense. Under these conditions, disturbance might also be expected to increase species diversity by providing patches of resource in which less-able competitors or ephemeral species may establish themselves temporarily. The general influence of disturbance, then, is similar to that of predation, in that both can lessen the influence of competition. Indeed, several investigators have found levels of disturbance and predation high enough to conclude that competition may play a relatively minor role in structuring the communities they studied. Perhaps competition exerts an important influence on the distribution of intertidal animals only in situations characterized by low rates of disturbance and predation.

In any event, predicting precisely which organism will occupy an area some time after it has been disturbed is a difficult exercise at best. Among the parameters that seem to be important are the magnitude of the disturbance, its periodicity, spatial scale, and timing. We will touch on these again in subsequent pages.

Refuges

Although wave shock, competition, and predation restrict the abundance and distribution of many intertidal organisms, many plants and animals also have partial refuges against these, and other, influences. Refuges take a variety of forms. There are temporal refuges, such as being active when one's predator is not, or timing one's larval settlement to occur during a season of lessened predation or competition. There are physical refuges, both those that are strictly abiotic and those that are biologically created: perhaps in the former case, a crevice that protects against desiccation, predation, and battering by logs and waves (Fig. 354); in the latter case, a dense stand of tube worms that protects against shifting sediments and predators. There are also spatial refuges, zones out of the reach of predators and competitors; for many animals, the upper levels of the shore seem to provide a refuge from predators and competitors occupying lower regions.

In addition to refuges such as these, animals or plants occasionally will grow fast enough, or escape from consumers long enough, to reach a size that makes them invulnerable to further attempts by consumers. At this point, the organism is often said to have "escaped in size" or found a "refuge" in size. Mussels that have escaped in size from *Pisaster* are shown in Figure 355, and many other examples of this have been described. Dayton (1971), for instance, found that patchy, intense settlements of the barnacle *Semibalanus cariosus* may on occasion temporarily overwhelm the ability of the predatory snail *Nucella* to consume all the barnacles in the spatial zone of overlap. A few lucky barnacles, escaping predation for a while, grew in that time to a size greater than *Nucella* could handle. These barnacles then had "escaped in size" from the snail; they were no longer vulnerable to this predator. Of course, the exact size that is "safe" depends on

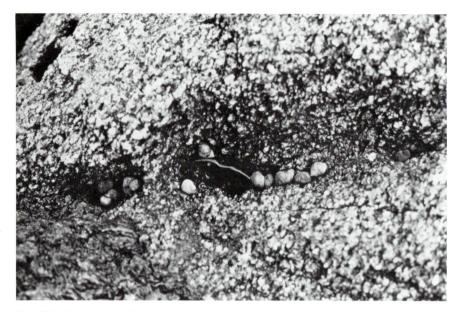

Fig. 354. Littorines in a high intertidal crevice, a refuge.

Fig. 355. The large mussels in this bed have grown sufficiently large that they are no longer vulnerable to predation by the seastar *Pisaster*; these mussels have gained a refuge in size, whereas smaller mussels higher on the rock are still vulnerable.

several factors, among them the size of the predators, the barnacle's intertidal position (lower on the shore the predator has a longer time available for drilling), and also the species of predator (*Semibalanus cariosus* never seems to grow large enough to escape *Pisaster ochraceus*).

Large size may provide a refuge against physical influences as well as consumers. Adult barnacles, for instance, are far less susceptible to desiccation stress than are barnacle spat. Consequently, a sustained chance period of favorable conditions may allow a group of young barnacles to attain a size at which they become much more resistant to desiccation stress. Environmental conditions that would have killed them as spat are now tolerable. They have attained a partial refuge from desiccation, simply by growing.

Finally, although we usually think of refuges in size as being refuges in large size, being small provides some refuge of its own. Smaller animals may be better able to hide in crevices or under boulders, or to escape from predators by climbing delicate fronds of algae. Smaller animals also tend to be passed over by many predators; tiny snails and protozoans have little to fear from a large seastar such as *Pisaster ochraceus*. Scale is always an important biological consideration.

Larval Settlement

A large percentage of marine invertebrates have a mobile larval or juvenile stage, which disperses members of the species. As we have seen, these young of a species are often more susceptible to physical and biological factors than adults are, and one might ask to what extent the characteristics and distributions of newly settled larvae or juveniles determine the future distribution of adults. Beyond the obvious (but important) observation that if the larvae of a sessile adult cannot survive in a location then no adults can occur there either, the answer is complex, depending on the species and on a number of other circumstances. Furthermore, any answer that might be given to our question should be regarded as tentative at best, since it is based on data that are meager indeed: uncomfortably few species have been examined in detail, and most of these primarily under laboratory conditions. Although field studies would be ideal, field studies of "settlement patterns," in practice, suffer greatly from severe technical difficulties. Recently settled larvae are small and usually occur only sporadically in a given location. As a consequence, conclusions about "settlement patterns" in the field most commonly are based on observations of animals that actually may be several weeks to several months old. One can sympathize with these technical problems, but the fact remains that during a few days, let alone a few weeks or months, many animals are likely to have died or moved, leaving us with a rather unreliable image of the true pattern at the time of settlement.

One aspect of larval settlement that has received considerable attention is the sensory cues used by larvae to choose a settlement site. The larvae of some species are highly selective in choosing sites, whereas others are much less so. The choices are most critical and the impact clearest, of course, for species with sessile adults. Serpulid worms and bryozoans, for example, may settle preferentially on certain algae, and some barnacles respond with increased settlement to a particular chemical adsorbed to the test of other barnacles of the same species. But the larvae of species with mobile adults can also be quite selective. Several opisthobranchs respond to the presence of their primary food, as does the lined chiton *Tonicella lineata* (Barnes & Gonor 1973). Selectivity is not restricted to organisms characteristic of hard substrata either. Larvae of *Urechis caupo*, the echiuran worm of mud-flat fame, respond to chemical substances produced by adults of the species (Suer 1982). Meadows and Campbell (1972) provide an excellent review of the settlement behavior of many marine invertebrates. They also discuss the many cues to which larvae may respond during the process of selecting a suitable settlement site. Among these are chemicals, particle size, light, gravity, pressure, current velocity, and surface properties such as texture, contour, angle, color, and reflectance. All of these factors may influence the distribution of larvae and thus may have a great impact on the distribution of adults.

Although some degree of selectivity seems nearly universal, larval distributions are seldom precisely the same as adult distributions (Fig. 356). Newly settled larvae in the extremes of vertical distribution may be killed by desiccation, temperature stress, competition, or predation, leading to a restricted adult distribution. Indeed, a few larvae almost always settle in areas that seem to doom them to failure. Although barnacle larvae are well known for using a wide variety of sophisticated cues, including the presence of conspecifics, to select an appropriate site for settlement, the primary distribution of *Balanus glandula* adults is considerably narrower than that of newly settled spat; young barnacles at the extremes are usually killed. Why then do these animals push, and very often exceed, the limits of success?

The answer seems to be, at least in part, that occasionally the long shot pays off, and when it does, the reproductive rewards are great. For example, larvae of the barnacle *Semibalanus balanoides* in Scotland (Connell 1961a) apparently survive to adulthood in the high intertidal only once every 3 to 5 years, but when chance favorable conditions do permit survival, those animals that are successful may be highly so, benefiting from reduced competition and predation; a similar situation seems to apply to the limpet *Collisella scabra* in California (Sutherland 1970). Recall, too, the possibility that larvae settling low in the intertidal, apparently within a predator's or competitor's reach, may occasionally survive long enough to escape in size. The probability of such success is admittedly low, but rare events do occur, and the potential rewards of such success are great (e.g., increased food and more-benign physical conditions). Certainly if one con-

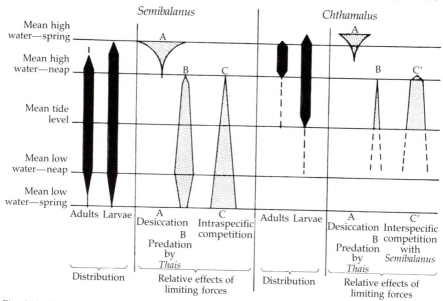

Fig. 356. The intertidal distribution of adults and newly settled larvae of *Semibalanus balanoides* and *Chthamalus stellatus*, with a diagrammatic representation of the relative effects of the principal limiting factors. *A*, desiccation; *B*, predation by *Thais lapillus*; *C*, intraspecific competition; *C′*, interspecific competition with *Semibalanus*.

siders a broad stretch of coast, with all of its various patches of physical and biological conditions, it seems likely that somewhere one or more of these "unlikely" successes is being quite successful indeed.

Intensity of Factors

Most of the factors influencing distribution and abundance in the intertidal vary along gradients of intensity and frequency. Wave shock, desiccation (and other factors related to tidal position), competition, predation, disturbance, and larval recruitment all vary along gradients, although for convenience we sometimes speak of these factors as if they were either "on" or "off." Nevertheless, variations in the intensity and frequency of parameters may profoundly influence the distribution and abundance of individual species, as well as broader aspects of community structure.

Having said this, it is not an easy matter to examine the effects of changes in intensity or frequency. The effect of a parameter may not follow a simple linear function; in fact, most likely it rarely does. Instead, there may be certain critical or threshold levels that trigger dramatic changes when they are crossed. Furthermore, the same apparent level of intensity or frequency of predation, for instance, may affect organisms differently.

barnacles apparently enhances the recruitment (or at least early survival) of the limpet *Collisella digitalis*, and similarly, recruitment of mussels also seems to be enhanced by the presence of barnacles. However, so far only in the case of the surfgrass *Phyllospadix*, which requires the presence of certain algae for recruitment, does the relationship seem to be obligate (Turner 1983).

Another factor affecting the intertidal landscape is chance—pure, but not so simple, chance. Primarily, chance events increase heterogeneity, both physical and temporal. A floating log or some other chance physical disturbance may clear a patch of space just when the larvae of a particular species are ready to settle from the plankton. At some other time of the year, a different species might have settled in the cleared space. An especially large settlement of larvae may swamp the ability of predators to consume them, allowing some prey to escape in size. Or predator populations may be devastated by a rare combination of low tides and extremely hot weather, again allowing prey to escape. As with other variables we have mentioned, these and other chance occurrences vary considerably in the frequency of occurrence and the scale of the impact. Some are rare; others are common. Some, such as a log battering the shore, affect only very local conditions; others, like unusual weather conditions, may influence many miles of shore. It is also worth noting here that small-scale fluctuations due to chance events tend to "even out" when averaged over a large scale. Local populations of the predatory snail *Nucella lamellosa*, for instance, fluctuate greatly in size; over a long stretch of shore, however, the fluctuations tend to cancel each other, and the total population size remains relatively stable over time (Spight 1974). Both the local fluctuations and the larger-scale stability are important considerations.

Superb examples of the combined influences of time, chance, and history were provided by two studies of disturbance in boulder fields, one on the Atlantic coast (Osman 1977) and another on the Pacific coast (Sousa 1979a, 1980). In boulder fields such as these (Fig. 359), rocks were periodically overturned by waves, usually killing the sessile occupants and clearing free space. The frequency of this disturbance was inversely proportional to the size of the rock (small rocks turned over more often) and directly related to the force of the waves. Once a rock had been cleared (turned over), the establishment and development of organisms on the rock, as well as the eventual outcome of the developmental progression, depended on many factors. In Osman's study, the seasonality of disturbance and the seasonality of larval abundances were extremely important in determining which species were first to colonize an area. The time of the disturbance determined when a substratum became available for initial colonization, and initial colonization was highly variable. What happened next, however, was fairly consistent: whichever species was there first, by chance, seemed to hold an advantage, and these early colonists resisted, rather than facilitated, the subsequent invasion of the space by other spe-

Fig. 359. A boulder field with rocks of various sizes. Several of the larger, more stable boulders have well-established populations of long-lived animals such as anemones and mussels, whereas many of the smaller rocks, more susceptible to being turned over, are covered with ephemeral algae.

cies. The final result in one of the boulder fields was that any of nine species could eventually dominate a rock depending in large measure on chance and the history of an area. Interestingly, in neither study was direct interference competition of primary importance in determining which species would dominate in the end. Rather, life span, reproductive strategy, and other life-history features were found to be critical. Long-lived species often "dominated" by default when short-lived species died.

One other point made in both studies is of interest here. Not only did disturbance determine when a substratum was cleared initially, but the frequency of disturbances determined how long it would be, on the average, before a space was cleared again. Since boulder size was inversely proportional to the average frequency of disturbance, the size range of boulders provided a gradient of disturbance frequency. How then did species diversity change along this gradient? Small rocks were disturbed so frequently that only those few species recently represented in the plankton were present on them. At the other extreme, large rocks, which were stable for long periods of time, were also usually dominated by one or a few species, presumably because other species had been killed by grazers, competition, or physical factors. In between, intermediate-sized rocks were stable long enough to acquire many colonists, but not long enough for any one to dominate completely. Thus, analogous to the situation with predator abundance, there was first an increase, then a decrease, in species diversity along a gradient of increasing disturbance. The maximum lay at some intermediate level of disturbance.

To summarize, then, chance may play an important role in determining what species will be found in any given location. The history of a site may also play a role, often through adult-larval interactions. Furthermore, since intertidal communities are dynamic entities, progressing in some cases through a series of developmental or historical stages, we should keep in mind that we are usually observing only one slice of a temporal sequence. Finally, changes occurring along various scales of time and space are an essential part of the natural scene. Such local and temporal variations are far from merely inconvenient deviations in more-general principles (e.g., competition, predation, etc.). Rather, the variation is itself a principle.

Behavior

Although elements of behavior have been mentioned in passing as various other influences on the distribution and abundance of organisms have been presented, the role of behavior is so general and important that it warrants emphasis in a separate section. This is especially true because most intertidal ecologists have paid surprisingly little attention to the role of behavior. Some of this unfortunate disregard is probably the result of historical precedent. Seminal experiments in marine ecology focused on factors influencing the distribution of sessile animals (primarily barnacles and mussels), animals not known for their behavioral prowess; when mobile animals were examined, they were generally treated as if they were sessile by confining them in small cages. The virtual absence of behavioral consideration in these studies apparently spawned the unfortunate impression that behavior was not an important influence on intertidal distributions.

To the contrary, modern ecologists recognize that behavior plays a critical and general role, either directly or indirectly, in determining distributions and abundances in the intertidal zone (Fig. 360). Organisms, as larvae, adults, or both, respond to various physical aspects of their environment, including wave shock, substratum type and slope, desiccation potential, temperature, light, salinity, and oxygen levels. They respond to such biological aspects of their environment as competitors, predators, and prey. Competitors respond to each other, consumers are attracted to their prey, prey respond to predators, and so forth. In short, behavior impinges on all the other factors that have been discussed. We should keep in mind, however, that behavior is likely to be the proximate expression of ultimate causes. Current behavior presumably reflects previous evolutionary selection by a host of influences, including all of those same physical and biological factors to which it is now related.

Fig. 360. Limpets (*Notoacmea scutum*, §25) and seastars (*Pisaster ochraceus*, §157) on a rock wall. These limpets avoid *Pisaster* and generally occur higher on the shore when this predator is present. They also move up and down with the seasons, presumably in response to physical aspects of their environment. (The white ruler is 15 cm long.)

Interaction and Integration

By necessity, the various influences on the distribution and abundance of organisms in the intertidal region have been identified and discussed as separate entities in the preceding pages. It should be clear by now, however, that no one factor stands alone. Reality for each organism is an interactive and continuously changing blend of all the influences that have been mentioned, and undoubtedly more. Understanding these influences and their interaction is an enormous task, one not made easier by ardent promoters of particular points of view. At times we seem to be floating in "principles" and points of view that gain additional weight as they are repeated but that are not diminished by facts to the contrary. Preoccupation with only a few selected principles and hypotheses can lead one to fit observations into a chosen scheme to the extent that what is really going on may be overlooked. However, scientific partisanship does have some potential value. It stimulates opposition, which in turn stimulates the design of critical tests, and as Dayton and Oliver (1980) have pointed out, science advances most rapidly by the falsification of hypotheses.

In this context, and with an eye toward integration, it might be useful to examine two current concepts: first, the generalization that upper limits of distribution are set by physical factors and lower limits by biological factors, and second, the concept of critical tide levels. Both relate to a debate in marine ecology over the relative importance of physical versus biological factors as influences on the distributions of organisms. The importance of physical factors was emphasized during the 1940's and 1950's, but over the past two decades, the influences of biological factors have been increasingly stressed. Unfortunately, in the process of recognizing the contributions of biological influences, those of the physical environment seem to have been relegated to secondary importance. The essential point is that physical and biological influences, far from being mutually exclusive, are concurrent aspects of the interrelated whole.

Are upper limits set by physical factors and lower limits by biological factors? To some extent yes, and to some extent no. Since most intertidal inhabitants are essentially marine organisms moving up into terrestrial conditions (and not the reverse), it seems intuitive that their advancing upper limits would be most affected by the increasingly harsh physical influences associated with the terrestrial environment. It also seems reasonable that lower limits would be affected more often by biological influences: not only are there simply more species in lower intertidal regions, but organisms presumably invaded the upper intertidal in evolutionary response to these biological interactions (i.e., competition and predation). The basic generalization, then, clearly has intuitive support.

It would be a mistake, however, to conclude that physical factors have no impact on the lower limits of organisms or that biological factors do not influence upper limits. Physical factors do influence individuals occupying lower regions of a species distribution, although they may not kill them outright. For example, *Fucus* and several other mid-intertidal algae, when held under a normal cycle of submersion and emersion, have lower photosynthetic rates when submerged than when exposed (Johnson et al. 1974); one might expect, therefore, faster growth of some individuals higher on the shore, at least within a certain vertical range. Increased submergence time, associated with being lower on the shore, also seems to have a deleterious effect on the barnacle *Semibalanus balanoides*: these barnacles, protected in cages from predators, died more often at lower levels (Connell 1961a). Increased wave action in the lower intertidal is apparently involved in setting the lower limit of adult *Littorina keenae*; large individuals apparently are unable to cling firmly enough to wave-swept rocks in the low zone. And scouring or burial by sand sets the lower limits of many plants and animals on some beaches (Fig. 361).

Similarly, biological factors can contribute to setting upper limits. Choat (1977), for instance, found that the upper limit of the distribution of the limpet *Collisella paradigitalis* was strongly influenced by competition with *C. digitalis*. And in general, one might expect the effects of terrestrial

Fig. 361. The lower distributional limit of *Anthopleura elegantissima* in this surge channel is likely set by sand scour and burial.

consumers (e.g., birds and insects) approaching from the high shore to be opposite those of marine predators and to influence primarily upper limits. One might also expect all the air-breathing, invertebrate invaders of the intertidal zone to be increasingly influenced by physical conditions at lower shore levels. Unfortunately, we know little of the activities or influences of such intertidal invaders from terrestrial environments, except in a few cases.

Nevertheless, the notion that upper limits of distributions are set by physical factors and lower limits by biological factors has fairly general applicability—at least for sessile species. For the majority of mobile species its applicability is far less certain, and perhaps even suspect. This is largely because behavior so crucially influences the distribution of mobile organisms. A case in point is Wolcott's study (1973) examining influences on the distribution of several species of high- and low-shore limpets. As mentioned, he found that physical factors, specifically desiccation, were important determinants of distribution only to the species living at very high levels on the shore, and then only occasionally. The reason was that most of the time the high-shore limpets maintained, by means of behavior, a position on the shore below the upper level at which desiccation would have killed them; the low-shore limpets always seemed to maintain a position far below the lethal level.

In many other studies as well, the laboratory tolerances of animals to temperature and desiccation considerably exceed the levels of these factors that normally occur in nature. Indeed, in contrast to the evidence gathered on sessile species, there is little evidence to suggest that the upper limits of distribution of mobile species are set directly and simply by physical factors such as desiccation, except for a few species living at the highest levels on the shore. Similarly, the lower limits of most mobile animals are at best imperfectly related to the influence of classic biological factors such as predation. Much has been written, for instance, about the consumption of the snail *Tegula funebralis* by the predator *Pisaster ochraceus*. Yet, there is no firm evidence that this predator actually determines the lower limit of the snail. Certainly, snails are eaten more frequently in the lower intertidal regions where the predator occurs, and it is thus reasonable to assume that snail abundance must be affected to some extent. But because the snails normally occur abundantly in the same zone as their predator, and frequently even lower on the shore, predation can hardly be said to determine the snail's distribution in the same way that it determines the distribution of sedentary mussels and barnacles. Behavior is again an important contributor to the snail's distribution; *Tegula*, as well as many other mobile animals, has evolved defensive responses triggered by the predator, and these responses are effective enough to allow co-occurrence with this and other predators.

The concept of critical tide levels has had its ups and downs over the past several decades, as the tide of sentiment shifted from emphasizing physical factors to emphasizing biological interactions. The notion was supported by several early investigations (e.g., Colman 1933; Hewatt 1937; Doty 1946; Evans 1947), but more recently it has encountered criticism and disregard (e.g., Connell 1972; Underwood 1978a). Despite these criticisms, however, the concept seems worth modifying and developing, not just discarding.

As often seems to be the case in ecology, part of the confusion and debate is due to different meanings being associated with a concept. There are two such meanings often associated with critical tide levels. To some scientists, these levels are relative tide levels (e.g., MLLW) at which some important parameter such as total emersion time changes abruptly or rapidly. To others, these are levels at which the distributional limits of many animals change abruptly and simultaneously. Despite the semantic confusion, the information that pertains to these two meanings is fairly clear. First, there are indeed abrupt or rapid changes in parameters (e.g., maximum continuous emersion time) that occur along a complete tidal transect (Fig. 362). These changes are more obvious along shores of the western coast of North America, where mixed tides generally occur, than along shores of Europe, where equal tides are the norm. Doty's classic

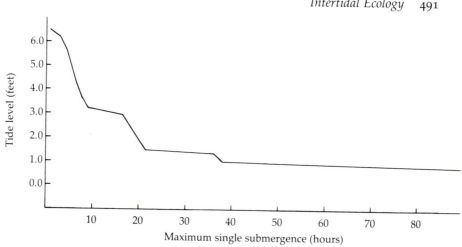

Fig. 362. Critical tide levels (after Doty 1946). Abrupt changes occur on the Pacific coast in the relationship between the tide level and various parameters of submergence and emersion, in this case the maximum single submergence experienced by an organism.

work (1946) on the Pacific coast recognizes six well-defined breaks in tidal parameters. (Doty, by the way, acknowledges the role played by R. Bolin in the development of his ideas, and Bolin, in turn, credits the influence of E. F. Ricketts.) Second, there are also distinct breaks in the distribution of many organisms in many areas (Fig. 363). This too is particularly evident along the Pacific coast (e.g., Doty 1946; Hewatt 1937).

The influence that some of these sudden changes in tidal parameters may have on distributions is intuitively clear. For example, an organism living just below the lowest extent of lower high tides will always be wetted twice a day, and it will be exposed to air for no more than about 7 to 8 hours at a stretch. An organism just above this point, however, will sometimes be wetted only once a day and may be out of water for 18 hours or more. All organisms whose tolerance limits fall somewhere between 8 and 18 hours of exposure would be affected by this change in height, and it would not take too many species to register a distributional break here. There is, of course, no reason to assume that all organisms should be affected similarly by any particular abrupt change. The relationships of organisms to tidal parameters are diverse and complex, and one should not expect the distributional breaks to be absolute. For one organism, the parameter of importance may be the maximum continuous period of exposure, which may be critical in regard to desiccation tolerance. For another, a rapid change in the total amount of submergence time averaged over many tidal cycles may be what is important, perhaps for this animal in regard to filter feeding. For yet another, the changes in average submergence may be less important than perhaps a minimum (not average) continuous submergence or exposure. As a concrete example of the last instance, the snail *Nucella* re-

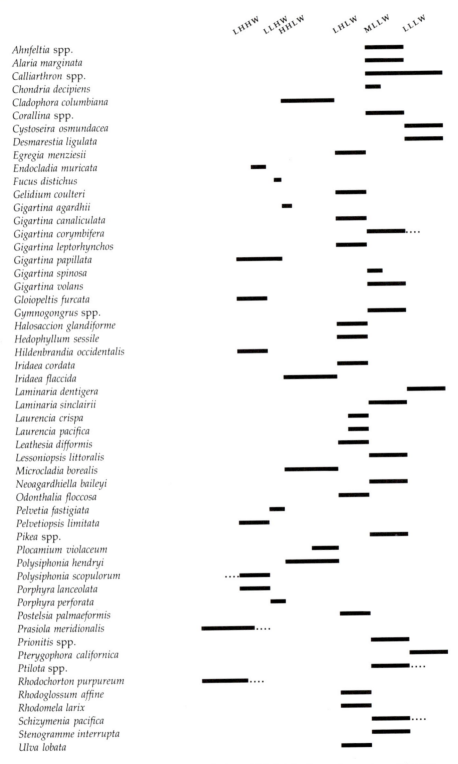

Fig. 363. Critical tide levels as breaks in the intertidal distributions of organisms. The vertical distribution of conspicuous algae on Pacific shores is indicated, with the critical tide levels indicated along the horizontal axis (after Doty 1946).

quires a certain minimum period of continuous submergence in order to bore through and consume a barnacle. For this snail, two 5-hour periods of submersion are not equivalent to one 10-hour period. When we add to all this theoretical discussion the various modifying influences that other factors such as competition and predation may have on individual distributions, and also the fact that physical and biological factors associated with tidal position are greatly influenced by local variations (e.g., wave action, slope of substratum, tide pools), it becomes difficult indeed to predict how breaks in the field distributions of animals should relate to abrupt changes in tidal parameters. Certainly, failure to find such well-defined, simultaneous breaks in the upper and lower limits of many organisms does not constitute successful refutation of the concept of critical tide levels, as some have suggested, if by critical tide levels one means tide levels at which important tidal parameters change rapidly. No more, we should add, than success in finding such breaks would prove that these breaks are *caused* by the changes occurring at those tide levels.

The difficulties and differences aside, critical tide levels are perhaps most constructively seen as a potentially important part of the arena in which both biological and physical influences interact. Defined as abrupt changes in important tidal parameters, they clearly exist, but we know little of their actual influence. They would seem to have great potential significance, however, and thus deserve more attention from ecologists.

To conclude this chapter on the various factors that influence the distributions of intertidal organisms, one example of the combined set influencing one organism seems appropriate for an integrated perspective. The distributions of intertidal organisms are determined, after all, by sets of interacting variables, not just one or two. The mussel *Mytilus californianus* will serve as our example, since we know more about this animal than most any other, and since on exposed rocky coasts it is so abundant (Fig. 364).

On exposed coasts, the mussel's upper limit of distribution seems to be set by several factors acting in concert. The list includes desiccation and nutritional stress as factors likely to be generally important. In addition, parasitism by a pea crab may also be important in some locations, and competition with limpets in others. The lower limit of distribution of the mussel is also influenced by multiple factors: most notably, predation by the seastar *Pisaster ochraceus*, and probably by several other predators (e.g., *Nucella* and crabs); and competition with algae and low intertidal invertebrates.

Competition is one of the mussel's strong suits, and some would say that *Mytilus californianus* is a "competitive dominant." This unadorned statement, however, tends to conceal several other factors interacting with the competitive process. Within its vertical distribution, for instance, the mussel's competitive prowess is influenced by the angle of the substratum (being less great on vertical surfaces) and by the amount of wave action.

Fig. 364. A small bed of *Mytilus californianus*, showing some of the factors influencing its distribution, among them predators (*Pisaster* in a dense aggregation along the waterline) and competitors (limpets and clumps of the goose barnacle *Pollicipes*). The patch in the bed also suggests disturbance by storm waves.

Indeed, on shores protected from wave action, *M. californianus* loses out entirely in competition with its congener *M. edulis*. The latter is apparently more mobile and crawls to the outside of mixed clumps of mussels, leaving *M. californianus* individuals inside the clump where they may die from the suffocating effects of siltation (Harger 1972). A more balanced and realistic picture might be gained by identifying the conditions, with appropriate modifiers, under which *M. californianus* is an efficient competitor—that is, the mussel is an efficient competitor for space on rocky shores, with strong wave action, within certain tidal limits, where predators are not abundant, and so on.

Finally, add the influences of larval settlement, history, chance, and disturbance to the mussel's circumstance. For example, larval recruitment of *Mytilus* is highly variable from year to year, and from place to place. And locally, recruitment success may be further affected by the present occupant of a site. In particular, recruitment is enhanced by the presence of barnacles, mussel byssal threads, and the alga *Endocladia*, but negatively affected by the grazing activities of limpets. Chance occurrences also have an influence: periodically, clumps of young mussels may escape in size from predation by *Pisaster*, and patches of adult mussels are at times ripped loose by disturbing agents such as storm waves and logs. The general effect is to increase heterogeneity in space and time.

It would seem evident, then, that all the factors discussed separately in this chapter play some role in determining the distribution of *Mytilus californianus*. The summary equation is a long one and the variables are many. Their influences are probably not simply additive, either—threshold values and synergistic interactions seem likely—and there is also an important dynamic element of time in the process. Most important, however, what is true of *Mytilus* would seem to be generally true of other organisms as well.

The diversity of life on the shore reflects different organisms emphasizing different variables of the equation. All organisms seem to do some things well and others not so well: high-intertidal barnacles are able to deal with desiccation better than mussels; mid-intertidal mussels have excellent competitive abilities; low-intertidal limpets are highly mobile and can move away from competitors, predators, and harsh physical conditions. But then, each benefit also seems to bear a cost that in the end balances the equation.

We have learned a good deal about the animals of the shore in the past several decades, but lest our readers surmise from this discussion that all of interest is now known, we hasten to add a strong statement to the contrary: we cannot claim even to have identified all of the important variables yet, let alone gathered all the data necessary to understand one single species, not to mention the many thousands of other species on the shore, many of them no doubt still to be discovered. Indeed, far more remains to be learned about intertidal life than we have learned to date. Certainly, the challenge of this void will keep us studying the processes of the shore for a long time to come.

That, and the natural feelings of exhilaration, joy, and satisfaction in making new acquaintances and watching old friends as they scurry or glide about their business on a foggy morning between Pacific tides.

And it is a strange thing that most of the feeling we call religious, most of the mystical outcrying which is one of the most prized and used and desired reactions of our species, is really the understanding and the attempt to say that man is related to the whole thing, related inextricably to all reality, known and unknowable. . . . It is advisable to look from the tide pool to the stars and then back to the tide pool again.

John Steinbeck
*The Log from the
Sea of Cortez (1951)*

Epilogue

Forty years after the first edition of *Between Pacific Tides* many things have changed on the shore. And much remains the same. Most of the species recorded as abundant in the first edition are still present, and still abundant, at least somewhere within their range. One must also assume that the essential elements of their biology, molded by millions of years of evolution, have changed relatively little, if at all. This is not to say, of course, that species cannot adapt to local or relatively recent changes in circumstance— witness, for example, the hermit crabs of Coyote Point (San Francisco Bay), which now occur almost exclusively in the discarded shells of snails introduced only recently from other shores. But neither the essential factors influencing animals' lives and distributions nor the great principles of community organization seem likely to have changed over a mere 40 years. There is something reassuring about all this, knowing that what was learned yesterday is likely still to be true today.

Although much remains the same on the shore, some changes are strikingly evident since Ed Ricketts first put pen to paper. Some are factual changes, whereas others reflect subtle shifts in our perceptions of the shore and its inhabitants. Perhaps not surprisingly, the concrete changes are most evident around centers of human population. Habitats have been created and destroyed as new jetties and marinas have been built and tidelands filled to make room for more animals of the two-legged variety, all in the name of progress. Few, if any, bays or estuaries on our Pacific coast do not show material effects of increased human presence.

On many shores, depletion in the abundance of certain species is evident. In some cases, the habitat or environmental conditions necessary for a species' success have been destroyed; in others, suitable habitats and environmental conditions remain, but increased levels of human predation have taken their toll. Densities of abalone, sea urchins, and various bivalves have been drastically reduced by overharvesting in some localities during the short time, geologically speaking, since the first edition was written. Reduced in some locations, too, are populations of animals not known for their culinary appeal, sea spiders and sand dollars, for example. And all along the Pacific coast, stretches of shore easily accessible to the public show the general wear and tear caused by the thousands of visitors whose feet trample the delicate forms and who collect the obvious, large ones, later to leave them stranded to die in the sun.

Not all of the changes associated with human activity have been in the direction of depletion. Additions to our fauna, in the form of species introduced from other shores, have also been documented. Occasionally, introductions have served some conscious purpose, but more often they have occurred unintentionally, as larvae were expelled in ship's ballast or as progeny were released by organisms fouling the bottoms of ships in port from foreign waters; adults, too, have been unintentionally introduced as hitchhikers with shipments of oysters imported for commercial purposes. Whatever the means of introduction, the number of exotic species now present on our Pacific shores is considerable and their ecological impact impressive. Carlton (1979) recorded over 100 introduced species on the Pacific coast, and noted that in the southern end of San Francisco Bay introduced species dominate benthic and fouling communities. In some cases the immigrants seem ominously to be displacing native species. The native snail *Cerithidia californica*, for example, is under pressure from a double dose of foreign imports. In Tomales Bay, *Cerithidia* apparently must compete for food with *Batillaria attramentaria*, a snail introduced from Japan; and in San Francisco Bay, the native snail already has lost ground to *Ilyanassa obsoleta*, an Atlantic import that preys on the eggs of *Cerithidia* and has driven the native snail from certain previously occupied habitats. The introduction of foreign species and the ecological impact of these imports on native species warrant careful attention. So far, species from other shores have been introduced only into our harbors and bays; none are known to have invaded the open coast, at least not yet.

In addition to material changes in habitats, species composition, and species abundances that have accompanied the passing of 40 years, changes are apparent, too, in the scientific and public perceptions of the shore and its inhabitants. Marine scientists at the time of the first edition of this book studied the natural histories of individual species and emphasized the physical factors (e.g., salinity, temperature, wave shock) influencing the distributions of seashore organisms. More recently there has been a tendency to focus efforts on what are seen as broad principles or organizing factors rather than on detailed information about specific organisms, and scientists studying the structure of marine communities during the last 40 years have come to place more emphasis on dynamic biological interactions (e.g., predation, competition) than on purely physical factors. These shifts are reflected in Chapter 12 of the present edition.

A great deal has been learned, in fact, about life on the shore. And along with the flood of information has come a lessening of ecological innocence, no doubt hastened by some well-publicized ecological disasters.

Significantly, our public perspective has become more global. Weather satellites in space daily transmit views of the continents and coastlines; scientists talk about great plates of the earth's crust moving relative to one another; species are introduced from one ocean to another; and so on. As a result of this new information and new perspective, a greater public awareness of seashore organisms has developed, and with it a greater concern for their continued existence, at least in some circles. But as our awareness and concern have increased, so has our reason to be concerned. The pressures of human activity on our Pacific coast have increased dramatically since Ricketts's day.

So, what of the future? What will be the state of the shore a decade from now? Clearly, the greatest cause for concern is the increase in human activity on or near the shore. The risk of disaster, never higher, can be expected to increase in proportion to the number of new oil rigs, power plants, toxic-waste dumps, and freshwater diversion projects. And even without a disaster, the effects of steadily increasing human activities— more housing developments, more fishing, more vacationers—will likely be evident, as they are today. We simply cannot continue to build homes on the shore, dump our wastes there, collect food and scientific specimens there, and expect that the shore will remain unchanged, pristine. There is no doubt that *Homo sapiens* is one of the most powerful organisms ever to reside on this planet. What remains to be discovered is what we will do with our power—create a park or produce a cesspool—for it is clear that not all our hopes for the shore are compatible. Choices and decisions will have to be made, and in this connection the current relationship, or lack of one, between scientists and political decision makers is decidedly less than ideal. Many politicians seem to follow scientific advice only when the advice matches their predilections, an unfortunate reality in any circumstance but one that is especially disconcerting when politicians control the purse strings of so many scientific projects. Scientists are not without blame in all of this, of course. Many fail to share their knowledge with decision makers, perhaps fearing that their science would somehow become tainted by utility, or that they would risk being drawn into an unwanted bureaucratic or legalistic role. Regardless of the motives of the participants, one cannot help feeling uneasy about the somewhat vacuous decisions that may result.

And what role will be played by our educational institutions? One might suppose they would be interested in expanding our knowledge of marine organisms, but instead, it seems that fewer investigations of the sort that might find their way into future revisions of this book are being supported by major universities. Alas, such basic, detailed work has fallen out of fashion, replaced (not supplemented, but replaced) by the fields of community ecology and cell biology. These are important fields of study, to be sure, but from where are to come the future likes of S. F. Light, J. W. Hedgpeth, R. I. Smith, and D. P. Abbott if not from our universities? We seem to be returning to a period when a large percentage of the scientists interested in seashore life work outside of major universities, as Ed Ricketts did. It is odd, and more than a little disconcerting, that at a time of increasing risk to our shore, we seem to have slowed our efforts to learn about the lives of the shore's inhabitants. If our universities do not value these organisms enough to study them, how can we expect the public to value their lives enough to preserve them?

Despite these concerns, one can find several reasons for hope. Over the past several decades much protective legislation has been enacted, and many reserves, state parks, and national seashores have been established where development and collecting are forbidden. We have also learned a good many things about life on the shore, and these should assist us in making intelligent decisions about our resources. Our knowledge can be used to guide future decisions, and some of it may allow us to reverse previous mistakes; aquaculture techniques, for instance, may enable us to restock depleted shores while we protect other less-damaged populations. Indeed, many of the shore's inhabitants have proved themselves impressively resilient to human disturbance, and some damaged shorelines have revived when simply left alone. Finally, there is hope in what seems to be a growing public realization that the earth is a relatively closed system, with finite limits of abuse, in which all parts are interrelated.

Although much has changed on the shore since the first edition of *Between Pacific Tides* was published, much more seems not to have changed. The species of the first edition are, by and large, the species of the present edition. The three factors—wave shock, type of bottom substratum, and tidal exposure—that served to organize the original book are still recognized as critical determinants of the distributions of intertidal organisms. Likewise, the lament of Gosse (p. 11) regarding the deleterious effects of human activity on the shore is with us still.

Thus, in many respects, this edition of *Between Pacific Tides* is similar to previous editions. The present book is considerably longer, of course, and many specifics are different—we have learned a great deal. But beyond this, the intentions of the present edition, and its essential message, remain the same as that of the original book by Ricketts and Calvin. As John Steinbeck wrote about the book in his Foreword to the revised edition (1948), "its main purpose is to stimulate curiosity, not to answer finally questions which are only temporarily answerable. . . . This book then says: 'There are good things to see in the tidepools and there are exciting and interesting thoughts to be generated from the seeing. Every new eye applied to the peep hole which looks out at the world may fish in some new beauty and some new pattern, and the world of the human mind must be enriched by such fishing.'"

There is no reason why our shore should not continue to be a rich source of sustenance for the human mind, body, and soul. It is still a treasure, full of interesting and beautiful organisms that can excite, intrigue, and delight, if only we apply our eyes to the peephole. All that is required of us is to keep our shore habitable for the organisms who live there and to allow them to prosper, along with ourselves.

Reference Matter

Outline of Major Taxa

Plants
Phylum "Protozoa"
Phylum Porifera. Sponges
Phylum Cnidaria. Jellyfishes, Corals, etc.
  Class Hydrozoa. Hydroids and
  Hydromedusae
      Order Trachylina
      Order Hydroida
        Suborder Anthomedusae
        Suborder Leptomedusae
        Suborder Limnomedusae
      Order Stylasterina. Hydrocorals
      Order Chondrophora
  Class Scyphozoa. Scyphomedusae,
  usually large jellyfish
      Order Stauromedusae
      Order Semaeostomeae
  Class Anthozoa. Sea anemones, Corals,
  Sea pens, etc.
    Subclass Alcyonaria
      Order Stolonifera
      Order Alcyonacea
      Order Pennatulacea
      Order Gorgonacea
    Subclass Zoantharia
      Order Actiniaria
      Order Ceriantharia
      Order Corallimorpharia
      Order Madreporaria
Phylum Ctenophora. Comb jellies
Phylum Platyhelminthes. Flatworms
  Class Trematoda. Flukes
  Class Cestoda. Tapeworms
  Class Turbellaria. Mostly free-living
  flatworms
      Order Acoela
      Order Rhabdocoelida

      Order Alloeocoela
      Order Tricladida
      Order Polycladida
Phylum Nemertea. Ribbon worms
Phylum Nematoda. Roundworms
Phylum Annelida. Segmented worms
  Class Hirudinea. Leeches
  Class Oligochaeta. Earthworms, etc.
  Class Polychaeta. Bristle worms, Fan
  worms, Clam worms, etc.
Phylum Sipuncula. Sipunculans, or Peanut
worms
Phylum Echiura. Echiuran worms
Phylum Arthropoda
  Subphylum Mandibulata
    Class Crustacea. Barnacles, Beach hop-
    pers, Shrimps, Lobsters, Crabs, etc.
      Subclass Branchiopoda
      Subclass Ostracoda
      Subclass Copepoda
      Subclass Cirripedia. Barnacles
        Order Acrothoracica
        Order Thoracica
        Order Rhizocephala
      Subclass Malacostraca
        Superorder Phyllocarida
          Order Leptostraca
        Superorder Hoplocarida
          Order Stomatopoda
        Superorder Peracarida
          Order Mysidacea
          Order Cumacea
          Order Tanaidacea (Chelifera)
          Order Isopoda. Pill bugs, Sow
          bugs, etc.
          Order Amphipoda. Beach hop-
          pers, Sand fleas, Skeleton
          shrimps, etc.

Superorder Eucarida
Order Euphausiacea
Order Decapoda. Shrimps, Lobsters, Crabs, etc.
  Infraorder Caridea. Shrimps
  Infraorder Palinura. Spiny lobsters
  Infraorder Thalassinidea. Ghost shrimps and Mud shrimps
  Infraorder Anomura. Hermit crabs, Porcelain crabs, and Mole crabs
  Infraorder Brachyura. True crabs: Shore crabs, Pea crabs, Spider crabs, Edible crabs, Fiddler crabs, etc.
Class Chilopoda. Centipedes
Order Geophilomorpha
Class Insecta (Hexapoda)
  Order Collembola. Springtails and Bristletails
  Order Thysanura. Silverfish
  Order Hemiptera. True bugs
  Order Dermaptera. Earwigs
  Order Diptera. Flies
  Order Coleoptera. Beetles
Subphylum Chelicerata
  Class Arachnida. Spiders, Mites, Pseudoscorpions, etc.
    Order Acarina. Mites
    Order Chelonethida. Pseudoscorpions
  Class Pycnogonida. Sea spiders
Phylum Mollusca
  Class Gastropoda. Snails, Limpets, Sea hares, Nudibranchs, etc.
    Subclass Prosobranchia
      Order Archaeogastropoda. Limpets, Abalones, Turbans
        Superfamily Pleurotomariacea. Abalones
        Superfamily Fissurellacea. Keyhole limpets
        Superfamily Patellacea. True limpets
        Superfamily Trochacea. Top shells and Turbans
      Order Mesogastropoda. Periwinkles, Slipper shells, Cowries, Moon snails, Horn shells, etc.
      Order Neogastropoda. Whelks, Rock snails, Olives, Cones, etc.
    Subclass Opisthobranchia. Nudibranchs, Sea hares, Sea slugs, etc.
      Order Cephalaspidea
      Order Anaspidea
      Order Sacoglossa
      Order Notaspidea
      Order Nudibranchia
    Subclass Pulmonata
  Class Bivalvia (Pelecypoda). Clams, Cockles, Mussels, Oysters, Shipworms, etc.
    Subclass Pteriomorpha
      Order Arcoida
      Order Mytiloida
      Order Pterioida
    Subclass Heterodonta
      Order Veneroida
      Order Myoida
  Class Polyplacophora. Chitons, or Sea cradles
  Class Cephalopoda. Octopods (or Octopuses), Squids, *Nautilus*
  Class Scaphopoda
  Class Aplacophora
  Class Monoplacophora
Phylum Phoronida
Phylum Brachiopoda
Phylum Ectoprocta (Bryozoa)
Phylum Entoprocta
Phylum Echinodermata
  Class Asteroidea. Seastars, or Starfish
  Class Ophiuroidea. Brittle stars, Serpent stars
  Class Echinoidea. Sea urchins, Sand dollars, and Heart urchins
  Class Holothuroidea. Sea cucumbers
  Class Crinoidea
Phylum Hemichordata (Enteropneusta). Acorn or Tongue worms, etc.
Phylum Chordata
  Subphylum Urochordata (Tunicata). Sea squirts, Compound ascidians, Tunicates
  Subphylum Cephalochordata. Lancelets
  Subphylum Vertebrata (Craniata)
    Class Agnatha
    Class Chondrichthyes. Sharks and Rays
    Class Osteichthyes. Bony fishes
    Class Reptilia
    Class Amphibia
    Class Aves. Birds
    Class Mammalia

Annotated Systematic Index

This index serves several purposes. First, it gives some idea of where a certain animal stands in the scheme of classification, to the level of orders and suborders (for an outline of this system, see the pages immediately preceding). The classification used here may not be universally accepted (none is), but it is not notably divergent from what might be called standard zoological opinion. Classification is, after all, simply an effort to present our ideas of how things should be arranged and how they may be related. In this sense, it reflects our ideas of evolution.

This index should also enable an interested reader to track down more information on any animal that is especially interesting, or to locate closely related species that are discussed in other sections of the text. In a few cases, species that have not been treated in the text are listed here; usually these are species about which we know a considerable amount, but they are less likely to be noticed than those treated in the text, except perhaps in special situations. All levels of classification, as well as such substantive material as occurs in the annotation, will be found listed in alphabetical order in the General Index. For books and articles of broader scope, the reader should consult the General Bibliography, pp. 593–609.

All names of plants and animals are listed according to their formal designation: the genus (plural "genera") and species in italics (boldface in the list here), followed by the name of the describer in roman. This is the designation by which a species (plural "species") is known. These names are supposed to look like Latin and be treated as such grammatically. There is, however, considerable confusion involved in determining "grammatically correct" forms for names; few people now learn enough Latin to know what ending the name of a species should have. Problems arise regularly, and although it is obvious that problems of nomenclature do not have much to do directly with biology, they do constitute a branch of international law. In zoology the guide for this sort of thing is the International Code of Zoological Nomenclature (approved by the XV International Congress of Zoology; published by the International Trust for Zoological Nomenclature, London, 1961). The text of this book is in French and English; of its 176 pages more than 60 are devoted primarily to these problems of name formation.

There are some differences between zoological and botanical naming conventions. Obvious patronymics are sometimes capitalized in botany (*Asparagus Sprengeri*) but never in zoology (*Hydractinia milleri*). In both fields the name of the describer is placed in parentheses if the generic name now in use differs from that in the original species description; thus, *Collisella scabra* (Gould) was described originally as *Patella scabra* by Gould. However, in zoology the name of the reviser responsible for the change is not indicated, whereas in botany it is; thus, for one of the algae, we have *Macrocystis pyrifera* (Linnaeus) C. A. Agardh.

Finally, a word or two is perhaps in order regarding the scope of the Annotated Systematic Index. *Between Pacific Tides* is a book about the animals, principally invertebrates, that reside more or less permanently in the intertidal zone. As such, the many species of vertebrates that periodically visit the shore during low or high tide are treated only tangentially, as they affect the more permanent, invertebrate residents. Plants too have been treated only incidentally, as sources of food and refuge for animal residents. Although an ideal treatment would have considered them along with the animals, this was not practical to do.

Plants

Although a detailed treatment of marine plants is outside the scope of this account, the seaweeds are one of the major sources of food in the intertidal and provide refuge in one way or another for many species of animals. Most marine plants are algae—red, green, or brown—but two genera of flowering plants also occur. A broad-leaved eelgrass, *Zostera marina* Linnaeus, is rooted in the muddy bottoms of quiet-water bays, and two species of narrow-leaved surfgrass, *Phyllospadix scouleri* Hooker and *P. torreyi* Watson, occur in more wave-swept rocky areas.

The literature on Pacific coast marine plants is large and steadily growing. The following references are useful as guides and background reading, and the book by Abbott and Hollenberg (1976) in particular contains a comprehensive bibliography of systematic literature on Pacific coast algae. In addition, selected papers dealing with algal ecology are cited in the section "Selected papers on marine ecology," in the General Bibliography; see especially the papers by Black (1976), Dayton (1975), Druehl and Green (1982), Gunnill (1980a,b; 1982a,b; 1983), Hay (1981), Hruby (1975; 1976), Johnson et al. (1974), Lubchenco and Cubit (1980), Paine (1979), Slocum (1980), Sousa (1979a,b; 1980), and Turner (1983a,b).

ABBOTT, I. A., and G. J. HOLLENBERG. 1976. Marine algae of California. Stanford, Calif.: Stanford University Press. 827 pp.

CHAPMAN, A. R. O. 1979. Biology of seaweeds: Levels of organization. Baltimore: University Park Press. 134 pp. ☐ An informative, yet relatively brief, account of seaweed biology from the level of the cell to the level of communities. Useful list of references.

CHAPMAN, V. J. 1964. Coastal vegetation. New York: Pergamon. 245 pp. ☐ In paperback. This is essentially an ecology of intertidal and maritime plants, including those of dunes and marshes. It is based primarily on British examples, but ecological concepts are not tied to species or localities.

DAWSON, E. Y. 1966a. Marine botany: An introduction. New York: Holt, Rinehart, & Winston. 371 pp. ☐ A good general text for non-botanists, with a strong flavor of the author's sense of the history of his subject.

———. 1966b. Seashore plants of northern California. Berkeley: University of California Press. 103 pp. ☐ Salt-marsh and dune plants are included in addition to the seaweeds. The northern California guide is concerned mainly with the Monterey and mid-California area; it does include plants found in Oregon, but only perfunctorily. The southern California guide goes as far north as Gaviota.

———. 1966c. Seashore plants of southern California. Berkeley: University of California Press. 101 pp.

DAYTON, P. K. 1973. Dispersion, dispersal, and persistence of the annual intertidal alga, *Postelsia palmaeformis* Ruprecht. Ecology 54: 433–38.

DOTY, M. S. 1947. The marine algae of Oregon. I, Chlorophyta and Phaeophyta. Farlowia 3: 1–65. II, Rhodophyta. Farlowia 3: 159–215.

GUBERLET, M. L. 1956. Seaweeds at ebb tide. Seattle: University of Washington Press. 182 pp. ☐ About the seaweeds of the Puget Sound area, with the usual hoked-up common names, such as

"loose color changer" and "red eyelet silk." Although there may be problems with some species because of a lack of detail, the work as a whole is useful, and is available as a comparatively inexpensive paperback.

LOBBAN, C. S., and M. J. WYNNE, eds. 1981. The biology of seaweeds. Berkeley: University of California Press. 786 pp.

ROUND, F. E. 1965. The biology of algae. New York: St. Martin's Press. 269 pp. ☐ An excellent, solid introduction to the algae, both freshwater and marine; as the dust cover says, it is "scholarly and general."

SCAGEL, R. F. 1967. Guide to common seaweeds of British Columbia. British Columbia Prov. Mus., Handbook 27. 330 pp. ☐ Probably the most useful guide for the Northwest.

SMITH, G. M. 1969. Marine algae of the Monterey Peninsula, California. Revised edition, including supplement by G. Hollenberg and I. Abbott. Stanford, Calif.: Stanford University Press. 752 pp.

Phylum "PROTOZOA"

Protozoans—acellular, or single-celled, organisms—are ordinarily too small for unaided eye observation, and have been for the most part omitted from the text. However, some are large enough to be seen with a hand lens. Although for some time protozoans have been considered a single phylum (the Protozoa) within the animal kingdom, a growing number of accounts now treat them as members of the kingdom Protista, a group that consists of several different unicellular phyla, together with most algal phyla. A small sample of references on various protozoans is offered below; there is no comprehensive reference, with the possible exception of works dealing with the Foraminifera. Forams, because of their long fossil history and the ease with which their remains may be recovered, are of great value to geologists, particularly in petroleum exploration.

ARNOLD, Z. M. 1980. Foraminifera: Shelled protozoans. *In* R. H. Morris, D. P. Abbott, and E. C. Haderlie, Intertidal invertebrates of California, pp. 9–20. Stanford, Calif.: Stanford University Press.

BURRESON, E. M. 1973. Symbiotic ciliates from description of *Parahypocoma rhamphisokarya* n. sp. Trans. Amer. Microsc. Soc. 92: 517–22. ☐ Three species of ciliates were frequently encountered in the pharyngeal basket of ascidians in the San Juan Islands and on the coast of central Oregon: *Trichophrya salparum* (a suctorian), *Euplotaspis cionaecola* (a hypotrich), and *Parahypocoma rhamphisokarya* n. sp. (a thigmotrich).

HAYNES, J. R. 1981. Foraminifera. New York: Halsted (Wiley). 433 pp.

HEDLEY, R. H., and C. G. ADAMS, eds. 1974–78. Foraminifera. Vol. 1 (1974), 276 pp.; Vol. 2 (1977), 260 pp.; Vol. 3 (1978), 290 pp. New York: Academic Press.

KOZLOFF, E. N. 1961. A new genus and two new species of ancistrocomid ciliates (Holotricha:

Thigmotricha) from sabellid polychaetes and from a chiton. J. Protozool. 8: 60–63.

———. 1966. *Phalacrocleptes verruciformis* gen. nov., sp. nov., an unciliated ciliate from the sabellid polychaete *Schizobranchia insignis* Bush. Biol. Bull. 130: 202–10.

LANKFORD, R. R., and F. B. PHLEGER. 1973. Foraminifera from the nearshore turbulent zone, western North America. J. Foram. Res. 3: 101–32.

LEVINE, N. D. 1981. New species of *Lankestria* (Apicomplexa, Eugregarinida) from ascidians on the central California coast. J. Protozool. 28: 363–70. □Describes 11 new species of gregarines from tunicates in Monterey Bay.

LIPPS, J. H., and M. G. ERSKIAN. 1969. Plastogamy in Foraminifera: *Glabratella ornatissima* (Cushman). J. Protozool. 16: 422–25. □Discusses reproduction of this protozoan from exposed, rocky-shore tide pools near Bodega Bay.

MATSUDO, H. 1966. A cytological study of a chonotrichous ciliate protozoan, *Lobochona prorates*, from the gribble. J. Morphol. 120: 359–90.

MURRAY, J. W. 1973. Distribution and ecology of living benthic foraminiferids. London: Heinemann Educational Books. 274 pp.

MYERS, E. H. 1940. Observations on the origin and fate of flagellated gametes in multiple tests of *Discorbis* (Foraminifera). J. Mar. Biol. Assoc. U.K. 24: 201–26. □A beautifully illustrated paper on a group of benthic tide-pool forams well represented on the California coast. Species occurring at La Jolla, Monterey, and Moss Beach are discussed.

NOBLE, A. E. 1929. Two species of the protozoan genus *Ephelota* from Monterey Bay, California. Univ. Calif. Publ. Zool. 33: 13–26.

RADIR, P. I. 1927. *Trichamoeba schaefferi*, a new species of large amoeba from Monterey Bay, California. Arch. Protistenkd. 59: 289–300.

Boveria subcylindrica Stevens, in the holothurian *Parastichopus*, §77

Ephelota gemmipara (Hertwig), suctorian occurring on the hydroid *Abietinaria*, §83

E. gigantea Noble, §323

Gonyaulax spp., §158

TAYLOR, D. L., and H. H. SELIGER, eds. 1979. Toxic dinoflagellate blooms. New York: Elsevier. 505 pp. □Proceedings of the Second International Conference on Toxic Dinoflagellate Blooms, held in Key Biscayne, Florida, in 1978. Many research papers address *Gonyaulax* and related forms.

Gromia oviformis Dujardin, §95

ARNOLD, Z. M. 1972. Observations on the biology of the protozoan *Gromia oviformis* Dujardin. Univ. Calif. Publ. Zool. 100: 1–168.

Licnophora macfarlandi Stevens, in *Parastichopus*, §77

Trichamoeba schaefferi Radir, §64

Phylum **PORIFERA**. Sponges

Sponges are notorious for great plasticity in response to local and geographic environmental variables, and this is undoubtedly one of the reasons the systematics of these animals is such a troublesome matter. Whatever the reason, a comprehensive revision of the sponges of the Pacific coast is sorely needed. The fact that the classification of sponges can still be debated at the class level makes it apparent that there is still much to learn about these organisms, as does the fact that specialists have changed, or otherwise called into doubt, the names of 10 of the 26 species mentioned in this volume since publication of the preceding edition.

BAKUS, G. J. 1966. Marine poeciloscleridan sponges of the San Juan Archipelago, Washington. J. Zool. (Lond.) 149: 415–531. □Concerns 23 species from the Friday Harbor area; proposes four new genera and seven new species. Discusses the zoogeography of Pacific coast species and reviews the development of sponge biology on this coast.

BAKUS, G. J., and D. P. ABBOTT. 1980. Porifera: The sponges. *In* R. H. Morris, D. P. Abbott, and E. C. Haderlie, Intertidal invertebrates of California, pp. 21–39. Stanford, Calif.: Stanford University Press.

BAKUS, G. J., and K. GREEN. 1977. Porifera of the Pacific coast of North America. Southern California Coastal Water Research Project, Taxonomic Standardization Program: Sponge Workshop. Photocopy. 22 pp.

BERGQUIST, P. R. 1978. Sponges. Berkeley: University of California Press. 268 pp. □An excellent book with relatively complete coverage of the basic aspects of sponge biology. Emphasizes recent studies and includes chapters on biochemistry, ecology, phylogeny, and structure and function.

BERGQUIST, P. R., W. HOFHEINZ, and G. OESTERHELT. 1980. Sterol composition and the classification of the Demospongiae. Biochem. Syst. Ecol. 8: 423–35.

BERGQUIST, P. R., M. E. SINCLAIR, and J. J. HOGG. 1970. Adaptation to intertidal existence: Reproductive cycles and larval behaviour in Demospongiae. Symp. Zool. Soc. Lond. 25: 247–71.

BOROJEVIC, R., W. G. FRY, W. C. JONES, C. LEVI, R. RASMONT, M. SARA, and J. VACELET. 1968. A reassessment of the terminology for sponges. Bull. Mus. Natl. Hist. Natur., ser. 2, 39: 1224–35.

DE LAUBENFELS, M. W. 1932. The marine and fresh-water sponges of California. Proc. U.S. Natl. Mus. 81: 1–140.

———. 1961. Porifera of Friday Harbor and vicinity. Pacific Sci. 15: 192–202. □Reports on 17 species collected and studied in the San Juan Islands, including 9 species covered in this volume.

FRY, W. G., ed. 1970. The biology of the Porifera. Symp. Zool. Soc. Lond. 25. London: Academic Press. 512 pp.

HARRISON, F. W., and R. R. COWDEN, eds. 1976. Aspects of sponge biology. New York: Academic Press. 354 pp. □Contains little information on local forms, but does provide accounts of various aspects of current research on sponges.

HARTMAN, W. D. 1975. Phylum Porifera. *In* R. I. Smith and J. T. Carlton, eds., Light's manual, pp. 32–64. Los Angeles: University of California Press. □Contains a helpful introduction to the terminology of spicules, the skeletal underpinnings of sponge taxonomy, and keys to the common sponges of central California.

PARKER, G. H. 1914. On the strength and the volume of the water currents produced by sponges. J. Exp. Zool. 16: 443–46.

REISWIG, H. M. 1971. Particle feeding in natural populations of three marine demosponges. Biol. Bull. 141: 568–91. □The three species are all tropical, but the investigation is relevant to local forms.

———. 1975a. The aquiferous systems of three marine Demospongiae. J. Morphol. 145: 493–502. □The three are *Haliclona permollis, Halichondria panicea,* and *Microciona prolifera.*

———. 1975b. Bacteria as food for temperate-water marine sponges. Can. J. Zool. 53: 582–89. □*Haliclona permollis* and *Suberites ficus* are the two sponges examined.

VOGEL, S. 1977. Current-induced flow through living sponges in nature. Proc. Natl. Acad. Sci. USA 74: 2069–71.

Antho lithophoenix (de Laubenfels) (formerly *Isociona*), §32

Aplysina fistularis (Pallas) (formerly *Verongia thiona* de Laubenfels), §98

Axocielita originalis (de Laubenfels) (formerly *Esperiopsis;* includes *Axocielita hartmani* Simpson), §140

Cliona celata Grant var. **californiana,** §217

EMSON, R. H. 1966. The reactions of the sponge *Cliona celata* to applied stimuli. Comp. Biochem. Physiol. 18: 805–27.

HATCH, W. I. 1980. The implication of carbonic anhydrase in the physiological mechanism of penetration of carbonate substrata by the marine burrowing sponge *Cliona celata* (Demospongiae). Biol. Bull. 159: 135–47.

RUTZLER, K., and G. RIEGER. 1973. Sponge burrowing: Fine structure of *Cliona lampa* penetrating calcareous substrata. Mar. Biol. 21: 144–62.

Halichondria panicea (Pallas), §97

FORESTER, A. J. 1979. The association between the sponge *Halichondria panicea* (Pallas) and scallop *Chlamys varia* (L.): A commensal-protective mutualism. J. Exp. Mar. Biol. Ecol. 36: 1–10.

PEATTIE, M. E., and R. HOARE. 1981. The sublittoral ecology of the Menai Strait: II. The sponge *Halichondria panicea* (Pallas) and its associated fauna. Estuarine Coastal Shelf Sci. 13: 621–35.

VOGEL, S. 1974. Current-induced flow through the sponge, *Halichondria.* Biol. Bull. 147: 443–56.

Haliclona sp. (? = *H. permollis* (Bowerbank)) (*H. permollis* (Bowerbank) of the 4th ed.; formerly *Reniera cinerea* (Grant)), §31

ELVIN, D. W. 1976. Seasonal growth and reproduction of an intertidal sponge, *Haliclona permollis* (Bowerbank). Biol. Bull. 151: 108–25.

———. 1979. The relationship of seasonal changes in the biochemical components to the reproductive behavior of the intertidal sponge, *Haliclona permollis.* Biol. Bull. 156: 47–61.

Haliclona sp. (formerly *H. rufescens* (Lambe), originally described as *Reniera*), §198

The Pacific coast representatives of the genus

Haliclona need further critical study to clarify their currently uncertain relationships. In addition to the two species of uncertain assignment covered in this book, there seems to be yet another, which is common on floats in San Francisco Bay and Bodega Harbor.

FELL, P. E. 1970. The natural history of *Haliclona ecbasis* de Laubenfels, a siliceous sponge of California. Pacific Sci. 24: 381–86.

Halisarca sacra de Laubenfels, §238

Hymeniacidon sinapium de Laubenfels, §§183, 225

Isodictya rigida (Lambe) (formerly *Neoesperiopsis*), §225

Leucandra heathi Urban (formerly *Leuconia*), §95

Leucilla nuttingi (Urban) (formerly *Rhabdodermella*), §94

Leucosolenia eleanor (Urban), §93

L. nautilia de Laubenfels, §229

Lissodendoryx firma (Lambe) (includes *L. noxiosa* de Laubenfels), §97

Mycale adhaerens (Lambe), §220

M. macginitiei de Laubenfels, §238

Myxilla incrustans (Esper) (formerly *Ectyodoryx parasitica* (Lambe), §220

BLOOM, S. A. 1975. The motile escape response of a sessile prey: A sponge-scallop mutualism. J. Exp. Mar. Biol. Ecol. 17: 311–21. □Concerns both *Myxilla incrustans* and *Mycale adhaerens.*

Ophlitaspongia pennata (Lambe), §32

Plocamia karykina de Laubenfels, §32

Spheciospongia confoederata de Laubenfels, §92

The generic placement of this sponge is at present uncertain.

Stelleta sp., §96

Suberites ficus (Johnson) (formerly *Choanites suberea* var. *lata* (Lambe)), §225

Tethya aurantia (Pallas) var. **californiana** de Laubenfels, §140

FISHELSON, L. 1981. Observations on the moving colonies of the genus *Tethya* (Demospongia, Porifera): I. Behavior and cytology. Zoomorphology 98: 89–99. □Young colonies of *T. seychellensis* (Red Sea) and *T. aurantia* (Mediterranean) can move from one site to another on 10- to 16-mm extensions of the body wall. Each of these "podia" bears an adhesive knob on its distal end. Movement may be induced by a stressful situation, such as burial by sand.

Tetilla mutabilis de Laubenfels, §298

Xestospongia vanilla de Laubenfels, §140

Phylum **CNIDARIA**. Jellyfishes, Corals, etc.

The cnidarians tend to be radially symmetrical animals, usually characterized by a single opening that serves both as mouth and anus and by a crown of tentacles that encircles the opening. The body is usually adapted either for a sedentary life, as a polyp, or for a mobile existence, as a medusa or jellyfish; some cnidarians have life cycles in which these body forms alternate, a fact that has caused some confusion among taxonomists; the two phases of one species have in some cases been given two separate species names. The group is diverse, including sea anemones, corals, sea pens, hydroids, and jellyfishes, and ancient, with many fossils in Cambrian rocks. Despite their often common occurrence, the natural history and taxonomy of many species, especially in the class Hydrozoa, are poorly known, and much remains to be done; detailed life-history studies are particularly needed.

BUCKLIN, A. 1982. The annual cycle of sexual reproduction in the sea anemone *Metridium senile*. Can. J. Zool. 60: 3241–48.

GLADFELTER, W. B. 1973. A comparative analysis of the locomotor system of medusoid Cnidaria. Helgol. Wiss. Meeresunters. 25: 228–72.

KRAMP, P. L. 1961. Synopsis of medusae of the world. J. Mar. Biol. Assoc. U.K. 40: 1–469.

MACKIE, G. O., ed. 1976. Coelenterate ecology and behavior. New York: Plenum. 744 pp. □This symposium volume contains a wealth of interesting papers on many aspects of cnidarian biology and ecology.

MILLER, R. L., and C. R. WYTTENBACH, eds. 1974. The developmental biology of the Cnidaria. Amer. Zool. 14: 440–866.

MILLS, C. E. 1981. Seasonal occurrence of planktonic medusae and ctenophores in the San Juan Archipelago (NE Pacific). Wasmann J. Biol. 39: 6–29.

MUSCATINE, L., and H. M. LENHOFF, eds. 1974. Coelenterate biology: Reviews and new perspectives. New York: Academic Press. 501 pp. □Contains review articles on bioluminescence, endosymbiosis, neurobiology, development, and several aspects of functional morphology.

REES, W. J., ed. 1966. The Cnidaria and their evolution. Symp. Zool. Soc. Lond. 16. New York: Academic Press. 449 pp. □Several good review articles on evolution within the Cnidaria.

Class HYDROZOA
Hydroids and Hydromedusae

ARAI, M. N., and A. BRINCKMANN-VOSS. 1980. Hydromedusae of British Columbia and Puget Sound, Canada. Can. Bull. Fish. Aquat. Sci. 204: 1–192.

FRASER, C. M. 1914. Some hydroids of the Vancouver Island region. Trans. Roy. Soc. Can., Biol. Sci. (3)8: 99–215.

————. 1937. Hydroids of the Pacific coast of Canada and the United States. University of Toronto Press. 207 pp. □Although this work is badly outdated and has many serious shortcomings (among them, Fraser's total neglect of medusa stages), it is, sadly, the only "comprehensive" taxonomic account of Pacific coast hydroids.

KRAMP, P. L. 1968. The Hydromedusae of the Pacific and Indian Oceans. Dana Report No. 72. Copenhagen: Carlsberg Foundation. 200 pp.

MILLS, C. E. 1981. Diversity of swimming behaviors in hydromedusae as related to feeding and utilization of space. Mar. Biol. 64: 185–89. □Concerns feeding behaviors of the medusa stage of *Pro-*

boscidactyla flavicirrata, *Stomotoca atra*, *Phialidium gregarium*, and *Polyorchis penicillatus*.

————. 1983. Vertical migration and diel activity patterns of hydromedusae: Studies in a large tank. J. Plankton Res. 5: 619–35.

Order TRACHYLINA

DONALDSON, S., G. O. MACKIE, and A. ROBERTS. 1980. Preliminary observations on escape swimming and giant neurons in *Aglantha digitale* (Hydromedusae: Trachylina). Can. J. Zool. 58: 549–52.

Order HYDROIDA
Suborder ANTHOMEDUSAE

Corymorpha palma Torrey, §288
BALL, E. E. 1973. Electrical activity and behavior in the solitary hydroid *Corymorpha palma*. I. Spontaneous activity in whole animals and in isolated parts. Biol. Bull. 145: 223–42.

WYMAN, R. 1965. Notes on the behavior of the hydroid, *Corymorpha palma*. Amer. Zool. 5: 491–97.

Eudendrium californicum Torrey, and other species, §79

Garveia annulata Nutting, §81

Hydractinia milleri Torrey, §82
KARLSON, R. H. 1981. A simulation study of growth inhibition and predator resistance in *Hydractinia echinata*. Ecol. Model. 13: 29–47.

Polyorchis penicillatus (Eschscholtz), §§193, 330
BRINCKMANN-VOSS, A. 1977. The hydroid of *Polyorchis penicillatus* (Eschscholtz) (Polyorchidae, Hydrozoa, Cnidaria). Can. J. Zool. 55: 93–96. □This is the only published account of a polyp stage of *Polyorchis*. The report was based on a single colony and is not universally accepted.

GLADFELTER, W. B. 1972. Structure and function of the locomotor system of *Polyorchis montereyensis* (Cnidaria, Hydrozoa). Helgol. Wiss. Meeresunters. 23: 38–79.

SKOGSBERG, T. 1948. A systematic study of the family Polyorchidae (Hydromedusae). Proc. Calif. Acad. Sci. (4)26: 101–24.

Sarsia tubulosa (M. Sars) (formerly *Syncoryne mirabilis* (L. Agassiz)), §333
The medusae have been long known as *Sarsia*, but the hydroid stage has been called *Syncoryne*.

MILLER, R. L. 1982. Identification of sibling species within the "*Sarsia tubulosa* complex" at Friday Harbor, Washington (Hydrozoa: Anthomedusae). J. Exp. Mar. Biol. Ecol. 62: 153–72. □There are at least three species of large *Sarsia* at Friday Harbor: *S. tubulosa*, *S. princeps*, and a common, but undescribed, species.

Tubularia crocea (L. Agassiz), §333
CAMPBELL, R. D., and F. CAMPBELL. 1968. *Tubularia* regeneration: Radial organization of tentacles, gonophores and endoderm. Biol. Bull. 134: 245–51.

MILLER, R. L., and C. J. BROKAW. 1970. Chemotactic turning behaviour of *Tubularia* spermatozoa. J. Exp. Biol. 52: 699–706.

MILLER, R. L., and C. Y. TSENG. 1974. Properties

Astropecten has difficulty crossing the sand-dollar bed and generally is kept seaward of the major *Renilla* population.

SATERLIE, R. A., and J. F. CASE. 1979. Development of bioluminescence and other effector responses in the pennatulid coelenterate *Renilla kollikeri*. Biol. Bull. 157: 506–23. □The first sign of bioluminescence appears shortly after settlement. Larval settlement in the laboratory is delayed if sand is not provided.

Stylatula elongata (Gabb), slender sea pen, §289
The known range of this species is Tomales Bay to San Diego, from low tide to 65 m. It is abundant in Tomales Bay.

Order GORGONACEA

Muricea californica Aurivillius and **M. fruticosa** (Verrill), §218
GRIGG, R. W. 1972. Orientation and growth form of sea fans. Limnol. Oceanogr. 17: 185–92. □Fan-shaped gorgonians, such as *Muricea californica* and *M. fruticosa*, orient at right angles to current flow.

———. 1977. Population dynamics of two gorgonian corals. Ecology 58: 278–90.

SATERLIE, R. A., and J. F. CASE. 1978. Neurobiology of the gorgonian coelenterates, *Muricea californica* and *Lophogorgia chilensis*: II. Morphology. Cell Tissue Res. 187: 379–96.

Subclass ZOANTHARIA

HAND, C. 1954. The sea anemones of central California. Part I. The corallimorpharian and athenarian anemones. Wasmann J. Biol. 12: 345–75.

———. 1955a. The sea anemones of central California. Part II. The endomyarian and mesomyarian anemones. Wasmann J. Biol. 13: 37–99.

———. 1955b. The sea anemones of central California. Part III. The acontiarian anemones. Wasmann J. Biol. 13: 189–251.

KOEHL, M. A. R. 1977a. Effects of sea anemones on the flow forces they encounter. J. Exp. Biol. 69: 87–105.

———. 1977b. Mechanical organization of cantilever-like sessile organisms: Sea anemones. J. Exp. Biol. 69: 127–42.

LINDBERG, W. J. 1976. Starvation behavior of the sea anemones, *Anthopleura xanthogrammica* and *Metridium senile*. Biologist 58: 81–88. □When starved, these anemones moved farther and were otherwise more active than when fed.

MARTIN, E. J. 1963. Toxicity of dialyzed extracts of some California anemones (Coelenterata). Pacific Sci. 17: 302–4.

SASSAMAN, C., and C. P. MANGUM. 1972. Adaptations to environmental oxygen levels in infaunal and epifaunal sea anemones. Biol. Bull. 143: 657–78.

SEBENS, K. P. 1981. The allometry of feeding, energetics, and body size in three sea anemone species. Biol. Bull. 161: 152–71. □The number of prey captured by *Anthopleura xanthogrammica*, *A. elegantissima*, and *Metridium senile* increased with increas-

ing feeding surface area. But only for *A. xanthogrammica* did prey size increase with predator size.

SPAULDING, J. G. 1974. Embryonic and larval development in sea anemones (Anthozoa: Actinaria). Amer. Zool. 14: 511–20.

TORREY, H. B. 1902. Anemones. Papers from the Harriman Alaska Expedition, XXX. Proc. Wash. Acad. Sci. 4: 373–410.

WATERS, V. L. 1973. Food preference of the nudibranch *Aeolidia papillosa*, and the effect of the defenses of the prey on predation. Veliger 15: 174–92.

Order ACTINIARIA

Anthopleura artemisia (Pickering), the burrowing anemone, §258
POWELL, D. C. 1964. Fluorescence in the sea anemone *Anthopleura artemisia*. Bull. Amer. Littoral Soc. 2: 17.

A. elegantissima (Brandt), the aggregated anemone, §24
CHIA, F.-S., and R. KOSS. 1979. Fine structural studies of the nervous system and the apical organ in the planula larva of the sea anemone *Anthopleura elegantissima*. J. Morphol. 160: 275–98.

FITT, W. K., and R. L. PARDY. 1981. Effects of starvation, and light and dark on the energy metabolism of symbiotic and aposymbiotic sea anemones, *Anthopleura elegantissima*. Mar. Biol. 61: 199–205.

FITT, W. K., R. L. PARDY, and M. M. LITTLER. 1982. Photosynthesis, respiration, and contribution to community productivity of the symbiotic sea anemone *Anthopleura elegantissima* (Brandt, 1835). J. Exp. Mar. Biol. Ecol. 61: 213–32. □The photosynthetic rate of the dinoflagellate *Symbiodinium microadriaticum* living inside *Anthopleura* is such that the anemone contributes to primary productivity at a rate comparable to that of many intertidal algae.

FRANCIS, L. 1973a. Clone specific segregation in the sea anemone *Anthopleura elegantissima*. Biol. Bull. 144: 64–72.

———. 1973b. Intraspecific aggression and its effect on the distribution of *Anthopleura elegantissima* and some related sea anemones. Biol. Bull. 144: 73–92.

———. 1976. Social organization within clones of the sea anemone *Anthopleura elegantissima*. Biol. Bull. 150: 361–76.

———. 1979. Contrast between solitary and clonal lifestyles in the sea anemone *Anthopleura elegantissima*. Amer. Zool. 19: 669–81. □The two forms of *Anthopleura elegantissima*, the clonal aggregated form and the larger solitary form, may actually be a sibling species pair.

HARRIS, L. G., and N. R. HOWE. 1979. An analysis of the defensive mechanisms observed in the anemone *Anthopleura elegantissima* in response to its nudibranch predator *Aeolidia papillosa*. Biol. Bull. 157: 138–52.

HOWE, N. R. 1976. Behavior of sea anemones evoked by the alarm pheromone anthopleurine. J. Comp. Physiol. 107: 67–76.

JENNISON, B. L. 1979. Gametogenesis and repro-

ductive cycles in the sea anemone *Anthopleura elegantissima* (Brandt, 1835). Can. J. Zool. 57: 403–11.
☐Gamete growth increases with increasing temperature, and spawning occurs during summer in central California.

LINDSTEDT, K. J. 1971. Biphasic feeding response in a sea anemone: Control by asparagine and glutathione. Science 173: 333–34.

LUBBOCK, R. 1980. Clone-specific cellular recognition in a sea anemone. Proc. Natl. Acad. Sci. U.S.A. 77: 6667–69.

PEARSE, V. B. 1974a. Modification of sea anemone behavior by symbiotic zooxanthellae: Phototaxis. Biol. Bull. 147: 630–40.

———. 1974b. Modification of sea anemone behavior by symbiotic zooxanthellae: Expansion and contraction. Biol. Bull. 147: 641–51.

SEBENS, K. P. 1982a. Asexual reproduction in *Anthopleura elegantissima* (Anthozoa: Actiniaria): Seasonality and spatial extent of clones. Ecology 63: 434–44.

———. 1982b. Recruitment and habitat selection in the intertidal sea anemones, *Anthopleura elegantissima* (Brandt) and *A. xanthogrammica* (Brandt). J. Exp. Mar. Biol. Ecol. 59: 103–24.

TAYLOR, P. R., and M. M. LITTLER. 1982. The roles of compensatory mortality, physical disturbance, and substrate retention in the development and organization of a sand-influenced, rocky-intertidal community. Ecology 63: 135–46. ☐Concerns *Anthopleura elegantissima*, its ability to withstand periodic burial under sand, and a competitor in the low intertidal, *Phragmatopoma californica*.

A. xanthogrammica (Brandt), the giant green or solitary anemone, §64

BATCHELDER, H. P., and J. J. GONOR. 1981. Population characteristics of the intertidal green sea anemone, *Anthopleura xanthogrammica*, on the Oregon Coast. Estuarine Coastal Shelf Sci. 13: 235–45.

BIGGER, C. H. 1980. Interspecific and intraspecific acrorhagial aggressive behavior among sea anemones: A recognition of self and not-self. Biol. Bull. 159: 117–34. ☐Documents one observation in the field of *A. xanthogrammica* expanding its acrorhagi; this anemone has previously been reported not to display such a response.

DAYTON, P. K. 1973. Two cases of resource partitioning in an intertidal community: Making the right prediction for the wrong reason. Amer. Natur. 107: 662–70. ☐*Anthopleura xanthogrammica* inadvertently benefits from the relationship between the seastar *Pycnopodia helianthoides* and its sea urchin prey when urchins attempting to flee from the seastar fall into the anemone's tentacles. A similar relationship is described among *Pisaster ochraceus*, *Mytilus californianus*, and *A. xanthogrammica*.

O'BRIEN, T. L. 1980. The symbiotic association between intracellular zoochlorellae (Chlorophyceae) and the coelenterate *Anthopleura xanthogrammica*. J. Exp. Zool. 211: 343–55.

SEBENS, K. P. 1981a. Recruitment in a sea anemone population: Juvenile substrate becomes adult prey. Science 213: 785–87.

———. 1981b. Reproductive ecology of the intertidal sea anemones *Anthopleura xanthogrammica* (Brandt) and *A. elegantissima* (Brandt): Body size, habitat, and sexual reproduction. J. Exp. Mar. Biol. Ecol. 54: 255–50.

———. 1982. The limits to indeterminate growth: An optimal size model applied to passive suspension feeders. Ecology 63: 209–22.

SIEBERT, A. E., JR. 1974. A description of the embryology, larval development, and feeding of the sea anemones *Anthopleura elegantissima* and *A. xanthogrammica*. Can. J. Zool. 52: 1383–88.

Charisea saxicola Torrey, §202

Edwardsiella californica McMurrich, §290

McMURRICH, J. P. 1913. Description of a new species of actinian of the genus *Edwardsiella* from southern California. Proc. U.S. Natl. Mus. 44: 551–53.

E. sipunculoides (Stimpson), §244

Epiactis prolifera Verrill, §37

DUNN, D. F. 1975. Reproduction of the externally brooding sea anemone *Epiactis prolifera* Verrill, 1869. Biol. Bull. 148: 199–218.

———. 1977a. Dynamics of external brooding in the sea anemone *Epiactis prolifera*. Mar. Biol. 39: 41–49.

———. 1977b. Locomotion by *Epiactis prolifera* (Coelenterata: Actiniaria). Mar. Biol. 39: 67–70.

———. 1977c. Variability of *Epiactis prolifera* (Coelenterata: Actiniaria) in the intertidal zone near Bodega Bay, California. J. Natur. Hist. 11: 457–63.

Haliplanella luciae (Verrill), §224

This widespread anemone has been variously called *Sagartia*, *Diadumene*, or *Aiptasiomorpha luciae*. Sexually reproducing phases are found along the coast of Japan, but elsewhere, reproduction is primarily by asexual division. This is a small, eurythermal, euryhaline form that has been the subject of many physiological studies, a sample of which is included below.

MINASIAN, L. L., JR. 1979. The effect of exogenous factors on morphology and asexual reproduction in laboratory cultures of the intertidal sea anemone, *Haliplanella luciae* (Verrill) (Anthozoa: Actiniaria) from Delaware. J. Exp. Mar. Biol. Ecol. 40: 235–46.

———. 1982. The relationship of size and biomass to fission rate in a clone of the sea anemone, *Haliplanella luciae* (Verrill). J. Exp. Mar. Biol. Ecol. 58: 151–62.

SHICK, J. M., and A. N. LAMB. 1977. Asexual reproduction and genetic population structure in the colonizing sea anemone *Haliplanella luciae*. Biol. Bull. 153: 604–17.

UCHIDA, T. 1932. Occurrence in Japan of *Diadumene luciae*, a remarkable actinian of rapid dispersal. J. Fac. Sci. Hokkaido Imp. Univ. (6)2(2): 69–82.

WILLIAMS, R. B. 1972. Chemical control of feeding behaviour in the sea anemone *Diadumene luciae* (Verrill). Comp. Biochem. Physiol. 41A: 361–71.

———. 1975. Catch-tentacles in sea anemones: Occurrence in *Haliplanella luciae* (Verrill) and a review of current knowledge. J. Natur. Hist. 9: 241–48.

Harenactis attenuata Torrey, §257

Metridium senile (Linnaeus), §321

BUCKLIN, A., and D. HEDGECOCK. 1982. Biochemical genetic evidence for a third species of *Metridium* (Coelenterata: Actiniaria). Mar. Biol. 66: 1–7. □Instead of one highly variable species, *M. senile*, the authors suggest that the genus *Metridium* contains three distinct species.

GOODWIN, M. H., and M. TELFORD. 1971. The nematocyst toxin of *Metridium*. Biol. Bull. 140: 389–99.

PURCELL, J. E. 1977a. Aggressive function and induced development of catch tentacles in the sea anemone *Metridium senile* (Coelenterata, Actiniaria). Biol. Bull. 153: 355–68.

———. 1977b. The diet of large and small individuals of the sea anemone *Metridium senile*. Bull. South. Calif. Acad. Sci. 76: 168–72.

PURCELL, J. E., and C. L. KITTING. 1982. Intraspecific aggression and population distributions of the sea anemone *Metridium senile*. Biol. Bull. 162: 345–59.

WALSH, P. J., and G. N. SOMERO. 1981. Temperature adaptation in sea anemones: Physiological and biochemical variability in geographically separate populations of *Metridium senile*. Mar. Biol. 62: 25–34.

Tealia coriacea (Cuvier), §65

T. crassicornis (Muller), §65
CHIA, F.-S., and J. G. SPAULDING. 1972. Development and juvenile growth of the sea anemone, *Tealia crassicornis*. Biol. Bull. 142: 206–18. □Spawning occurred in the laboratory from April to June. Growth was slow (to 4 cm in diameter within 18 months) and dependent on food supply. Development is described.

T. lofotensis (Danielssen), §65
SEBENS, K. P., and G. LAAKSO. 1977. The genus *Tealia* (Anthozoa: Actiniaria) in the waters of the San Juan Archipelago and the Olympic Peninsula. Wasmann J. Biol. 35: 152–68.

Order CERIANTHARIA

TORREY, H. B., and F. L. KLEEBERGER. 1909. Three species of *Cerianthus* from southern California. Univ. Calif. Publ. Zool. 2: 115–25.

Pachycerianthus aestuari (Torrey and Kleeberger), §256
ARAI, M. N. 1971. *Pachycerianthus* (Ceriantharia) from British Columbia and Washington. J. Fish. Res. Board Can. 28: 1677–80.

P. fimbriatus McMurrich, §256
ARAI, M. N., and G. L. WALDER. 1973. The feeding response of *Pachycerianthus fimbriatus* (Ceriantharia). Comp. Biochem. Physiol. 44A: 1085–92.

Order CORALLIMORPHARIA

Corynactis californica Carlgren, §40
CARLGREN, O. 1936. Some west American sea anemones. J. Wash. Acad. Sci. 26: 16–23. □Describes *Corynactis californica* and includes information on other species. Based on material collected by Ed Ricketts.

Order MADREPORARIA

Astrangia lajollaensis Durham, §39
BOSCHMA, H. 1925. On the feeding reactions and digestion in the coral polyp *Astrangia danae*, with notes on its symbiosis with zooxanthellae. Biol. Bull. 49: 407–39.

FADLALLAH, Y. H. 1982. Reproductive ecology of the coral *Astrangia lajollaensis*: Sexual and asexual patterns in a kelp forest habitat. Oecologia 55: 379–88. □Data on reproduction of subtidal *A. lajollaensis* at Pacific Grove; the coral's occurrence there extends the known range 200–300 km to the north.

Balanophyllia elegans Verrill, §39
FADLALLAH, Y. H., and J. S. PEARSE. 1982. Sexual reproduction in solitary corals: Overlapping oogenic and brooding cycles, and benthic planulas in *Balanophyllia elegans*. Mar. Biol. 71: 223–31.

GERRODETTE, T. 1981. Dispersal of the solitary coral *Balanophyllia elegans* by demersal planular larvae. Ecology 62: 611–19. □Dispersal is usually slow. The mean larval dispersal distance is less than 0.5 meter from the parent.

Caryophyllia alaskensis Vaughan, §39

Phylum **CTENOPHORA**. Comb jellies

Although resembling cnidarian jellyfish in that much of the body consists of jellylike material, these animals have a number of characteristics that warrant their placement in a separate phylum. Probably the most apparent to the shore observer will be the mode of locomotion. Instead of pulsating jerkily through the water, ctenophores glide smoothly along, propelled by means of eight rows of ciliated plates (comb plates, or ctenes).

HORRIDGE, G. A. 1974. Recent studies on the Ctenophora. *In* L. Muscatine and H. M. Lenhoff, eds., Coelenterate biology, pp. 439–68. New York: Academic Press.

Pleurobrachia bachei (A. Agassiz), §193

Phylum **PLATYHELMINTHES**. Flatworms

The flatworms are divided taxonomically into three classes: Trematoda, Cestoda, and Turbellaria. However, adult members of the Trematoda (flukes) and Cestoda (tapeworms) are parasites, mainly inside vertebrates, and thus are unlikely to be encountered by users of this book. Only members of the Turbellaria, which are mainly free-living, will be treated further.

Identification of turbellarian flatworms is often difficult. In some cases, color patterns may be distinctive, but more often, coloration is extremely variable (changing with diet) and unreliable. Specimens of the polyclad *Freemania litoricola*, for instance, may be a rather uniform tan, reddish brown, or dark brown; more often they may be mottled with splotches of various shades of brown, tan, or black. Positive identification usually must be made by an expert, and often based on microscopic examination of the reproductive organs. The difficulties attendant in identifying flatworms, coupled with the retiring habits of most of them, combine to produce a group of common animals about which surprisingly little in the way of natural history is known.

Class TURBELLARIA
Free-living flatworms

HADERLIE, E. C. 1975. Phylum Platyhelminthes. *In* R. I. Smith and J. T. Carlton, eds., Light's manual, pp. 100–111. Berkeley: University of California Press.

——. 1980. Platyhelminthes: The flatworms. *In* R. H. Morris, D. P. Abbott, and E. C. Haderlie, Intertidal invertebrates of California, pp. 76–83. Stanford, Calif.: Stanford University Press.

HOLLEMAN, J. J. 1972. Marine turbellarians of the Pacific coast. Proc. Biol. Soc. Wash. 85: 405–12.

HYMAN, L. H. 1951. The invertebrates: Platyhelminthes and Rhynchocoela. Vol. II. New York: McGraw-Hill. 550 pp.

——. 1959. Some Turbellaria from the coast of California. Amer. Mus. Novit. 1943: 1–17. □Concerns one acoel, *Childia groenlandica*, and six species of polyclads, including three new species.

KARLING, T. G. 1962–66. Marine Turbellaria from the Pacific coast of North America. I, Plagiostomidae. Ark. Zool. 15 (1962): 113–41. II, Pseudostomidae and Cylindrostomidae. Ark. Zool. 15 (1962): 181–209. III, Otoplanidae. Ark. Zool. 16 (1964): 527–41. IV, Coelogynoporidae and Monocelididae. Ark. Zool. 18 (1966): 493–528.

KOOPOWITZ, H. 1975. Electrophysiology of the peripheral nerve net in the polyclad flatworm *Freemania litoricola*. J. Exp. Biol. 62: 469–79. □Cited here as an example of recent work on the electrophysiology of flatworms.

MARTIN, G. G. 1978a. A new function of rhabdites: Mucus production for ciliary gliding. Zoomorphologie 91: 235–48. □Concerns *Alloioplana californica*, *Monocelis cincta*, and *Polychoerus carmelensis*, and the structure, synthesis, and secretion of rhabdites, thought to form the mucus used in locomotion by these flatworms.

——. 1978b. Ciliary gliding in lower invertebrates. Zoomorphologie 91: 249–61. □Describes ultrastructure of the ciliated and mucus-secreting cells on the locomotor surface of *Polychoerus carmelensis* and *Monocelis cincta*; discusses locomotion of these forms.

RISER, N. W., and M. P. MORSE, eds. 1974. Biology of the Turbellaria. New York: McGraw-Hill. 530 pp. □An international symposium in memoriam of Libbie Hyman. Contains several papers of general interest on the morphology, phylogeny, physiology, and ecology of turbellarians.

SCHOCKAERT, E. R., and I. R. BALL, eds. 1981. The biology of the Turbellaria. The Hague: Junk. 302 pp.

Order ACOELA

KOZLOFF, E. N. 1965. New species of acoel turbellarians from the Pacific coast. Biol. Bull. 129: 151–66. □Three species from San Juan Island, collected intertidally.

Polychoerus carmelensis Costello, §59

ARMITAGE, K. B. 1961. Studies of the biology of *Polychoerus carmelensis* (Turbellaria, Acoela). Pacific Sci. 15: 203–10.

COSTELLO, H. M., and D. P. COSTELLO. 1938a. Copulation in the acoelous turbellarian *Polychoerus carmelensis*. Biol. Bull. 75: 85–98.

——. 1938b. A new species of *Polychoerus* from the Pacific coast. Ann. Mag. Natur. Hist. (11)1: 148–55.

——. 1939. Egg laying in the acoelous turbellarian *Polychoerus carmelensis*. Biol. Bull. 76: 80–89.

SCHWAB, R. G. 1967. Overt responses of *Polychoerus carmelensis* (Turbellaria: Acoela) to abrupt changes in ambient water temperature. Pacific Sci. 21: 85–90.

Order RHABDOCOELIDA

Syndesmis franciscana (Lehman), §73

GIESE, A. C. 1958. Incidence of *Syndesmis* in the gut of two species of sea urchins. Anat. Rec. 132: 441–42.

LEHMAN, H. E. 1946. A histological study of *Syndisyrinx franciscanus* gen. et sp. nov., an endoparasitic rhabdocoel of the sea urchin, *Strongylocentrotus franciscanus*. Biol. Bull. 91: 295–311.

Order ALLOEOCOELA

Most of what is known about alloeocoels of the Pacific coast is to be found in Karling's works (1962–66) cited above.

Monocelis sp., §60

Karling, cited in Haderlie (1975), says that the *Monocelis* sp. listed here is *M. cincta*.

MARTIN, G. G. 1978. The duo-gland adhesive system of the archiannelids *Protodrilus* and *Saccocirrus* and the turbellarian *Monocelis*. Zoomorphologie 91: 63–75. □*Monocelis cincta* attaches to the substratum by a crescent-shaped adhesive area at the posterior margin of the ventral surface. The area contains two types of glands: one thought to contain adhesive; the other, granules to break the attachment.

Order TRICLADIDA

HOLLEMAN, J. J., and C. HAND. 1962. A new species, genus, and family of marine flatworms (Turbellaria: Tricladida, Maricola) commensal with mollusks. Veliger 5: 20–22. □Describes *Nexilis epichitonius*, about 3 mm long, occurring on the mantle of *Mopalia hindsi*; also found on *Nucella emarginata*.

HOLMQUIST, C., and T. G. KARLING. 1972. Two new species of interstitial marine triclads from the North American Pacific coast, with comments on evolutionary trends and systematics in Tricladida (Turbellaria). Zool. Scr. 1: 175–84. □Describes two new species of triclads, *Pacificides psammophilus* and *Oregoniplana opisthopora*, inhabitants of the surf zone of sandy beaches; brings the number of triclads known from the North American Pacific coast to five.

HYMAN, L. H. 1954. A new marine triclad from the coast of California. Amer. Mus. Novit. 1679: 1–5. ☐Describes *Procerodes pacifica*, collected near San Diego.

———. 1956. North American triclad Turbellaria. 15. Three new species. Amer. Mus. Novit. 1808: 1–14. ☐Describes *Nesion arcticum*, from Alaska.

Order POLYCLADIDA

The works of Hyman (1953, 1955) still contain the most comprehensive and valuable accounts of most of the described polyclads from the Pacific coast of North America.

BOONE, E. S. 1929. Five new polyclads from the California coast. Ann. Mag. Natur. Hist. (10)3: 33–46. ☐Includes *Notoplana acticola* (as *Leptoplana acticola*).

CHING, H. L. 1977. Redescription of *Eurylepta leoparda* Freeman, 1933 (Turbellaria: Polycladida), a predator of the ascidian *Corella willmeriana* Herdman, 1898. Can. J. Zool. 55: 338–42. ☐Redescribed, since morphology was known from only two whole mounts; reported from British Columbia and Washington.

FREEMAN, D. 1933. The polyclads of the San Juan region of Puget Sound. Trans. Amer. Microsc. Soc. 52: 107–46. ☐Includes *Freemania litoricola* (as *Notoplana segnis*).

HEATH, H., and E. A. McGREGOR. 1912. New polyclads from Monterey Bay, California. Proc. Acad. Natur. Sci. Phila. 64: 455–88. ☐Includes *Alloioplana californica* (as *Planocera californica*), *Notoplana inquieta* (as *Leptoplana inquieta*), and *Freemania litoricola* (as *Phylloplana litoricola*).

HURLEY, A. C. 1976. The polyclad flatworm *Stylochus tripartitus* Hyman as a barnacle predator. Crustaceana 31: 110–11. ☐Other species in this genus are predators on oysters and mussels, as well as on barnacles.

HYMAN, L. H. 1953. The polyclad flatworms of the Pacific coast of North America. Bull. Amer. Mus. Natur. Hist. 100: 269–392.

———. 1955. The polyclad flatworms of the Pacific coast of North America: Additions and corrections. Amer. Mus. Novit. 1704: 1–11.

Alloioplana californica (Heath and McGregor), §3

Freemania litoricola (Heath and McGregor), §235

PHILLIPS, D. W., and M. L. CHIARAPPA. 1980. Defensive responses of gastropods to the predatory flatworms *Freemania litoricola* (Heath and McGregor) and *Notoplana acticola* (Boone). J. Exp. Mar. Biol. Ecol. 47: 179–89.

Kaburakia excelsa Bock, §235

Notoplana acticola (Boone), §17

THUM, A. B. 1974. Reproductive ecology of the polyclad turbellarian *Notoplana acticola* (Boone,

1929) on the central California coast. *In* N. W. Riser and M. P. Morse, eds., Biology of the Turbellaria, pp. 431–45. New York: McGraw-Hill.

N. inquieta (Heath and McGregor), §243

HURLEY, A. C. 1975. The establishment of populations of *Balanus pacificus* Pilsbry (Cirripedia) and their elimination by predatory Turbellaria. J. Anim. Ecol. 44: 521–32.

Thysanozoon sp. and undetermined forms, §138

Phylum NEMERTEA. Ribbon worms

The nemerteans, or ribbon worms, are like flatworms in some respects. Their soft, unsegmented, often flattened bodies glide across the substratum by means of cilia. Unlike flatworms, however, the nemerteans are characteristically highly contractile, slender worms that sport near their anterior end a distinctive eversible proboscis that can be extended to great lengths. A posterior anus separate from the mouth opening is another distinguishing feature. Nemerteans, like the polyclad flatworms, are primarily predators.

COE, W. R. 1901. The nemerteans. Papers from the Harriman Alaska Expedition, XX. Proc. Wash. Acad. Sci. 3: 1–110.

———. 1905. Nemerteans of the west and northwest coasts of America. Bull. Mus. Comp. Zool. Harv. Univ. 47: 1–319.

———. 1929. Regeneration in nemerteans. J. Exp. Zool. 54: 411–60.

———. 1930. Asexual reproduction in nemerteans. Physiol. Zool. 3: 297–308.

———. 1940. Revision of the nemertean fauna of the Pacific coasts of North, Central and northern South America. Allan Hancock Pacific Exped. 2: 247–323.

———. 1944. Geographical distribution of the nemerteans of the Pacific coast of North America, with descriptions of two new species. J. Wash. Acad. Sci. 34: 27–32.

JENNINGS, J. B., and R. GIBSON. 1969. Observations on the nutrition of seven species of rhynchocoelan worms. Biol. Bull. 136: 405–43.

STRICKER, S. A., and R. A. CLONEY. 1982. Stylet formation in nemerteans. Biol. Bull. 162: 387–403. ☐Describes formation of the first stylets produced by larvae and the replacement stylets of adults. Nine species are examined.

WILLMER, E. N. 1974. Nemertines as possible ancestors of the vertebrates. Biol. Rev. 49: 321–63. ☐An interesting, detailed, speculative account, well worth reading.

Amphiporus bimaculatus Coe, §48

A. imparispinosus Griffin, §165

Carcinonemertes epialti Coe, §152

KURIS, A. M. 1978. Life cycle, distribution and abundance of *Carcinonemertes epialti*, a nemertean egg predator of the shore crab, *Hemigrapsus oregonensis*, in relation to host size, reproduction and molt cycle. Biol. Bull. 154: 121–37.

RoE, P. 1979. Aspects of development and occurrence of *Carcinonemertes epialti* (Nemertea) from shore crabs in Monterey Bay, California. Biol. Bull. 156: 130–40.

STRICKER, S. A., and C. G. REED. 1981. Larval morphology of the nemertean *Carcinonemertes epialti* (Nemertea: Hoplonemertea). J. Morphol. 169: 61–70.

C. errans Wickham, §152

RoE, P., J. H. CROWE, L. M. CROWE, and D. E. WICKHAM. 1981. Uptake of amino acids by juveniles of *Carcinonemertes errans* (Nemertea). Comp. Biochem. Physiol. 69A: 423–27. □Juveniles have no apparent means of feeding, and the implication here is that they may acquire nutrients in the form of dissolved organic materials, transported directly across the body wall, from the surrounding water.

WICKHAM, D. E. 1978. A new species of *Carcinonemertes* (Nemertea: Carcinonemertidae) with notes on the genus from the Pacific coast. Proc. Biol. Soc. Wash. 91: 197–202.

———. 1979. Predation by the nemertean *Carcinonemertes errans* on eggs of the Dungeness crab *Cancer magister*. Mar. Biol. 55: 45–53.

———. 1980. Aspects of the life history of *Carcinonemertes errans* (Nemertea: Carcinonemertidae), an egg predator of the crab *Cancer magister*. Biol. Bull. 159: 247–57.

Cerebratulus sp., of Elkhorn Slough, §312
Cerebratulus montgomeryi Coe, §239
Emplectonema gracile (Johnston), §165

Lineus vegetus Coe, §139

COE, W. R. 1931. A new species of nemertean (*Lineus vegetus*) with asexual reproduction. Zool. Anz. 94: 54–60.

FISHER, F. M., and J. A. OAKS. 1978. Evidence for a nonintestinal nutritional mechanism in the rhynchocoelan, *Lineus ruber*. Biol. Bull. 154: 213–25.

Malacobdella grossa (Müller), §189

ADDICOTT, W. O. 1968. Additional Pacific coast *Malacobdella grossa*. Nautilus 81: 144.

GIBSON, R., and J. B. JENNINGS. 1969. Observations on the diet, feeding mechanisms, digestion and food reserves of the entocommensal rhynchocoelan *Malacobdella grossa*. J. Mar. Biol. Assoc. U.K. 49: 17–32.

JENNINGS, J. B. 1968. A new astomatous ciliate from the entocommensal rhynchocoelan *Malacobdella grossa* (O. F. Müller). Arch. Protistenkd. 110: 422–25. □The commensal is itself host to various parasites, here a protozoan.

Micrura sp., of Elkhorn Slough, §312
M. verrilli Coe, §139

Paranemertes peregrina Coe, §§139, 198, 312

AMERONGEN, H. M., and F.-S. CHIA. 1982. Behavioural evidence for a chemoreceptive function of the cerebral organs in *Paranemertes peregrina* Coe (Hoplonemertea: Monostilifera). J. Exp. Mar. Biol. Ecol. 64: 11–16. □Ablation of the cerebral organs abolished the worm's ability to follow prey trails.

GIBSON, R. 1970. The nutrition of *Paranemertes peregrina* (Rhynchocoela: Hoplonemertea). II. Observations on the structure of the gut and proboscis, site and sequence of digestion, and food reserves. Biol. Bull. 139: 92–106.

RoE, P. 1970. The nutrition of *Paranemertes peregrina* (Rhynchocoela: Hoplonemertea). I. Studies on food and feeding behavior. Biol. Bull. 139: 80–91.

———. 1976. Life history and predator-prey interactions of the nemertean *Paranemertes peregrina* Coe. Biol. Bull. 150: 80–106. □The most complete study of feeding ecology and natural history of any of our free-living nemerteans.

———. 1979. A comparison of aspects of the biology of *Paranemertes peregrina* (Nemertea) from Bodega Harbor, California, and Washington State. Pacific Sci. 33: 281–87.

STRICKER, S. A., and R. A. CLONEY. 1981. The stylet apparatus of the nemertean *Paranemertes peregrina*: Its ultrastructure and role in prey capture. Zoomorphology 97: 205–23.

Procephalothrix major Coe (formerly *Cephalothrix*), §139

Tubulanus polymorphus Renier (includes *Carinella rubra* Griffin of previous editions), §§139, 239

Phylum **NEMATODA**. Roundworms

Nematodes are abundant in sediment and among algal holdfasts, and some of the largest free-living species are found on the shore. However, specific identification is difficult, requiring a specialist, and these worms are often ignored, as the paucity of references attests.

CHITWOOD, B. G. 1960. A preliminary contribution on the marine nemas (Adenophorea) of northern California. Trans. Amer. Microsc. Soc. 79: 347–84.

HOPE, W. D., and D. G. MURPHY. 1972. A taxonomic hierarchy and checklist of the genera and higher taxa of marine nematodes. Smithsonian Contr. Zool., No. 137. 101 pp. □Little text, but an extensive bibliography.

WIESER, W. 1959. Free-living nematodes and other small invertebrates of Puget Sound beaches. Seattle: University of Washington Press. 179 pp.

Phylum **ANNELIDA**. Segmented worms

The phylum consists (at least in most accounts) of three classes of segmented worms: Hirudinea (leeches), Oligochaeta, and Polychaeta. The polychaetes, of course, are familiar to any visitor of the shore. However, marine members of the other two taxa are still a novelty to most Pacific coast biologists, and will be represented here by only a small sample of the many interesting papers and books residing on library shelves.

BRINKHURST, R. O. 1982. Evolution in the Annelida. Can. J. Zool. 60: 1043–59. □Mostly concerns the Oligochaeta.

DALES, R. P. 1967. Annelids. London: Hutchinson. 200 pp. □A useful introduction.

HERMANS, C. O. 1969. The systematic position of the Archiannelida. Syst. Zool. 18: 85–102.

MILL, P. J., ed. 1978. Physiology of annelids. London: Academic Press. 683 pp. ☐Contributed chapters on many topics of general importance.

Class HIRUDINEA. Leeches

KNIGHT-JONES, E. W. 1962. The systematics of marine leeches. *In* K. H. Mann, Leeches (Hirudinea): Their structure, physiology, ecology and embryology, pp. 169–86. New York: Pergamon.

Class OLIGOCHAETA. Earthworms, etc.

BRINKHURST, R. O., and H. R. BAKER. 1979. A review of the marine Tubificidae (Oligochaeta) of North America. Can. J. Zool. 57: 1553–69.

BRINKHURST, R. O., and D. G. COOK, eds. 1980. Aquatic oligochaete biology. New York: Plenum. 529 pp.

BRINKHURST, R. O., and M. L. SIMMONS. 1968. The aquatic Oligochaeta of the San Francisco Bay system. Calif. Fish Game 54: 180–94.

CHAPMAN, P. M., and R. O. BRINKHURST. 1980. Salinity tolerance in some selected aquatic oligochaetes. Int. Rev. Gesamten Hydrobiol. 65: 499–505.

COATES, K., and D. V. ELLIS. 1981. Taxonomy and distribution of marine Enchytraeidae (Oligochaeta) in British Columbia. Can. J. Zool. 59: 2129–50.

GIERE, O., and O. PFANNKUCHE. 1982. Biology and ecology of marine Oligochaeta. Oceanogr. Mar. Biol. Ann. Rev. 20: 173–308.

Class POLYCHAETA
Bristle worms, Fan worms, Clam worms, etc.

For better or worse—depending on one's inclination, interest, and ability—the polychaetes, or bristle worms, are among the most numerous and significant animals of the sea. Some are pelagic, living on the high seas, and a few are in the great deeps. There are many in the shallow sandy and muddy bottoms of the seas everywhere, and on intertidal mud flats they occur in nearly countless variety. Since it takes a bit of hard work, aided by literature at times obscurely couched in a rather special terminology, to identify polychaetes, they are considered difficult subjects for study. But if a diligent observer of the life of mud flats keeps at it for a while, he or she may find perhaps a hundred species on a stretch of mud flat 20 meters square. Many of them are beautifully iridescent.

Thousands of papers have been published about polychaete worms. Olga Hartman, doyenne of Pacific coast polychaetology, has done yeoman service in trying to keep us abreast of the literature. Her *Catalogue of the polychaetous annelids of the world* is a basic guide to the systematic literature. Parts I and II of this appeared in 1959 (Occasional Papers of the Allan Hancock Foundation, No. 23, Parts I and II, 628 pp.), and the supplement and index appeared as Part III (197 pp.) at the end of 1965. Her summary of the literature, in which many topics are indexed to provide a key to the unwieldy mass of papers on polychaetes, appeared in 1951 and included more than 1,300 authors and 4,000 titles (*The literature of the polychaetous annelids, Part I: Bibliography and subject analysis.* Los Angeles: privately printed, 290 pp.); this work, of course, is now rather out of date, since the flood of papers has not subsided. In 1968 the first volume of her two-volume comprehensive key to California species appeared, *Atlas of the errantiate polychaetous annelids from California* (Los Angeles: Allan Hancock Foundation, University of Southern California, 828 pp.), followed in 1969 by the second, *Atlas of the sedentariate polychaetous annelids from California* (Los Angeles: Allan Hancock Foundation, University of Southern California, 812 pp.).

In addition to Hartman's monumental works, other treatments of systematic importance include those of Blake (1975) and Fauchald (1977).

BANSE, K. 1979. Sabellidae (Polychaeta) principally from the northeast Pacific Ocean. J. Fish. Res. Board Can. 36: 869–82.

BANSE, K., and K. D. HOBSON. 1968. Benthic polychaetes from Puget Sound, Washington, with remarks on four other species. Proc. U.S. Natl. Mus. 125: 1–53. ☐Describes 8 new species, gives 25 new records, and discusses 35 other forms.

BERKELEY, E., and C. BERKELEY. 1932. On a collection of littoral Polychaeta from the west coast of Vancouver Island. Contr. Can. Biol. Fish., n.s. 7: 309–18.

————. 1935. Some notes on the polychaetous annelids of Elkhorn Slough, Monterey Bay, California. Amer. Midl. Natur. 16: 766–75.

————. 1948. Annelida: Polychaeta Errantia. Can. Pacific Fauna 9b(1): 1–100.

————. 1952. Annelida: Polychaeta Sedentaria. Can. Pacific Fauna 9b(2): 1–139.

BERRILL, N. J. 1977. Functional morphology and development of segmental inversion in sabellid polychaetes. Biol. Bull. 153: 453–67. ☐Form and function are described and analyzed particularly with respect to the ability of sabellid worms to deal with predators, through behavior and the regeneration of lost segments.

BLAKE, J. A. 1975a. The larval development of Polychaeta from the California coast. III. Eighteen species of Errantia. Ophelia 14: 23–84.

————. 1975b. Phylum Annelida: Class Polychaeta. *In* R. I. Smith and J. T. Carlton, eds., Light's manual, pp. 151–243. Berkeley: University of California Press.

————. 1981. A new coralline boring species of *Polydora* (Polychaeta: Spionidae) from northern California. Bull. South. Calif. Acad. Sci. 80: 32–35. ☐Describes *Polydora bifurcata*, a borer in coralline algae (*Lithophyllum*) from Tomales Point.

BONAR, D. B. 1972. Feeding and tube construction in *Chone mollis* Bush (Polychaeta, Sabellidae). J. Exp. Mar. Biol. Ecol. 9: 1–18. ☐*Chone mollis* is found in muddy sands of estuaries from northern

Oregon to Mexico. Tube formation is a consequence of burrowing, and a new tube can be quickly built, an advantage in an environment subject to rapid sedimentation.

BRENCHLEY, G. A. 1976. Predator detection and avoidance: Ornamentation of tube-caps of *Diopatra* spp. (Polychaeta: Onuphidae). Mar. Biol. 38: 179–88. □Presents data consistent with the hypothesis that the ornamentation of above-sediment portions of the tube facilitates the detection and, thus, avoidance of surface predators.

CLARK, R. B. 1965. Endocrinology and the reproductive biology of polychaetes. Oceanogr. Mar. Biol. Ann. Rev. 3: 211–55.

FAUCHALD, K. 1975. Polychaete phylogeny: A problem in protostome evolution. Syst. Zool. 23: 493–506.

———. 1977. The polychaete worms: Definitions and keys to the orders, families, and genera. Los Angeles Co. Mus. Natur. Hist., Sci. Ser. 28. 190 pp.

———. 1983. Life diagram patterns in benthic polychaetes. Proc. Biol. Soc. Wash. 96: 160–77.

FAUCHALD, K., and P. A. JUMARS. 1979. The diet of worms: A study of polychaete feeding guilds. Oceanogr. Mar. Biol. Ann. Rev. 17: 193–284.

GAFFNEY, P. M. 1973. Setal variation in *Halosydna brevisetosa*, a polynoid polychaete. Syst. Zool. 22: 171–75. □Reports considerable intraspecific variation in the setae of *H. brevisetosa*, and questions the constancy and diagnostic utility of setal characters in polychaete taxonomy.

HANNAN, C. A. 1981. Polychaete larval settlement: Correspondence of patterns in suspended jar collectors and in adjacent natural habitat in Monterey Bay, California. Limnol. Oceanogr. 26: 159–71. □Larval settlement patterns of *Armandia brevis*, *Capitella* spp., and two other polychaetes are examined and discussed.

HEACOX, A. E. 1980. Reproduction and larval development of *Typosyllis pulchra* (Berkeley and Berkeley) (Polychaeta: Syllidae). Pacific Sci. 34: 245–59.

HICKOK, J. F., and D. DAVENPORT. 1957. Further studies in the behavior of commensal polychaetes. Biol. Bull. 113: 397–406. □Three species of commensal polychaetes, *Ophiodromus pugettensis*, *Arctonoe fragilis*, and *A. vittata*, are attracted by chemicals released by their respective hosts.

HOBSON, K. D., and K. BANSE. 1981. Sedentariate and archiannelid polychaetes of British Columbia and Washington. Can. Bull. Fish. Aquat. Sci. 209: 144 pp. □A checklist of 273 species with keys. About 45 percent of the benthic polychaete species of British Columbia and Washington also occur in southern California.

JORGENSEN, N. O. G., and E. KRISTENSEN. 1980. Uptake of amino acids by three species of *Nereis* (Annelida: Polychaeta): I. Transport kinetics and net uptake from natural concentrations. Mar. Ecol. Prog. Ser. 3: 329–40.

KENNEDY, B., and H. KRYVI. 1980. Autotomy in a polychaete: Abscission zone at the base of the tentacular crown of *Sabella penicillus*. Zoomorphology 96: 33–43.

LEVIN, L. A. 1981. Dispersion, feeding behavior and competition in two spionid polychaetes. J. Mar. Res. 39: 99–117. □*Pseudopolydora paucibranchiata* (Okuda) and *Streblospio benedicti* (Webster) on intertidal mud flats of Mission Bay, San Diego County.

MAY, D. R. 1972. The effects of oxygen concentration and anoxia on respiration of *Abarenicola pacifica* and *Lumbrineris zonata* (Polychaeta). Biol. Bull. 142: 71–83. □Measurements of oxygen concentration in burrows of *A. pacifica* at low tide indicate that water in these burrows is never completely anoxic; nevertheless, specimens in the laboratory were healthy after three days of anoxia.

ORRHAGE, L. 1980. On the structure and homologues of the anterior end of the polychaete families Sabellidae and Serpulidae. Zoomorphology 96: 113–68.

PETTIBONE, M. H. 1953. Some scale-bearing polychaetes of Puget Sound and adjacent waters. Seattle: University of Washington Press. 89 pp.

REISH, D. J. 1971. Seasonal settlement of polychaetous annelids on test panels in Los Angeles–Long Beach harbors 1950–1951. J. Fish. Res. Board Can. 28: 1459–67.

REISH, D. J., and M. C. ALOSI. 1968. Aggressive behavior in the polychaetous annelid family Nereidae. Bull. South. Calif. Acad. Sci. 67: 21–28. □Concerns encounters between *Neanthes arenaceodentata*, *N. limnicola*, *N. succinea*, *Nereis grubei*, *N. latescens*, and *Platynereis bicanaliculata*. In all cases but one, individuals were aggressive, and they tended to be more aggressive against members of their own species than against members of other species. Only *Neanthes limnicola* showed little aggressiveness.

REISH, D. J., and K. FAUCHALD, eds. 1977. Essays on polychaetous annelids in memory of Dr. Olga Hartman. Los Angeles: Allan Hancock Foundation, University of Southern California. 604 pp.

ROE, P. 1975. Aspects of life history and of territorial behavior in young individuals of *Platynereis bicanaliculata* and *Nereis vexillosa* (Annelida, Polychaeta). Pacific Sci. 29: 341–48.

SCHROEDER, P. C., and C. O. HERMANS. 1975. The polychaete annelids. *In* A. C. Giese and J. S. Pearse, eds., Reproduction of marine invertebrates, Vol. III, pp. 1–213. New York: Academic Press.

WILSON, W. H., JR. 1980. A laboratory investigation of the effect of a terebellid polychaete on the survivorship of nereid polychaete larvae. J. Exp. Mar. Biol. Ecol. 46: 73–80. □The feeding activities of *Eupolymnia heterobranchia* lower the survivorship of larvae of *Nereis vexillosa*.

WOODIN, S. A. 1974. Polychaete abundance patterns in a marine soft-sediment environment: The importance of biological interactions. Ecol. Monogr. 44: 171–87. □Concerns the abundances and interactions of the four numerically important large polychaetes on a tidal flat in the San Juan Islands: *Armandia brevis*, *Axiothella rubrocincta*, *Lumbrineris inflata*, and *Platynereis bicanaliculata*.

———. 1982. Browsing: Important in marine sedimentary environments? Spionid polychaete examples. J. Exp. Mar. Biol. Ecol. 60: 35–45. □Several species of spionid polychaetes showed evidence of tentacle loss, presumably the result of browsing by visual predators; from 6 to 16 percent of individual worms were so affected.

Abarenicola pacifica Healy and Wells, §310

HOBSON, K. D. 1967. The feeding and ecology of two North Pacific *Abarenicola* species (Arenicolidae, Polychaeta). Biol. Bull. 133: 343–54. □The two are *Abarenicola pacifica* and *A. claparedii vagabunda*, studied in the San Juan Islands.

HYLLEBERG, J. 1975. Selective feeding by *Abarenicola pacifica* with notes on *Abarenicola vagabunda* and a concept of gardening in lugworms. Ophelia 14: 113–37.

OGLESBY, L. C. 1973. Salt and water balance in lugworms (Polychaeta: Arenicolidae), with particular reference to *Abarenicola pacifica* in Coos Bay, Oregon. Biol. Bull. 145: 180–99. □Although *Abarenicola pacifica* can tolerate salinities as low as 23 percent seawater in the laboratory, it is unlikely to survive more than brief exposure to salinities lower than 50 percent in the field.

SWINBANKS, D. D. 1981. Sediment reworking and the biogenic formation of clay laminae by *Abarenicola pacifica*. J. Sediment. Petrol. 51: 1137–45. □The worms of Boundary Bay, British Columbia, attain densities of 200 individuals per square meter and annually rework about a million cubic meters of sand, completely reworking the substratum they live in to a depth of 10 cm in 100 days.

WILSON, W. H., JR. 1981. Sediment-mediated interactions in a densely populated infaunal assemblage: The effects of the polychaete *Abarenicola pacifica*. J. Mar. Res. 39: 735–48.

A. vagabunda oceanica Healy and Wells (= *A. claparedii oceanica*), §310

A. v. vagabunda Healy and Wells (= *A. claparedii vagabunda*), §310

See additional references listed under *Arenicola*.

HEALY, E. A., and G. P. WELLS. 1959. Three new lugworms (Arenicolidae, Polychaeta) from the North Pacific area. Proc. Zool. Soc. Lond. 133: 315–35. □The three are described as *Abarenicola pacifica*, *A. vagabunda vagabunda*, and *A. vagabunda oceanica*.

Anaitides medipapillata (Moore), §153

Arabella iricolor (Montagu), §153

Arctonoe fragilis (Baird) (formerly *Polynoe*), in ambulacral grooves of several species of seastars, §207

A. pulchra (Johnson) (originally *Polynoe*), a common commensal of *Cryptochiton, Megathura,* and *Parastichopus*, §77

A. vittata (Grube) (includes *Lepidonotus lordi* Baird), a common commensal of *Diodora aspera, Cryptochiton stelleri,* and *Dermasterias imbricata*, §§76, 211

DAVENPORT, D. 1950. Studies in the physiology of commensalism. I. The polynoid genus *Arctonoe*. Biol. Bull. 98: 81–93.

DAVENPORT, D., and J. F. HICKOK. 1951. Studies in the physiology of commensalism. 2. The polynoid genera *Arctonoe* and *Halosydna*. Biol. Bull. 100: 71–83. □A classic study, primarily concerning *Arctonoe fragilis* and *A. pulchra*, showing that poly-

noid worms are attracted to their hosts by chemicals.

DIMOCK, R. V. 1974. Intraspecific aggression and the distribution of a symbiotic polychaete on its hosts. *In* W. B. Vernberg, ed., Symbiosis in the sea, pp. 29–44. Columbia: University of South Carolina Press. □Concerns *Arctonoe pulchra*.

DIMOCK, R. V., and D. DAVENPORT. 1971. Behavioral specificity and the induction of host recognition in a symbiotic polychaete. Biol. Bull. 141: 472–84. □*Arctonoe pulchra*.

GERBER, H. S., and J. F. STOUT. 1968. Sensory basis of the symbiotic relationship of *Arctonoe vittata* (Grube) (Polychaeta, Polynoidae) to the keyhole limpet *Diadora aspera*. Physiol. Zool. 41: 169–79.

WAGNER, R. H., D. W. PHILLIPS, J. D. STANDING, and C. HAND. 1979. Commensalism or mutualism: Attraction of a sea star towards its symbiotic polychaete. J. Exp. Mar. Biol. Ecol. 39: 205–10. □Not only are scale worms attracted to their hosts, but in the case of *Dermasterias imbricata* and *Arctonoe vittata*, the host seastar is attracted to the worm.

WEBSTER, S. K. 1968. An investigation of the commensals of *Cryptochiton stelleri* (Middendorff, 1847) in the Monterey Peninsula area, California. Veliger 11: 121–25. □*Arctonoe vittata*, one of the commensals, is attracted to its host.

Arenicola brasiliensis Nonato, lugworm, §310

WELLS, G. P. 1963. Barriers and speciation in lugworms. *In* J. P. Harding and N. Tebble, eds., Speciation in the sea, pp. 79–98. Syst. Assoc. Publ. 5: 1–199.

A. cristata Stimpson, §310

See additional references listed under *Abarenicola*.

DeVLAS, J. 1979. Secondary production by tail regeneration in a tidal flat population of lugworms (*Arenicola marina*), cropped by flatfish. Neth. J. Sea Res. 13: 362–93.

WELLS, G. P. 1962. The warm-water lugworms of the world (Arenicolidae, Polychaeta). Proc. Zool. Soc. Lond. 138: 331–53. □Concerns *Arenicola cristata* and four other species of the genus.

Armandia brevis (Moore) (formerly *A. bioculata* Hartman), §265

See also Woodin (1974) and Hannan (1981) above.

HERMANS, C. O. 1966. The natural history and larval anatomy of *Armandia brevis* (Polychaeta: Opheliidae). Ph.D. dissertation, Zoology, University of Washington, Seattle. 184 pp.

———. 1978. Metamorphosis in the opheliid polychaete *Armandia brevis*. In F.-S. Chia and M. E. Rice, eds., Settlement and metamorphosis of marine invertebrate larvae, pp. 113–26. New York: Elsevier.

Axiothella rubrocincta (Johnson) (formerly *Clymenella*), §293

KUDENOV, J. D. 1978. The feeding ecology of *Axiothella rubrocincta* (Johnson) (Polychaeta: Maldanidae). J. Exp. Mar. Biol. Ecol. 31: 209–21.

———. 1982. Rates of seasonal sediment reworking in *Axiothella rubrocincta* (Polychaeta: Maldanidae). Mar. Biol. 70: 181–86. □*Axiothella* in Tomales Bay rework about 5 grams of dry sediment per day.

Boccardia columbiana Berkeley, §152

B. proboscidea Hartman, §152

BLAKE, J. A., and J. W. EVANS. 1973. *Polydora* and related genera as borers in mollusk shells and other calcareous substrates. Veliger 15: 235–49.

BLAKE, J. A., and K. H. WOODWICK. 1971. A review of the genus *Boccardia* Carazzi (Polychaeta: Spionidae) with descriptions of two new species. Bull. South. Calif. Acad. Sci. 70: 31–42. □Describes *Boccardia berkeleyorum* from central and northern California.

DEAN, D., and J. A. BLAKE. 1966. Life-history of *Boccardia hamata* (Webster) on the east and west coasts of North America. Biol. Bull. 130: 316–30.

LIGHT, W. J. 1978. Spionidae: Polychaeta: Annelida. 1. Invertebrates of the San Francisco Bay estuary system. Pacific Grove, Calif.: Boxwood Press. 211 pp. □Concerns the morphology and taxonomy of members of the polychaete family Spionidae. Includes a key to eight species of *Boccardia* in California.

TAGHON, G. L., A. R. M. NOWELL, and P. A. JUMARS. 1980. Induction of suspension feeding in spionid polychaetes by high particulate fluxes. Science 210: 562–64. □As the velocity and the concentration of particulate matter in a water current was increased experimentally, the number of spionids suspension feeding increased relative to the number deposit feeding; three species were examined, including *Boccardia proboscidea.*

Capitella capitata (Fabricius), §293

GRASSLE, J. P., and J. F. GRASSLE. 1976. Sibling species in the marine pollution indicator *Capitella* (Polychaeta). Science 192: 567–69.

WARREN, L. M. 1976. A population study of the polychaete *Capitella capitata* at Plymouth. Mar. Biol. 38: 209–16.

Chaetopterus variopedatus (Renier) (includes *C. pergamentaceus* Cuvier), §309

BROWN, S. C. 1975. Biomechanics of water-pumping by *Chaetopterus variopedatus* Renier: Skeletomusculature and kinematics. Biol. Bull. 149: 136–50. □Describes the unusual method by which *Chaetopterus* pumps water through its burrow. Here, water is pumped by the coordinated activities of three segments, each acting as a "piston" to drive a quantity of water along the tube. This mechanism is unlike those employed by the majority of tube-dwelling worms, who use peristaltic or undulatory movements.

———. 1977. Biomechanics of water-pumping by *Chaetopterus variopedatus* Renier: Kinetics and hydrodynamics. Biol. Bull. 153: 121–32.

BROWN, S. C., and J. S. ROSEN. 1978. Tube-cleaning behaviour in the polychaete annelid *Chaetopterus variopedatus* (Renier). Anim. Behav. 26: 160–66.

FLOOD, P. R., and A. FIALA-MEDIONI. 1982. Structure of the mucous feeding filter of *Chaetopterus variopedatus* (Polychaeta). Mar. Biol. 72: 27–33.

MacGINITIE, G. E. 1939. The method of feeding of *Chaetopterus.* Biol. Bull. 77: 115–18.

Cirriformia luxuriosa (Moore) (formerly *Cirratulus*), §16

YOSHIYAMA, R. M., and J. D. S. DARLING. 1982. Grazing by the intertidal fish *Anoplarchus purpurescens* upon a distasteful polychaete worm. Environ. Biol. Fishes 7: 39–45. □*Anoplarchus* is the only intertidal fish that regularly eats large quantities of *Cirriformia luxuriosa* tentacles. Could the worm be distasteful?

C. spirabrancha (Moore) (originally *Cirratulus spirabranchus*), §293

BLAKE, J. A. 1975. The larval development of Polychaeta from the northern California coast. 1. *Cirriformia spirabrancha* (family Cirratulidae). Trans. Amer. Microsc. Soc. 94: 179–88.

DICE, J. F. 1969. Osmoregulation and salinity tolerance in the polychaete annelid *Cirriformia spirabrancha* (Moore, 1904). Comp. Biochem. Physiol. 28: 1331–43. □*Cirriformia* can tolerate salinities as low as 23 percent seawater and as high as 220 percent seawater for at least 6 hours.

Cistenides brevicoma (Johnson), §281

Dodecaceria fewkesi Berkeley and Berkeley (formerly *D. fistulicola* Ehlers and *D. pacifica* (Fewkes)), §63

BERKELEY, E., and C. BERKELEY. 1954. Notes on the life-history of the polychaete *Dodecaceria fewkesi* (nom. n.). J. Fish. Res. Board Can. 11: 326–34.

GIBSON, P. H. 1978. Systematics of *Dodecaceria* (Annelida: Polychaeta) and its relation to the reproduction of its species. Zool. J. Linn. Soc. 63: 275–87. □Discusses the confused state of the systematics of the genus *Dodecaceria,* including uncertainty about whether *D. fistulicola* and *D. fewkesi* are indeed identical as indicated above; suggests that reproductive features are more useful taxonomic characteristics than morphological features in this genus.

Eudistylia polymorpha (Johnson), §129

E. vancouveri (Kinberg), §129

Eupolymnia heterobranchia (Johnson) and other terebellids, §53

See footnote, p. 80: this may not be a legitimate species.

Euzonus mucronata (Treadwell) (formerly *Thoracophelia*), §§155, 191

Two additional species, *Euzonus dillonensis* (Hartman) and *E. williamsi* (Hartman), occur in clean sand at Dillon Beach.

DALES, R. P. 1952. The larval development and ecology of *Thoracophelia mucronata* (Treadwell). Biol. Bull. 102: 232–42.

McCONNAUGHEY, B. H., and D. L. FOX. 1949. The anatomy and biology of the marine polychaete *Thoracophelia mucronata* (Treadwell) Opheliidae. Univ. Calif. Publ. Zool. 47: 319–40.

Glycera americana Leidy (including *Glycera rugosa* Johnson), proboscis worm, §280

G. robusta Ehlers, the large proboscis worm, §311

GIBBS, P. E., and G. W. BRYAN. 1980. Copper—the major metal component of glycerid polychaete jaws. J. Mar. Biol. Assoc. U.K. 60: 205–14.

HOFFMAN, R. J., and C. P. MANGUM. 1970. The function of coelomic cell hemoglobin in the polychaete *Glycera dibranchiata*. Comp. Biochem. Physiol. 36: 211–28. □The worm is able to survive low oxygen conditions for at least 48 hours. It is likely that hemoglobin functions as an oxygen storage compound when ventilation of the burrow ceases during low tide.

Grubeopolynoe tuta (Grube) (formerly *Hololepidella*), §234

Halosydna brevisetosa Kinberg (includes *H. johnsoni* (Darboux)), §49

LWEBUGA-MUKASA, J. 1970. The role of elytra in the movement of water over the surface of *Halosydna brevisetosa* (Polychaeta: Polynoidae). Bull. South. Calif. Acad. Sci. 69: 154–60.

Hemipodus borealis Johnson, §234

Hesperonoe adventor (Skogsberg) (formerly *Harmothoe*), commensal with *Urechis caupo*, §306

FISHER, W. K. 1946. Echiuroid worms of the north Pacific Ocean. Proc. U.S. Natl. Mus. 96: 215–92. □An account of the relations of this worm with its host and other commensals is given on pp. 277–78.

H. complanata (Johnson) (formerly *Harmothoe*), commensal with *Callianassa* and *Upogebia*, §292

Hydroides pacificus Hartman, not treated

This warm-water species is found in southern California harbors, where it encrusts piling and other objects and may foul yacht rudders and propellers with masses of calcareous tubes. *Hydroides pacificus* has been recorded from southern California as *H. norvegicus* Gunnerus, and although described as new in 1969, the former may not be distinct from the latter, a well-known cosmopolitan species.

REISH, D. J. 1961. The relationship of temperature and dissolved oxygen to the seasonal settlement of the polychaetous annelid *Hydroides norvegica* (Gunnerus). Bull. South. Calif. Acad. Sci. 60: 1–11.

SCHELTEMA, R. S., I. P. WILLIAMS, M. A. SHAW, and C. LOUDON. 1981. Gregarious settlement by the larvae of *Hydroides dianthus* (Polychaeta: Serpulidae). Mar. Ecol. Prog. Ser. 5: 69–74.

Lumbrineris erecta (Moore), §153

L. zonata (Johnson), §§130, 293

Marphysa stylobranchiata Moore, §129

Mesochaetopterus taylori Potts, §265

BARNES, R. D. 1965. Tube-building and feeding in chaetopterid polychaetes. Biol. Bull. 129: 217–33. □Compares and contrasts the tube-building and feeding methods of various chaetopterids, among them *Mesochaetopterus taylori*.

Naineris dendritica (Kinberg) (formerly *N. laevigata* (Grube)), §191

PARKINSON, G. T. 1978. Aspects of feeding, burrowing, and distribution of *Haploscoloplos elongatus* (Polychaeta: Orbiniidae) at Bodega Harbor, California. Pacific Sci. 32: 149–55. □The distribution of *Haploscoloplos elongatus* (an orbinid, like *Naineris*) in Bodega Harbor is aggregated and correlated with a sandy mud substratum. The worm employs its prostomium in burrowing, not its proboscis, which is used solely for deposit feeding.

Neanthes brandti (Malmgren) (formerly *Nereis*), §164

N. limnicola (Johnson) (formerly *N. lighti* Hartman), not treated

This species inhabits brackish to fresh water of estuarine streams, and thus is unlikely to be encountered by most intertidal visitors. It has sparked interest (and publications), however, because it has features unusual for a polychaete: viviparous reproduction and the freshwater habitat. A sample of research papers is given below.

BASKIN, D. G., and D. K. GOLDING. 1970. Experimental studies on the endocrinology and reproductive biology of the viviparous polychaete annelid, *Nereis limnicola* Johnson. Biol. Bull. 139: 461–75.

OGLESBY, L. C. 1968. Responses of an estuarine population of the polychaete *Nereis limnicola* to osmotic stress. Biol. Bull. 134: 118–38.

SMITH, R. I. 1950. Embryonic development in the viviparous nereid polychaete, *Neanthes lighti* Hartman. J. Morphol. 87: 417–65.

N. succinea (Frey and Leuckart) (formerly *Nereis*), not treated

CAMMEN, L. M. 1980. The significance of microbial carbon in the nutrition of the deposit feeding polychaete *Nereis succinea*. Mar. Biol. 61: 9–20.

KUHL, D. L., and L. C. OGLESBY. 1979. Reproduction and survival of the pileworm *Nereis succinea* in higher Salton Sea salinities. Biol. Bull. 157: 153–65. □The worm is the major benthic detritivore in this inland salt lake (present salinity of 36 parts per thousand). Immature worms can survive extended periods at salinities as high as 65 parts per thousand, and reproduction is successful at salinities at least as high as 45 parts per thousand.

N. virens (Sars) (formerly *Nereis*), §311

BASS, N. R., and A. E. BRAFIELD. 1972. The lifecycle of the polychaete *Nereis virens*. J. Mar. Biol. Assoc. U.K. 52: 701–26. □Study of a population in the Thames estuary, Britain. Spawning generally occurs in early May; both lunar period and temperature seem to be involved in initiating spawning.

COMMITO, J. A. 1982. Importance of predation by infaunal polychaetes in controlling the structure of a soft-bottom community in Maine, USA. Mar. Biol. 68: 77–81. □Adult *Neanthes virens* fed on the amphipod *Corophium volutator*. When the polychaete was removed experimentally, there was an increase in amphipod abundance.

DEAN, D. 1978. Migration of the sandworm *Nereis virens* during winter nights. Mar. Biol. 45: 165–73. □During winter months, these worms were found swimming in surface waters at night in an estuary in Maine. Since none of the swimming worms contained ripe gametes, it was concluded that swimming during winter months was unrelated to reproduction.

Neoamphitrite robusta (Johnson) (formerly *Terebella* or *Amphitrite*), §§53, 54, 232

BROWN, P. L., and D. V. ELLIS. 1971. Relation between tube-building and feeding in *Neoamphitrite robusta* (Polychaeta: Terebellidae). J. Fish. Res. Board Can. 28: 1433–35.

DALES, R. P. 1955. Feeding and digestion in

terebellid polychaetes. J. Mar. Biol. Assoc. U.K. 34: 55–79.

Nephtys caecoides Hartman, §265
BANKS, R. C. 1962. Observations on the polychaete genus *Nephtys* near Bolinas, California (Annelida: Nephtyidae). Wasmann J. Biol. 20: 107–14. □Concerns habitat and morphological differences between *Nephtys caecoides* and *N. californiensis*. The two species are sympatric in some areas.
CLARK, R. B., and E. C. HADERLIE. 1962. The distribution of *Nephtys californiensis* and *N. caecoides* on the California coast. J. Anim. Ecol. 31: 339–57.

N. californiensis Hartman, §§155, 265

Nereis grubei (Kinberg) (formerly *N. mediator* Chamberlin), the small mussel worm, §§163, 320
REISH, D. J. 1954. The life history and ecology of the polychaetous annelid *Nereis grubei* (Kinberg). Allan Hancock Found. Occ. Pap. 14: 1–75.
SCHROEDER, P. C. 1967. Morphogenesis of epitokous setae during normal and induced metamorphosis in the polychaete annelid *Nereis grubei* (Kinberg). Biol. Bull. 133: 426–37.
———. 1968. On the life history of *Nereis grubei* (Kinberg), a polychaete annelid from California. Pacific Sci. 22: 476–81. □Concerns a population at Pescadero Point (Monterey County). During spring and summer, the worms are most readily found in the sandy substratum beneath the alga *Gastroclonium*. During the winter, however, they are more abundant in holdfasts of *Egregia*; presumably they are protected there from violent winter surf.

N. procera Ehlers, §274

N. vexillosa Grube, the large mussel worm, §163
JOHNSON, M. W. 1943. Studies on the life history of the marine annelid *Nereis vexillosa*. Biol. Bull. 84: 106–14.
WOODIN, S. A. 1977. Algal "gardening" behavior by nereid polychaetes: Effects on soft-bottom community structure. Mar. Biol. 44: 39–42. □Describes the attachment of drift algae to the tubes of *Nereis vexillosa* and *Platynereis bicanaliculata* living on a mud flat in the San Juan Islands. The attached algae are eaten by the worms and provide some protection from physical stresses.

Notomastus tenuis Moore, §293

Odontosyllis phosphorea Moore, and similar syllids, §151
BANSE, K. 1972. On some species of Phyllodocidae, Syllidae, Nephtyidae, Goniadidae, Apistobranchidae, and Spionidae (Polychaeta) from the northeast Pacific Ocean. Pacific Sci. 26: 191–222.

Ophiodromus pugettensis (Johnson) (formerly *Podarke*), §26
LANDE, R., and D. J. REISH. 1968. Seasonal occurrence of the commensal polychaetous annelid *Ophiodromus pugettensis* on the starfish *Patiria miniata*. Bull. South. Calif. Acad. Sci. 67: 104–11. □The number of *Ophiodromus* on *Patiria* varied seasonally, being highest in winter when water temperature was lowest. There was considerable movement of worms on and off the seastar hosts.
SHAFFER, P. L. 1979. The feeding biology of *Podarke*

pugettensis (Polychaeta: Hesionidae). Biol. Bull. 156: 343–55.

Pareurythoe californica (Johnson) (originally *Eurythoe*), §311

Pectinaria californiensis Hartman, §281
WATSON, A. T. 1928. Observations on the habits and life history of *Pectinaria (Lagis) koreni* Mgr. Proc. Trans. Liverpool Biol. Soc. 42: 25–60.

Pherusa papillata (Johnson) (formerly *Stylarioides*), §233
SPIES, R. B. 1975. Structure and function of the head in flabelligerid polychaetes. J. Morphol. 147: 187–208. □Based on detailed studies of *Flabelliderma commensalis*, with comparisons to other flabelligerid polychaetes, including *Pherusa neopapillata* (but not *P. papillata*).

Phragmatopoma californica (Fewkes) (formerly *Sabella*), §166
DALES, R. P. 1952. The development and structure of the anterior region of the body in the Sabellariidae, with special reference to *Phragmatopoma californica*. J. Microsc. Sci. 93: 435–52.
ECKELBARGER, K. J. 1977. Larval development of *Sabellaria floridensis* from Florida and *Phragmatopoma californica* from southern California (Polychaeta: Sabellariidae), with a key to the sabellariid larvae of Florida and a review of development in the family. Bull. Mar. Sci. 27: 241–55.
ROY, P. A. 1974. Tube dwelling behavior in the marine annelid *Phragmatopoma californica* (Fewkes) (Polychaeta: Sabellariidae). Bull. South. Calif. Acad. Sci. 73: 117–25. □Concerns the worm's locomotion and ciliary currents within its tube, morphology, and elimination of feces.
See also Taylor and Littler (1982), under *Anthopleura elegantissima*, p. 515.

Phyllochaetopterus prolifica Potts, §118

Platynereis bicanaliculata (Baird) (*P. agassizi* (Ehlers) in previous editions), §§153, 330
See Roe (1975) and Woodin (1974).

Sabellaria cementarium Moore, §166

Salmacina tribranchiata (Moore), §183
Listed as *Salmacina* sp. or *Filograna* sp. in previous editions; *Salmacina tribranchiata* is the most common species of this genus along the Pacific coast.

Serpula vermicularis Linnaeus (includes *S. columbiana* Johnson), §§129, 213

Spirorbids, small serpulid worms, §22
Previous editions list *Spirorbis spirillum* or *Spirorbis* sp. here. However, the taxonomy of the several species of spirorbids occurring along the Pacific coast is rather confused or, at the very least, far more complicated than earlier supposed. Designating, for the present, these species broadly as "spirorbids," though unsatisfactory in some ways, will perhaps keep us from thinking we know more than we do.
BAILEY, J. H. 1969. Methods of brood protection as a basis for reclassification of the Spirorbinae (Serpulidae). Zool. J. Linn. Soc. 48: 387–407.
KNIGHT-JONES, P. 1978. New Spirorbidae (Polychaeta: Sedentaria) from the east Pacific, Atlantic, Indian and Southern Oceans. Zool. J. Linn. Soc. 64:

201–40. ☐Describes 18 new species, 8 from California, and discusses the taxonomy of "spirorbids."

O'CONNOR, R. J., and P. LAMONT. 1978. The spatial organization of an intertidal *Spirorbis* community. J. Exp. Mar. Biol. Ecol. 32: 143–69. ☐Concerns species occurring along the shore of Northern Ireland but provides clues about the likely relationships of spirorbids on our Pacific coast.

POTSWALD, H. E. 1968. The biology of fertilization and brood protection in *Spirorbis* (*Laeospira*) *morchi*. Biol. Bull. 135: 208–22.

WILLIAMS, G. B. 1964. The effects of extracts of *Fucus serratus* in promoting the settlement of larvae of *Spirorbis borealis* (Polychaeta), J. Mar. Biol. Assoc. U.K. 44: 397–414.

Terebella californica Moore, §§53, 311

SCOTT, J. W. 1911. Further experiments on the methods of egg laying in *Amphitrite*. Biol. Bull. 20: 252–65.

WELSH, J. H. 1934. The structure and reactions of the tentacles of *Terebella magnifica* W. Biol. Bull. 66: 339–45.

Thelepus crispus Johnson, §53

Phylum SIPUNCULA
Sipunculans, or Peanut worms

The sipunculans, a small group of unsegmented, cylindrical worms, bear some superficial resemblance to certain polychaete annelids and perhaps to certain apodous sea cucumbers. Indeed, up until the 1940's they were often placed in a class of the Annelida, the "Gephyrea," along with the echiurans. However, the fundamental differences between sipunculans and other worms are now seen as warranting their placement in a separate phylum. When relaxed, the body of a sipunculan consists of a slender anterior introvert (which can be retracted, or introverted, into the body) and a more bulbous posterior trunk. With the introvert retracted, some of the locally common sipunculans become roughly peanut-shaped, giving rise to their common name.

FISHER, W. K. 1952. The sipunculid worms of California and Baja California. Proc. U.S. Natl. Mus. 102: 371–450. ☐Concerns the distribution, anatomy, and taxonomy of all the sipunculans treated in this book. Essential reference for identification of species on the Pacific coast.

HANSEN, M. D. 1978. Food and feeding behavior of sediment feeders as exemplified by sipunculids and holothurians. Helgol. Wiss. Meeresunters. 31: 191–221.

HYMAN, L. H. 1959. The invertebrates: Smaller coelomate groups. Vol. V. New York: McGraw-Hill. (Sipunculida, pp. 610–96.)

NICOSIA, S. V. 1979. Lectin-induced mucus release in the urn cell complex of the marine invertebrate *Sipunculus nudus* (Linnaeus). Science 206: 698–700. ☐Cited here as an example of recent work on the physiology of sipunculans. Urn cell complexes play an important role in the immune response of sipunculans; they secrete mucus that traps foreign cells and leads to their destruction.

PILGER, J. F. 1982. Ultrastructure of the tentacles of *Themiste lageniformis* (Sipuncula). Zoomorphology 100: 143–56. ☐The only ultrastructural work done to date on sipunculan tentacles. Several species of this genus occur on our shores.

RICE, M. E. 1967. A comparative study of the development of *Phascolosoma agassizii*, *Golfingia pugetensis*, and *Themiste pyroides* with a discussion of developmental patterns in the Sipuncula. Ophelia 4: 143–71. ☐Describes the breeding season and development of these three sipunculans from the San Juan Islands.

———. 1975. Unsegmented coelomate worms. *In* R. I. Smith and J. T. Carlton, eds., Light's manual, pp. 128–34. Berkeley: University of California Press.

———. 1976. Larval development and metamorphosis in Sipuncula. Amer. Zool. 16: 563–71.

———. 1980. Sipuncula and Echiura. *In* R. H. Morris, D. P. Abbott, and E. C. Haderlie, Intertidal invertebrates of California, pp. 490–98. Stanford, Calif.: Stanford University Press.

RICE, M. E., and M. TODOROVIC, eds. 1975–1976. Proceedings of the International Symposium on the Biology of Sipuncula and Echiura. Vol. 1 (1975), 355 pp.; Vol. 2 (1976), 204 pp. Belgrade: Naucno Delo Press.

STEPHEN, A. C., and S. J. EDMUNDS. 1972. The phyla Sipuncula and Echiura. London: British Museum (Natur. Hist.). 528 pp.

Golfingia hespera (Chamberlin), §293

FISHER, W. K. 1950. The sipunculid genus *Phascolosoma*. Ann. Mag. Natur. Hist. (12)3: 547–52. ☐Places *Phascolosoma agassizii* and *Golfingia hespera* in their correct genera after erroneous use of the name *Phascolosoma* by authors. Fisher notes that the generic name *Golfingia* was coined by E. R. Lankester in honor of an afternoon golfing at St. Andrews; "scarcely, however, were good taste and euphony advanced thereby."

Phascolosoma agassizii Keferstein (formerly *Golfingia* or *Physcosoma*), §52

RICE, M. E. 1973. Morphology, behavior, and histogenesis of the pelagosphera larva of *Phascolosoma agassizii* (Sipuncula). Smithsonian Contr. Zool. 132: 1–51.

TOWLE, A., and A. C. GIESE. 1967. The annual reproductive cycle of the sipunculid *Phascolosoma agassizii*. Physiol. Zool. 40: 229–37.

Siphonosoma ingens (Fisher) (formerly *Siphonomecus*), §308

Sipunculus nudus Linnaeus, §308

Themiste perimeces (Fisher) (formerly *Dendrostoma*), §282

FISHER, W. K. 1928. New Sipunculoidea from California. Ann. Mag. Natur. Hist. (10) 1: 194–99. ☐Concerns *Themiste perimeces* (as *Dendrostoma*).

T. pyroides (Chamberlin) (formerly *Dendrostoma*), §145

T. zostericola (Chamberlin) (formerly Dendrostoma), §145

PEEBLES, F., and D. L. FOX. 1933. The structure, functions, and general reactions of the marine sipunculid worm *Dendrostoma zostericola*. Bull. Scripps Inst. Oceanogr., Tech. Ser. 3: 201–24.

Phylum **ECHIURA**. Echiuran worms

The echiurans are a second small phylum of unsegmented, largely sedentary worms, whose sausage-shaped bodies differ in many details of anatomy from sipunculans and other worms, enough to warrant placement in a separate phylum.

FISHER, W. K. 1946. Echiuroid worms of the North Pacific Ocean. Proc. U.S. Natl. Mus. 96: 215–92. ☐The most comprehensive treatment of Pacific coast echiurans.

NEWBY, W. W. 1940. The embryology of the echiuroid worm *Urechis caupo*. Mem. Amer. Philos. Soc. 16. 219 pp.

STEPHEN, A. C., and S. J. EDMONDS. 1972. The phyla Sipuncula and Echiura. London: British Museum (Natur. Hist.). 528 pp.

Echiurus echiurus alaskensis Fisher, §306

GISLÈN, T. 1940. Investigations on the ecology of *Echiurus*. Lunds Universitets Årsskrift, N.F. Avd. 2, Bd. 36, Nr. 10, pp. 3–39. ☐Concerns the biology and ecology of *Echiurus echiurus* in Swedish waters.

Listriolobus pelodes Fisher, §307

PILGER, J. F. 1980. The annual cycle of oogenesis, spawning, and larval settlement of the echiuran *Listriolobus pelodes* off southern California. Pacific Sci. 34: 129–42. ☐Concerns the natural history and reproductive biology of *Listriolobus* in a subtidal population, 60 meters deep, off Palos Verdes, California.

Ochetostoma octomyotum Fisher, §307

Urechis caupo Fisher and MacGinitie, §306

CHAPMAN, G. 1968. The hydraulic system of *Urechis caupo* Fisher and MacGinitie. J. Exp. Biol. 49: 657–67.

FISHER, W. K., and G. E. MACGINITIE. 1928a. A new echiuroid worm from California. Ann. Mag. Natur. Hist. (10)1: 199–204.

———. 1928b. The natural history of an echiuroid worm. Ann. Mag. Natur. Hist. (10)1: 204–13.

LAWRY, J. V. 1966. Neuromuscular mechanisms of burrow irrigation in the echiuroid worm *Urechis caupo* Fisher and MacGinitie. I. Anatomy of the neuromuscular system and activity of intact animals. II. Neuromuscular activity of dissected preparations. J. Exp. Biol. 45: 343–56, 357–68. ☐*Urechis* shows no endogenous periodicity or regular rhythm, but pumps more or less continuously in artificial burrows in the laboratory. In nature, when the burrow is uncovered at low tide, the worm apparently ceases pumping.

MACGINITIE, G. E. 1935. Normal functioning and experimental behavior of the egg and sperm collectors of the echiuroid, *Urechis caupo*. J. Exp. Zool. 70: 341–55.

———. 1945. The size of the mesh openings in mucous feeding nets of marine animals. Biol. Bull. 88: 107–11. ☐Examines properties of the mucous nets of *Urechis caupo* and the polychaete *Chaetopterus variopedatus*.

PRITCHARD, A., and F. N. WHITE. 1981. Metabolism and oxygen transport in the innkeeper *Urechis caupo*. Physiol. Zool. 54: 44–54.

SUER, A. L. 1982. Larval settlement, growth, and reproduction of the marine echiuran *Urechis caupo*. Ph.D. dissertation, Zoology, University of California, Davis. 198 pp.

SUER, A. L., and D. W. PHILLIPS. 1983. Rapid, gregarious settlement of the larvae of the marine echiuran *Urechis caupo* Fisher and MacGinitie 1928. J. Exp. Mar. Biol. Ecol. 67: 243–59.

Phylum **ARTHROPODA**

The arthropods are an enormous group of animals, more species having been described for it than for all the rest of the animal kingdom combined. But beyond their sheer numbers, they have succeeded in virtually every habitat that supports life, and in many habitats arthropods are among the most prominent residents. Despite their diversity, arthropods can often be easily identified as such by their firm, jointed exoskeleton and jointed appendages.

Although the arthropods are presented here as one phylum, there is a growing suspicion that "arthropodization" may have begun independently in two or three lines leading from annelidan ancestors. If this were true, the arthropods would be a polyphyletic group, and each of the major lines would warrant status as a separate phylum. The crustaceans are seen by many as the result of one such line; the myriapods (centipedes, millipedes, etc.) and insects as another; and the chelicerates (spiders, mites, pycnogonids, etc.) as a third. Of the general accounts of arthropod biology listed below, several deal with the question of arthropod phylogeny.

ANDERSON, D. T. 1973. Embryology and phylogeny in annelids and arthropods. New York: Pergamon. 495 pp.

CLARKE, K. U. 1973. The biology of the Arthropoda. London: Arnold. 270 pp.

GUPTA, A. P., ed. 1979. Arthropod phylogeny. New York: Van Nostrand Reinhold. 762 pp.

MANTON, S. M. 1973. Arthropod phylogeny—a modern synthesis. J. Zool. (Lond.) 171: 111–30.

———. 1977. The Arthropoda: Habits, functional morphology, and evolution. Oxford: Oxford University Press (Clarendon Press). 527 pp.

Subphylum **MANDIBULATA**

Class CRUSTACEA. Barnacles, Beach hoppers, Shrimps, Lobsters, Crabs, etc.

Crustaceans are by far the dominant arthropods in the sea and on the shore. The group is extremely diverse, including such varied forms as barnacles, shrimps, crabs, and sow bugs. As is often the case, great diversity produces a rather complex taxonomy, and that of the crustaceans seems at times to exceed the limits of reason. But the diversity and complexity are real and should be appreciated. In addition to cataloguing this diversity, taxonomy expresses hypotheses about evolutionary relationships, and these hypotheses are worthy of reflection.

BLISS, D. E., and L. H. MANTEL, eds. 1968. Terrestrial adaptations in Crustacea. Amer. Zool. 8: 307–685. ☐A symposium volume. See especially the papers by Edney (isopods), Hurley (amphipods), Bliss (decapods), Warburg (isopods), and Barnwell (fiddler crabs).

GREEN, J. 1963. A biology of Crustacea. London: Witherby. 180 pp.

GUNNILL, F. C. 1982. Macroalgae as habitat patch islands for *Scutellidium lamellipes* (Copepoda: Harpacticoida) and *Ampithoe tea* (Amphipoda: Gammaridae). Mar. Biol. 69: 103–16. ☐These two occur in patches formed by the brown alga *Pelvetia fastigiata* in the mid-intertidal zone near La Jolla.

KAESTNER, A. 1970. Invertebrate zoology. III. Crustacea. (Translated and adapted from the second German edition by H. W. Levi and L. R. Levi.) New York: Interscience. 523 pp.

LOCKWOOD, A. P. M. 1967. Aspects of the physiology of Crustacea. San Francisco: Freeman. 328 pp.

MACGINITIE, G. E. 1937. Notes on the natural history of several marine Crustacea. Amer. Midl. Natur. 18: 1031–37.

MCLAUGHLIN, P. A. 1980. Comparative morphology of recent Crustacea. San Francisco: Freeman. 177 pp.

MOORE, R. C., and L. McCORMICK. 1969. General features of Crustacea. *In* R. C. Moore, ed., Treatise on invertebrate paleontology, part R, Arthropoda 4, Vol. 1, pp. R57–120. New York: Geologic Society of America. 398 pp.

SCHMITT, W. L. 1965. Crustaceans. Ann Arbor: University of Michigan Press. 204 pp.

SNODGRASS, R. E. 1956. Crustacean metamorphoses. Smithsonian Misc. Coll. 131(10): 1–78.

WATERMAN, T. H., ed. 1960. The physiology of Crustacea. 1. Metabolism and growth. New York: Academic Press. 670 pp.

——. 1961. The physiology of Crustacea. 2. Sense organs, integration, and behavior. New York: Academic Press. 681 pp.

WENNER, A. M. 1972. Sex ratio as a function of size in marine Crustacea. Amer. Natur. 106: 321–50.

Subclass BRANCHIOPODA

The branchiopods, almost exclusively freshwater animals, are divided into four orders: Anostraca (the fairy shrimps), Notostraca (the tadpole shrimps), Conchostraca (the clam shrimps), and Cladocera (the water fleas). None is likely to be encountered intertidally, although one anostracan at least, the brine shrimp *Artemia salina* (Linnaeus), is well known to aquarists, who use it to feed their fish. In addition, two genera of cladocerans (*Evadne* and *Podon*) are common in marine plankton tows.

BAKER, H. M. 1938. Studies on the Cladocera of Monterey Bay. Proc. Calif. Acad. Sci. (4)23: 311–65.

GELLER, W., and H. MULLER. 1981. The filtration apparatus of Cladocera: Filter mesh-sizes and their implications on food selectivity. Oecologia 49: 316–21.

LITTLEPAGE, J. L., and M. N. McGINLEY. 1965. A bibliography of the genus *Artemia* (*Artemia salina*)

1812–1962. San Francisco Aq. Soc. Spec. Publ. 1. 73 pp.

Subclass OSTRACODA

The Ostracoda are often numerous, and several species reside among seaweeds and bottom litter at the shore, but no attempt has been made to consider them. Unfortunately, there is no easily accessible reference, and no easy way into the literature. Local intertidal species have not been studied in detail.

BATE, R. H., E. ROBINSON, and L. M. SHEPPARD, eds. 1982. Fossil and recent ostracods. New York: Halsted. 494 pp.

BENSON, R. H. 1966. Recent marine podocopid ostracodes. Oceanogr. Mar. Biol. Ann. Rev. 4: 213–32.

LUCAS, V. Z. 1931. Some Ostracoda of the Vancouver Island region. Contr. Can. Biol. Fish. 7: 397–416.

NEALE, J. W., ed. 1969. The taxonomy, morphology, and ecology of Recent Ostracoda. Edinburgh: Oliver and Boyd. 553 pp. ☐A symposium volume.

SKOGSBERG, T. 1928. Studies on marine ostracods. Part II. External morphology of the genus *Cythereis*, with descriptions of twenty-one new species. Occ. Pap. Calif. Acad. Sci. 15: 1–155. ☐Includes five new species from Pacific Grove.

——. 1950. Two new species of marine Ostracoda (Podocopa) from California. Proc. Calif. Acad. Sci. (4)26: 485–505.

WATLING, L. 1970. Two species of Cytherinae (Ostracoda) from central California. Crustaceana 19: 251–63.

——. 1972. A new species of *Acetabulastoma* Shornikov from central California with a review of the genus. Proc. Biol. Soc. Wash. 85: 481–88. ☐Both papers by Watling concern forms collected at Tomales Bay.

Subclass COPEPODA

The subclass Copepoda is one of the largest crustacean groups, and there are hundreds of species along our Pacific coast. However, they tend to be minute and difficult to identify, so only a few that occupy distinctive habitats have been mentioned.

EPP, R. W., and W. M. LEWIS, JR. 1981. Photosynthesis in copepods. Science 214: 1349–50. ☐Grazing copepods consume algae that are not digested. One of these copepods, *Acanthocyclops vernalis*, was shown to produce oxygen in the light, presumably compliments of the ingested algae.

FRASER, J. H. 1936. The distribution of rock pool Copepoda according to tidal level. J. Anim. Ecol. 5: 23–28.

HUMES, A. G., and J. H. STOCK. 1973. A revision of the family Lichomolgidae Kossman, 1877, cyclopoid copepods mainly associated with marine invertebrates. Smithsonian Contr. Zool. 127. 368 pp.

ILLG, P. L. 1958. North American copepods of the family Notodelphyidae. Proc. U.S. Natl. Mus. 107: 463–649.

ILLG, P. L., and P. L. DUDLEY. 1980. The family Ascidicolidae and its subfamilies (Copepoda, Cyclopoida), with descriptions of new species. Mem. Mus. Natl. Hist. Natur., n.s., Ser. A. Zool. 117: 1–192.

JOHNSON, M. W., and J. B. OLSEN. 1948. The life history and biology of a marine harpacticoid copepod, *Tisbe furcata* (Baird). Biol. Bull. 95: 320–32.

KOEHL, M. A. R., and J. R. STRICKLER. 1981. Copepod feeding currents: Food capture at low Reynolds number. Limnol. Oceanogr. 26: 1062–73. □This work necessitates a rewrite of the textbook version of "filter feeding" in copepods. Instead of pumping large volumes of particle-laden water through filtering sieves, the copepod scans large volumes of water and then uses its bristled mouthparts more like paddles than filters to selectively capture particles.

LANG, K. 1965. Copepoda Harpacticoidea from the California Pacific coast. Kungl. Sven. Vetenskapakad. Handl., Ser. 4, Vol. 10(2). 560 pp. □Based principally on collections made at Dillon Beach and Pacific Grove. Nearly a hundred species recorded.

LOPEZ, G. W. 1982. Short-term population dynamics of *Tisbe cucumariae* (Copepoda: Harpacticoida). Mar. Biol. 68: 333–41.

MONK, C. R. 1941. Marine harpacticoid copepods from California. Trans. Amer. Microsc. Soc. 60: 75–99.

WILSON, C. B. 1935. Parasitic copepods from the Pacific coast. Amer. Midl. Natur. 16: 776–97.

Clausidium vancouverense (Haddon), on *Callianassa* and *Upogebia*, §292

GOODING, R. V. 1957. "*Callianassa pugettensis*" (Decapoda, Anomura), type host of the copepod *Clausidium vancouverense* (Haddon). With a note on *Hemicyclops pugettensis* Light and Hartman, another copepod associated with callianassids. Ann. Mag. Natur. Hist. (12)10: 695–700.

LIGHT, S. F., and O. HARTMAN. 1937. A review of the genera *Clausidium* Kossmann and *Hemicyclops* Boeck (Copepoda, Cyclopoida), with the description of a new species from the northeast Pacific. Univ. Calif. Publ. Zool. 41: 173–88. □See Gooding (1960) below for a revision of the genus *Hemicyclops*.

Diathrodes cystoecus Fahrenbach, not treated

FAHRENBACH, W. H. 1962. The biology of a harpacticoid copepod. La Cellule 62: 301–76. □Originally described from *Halosaccion* at Moss Beach, this copepod is always found in or on red algae, often in the bladders of *Halosaccion*. It occurs from San Pedro to the Queen Charlotte Islands.

Hemicyclops thysanotus Wilson (formerly *H. callianassae* Wilson), on *Callianassa*, *Upogebia*, and *Hermissenda*, §292

GOODING, R. V. 1960. North and South American copepods of the genus *Hemicyclops* (Cyclopoida: Clausiidae). Proc. U.S. Natl. Mus. 112: 159–95. □Synonymizes *Hemicyclops thysanotus*, *H. callianassae*, and *H. pugettensis*.

Lamippe sp. or a member of the family Lamippidae (generic and specific placement uncertain), parasitic in *Renilla* polyps, §248

Mytilicola orientalis Mori, §219

BERNARD, F. R. 1969. The parasitic copepod *Mytilicola orientalis* in British Columbia bivalves. J. Fish. Res. Board Can. 26: 190–91.

BRADLEY, W., and A. E. SIEBERT, JR. 1978. Infection of *Ostrea lurida* and *Mytilus edulis* by the parasitic copepod *Mytilicola orientalis* in San Francisco Bay, California. Veliger 21: 131–34.

DARE, P. J. 1982. The susceptibility of seed oysters of *Ostrea edulis* and *Crassostrea gigas* to natural infestation by the copepod *Mytilicola intestinalis*. Aquaculture 26: 201–12.

DAVEY, J. T., J. M. GEE, and S. L. MOORE. 1978. Population dynamics of *Mytilicola intestinalis* in *Mytilus edulis* in southwest England. Mar. Biol. 45: 319–27.

Scolecodes huntsmani (Henderson) (formerly *Scolecimorpha*), in the ascidian *Styela gibbsii*, §216

Tigriopus californicus (Baker), §8

DETHIER, M. N. 1980. Tidepools as refuges: Predation and the limits of the harpacticoid copepod *Tigriopus californicus* (Baker). J. Exp. Mar. Biol. Ecol. 42: 99–111. □Suggests that the high tide-pool habitat of *Tigriopus* is a refuge from potential predators such as sculpins and anemones.

HARRIS, R. P. 1973. Feeding, growth, reproduction and nitrogen utilization by the harpacticoid copepod, *Tigriopus californicus*. J. Mar. Biol. Assoc. U.K. 53: 785–800.

KONTOGIANNIS, J. E. 1973. Acquisition and loss of heat resistance in adult tide-pool copepod *Tigriopus californicus*. Physiol. Zool. 46: 50–54.

LEAR, D. W., JR., and C. H. OPPENHEIMER, JR. 1962. Consumption of microorganisms by the copepod *Tigriopus californicus*. Limnol. Oceanogr. 7(suppl.): lxiii–lxv.

Subclass CIRRIPEDIA. Barnacles

A comprehensive bibliography of barnacles would occupy a volume; we offer only a small sample. Barnacles, as components of communities and zones, also occupy a place in many papers. Some of these are cited in the marine-ecology section of the General Bibliography—see, for example, papers by Connell (1961a,b), Dayton (1971), and Paine (1974).

BARNES, H. 1959. Stomach contents and microfeeding of some common cirripedes. Can. J. Zool. 37: 231–36.

BOUSFIELD, E. L. 1955. Ecological control of the occurrence of barnacles in the Miramichi Estuary. Bull. Natl. Mus. Can. 137: 1–69.

BRANSCOMB, E. S., and K. VEDDER. 1982. A description of the naupliar stages of the barnacles *Balanus glandula* Darwin, *Balanus cariosus* Pallas, and *Balanus crenatus* Bruguière (Cirripedia, Thoracica). Crustaceana 42: 83–95.

CHENG, L., and R. A. LEWIN. 1976. Goose barnacles (Cirripedia: Thoracica) on flotsam beached at La Jolla, California. U.S. Natl. Mar. Fish. Serv., Fish. Bull. 74: 212–17.

CORNWALL, I. E. 1951. The barnacles of California (Cirripedia). Wasmann J. Biol. 9: 311–46.

———. 1955. The barnacles of British Columbia. British Columbia Prov. Mus. Handbook 7: 1–69.

CRISP, D. J., and A. J. SOUTHWARD. 1961. Different types of cirral activity of barnacles. Phil. Trans. Roy. Soc. Lond. B243: 271–307.

FOSTER, B. A. 1969. Tolerance of high temperatures by some intertidal barnacles. Mar. Biol. 4: 326–32.

———. 1971. Desiccation as a factor in the intertidal zonation of barnacles. Mar. Biol. 8: 12–29.

HAVEN, S. B. 1973. Occurrence and identification of *Balanus balanoides* (Crustacea: Cirripedia) in British Columbia. Syesis 6: 97–99.

HENRY, D. P. 1940. The Cirripedia of Puget Sound, with a key to the species. Univ. Wash. Publ. Oceanogr. 4: 1–48.

———. 1942. Studies on the sessile Cirripedia of the Pacific coast of North America. Univ. Wash. Publ. Oceanogr. 4: 95–134.

HENRY, D. P., and P. A. McLAUGHLIN. 1975. The barnacles of the *Balanus amphitrite* complex (Cirripedia, Thoracica). Zool. Verhandl. (Leiden) 141: 1–254.

HINES, A. H. 1978. Reproduction in three species of intertidal barnacles from central California. Biol. Bull. 154: 262–81.

———. 1979. The comparative reproductive ecology of three species of intertidal barnacles. *In* S. E. Stancyk, ed., Reproductive ecology of marine invertebrates, pp. 213–34. Belle W. Baruch Library in Marine Science 9. Columbia: University of South Carolina Press. □This paper and the one above concern *Balanus glandula, Chthamalus fissus,* and *Tetraclita rubescens.*

LEWIS, C. A. 1978. A review of substratum selection in free-living and symbiotic cirripeds. *In* F.-S. Chia and M. E. Rice, eds., Settlement and metamorphosis of marine invertebrate larvae, pp. 207–18. New York: Elsevier.

McLAUGHLIN, P. A., and D. P. HENRY. 1972. Comparative morphology of complemental males in four species of *Balanus* (Cirripedia, Thoracica). Crustaceana 22: 13–30.

NEWMAN, W. A. 1967. On physiology and behaviour of estuarine barnacles. Symp. Crustacea, Proc. Mar. Biol. Assoc. India, Part 3: 1038–66. □Concerns the osmoregulation and behavior of *Balanus glandula, B. improvisus,* and *B. amphitrite,* and their distribution within the San Francisco Bay estuarine system.

———. 1979. Californian transition zone: Significance of short-range endemics. *In* J. Gray and A. J. Boucot, eds., Historical biogeography, plate tectonics, and the changing environment, pp. 399–416. 37th Ann. Biol. Colloq. Corvallis: Oregon State University Press.

NEWMAN, W. A., and A. Ross. 1976. Revision of the balanomorph barnacles; including a catalog of the species. San Diego Natur. Hist. Mus. Mem. 9: 1–108.

PILSBRY, H. A. 1916. The sessile barnacles (Cirripedia) contained in the collections of the U.S. National Museum; including a monograph of the American species. Bull. U.S. Natl. Mus. 93: 1–366.

SOUTHWARD, A. J., and D. J. CRISP. 1965. Activity rhythms of barnacles in relation to respiration and feeding. J. Mar. Biol. Assoc. U.K. 45: 161–85.

STANDING, J. D. 1980. Common inshore barnacle cyprids of the Oregonian Faunal Province (Crustacea: Cirripedia). Proc. Biol. Soc. Wash. 93: 1184–1203.

STRATHMANN, R. R., E. S. BRANSCOMB, and K. VEDDER. 1981. Fatal errors in set as a cost of dispersal and the influence of intertidal flora on set of barnacles. Oecologia 48: 13–18.

Order ACROTHORACICA

TOMLINSON, J. T. 1955. The morphology of an acrothoracican barnacle, *Trypetesa lateralis.* J. Morphol. 96: 97–122. □This small barnacle forms a slitlike burrow in gastropod shells (especially *Tegula funebralis* and *Tegula brunnea*); its range is Point Arena to Point Conception.

———. 1969. The burrowing barnacles (Cirripedia: order Acrothoracica). Bull. U.S. Natl. Mus. 296: 1–162.

Order THORACICA

Balanus glandula Darwin, §§4, 194

BARNES, H., and M. BARNES. 1956. The general biology of *Balanus glandula* Darwin. Pacific Sci. 10: 415–22.

BERGEN, M. 1968. The salinity tolerance limits of the adults and early-stage embryos of *Balanus glandula* Darwin, 1854 (Cirripedia, Thoracica). Crustaceana 15: 229–34.

CONNELL, J. H. 1970. A predator-prey system in the marine intertidal region. I. *Balanus glandula* and several predatory species of *Thais.* Ecol. Monogr. 40: 49–78. □Examines predators, growth, and mortality rates.

WU, R. S. S. 1980. Effects of crowding on the energetics of the barnacle *Balanus glandula.* Can. J. Zool. 58: 559–66.

———. 1981. The effect of aggregation on breeding in the barnacle *Balanus glandula* Darwin. Can. J. Zool. 59: 890–92.

WU, R. S. S., and C. D. LEVINGS. 1979. Energy flow and population dynamics of the barnacle *Balanus glandula.* Mar. Biol. 54: 83–89. □Cites two other papers on the energy budgets of *Balanus glandula* by Wu.

B. nubilus Darwin (often misspelled *B. nubilis* following an error by Pilsbry), §317

BARNES, H., and M. BARNES. 1959. The naupliar stages of *Balanus nubilus* Darwin. Can. J. Zool. 37: 15–23.

HOYLE, G., P. A. McNEILL, and A. I. SELVERSTON. 1973. Ultrastructure of barnacle giant muscle fibers. J. Cell. Biol. 56: 74–91. □*Balanus nubilus* contains the largest individual muscle fibers known to science and is a popular subject for physiological research.

Chthamalus dalli Pilsbry, §194
KORN, O. M., and I. I. OVSYANNIKOVA. 1979. Larval development of the barnacle *Chthamalus dalli.* Biologia Morya. (Vladivost.) 5: 423–30.

C. fissus Darwin, §5
HEDGECOCK, D. 1979. Biochemical genetic variation and evidence of speciation in *Chthamalus* barnacles of the tropical eastern Pacific Ocean. Mar. Biol. 54: 207–14.

Lepas anatifera (Linnaeus), §193
Megabalanus californicus (Pilsbry) (formerly *Balanus tintinnabulum californicus* Pilsbry), §317
Pollicipes polymerus Sowerby (formerly *Mitella*), §160
BARNES, H., and E. S. REESE. 1960. The behaviour of the stalked intertidal barnacle *Pollicipes polymerus* J. B. Sowerby, with special reference to its ecology and distribution. J. Anim. Ecol. 29: 169–85.
CIMBERG, R. L. 1981. Variability in brooding activity in the stalked barnacle *Pollicipes polymerus.* Biol. Bull. 160: 31–42. □Seasonal patterns of brooding revealed two physiological races, distributed north and south of Point Conception. The northern one broods maximally at 14°C or less; the southern one at warmer temperatures (20°C).
FYHN, H. J., J. A. PETERSEN, and K. JOHANSEN. 1972. Eco-physiological studies of an intertidal crustacean, *Pollicipes polymerus* (Cirripedia, Lepadomorpha). I. Tolerance to body temperature change, desiccation and osmotic stress. J. Exp. Biol. 57: 83–102.
HILGARD, G. H. 1960. A study of reproduction in the intertidal barnacle, *Mitella polymerus,* in Monterey Bay, California. Biol. Bull. 119: 169–88.
HOWARD, G. K., and H. C. SCOTT. 1959. Predaceous feeding in two common gooseneck barnacles. Science 129: 717–18. □*Lepas anatifera* and *Pollicipes polymerus.*
LEWIS, C. A. 1975. Development of the gooseneck barnacle *Pollicipes polymerus* (Cirripedia: Lepadomorpha): Fertilization through settlement. Mar. Biol. 32: 141–53.
———. 1981. Juvenile to adult shift in feeding strategies in the pedunculate barnacle *Pollicipes polymerus* (Sowerby) (Cirripedia, Lepadomorpha). Crustaceana 41: 14–20. □In calm water (laboratory tanks), juvenile *Pollicipes* exhibit cirral beating somewhat like balanoid barnacles. In fast currents, juvenile and adult *Pollicipes* simply extend their cirri to capture food. Juveniles consume smaller food particles than adults.
LEWIS, C. A., and F.-S. CHIA. 1981. Growth, fecundity, and reproductive biology in the pedunculate cirripede *Pollicipes polymerus* at San Juan Island, Washington. Can. J. Zool. 59: 893–901.
PETERSEN, J. A., H. J. FYHN, and K. JOHANSEN. 1974. Eco-physiological studies of an intertidal crustacean, *Pollicipes polymerus* (Cirripedia, Lepadomorpha): Aquatic and aerial respiration. J. Exp. Biol. 61: 309–20. □Barnacles in air had a higher respiratory rate than those submerged in water.

Semibalanus cariosus (Pallas) (formerly *Balanus*), §194

Solidobalanus hesperius (Pilsbry) (formerly *Balanus*), §4
Tetraclita rubescens Darwin (formerly *T. squamosa rubescens* Darwin), §168
VILLALOBOS, C. R. 1979. Variations in population structure in the genus *Tetraclita* (Crustacea: Cirripedia) between temperate and tropical populations. I. Fecundity, recruitment, mortality and growth in *T. rubescens.* II. The age structure of *T. rubescens.* Rev. Biol. Trop. 27: 279–91, 293–300.

Order RHIZOCEPHALA

BOSCHMA, H. 1953. The Rhizocephala of the Pacific. Zool. Meded. (Leiden) 32: 185–201. □Cites several earlier works of this author.
HOEG, J. T. 1982. The anatomy and development of the rhizocephalan barnacle *Clistosaccus paguri* Lilljeborg and relation to its host *Pagurus bernhardus* (L.). J. Exp. Mar. Biol. Ecol. 58: 87–125.
REINHARD, E. G. 1942. Studies on the life history and host-parasite relationships of *Peltogaster paguri.* Biol. Bull. 83: 401–15.
———. 1944. Rhizocephalan parasites of hermit crabs from the northwest Pacific. J. Wash. Acad. Sci. 34: 49–58.
———. 1956. Parasitological reviews: Parasitic castration of Crustacea. Parasitology 5: 79–107.
REISCHMAN, P. G. 1959. Rhizocephala of the genus *Peltogasterella* from the coast of the State of Washington to the Bering Sea. Koninkl. Ned. Akad. Wet. Verh., Ser. C 62: 409–35.
RITCHIE, L. E., and J. T. HOEG. 1981. The life history of *Lernaeodiscus porcellanae* (Cirripedia: Rhizocephala) and co-evolution with its porcellanid host. J. Crust. Biol. 1: 334–47. □Found on and in *Petrolisthes cabrilloa* at La Jolla.
YANAGIMACHI, R. 1961. The life-cycle of *Peltogasterella* (Cirripedia, Rhizocephala). Crustaceana 2: 183–86.

Heterosaccus californicus Boschma, on spider crabs, §327
Loxothylacus panopaei (Gissler), on xanthid crabs, §231
Peltogasterella gracilis (Boschma) (probably the *Peltogaster* sp. on hermit crabs in earlier editions), §28
Loxothylacus, Peltogasterella, and other rhizocephalans increase in abundance to the north. They are rare at Monterey, but abundant at Sitka, where *Lophopanopeus* and *Cancer oregonensis* are commonly infected.

Subclass MALACOSTRACA

The great subclass Malacostraca includes most of the large and conspicuous crustaceans of this book and of oceans all over the world. Unlike the branchiopods, copepods, and barnacles, which vary substantially in the number of segments that compose the body, the malacostracans vary remarkably little in basic plan considering the large size of the group. Typically, there is a head, a thorax consisting of eight segments, and an abdomen of six or seven segments plus a terminal telson. Settling on this basic plan seems not to have constrained them much, as the group includes such varied forms as isopods (e.g., sow bugs), amphipods (e.g., beach hoppers), lobsters, true crabs, hermit crabs, and shrimps of various sorts.

Superorder PHYLLOCARIDA

Order LEPTOSTRACA

CANNON, H. G. 1927. On the feeding mechanism of *Nebalia bipes*. Trans. Roy. Soc. Edinb. 55: 355–69.

CLARK, A. E. 1932. *Nebaliella caboti*, n. sp., with observations on other Nebaliacea. Trans. Roy. Soc. Can., Ser. 3, Sec. 5, 26: 217–35.

LaFOLLETTE, R. 1914. A *Nebalia* from Laguna Beach. J. Entom. Zool. 6: 204–6.

MENZIES, R. J., and J. L. MOHR. 1952. The occurrence of the wood-boring crustacean *Limnoria* and the Nebaliacea in Morro Bay, California. Wasmann J. Biol. 10: 81–86.

Nebalia pugettensis (Clark) (formerly *Epinebalia*), §204

Superorder HOPLOCARIDA

Order STOMATOPODA

Caldwell and Dingle (1976), in an interesting, general account, divided the stomatopods into two functional groups, the smashers and the spearers, based on differences in the raptorial forelimbs. The smashers have a smooth dactylus with a blunt heel that is used to strike a crushing blow to hardshelled prey such as clams and oysters. The spearers have sharp spines on the dactylus that are used to impale soft-bodied prey, including fish in some cases. Both species listed below are spearers. A third species, a smasher by the name of *Hemisquilla ensigera californiensis* (Stephenson), occupies subtidal burrows in shallow waters from Point Conception to the Gulf of California.

BURROWS, M. 1969. The mechanics and neural control of the prey capture strike in the mantid shrimps *Squilla* and *Hemisquilla*. Z. Vergl. Physiol. 62: 361–81.

CALDWELL, R. L., and H. DINGLE. 1976. Stomatopods. Sci. Amer. 234: 80–89.

MANNING, R. B. 1967. *Nannosquilla anomala*, a new stomatopod crustacean from California. Proc. Biol. Soc. Wash. 80: 147–50.

———. 1968. A revision of the family Squillidae (Crustacea, Stomatopoda), with the description of eight new genera. Bull. Mar. Sci. 18: 105–42.

———. 1971. Eastern Pacific expeditions of the New York Zoological Society. Stomatopod Crustacea. Zoologica 56: 95–113.

SCHMITT, W. L. 1940. The stomatopods of the west coast of America based on collections made by the Allan Hancock Expeditions, 1933–38. Allan Hancock Pacific Exped. 5: 129–225. □Monographic for the region covered, chiefly in the Panamic area. Includes species occurring in California.

STEPHENSON, W. 1967. A comparison of Australasian and American specimens of *Hemisquilla ensig-*

era (Owen, 1832) (Crustacea: Stomatopoda). Proc. U.S. Natl. Mus. 120(3564): 1–18.

Pseudosquillopsis marmorata (Lockington) (*Pseudosquilla lessonii* (Guèrin) in previous editions), §146

MANNING, R. B. 1969. The postlarvae and juvenile stages of two species of *Pseudosquillopsis* (Crustacea, Stomatopoda) from the Eastern Pacific region. Proc. Biol. Soc. Wash. 82: 525–37.

Schmittius politus (Manning) (*Squilla polita* Bigelow in previous editions), §146

Superorder PERACARIDA

Order MYSIDACEA

BANNER, A. H. 1948–1950. A taxonomic study of the Mysidacea and Euphausiacea (Crustacea) of the northeastern Pacific. Part 1, Trans. Roy. Can. Inst. 26 (1948): 345–99 (includes *Archaeomysis*). Part 2, Trans. Roy. Can. Inst. 27 (1948): 65–125 (includes *Neomysis* and *Acanthomysis*). Part 3, Trans. Roy. Can. Inst. 28 (1950): 1–63 (includes keys).

———. 1954. New records of Mysidacea and Euphausiacea from the northeastern Pacific and adjacent areas. Pacific Sci. 8: 125–39. □Includes *Acanthomysis*, *Archaeomysis*, and *Neomysis*.

CANNON, H. G., and S. M. MANTON. 1927. On the feeding mechanism of a mysid crustacean, *Hemimysis lamornae*. Trans. Roy. Soc. Edinb. 55: 219–53.

MAUCHLINE, J. 1980. The biology of mysids and euphausids. Advances in Marine Biology, 18. New York: Academic Press. 681 pp. □Presented as two separate reviews, one on mysids and one on euphausids. Both are excellent and contain extensive bibliographies.

TATTERSALL, W. M. 1951. A review of the Mysidacea of the United States National Museum. Bull. U.S. Natl. Mus. 201. 292 pp.

Acanthomysis costata (Holmes) (originally *Mysis*), §61

A. sculpta (Tattersall), §61

Archaeomysis maculata (Holmes) (originally *Callomysis*), §187

Mysidopsis californica Tattersall, §314

Order CUMACEA

None of these usually small (less than 4 mm), grotesque, shrimplike crustaceans is treated here, although a few species (e.g., *Cumella vulgaris* Hart) are sometimes common inshore in the region involved.

GLADFELTER, W. B. 1975. Quantitative distribution of shallow-water Cumacea from the vicinity of Dillon Beach, California, with descriptions of five new species. Crustaceana 29: 241–51.

HART, J. F. L. 1930. Some Cumacea of the Vancouver Island region. Contr. Can. Biol. Fish. 6: 1–18.

LIE, U. 1969. Cumacea from Puget Sound and off the northwestern coast of Washington, with descriptions of two new species. Crustaceana 17: 19–30.

WIESER, W. 1956. Factors influencing the choice of substratum in *Cumella vulgaris* Hart (Crustacea, Cumacea). Limnol. Oceanogr. 1: 274–85.

ZIMMER, C. 1936. California Crustacea of the order Cumacea. Proc. U.S. Natl. Mus. 83: 423–39.

Order TANAIDACEA (CHELIFERA)

These small, chelate, isopod-like animals, especially common in mussel beds, are usually treated with the isopods in systematic papers. *Leptochelia* is a genus commonly represented on this coast.

GARDINER, L. F. 1973. New species of the genera *Synapseudes* and *Cycloapseudes* with notes on morphological variation, postmarsupial development, and phylogenetic relationships within the family Metapseudidae (Crustacea: Tanaidacea). Zool. J. Linn. Soc. 53: 25–58.

HIGHSMITH, R. C. 1982. Induced settlement and metamorphosis of sand dollar (*Dendraster excentricus*) larvae in predator-free sites: Adult sand dollar beds. Ecology 63: 329–37. □ *Leptochelia dubia* is a predator on juvenile sand dollars.

———. 1983. Sex reversal and fighting behavior: Coevolved phenomena in a tanaid crustacean. Ecology 64: 719–26.

HOWARD, A. D. 1952. Molluscan shells occupied by tanaids. Nautilus 65: 75–76.

LANG, K. 1957. Tanaidacea from Canada and Alaska. Contr. Dept. Pêche. Quebec No. 52: 1–54.

———. 1961. Further notes on *Pancolus californiensis* Richardson. Ark. Zool. (2) 13: 573–77.

MENDOZA, J. A. 1982. Some aspects of the autecology of *Leptochelia dubia* (Krøyer, 1842) (Tanaidacea). Crustaceana 43: 225–40.

MENZIES, R. J. 1953. The apseudid Chelifera of the eastern tropical and north temperate Pacific Ocean. Bull. Mus. Comp. Zool. Harv. Univ. 107: 443–96.

SIEG, J., and R. N. WINN. 1981. The Tanaidae (Crustacea; Tanaidacea) of California, with a key to the world genera. Proc. Biol. Soc. Wash. 94: 315–43.

See also Hatch (1947), under Isopoda.

Order ISOPODA. Pill bugs, Sow bugs, etc.

BOWMAN, T. E., N. L. BRUCE, and J. D. STANDING. 1981. Recent introduction of the cirolanid isopod crustacean *Cirolana arcuata* into San Francisco Bay. J. Crust. Biol. 1: 545–57. □ *Cirolana arcuata* was introduced from Australia and New Zealand.

BOYLE, P. J., and R. MITCHELL. 1978. Absence of microorganisms in crustacean digestive tracts. Science 200: 1157–59. □ The guts of *Limnoria lignorum* and *L. tripunctata*, as well as those of the terrestrial isopod *Oniscus asellus* and the marine wood-inhabiting amphipod *Chelura terebrans*, are devoid of microorganisms, suggesting that these crustaceans produce their own cellulolytic enzymes.

BRUSCA, G. J. 1966. Studies on the salinity and humidity tolerances of five species of isopods in a transition from marine to terrestrial life. Bull. South. Calif. Acad. Sci. 65: 146–54. □ Concerns *Alloniscus perconvexus, Cirolana harfordi, Idotea wosnesenskii, Ligia occidentalis,* and *Porcellio scaber.* In general, the five species demonstrate differences in structure, behavior, and tolerance appropriate to the shore level occupied.

BRUSCA, R. C., and B. R. WALLERSTEIN. 1977. The marine isopod Crustacea of the Gulf of California. I. Family Idoteidae. Amer. Mus. Novit. 2634: 1–17.

DANFORTH, C. G. 1970. Epicaridea (Crustacea: Isopoda) of North America. Ann Arbor: University Microfilms. 190 pp.

GEORGE, R. Y., and J.-O. STROMBERG. 1968. Some new species and new records of marine isopods from San Juan Archipelago, Washington, U.S.A. Crustaceana 14: 225–54. □ Includes a checklist of isopods recorded from the San Juans.

HATCH, M. H. 1947. The Chelifera and Isopoda of Washington and adjacent regions. Univ. Wash. Publ. Biol. 10: 155–274.

KURIS, A. M., G. O. POINAR, and R. T. HESS. 1980. Post-larval mortality of the endoparasitic isopod castrator *Portunion conformis* (Epicaridea: Entoniscidae) in the shore crab, *Hemigrapsus oregonensis,* with a description of the host response. Parasitology 80: 211–32.

MENZIES, R. J. 1950. The taxonomy, ecology and distribution of northern California isopods of the genus *Idothea* with the description of a new species. Wasmann J. Biol. 8: 155–95. □ *Idotea schmitti* is the new species.

———. 1962. The marine isopod fauna of Bahía de San Quintín, Baja California, Mexico. Pacific Natur. 3: 337–48.

MENZIES, R. J., and M. A. MILLER. 1972. Systematics and zoogeography of the genus *Synidotea* (Crustacea: Isopoda) with an account of Californian species. Smithsonian Contr. Zool. 102: 1–33.

MILLER, M. A. 1938. Comparative ecological studies on the terrestrial isopod Crustacea of the San Francisco Bay region. Univ. Calif. Publ. Zool. 43: 113–42. □ Includes information on *Ligia occidentalis, Ligia pallasii, Alloniscus perconvexus,* and others.

———. 1975. Phylum Arthropoda: Crustacea, Tanaidacea and Isopoda. *In* R. I. Smith and J. T. Carlton, eds., Light's manual, pp. 277–312. Berkeley: University of California Press.

MILLER, M. A., and W. L. LEE. 1970. A new idoteid isopod, *Idotea (Pentidotea) kirchanskii,* from central California (Crustacea). Proc. Biol. Soc. Wash. 82: 789–98.

RICHARDSON, H. 1905. Monograph on the isopods of North America. Bull. U.S. Natl. Mus. 54: 1–727.

SCHULTZ, G. A. 1969. How to know the marine isopod crustaceans. Dubuque, Iowa: Brown. 359 pp.

Alloniscus perconvexus Dana, §184
Argeia pugettensis Dana, on crangonid shrimps, §254
Bopyrina striata Nierstrasz and Brender à Brandis, §230

Cirolana harfordi (Lockington), §21

ENRIGHT, J. T. 1965. Entrainment of a tidal rhythm. Science 147: 864–67. □ The endogenous activity of *Excirolana chiltoni* related to tidal rhythm has been synchronized by simulating the wave action on the beach, which suggests that the stimulus for movement up and down the beach is mechanical in this case.

JOHNSON, W. S. 1976a. Biology and population dynamics of the intertidal isopod *Cirolana harfordi*. Mar. Biol. 36: 343–50.

―――. 1976b. Population energetics of the intertidal isopod *Cirolana harfordi*. Mar. Biol. 36: 351–57.

Colidotea rostrata (Benedict), among spines of urchins, §174

Gnorimosphaeroma oregonensis (Dana), **G. luteum** Menzies, and **G. noblei** Menzies, §198

ERIKSEN, C. H. 1968. Aspects of the limno-ecology of *Corophium spinicorne* Stimpson (Amphipoda) and *Gnorimosphaeroma oregonensis* (Dana) (Isopoda). Crustaceana 14: 1–12.

HOESTLANDT, H. 1973. Étude systématique et génétique de trois espèces pacifiques nord-américaines du genre *Gnorimosphaeroma* Menzies (Isopodes: Flabellifères). I. Considérations générales et systématique. Arch. Zool. Exp. Gén. 114: 349–95.

RIEGEL, J. A. 1959. A revision in the sphaeromid genus *Gnorimosphaeroma* Menzies (Crustacea: Isopoda) on the basis of morphological, physiological and ecological studies on two of its "subspecies." Biol. Bull. 117: 154–62.

STANDING, J. D., and D. D. BEATTY. 1978. Humidity behavior and reception in the sphaeromatid isopod *Gnorimosphaeroma oregonensis* (Dana). Can. J. Zool. 56: 2004–14. □This isopod orients behaviorally in experimental humidity gradients, aggregating in high-humidity areas.

Idotea montereyensis Maloney, §205

Idotea is sometimes spelled *Idothea* in older works.

LEE, W. L. 1966a. Pigmentation of the marine isopod *Idothea montereyensis*. Comp. Biochem. Physiol. 18: 17–36.

―――. 1966b. Color change and the ecology of the marine isopod *Idothea* (*Pentidotea*) *montereyensis* Maloney, 1933. Ecology 47: 930–41.

―――. 1972. Chromatophores and their role in color change in the marine isopod *Idotea montereyensis* (Maloney). J. Exp. Mar. Biol. Ecol. 8: 201–15.

I. urotoma Stimpson (includes *I. rectilineata* Lockington), §§46, 205

I. (Pentidotea) resecata Stimpson, §§205, 278

LEE, W. L., and B. M. GILCHRIST. 1972. Pigmentation, color change and the ecology of the marine isopod *Idotea resecata* (Stimpson, 1857). J. Exp. Mar. Biol. Ecol. 10: 1–27.

MENZIES, R. J., and R. J. WAIDZUNAS. 1948. Postembryonic growth changes in the isopod *Pentidotea resecata* (Stimpson) with remarks on their taxonomic significance. Biol. Bull. 95: 107–13.

I. (P.) schmitti Menzies (formerly *Pentidotea whitei* (Stimpson)), §172

I. (P.) stenops (Benedict), §172

I. (P.) wosnesenskii (Brandt), §165

Ligia occidentalis Dana, §2

ARMITAGE, K. B. 1960. Chromatophore behavior in the isopod *Ligia occidentalis* Dana, 1853. Crustaceana 1: 193–207.

WILSON, W. J. 1970. Osmoregulatory capabilities in isopods: *Ligia occidentalis* and *Ligia pallasii*. Biol. Bull. 138: 96–108.

L. pallasii Brandt, §162

CAREFOOT, T. H. 1973a. Feeding, food preference, and the uptake of food energy by the supralittoral isopod *Ligia pallasii*. Mar. Biol. 18: 228–36.

―――. 1973b. Studies on the growth, reproduction, and life cycle of the supralittoral isopod *Ligia pallasii*. Mar. Biol. 18: 302–11.

―――. 1979. Microhabitat preferences of young *Ligia pallasii* Brandt (Isopoda). Crustaceana 36: 209–14.

Limnoria spp., §329

Three species of wood-boring *Limnoria* occur on the Pacific coast: *L. lignorum* Rathke from Alaska to Point Arena, *L. quadripunctata* Holthuis from Crescent City to La Jolla, and *L. tripunctata* Menzies from San Francisco Bay to Mazatlán, Mexico. The last two species occur in San Francisco Bay; the species of the piling report, at least as illustrated, is *L. quadripunctata*, from the colder waters of the bay; *L. tripunctata* occurs in slightly warmer waters, and is capable of boring into creosoted timbers. All three species are widely distributed elsewhere in the world. In addition to the wood-boring *Limnoria*, one species on our coast (*L. algarum* Menzies) acts as a parasite, boring into the living holdfasts of the large kelps; its range is Cape Arago to Bahía Tortugas, Baja California.

BECKMAN, C., and R. MENZIES. 1960. The relationship of reproductive temperature and the geographical range of the marine wood-borer *Limnoria tripunctata*. Biol. Bull. 118: 9–16.

JOHNSON, M. W., and R. J. MENZIES. 1956. The migratory habits of the marine gribble *Limnoria tripunctata* Menzies in the San Diego harbor, California. Biol. Bull. 110: 54–68.

MENZIES, R. J. 1954. The comparative biology of reproduction in the wood-boring isopod crustacean *Limnoria*. Bull. Mus. Comp. Zool. Harv. Univ. 112: 364–88.

―――. 1957. The marine borer family Limnoriidae (Crustacea, Isopoda). Bull. Mar. Sci. Gulf Caribb. 7: 101–200.

MENZIES, R. J., J. MOHR, and C. M. WAKEMAN. 1963. The seasonal settlement of wood-borers in Los Angeles–Long Beach harbors. Wasmann J. Biol. 21: 97–120.

REISH, D. J., and W. M. HETHERINGTON. 1969. The effects of hyper- and hypo-chlorinities on members of the wood-boring genus *Limnoria*. Mar. Biol. 2: 137–39.

SLEETER, T. D., P. J. BOYLE, A. M. CUNDELL, and R. MITCHELL. 1978. Relationship between marine microorganisms and the wood-boring isopod *Limnoria tripunctata*. Mar. Biol. 45: 329–36.

Sphaeroma quoyana H. Milne-Edwards (*S. pentodon* Richardson in previous editions, and in the first two references below), §336

Sphaeroma quoyana is often accompanied by a tiny commensal isopod, *Iais californica* (Richardson), which crawls about the host's ventral surface.

BARROWS, A. L. 1919. The occurrence of a rock-boring isopod along the shore of San Francisco Bay, California. Univ. Calif. Publ. Zool. 19: 299–316.

RIEGEL, J. A. 1959. Some aspects of osmoregulation in two species of sphaeromid isopod Crustacea. Biol. Bull. 116: 272–84. □Discusses *Gnorimosphaeroma oregonensis, G. luteum* (as subspecies), and *Sphaeroma quoyana* (as *S. pentodon*).

ROTRAMEL, G. 1972. *Iais californica* and *Sphaeroma quoyanum*, two symbiotic isopods introduced to California (Isopoda, Janiridae and Sphaeromatidae). Crustaceana 3 (suppl.): 193–97.

Tylos punctatus Holmes and Gay, §184

HAMNER, W. M., M. SMYTH, and E. D. MULFORD. 1968. Orientation of the sand-beach isopod *Tylos punctatus*. Anim. Behav. 16: 405–9. □The animal can orient to slopes as little as one degree, moving downhill when the sand surface is dry and uphill when it is wet.

———. 1969. The behavior and life history of a sand-beach isopod, *Tylos punctatus*. Ecology 50: 442–53.

HAYES, W. B. 1974. Sand-beach energetics: Importance of the isopod *Tylos punctatus*. Ecology 55: 838–47.

———. 1977. Factors affecting the distribution of *Tylos punctatus* (Isopoda, Oniscoidea) on beaches in southern California and northern Mexico. Pacific Sci. 31: 165–86. □Records population densities of 1,000 to 100,000 animals per meter of beach frontage at five sites between Los Angeles and Punta Banda, Baja California. Distribution on the beach appears related to sand moisture content and not sand coarseness.

HOLANOV, S. H., and J. R. HENDRICKSON. 1980. The relationship of sand moisture to burrowing depth of the sand-beach isopod, *Tylos punctatus* Holmes and Gay. J. Exp. Mar. Biol. Ecol. 46: 81–88. □Suggests that *Tylos punctatus* burrows down into the sand until it reaches sand with at least 1 percent moisture. Most of the animals were buried 5 to 20 cm deep.

Order AMPHIPODA. Beach hoppers, Sand fleas, Skeleton shrimps, etc.

A sweeping revision and compendium of amphipod systematics and distributional ecology has been started by E. L. Bousfield and coworkers at the National Museum of Natural Sciences, Ottawa, Canada. Several new generic names have been proposed already (see the papers below) and more are likely on the way. That the above-mentioned compendium is said to include more than 600 coastal marine species from Alaska to California should be warning enough that our list is but a small sample of the Pacific coast's rich amphipod fauna. But a much longer list would be difficult to justify for a book with our scope and intent. Generic and species placements often depend on fine details in the structure of appendages and other body parts, and most require the assistance of a specialist for any degree of certainty.

BARNARD, J. L. 1952. Some Amphipoda from central California. Wasmann J. Biol. 10: 9–36.

———. 1954. Marine Amphipoda of Oregon. Oregon State Monogr., Stud. Zool. 8: 1–103.

———. 1965. Marine Amphipoda of the family Ampithoidae from southern California. Proc. U.S. Natl. Mus. 118: 1–46.

———. 1969a. Gammaridean Amphipoda of the rocky intertidal of California: Monterey Bay to La Jolla. Bull. U.S. Natl. Mus. 258: 1–230.

———. 1969b. The families and genera of marine gammaridean Amphipoda. Bull. U.S. Natl. Mus. 271: 1–535.

———. 1973. Revision of Corophiidae and related families (Amphipoda). Smithsonian Contr. Zool. 151. 27 pp.

BOUSFIELD, E. L. 1979. The amphipod superfamily Gammaroidea in the northeastern Pacific region: Systematics and distributional ecology. Bull. Biol. Soc. Wash. 3: 297–357.

———. 1982. The amphipod superfamily Talitroidea in the northeastern Pacific region. 1. Family Talitridae: Systematics and distributional ecology. Natl. Mus. Natur. Sci. (Ottawa), Publ. Biol. Oceanogr. 11. 73 pp.

BOUSFIELD, E. L., and J. T. CARLTON. 1967. New records of Talitridae (Crustacea: Amphipoda) from the central California coast. Bull. South. Calif. Acad. Sci. 66: 277–84.

BOWERS, D. E. 1963. Field identification of five species of Californian beach hoppers (Crustacea: Amphipoda). Pacific Sci. 17: 315–20.

CAINE, E. A. 1976. Cleansing mechanism of caprellid amphipods (Crustacea) from North America. Mar. Behav. Physiol. 4: 161–69.

———. 1977. Feeding mechanisms and possible resource partitioning of the Caprellidae (Crustacea: Amphipoda) from Puget Sound, U.S.A. Mar. Biol. 42: 331–36.

———. 1978. Habitat adaptations of North American caprellid Amphipoda (Crustacea). Biol. Bull. 155: 288–96. □Discusses adaptations associated with four habitat categories: cosmopolitan substrata, seastar epibionts, gorgonian and bryozoan epibionts, and free on the bottom.

———. 1979. Population structures of two species of caprellid amphipods (Crustacea). J. Exp. Mar. Biol. Ecol. 40: 103–14.

———. 1980. Ecology of two littoral species of caprellid amphipods (Crustacea) from Washington, U.S.A. Mar. Biol. 56: 327–35. □Concerns two species of caprellid common in the San Juan Islands: *Caprella laeviuscula*, a periphyton scraper/filter-feeder common on the eelgrass *Zostera*, and *Deutella californica*, a predator (eating mostly nematodes and copepods), common on *Obelia* and other hydroids.

CARTER, J. W. 1982. Natural history observations on the gastropod shell-using amphipod *Photis conchicola* Alderman, 1936. J. Crust. Biol. 2: 328–41.

CARTER, J. W., and D. W. BEHRENS. 1980. Gastropod mimicry by another pleustid amphipod in central California. Veliger 22: 376–77. □The amphipod *Pleustes depressa* resembles the gastropod *Alia carinata* and may derive some benefit in the form of reduced predation.

CONLAN, K. E., and E. L. BOUSFIELD. 1982a. The

amphipod superfamily Corophioidea in the northeastern Pacific region. Family Ampithoidae: Systematics and distributional ecology. Natl. Mus. Natur. Sci. (Ottawa), Publ. Biol. Oceanogr. 10: 41–75.

————. 1982b. The superfamily Corophioidea in the North Pacific region. Family Aoridae: Systematics and distributional ecology. Natl. Mus. Natur. Sci. (Ottawa), Publ. Biol. Oceanogr. 10: 77–101.

DICKINSON, J. J. 1982. The systematics and distributional ecology of the family Ampeliscidae (Amphipoda: Gammaridea) in the northeastern Pacific region. I. The genus *Ampelisca*. Natl. Mus. Natur. Sci. (Ottawa), Publ. Biol. Oceanogr. 10: 1–39.

DOUGHERTY, E. C., and J. E. STEINBERG. 1953. Notes on the skeleton shrimps (Crustacea, Caprellidae) of California. Proc. Biol. Soc. Wash. 66: 39–50.

FIELD, L. H. 1974. A description and experimental analysis of Batesian mimicry between a marine gastropod and an amphipod. Pacific Sci. 28: 439–47. □The amphipod *Stenopleustes* sp. apparently mimics species of the gastropod *Lacuna* in appearance and locomotory behavior on eelgrass. Fish will readily eat the amphipod if it is noticed.

HUGHES, J. E. 1982. Life-history of the sandy-beach amphipod *Dogielinotus loquax* (Crustacea: Dogielinotidae) from the outer coast of Washington, U.S.A. Mar. Biol. 71: 167–75.

JARRETT, N. E., and E. L. BOUSFIELD. 1982. Studies on the amphipod family Lysianassidae in the northeastern Pacific region. *Hippomedon* and related genera: Systematics and distributional ecology. Natl. Mus. Natur. Sci. (Ottawa), Publ. Biol. Oceanogr. 10: 103–28.

JOHNSON, S. E. 1973. The ecology of *Oligochinus lighti* J. L. Barnard, 1969, a gammarid amphipod from the high rocky intertidal region of Monterey Bay, California. Ph.D. dissertation, Biological Sciences, Stanford University, Stanford, Calif. 280 pp.

KITRON, V. D. 1980. The pattern of infestation of the beach-hopper amphipod *Orchestoidea corniculata*, by a parasitic mite. Parasitology 81: 235–49. □Near Santa Barbara, the prevalence and intensity of infestation by *Gammaridacarus orchestoideae* were highest in the winter, and females were mitier than males.

LAUBITZ, D. R. 1970. Studies on the Caprellidae (Crustacea, Amphipoda) of the American North Pacific. Natl. Mus. Can. Publ. Biol. Oceanogr. 1: 1–89.

————. 1977. A revision of the genera *Dulichia* Kroyer and *Paradulichia* Boeck (Amphipoda, Podoceridae). Can. J. Zool. 55: 942–82.

MARELLI, D. C. 1981. New records for Caprellidae in California, and notes on a morphological variant of *Caprella verrucosa* Boeck, 1871. Proc. Biol. Soc. Wash. 94: 654–62. □Includes *Caprella scaura*, new to North America (collected in San Francisco Bay and Elkhorn Slough), and extends the known range of *Caprella alaskana* south to San Francisco Bay.

MARTIN, D. M. 1977. A survey of the family Caprellidae (Crustacea, Amphipoda) from selected sites along the northern California coast. Bull. South. Calif. Acad. Sci. 76: 146–67. □Treats 19 species of caprellids, collected in northern California from the Oregon border to Fort Bragg; another 5 species had been recorded from California earlier.

McCAIN, J. C. 1970. Familial taxa within the Caprellidae (Crustacea: Amphipoda). Proc. Biol. Soc. Wash. 82: 837–42.

McCLOSKEY, L. R. 1970. A new species of *Dulichia* (Amphipoda, Podoceridae) commensal with a sea urchin. Pacific Sci. 24: 90–98. □Describes *Dulichia rhabdoplastis*, an apparently obligate commensal, living among the spines of the red sea urchin.

MILLS, E. L. 1961. Amphipod crustaceans of the Pacific coast of Canada, I. Family Atylidae. Bull. Natl. Mus. Canada 172: 13–33.

————. 1962. Amphipod crustaceans of the Pacific coast of Canada, II. Family Oedicerotidae. Natur. Hist. Papers, Natl. Mus. Can. 15: 1–21.

REISH, D. J., and J. L. BARNARD. 1967. The benthic Polychaeta and Amphipoda of Morro Bay, California. Proc. U.S. Natl. Mus. 120 (3565): 1–26.

SHOEMAKER, C. R. 1938. Three new species of the amphipod genus *Ampithoe* from the west coast of America. J. Wash. Acad. Sci. 28: 15–25.

————. 1964. Seven new amphipods from the west coast of North America with notes on some unusual species. Proc. U.S. Natl. Mus. 115: 391–429.

Ampithoe spp., §42

At least nine species of *Ampithoe* and five species of the closely related *Peramphithoe* have been recorded from our coast.

Anisogammarus confervicolus (Stimpson) (also currently cited as *Eogammarus*), §314

LEVINGS, C. D. 1980. The biology and energetics of *Eogammarus confervicolus* (Stimpson) (Amphipoda, Anisogammaridae) at the Squamish River estuary, British Columbia. Can. J. Zool. 58: 1652–63. □The diet of this amphipod consists primarily of benthic algae (e.g., *Enteromorpha*) and plant debris. It is in turn eaten by juvenile salmon. The minimum length of females with eggs is about 7 mm; mature females were most abundant from October to December and from July to August.

Aoroides columbiae Walker, §153

Caprella californica Stimpson, §272

KEITH, D. E. 1969. Aspects of feeding in *Caprella californica* Stimpson and *Caprella equilibra* Say (Amphipoda). Crustaceana 16: 119–24. □These opportunistic omnivores are capable of capturing live prey, consuming floating detritus, and scraping diatoms, dinoflagellates, and protozoans from surfaces.

————. 1971. Substrate selection in caprellid amphipods of southern California, with emphasis on *Caprella californica* Stimpson and *Caprella equilibra* Say (Amphipoda). Pacific Sci. 25: 387–94. □*Caprella californica* prefers the bryozoan *Bugula neritina* over two other natural substrata, the algae *Ulva* and *Polysiphonia*. *Caprella equilibra* was less selective.

C. equilibra Say, §89

There are many species of small caprellids on the Pacific coast. One of them is *C. equilibra*. For others, see Laubitz (1970) and Martin (1977) above.

Elasmopus rapax Costa, §165
In addition, two forms formerly considered subspecies are now commonly elevated to the species level: *Elasmopus mutatus* Barnard and *E. serricatus* Barnard.

Eurystheus thompsoni (Walker) (formerly *E. tenuicornis* (Holmes)), §153
Hyale spp., §153

Megalorchestia benedicti (Shoemaker) (formerly *Orchestoidea*), §154
SHOEMAKER, C. R. 1930. Descriptions of two new amphipod crustaceans (Talitridae) from the United States. J. Wash. Acad. Sci. 20: 107–14. □Describes *Megalorchestia benedicti*.

M. californiana Brandt (formerly *Orchestoidea*), §§154, 185

M. corniculata (Stout) (formerly *Orchestoidea*), §154
BOWERS, D. E. 1964. Natural history of two beachhoppers of the genus *Orchestoidea* (Crustacea: Amphipoda) with reference to their complemental distribution. Ecology 45: 677–96. □Concerns *Megalorchestia corniculata* and *M. californiana*, mostly in central California.
CRAIG, P. C. 1971. An analysis of the concept of lunar orientation in *Orchestoidea corniculata* (Amphipoda). Anim. Behav. 19: 368–74.
———. 1973. Behaviour and distribution of the sand-beach amphipod *Orchestoidea corniculata*. Mar. Biol. 23: 101–9.
ERCOLINI, A., and F. SCAPINI. 1974. Sun compass and shore slope in the orientation of littoral amphipods (*Talitrus saltator* Montagu). Monit. Zool. Ital. 8: 85–115.
SCURLOCK, D. 1975. Infestation of the sandy beach amphipod *Orchestoidea corniculata* by *Gammaridacarus brevisternalis* (Acari: Laelaptidae). Bull. South. Calif. Acad. Sci. 74: 5–9. □The mites, up to 33 of them on a large specimen, are attached to the ventral surface of the hosts, *Megalorchestia corniculata* and *M. californiana*. Mites left hosts within 2 to 9 hours after the amphipod's death, and attached to new, living hosts. Some mites were found in decomposing beach wrack.

Melita sulca (Stout) (formerly listed here as *Melita palmata* (Montagu)), §20
Metacaprella kennerlyi (Stimpson), §89

Polycheria osborni Calman (*Polycheria antarctica* (Stebbing) in previous editions), §100
SKOGSBERG, T., and G. H. VANSELL. 1928. Structure and behavior of the amphipod, *Polycheria osborni*. Proc. Calif. Acad. Sci. (4)17: 267–95.

Traskorchestia sp., §50

T. traskiana (Stimpson) (formerly *Orchestia*), §2
PAGE, H. M. 1979. Relationship between growth, size, molting, and number of antennal segments in *Orchestia traskiana* Stimpson (Amphipoda, Talitridae). Crustaceana 37: 247–52. □This amphipod adds, on average, one segment to the second antennae each time it molts, at least up to 16 segments for males and 13 for females.

Superorder EUCARIDA

Order EUPHAUSIACEA

Thysanoessa gregaria G. O. Sars, §330
According to Banner (Part III, 1950, listed under Mysidacea, p. 532), records for this species from off northern California to British Columbia actually refer to *Thysanoessa longipes* Brandt; and he implied that *T. gregaria* may not occur here at all. See his paper for a consideration of other euphausiids commonly found near shore, especially along the coasts of Washington and British Columbia.
See also Mauchline (1980), under Mysidacea, for a recent general treatment of the group.

Order DECAPODA
Shrimps, Lobsters, Crabs, etc.

BAUER, R. T. 1981. Grooming behavior and morphology in the decapod Crustacea. J. Crust. Biol. 1: 153–73.
BOOLOOTIAN, R. A., A. C. GIESE, A. FARMANFARMAIAN, and J. TUCKER. 1959. Reproductive cycles of five west coast crabs. Physiol. Zool. 32: 213–20. □Concerns three brachyurans (*Hemigrapsus nudus*, *Pachygrapsus crassipes*, and *Pugettia producta*) and two anomurans (*Emerita analoga* and *Petrolisthes cinctipes*) in Monterey Bay.
FEDER, H. M., and A. J. PAUL. 1980. Food of the king crab, *Paralithodes camtschatica*, and the Dungeness crab, *Cancer magister*, in Cook Inlet, Alaska. Proc. Natl. Shellfish. Assoc. 70: 240–46.
GLAESSNER, M. F. 1969. Decapoda. *In* R. C. Moore, ed., Treatise on invertebrate paleontology, part R, Arthropoda 4, vol. 2, pp. R399–533 (and bibliography, pp. R552–66). New York: Geological Society of America; Lawrence: University of Kansas Press. 253 pp.
GROSS, W. J. 1957. An analysis of response to osmotic stress in selected decapod Crustacea. Biol. Bull. 112: 43–62. □*Emerita analoga* and *Pachygrapsus crassipes* are principal subjects.
HAIG, J., and M. K. WICKSTEN. 1975. First records and range extensions of crabs in California waters. Bull. South. Calif. Acad. Sci. 74: 100–104.
HART, J. F. L. 1930. Some decapods from the south-eastern shores of Vancouver Island. Can. Field Natur. 44: 101–9.
———. 1940. Reptant decapod Crustacea of the west coasts of Vancouver and Queen Charlotte Islands, British Columbia. Can. J. Res. D18: 86–105.
———. 1968. Crab-like Anomura and Brachyura (Crustacea: Decapoda) from southeastern Alaska and Prince William Sound. Natl. Mus. Can. Natur. Hist. Pap. 38: 1–6.
———. 1971a. New distribution records of reptant decapod Crustacea, including descriptions of three new species of *Pagurus*, from the waters adjacent to British Columbia. J. Fish. Res. Board Can. 28: 1527–44.
———. 1971b. Key to planktonic larvae of families of decapod Crustacea of British Columbia. Syesis 4: 227–34.
———. 1980. New records and extensions of

range of reptant decapod Crustacea from the north-eastern Pacific Ocean. Can. J. Zool. 58: 767–69.

————. 1982. Crabs and their relatives of British Columbia. British Columbia Prov. Mus., Handbook 40. 267 pp.

HAZLETT, B. A. 1971. Antennule chemosensitivity in marine decapod Crustacea. J. Anim. Morph. Physiol. 18: 1–10.

KNUDSEN, J. W. 1964. Observations of the reproductive cycles and ecology of the common Brachyura and crablike Anomura of Puget Sound, Washington. Pacific Sci. 18: 3–33.

SCHMITT, W. L. 1921. The marine decapod Crustacea of California. Univ. Calif. Publ. Zool. 23: 1–470.

WAY, W. F. 1917. Brachyura and crab-like Anomura of Friday Harbor, Washington. Publ. Puget Sound Biol. Sta. 1: 349–96.

Infraorder CARIDEA. Shrimps

BAUER, R. T. 1978. Antifouling adaptations of caridean shrimps: Cleaning of the antennal flagellum and general body grooming. Mar. Biol. 49: 69–82.

————. 1979. Antifouling adaptations of marine shrimp (Decapoda: Caridea): Gill cleaning mechanisms and grooming of brooded embryos. Zool. J. Linn. Soc. 65: 281–303.

BREED, G. M., and R. E. OLSON. 1977. Biology of the microsporidan parasite *Pleistophora crangoni* n. sp. in three species of crangonid sand shrimps. J. Invert. Pathol. 30: 387–405.

BUTLER, T. H. 1980. Shrimps of the Pacific coast of Canada. Can. Bull. Fish. Aquat. Sci. 202: 1–280.

CHACE, F. A. 1951. The grass shrimps of the genus *Hippolyte* from the west coast of North America. J. Wash. Acad. Sci. 41: 35–39.

COOMBS, E. F., and J. A. ALLEN. 1978. The functional morphology of the feeding appendages and gut of *Hippolyte varians* (Crustacea: Natantia). Zool. J. Linn. Soc. 64: 261–82. ☐A detailed account of appendages and alimentary system, with comparisons between *Hippolyte varians* and *Pandalus danae*.

EDWARDS, R. R. C. 1978. The fishery and fisheries biology of penaeid shrimp on the Pacific coast of Mexico. Oceanogr. Mar. Biol. Ann. Rev. 16: 145–80.

HART, J. F. L. 1964. Shrimps of the genus *Betaeus* on the Pacific coast of North America with descriptions of three new species. Proc. U.S. Natl. Mus. 115: 431–66.

KURIS, A. M., and J. T. CARLTON. 1977. Description of a new species, *Crangon handi*, and new genus, *Lissocrangon*, of crangonid shrimps (Crustacea: Caridea) from the California coast, with notes on adaptation in body shape and coloration. Biol. Bull. 153: 540–59.

PIKE, R. B., and D. I. WILLIAMSON. 1961. The larvae of *Spirontocaris* and related genera (Decapoda: Hippolytidae). Crustaceana 2: 187–208.

SIEGFRIED, C. A. 1982. Trophic relations of *Crangon franciscorum* Stimpson and *Palaemon macrodactylus* Rathbun: Predation on the opossum shrimp, *Neomysis mercedis* Holmes. Hydrobiologia 88: 129–39.

STANDING, J. D. 1981. Occurrences of shrimps (Natantia: Penaeidea and Caridea) in central California and Oregon. Proc. Biol. Soc. Wash. 94: 774–86.

WORD, J. Q., and D. K. CHARWAT. 1976. Invertebrates of southern California coastal waters. Vol. 2. Natantia. El Segundo, Calif.: Southern California Coastal Water Research Project. 238 pp.

Alpheus californiensis Holmes (formerly *Crangon*), §278

KNOWLTON, R. E., and J. M. MOULTON. 1963. Sound production in the snapping shrimps *Alpheus* (*Crangon*) and *Synalpheus*. Biol. Bull. 125: 311–31.

RITZMANN, R. 1973. Snapping behavior of the shrimp *Alpheus californiensis*. Science 181: 459–60.

————. 1974. Mechanisms for the snapping behavior of two alpheid shrimp, *Alpheus californiensis* and *Alpheus heterochelis*. J. Comp. Physiol. 95: 217–36.

SCHEIN, H. 1975. Aspects of the aggressive and sexual behavior of *Alpheus heterochaelis* Say. Mar. Behav. Physiol. 3: 83–96.

————. 1977. The role of snapping in *Alpheus heterochaelis* Say, 1818, the big-clawed snapping shrimp. Crustaceana 33: 182–88. ☐Snapping is an aggressive act, in this case, committed intraspecifically.

A. clamator Lockington (formerly *Crangon dentipes* (Guèrin)), §143

Betaeus ensenadensis Glassell, commensal with *Callianassa*, §292

B. harfordi (Kingsley), commensal with abalone, §75

B. longidactylus Lockington, §46

B. macginitieae Hart, associated with sea urchins, §174

ACHE, B. W. 1975. Antennular mediated host location by symbiotic crustaceans. Mar. Behav. Physiol. 3: 125–30.

ACHE, B. W., and D. DAVENPORT. 1972. The sensory basis of host recognition by symbiotic shrimps, genus *Betaeus*. Biol. Bull. 143: 94–111. ☐ *Betaeus harfordi* and *B. macginitieae* employ chemical cues to locate their hosts, abalones and sea urchins, respectively; the second shrimp may also use visual cues.

B. setosus Hart, §326

Crangon nigricauda Stimpson and **C. nigromaculata** Lockington, similar in appearance and habitat (both formerly species of *Crago*), §254

ISRAEL, H. R. 1936. A contribution toward the life histories of two California shrimps, *Crago franciscorum* (Stimpson) and *Crago nigricauda* (Stimpson). Calif. Dept. Fish Game, Fish Bull. 46: 1–28.

KRYGIER, E. E., and H. F. HORTON. 1975. Distribution, reproduction, and growth of *Crangon nigricauda* and *Crangon franciscorum* in Yaquina Bay, Oregon. Northwest Sci. 49: 216–40. ☐Both shrimps exhibit a tolerance for wide temperature and salinity ranges, but of the two, *Crangon nigricauda* prefers cooler temperatures and higher salinity. The spawning season for both is from December to mid-August. Females live about 1½ years, males less.

Heptacarpus brevirostris (Dana), §150
All species of *Heptacarpus* were listed as *Spirontocaris* in previous editions.

H. camtschatica (Stimpson), §278
H. palpator (Owen), §150
H. paludicola Holmes, §§62, 278
BAUER, R. T. 1979. Sex attraction and recognition in the caridean shrimp *Heptacarpus paludicola* Holmes (Decapoda: Hippolytidae). Mar. Behav. Physiol. 6: 157–74.

H. pictus (Stimpson), §62
BAUER, R. T. 1976. Mating behaviour and spermatophore transfer in the shrimp *Heptacarpus pictus* (Stimpson) (Decapoda: Caridea: Hippolytidae). J. Natur. Hist. 10: 415–40.

———. 1981. Color patterns of the shrimps *Heptacarpus pictus* and *H. paludicola* (Caridea: Hippolytidae). Mar. Biol. 64: 141–52. □In both species there are five basic color morphs: green, banded (with brown and pink bands), striped (with a middorsal longitudinal stripe of variable color), speckled, and transparent. Color is lost at night, and probably serves as camouflage against predatory fish.

———. 1982. Polymorphism of colour pattern in the caridean shrimps *Heptacarpus pictus* and *H. paludicola*. Mar. Behav. Physiol. 8: 249–65.

H. stimpsoni Holthuis (known also as *H. cristatus* (Stimpson)), §278
H. taylori (Stimpson), §150
Hippolyte californiensis Holmes, §§230, 278
H. clarki Chace, §278

Lysmata californica (Stimpson) (formerly *Hippolysmata*), §134
LIMBAUGH, C., H. PEDERSON, and F. A. CHACE, JR. 1961. Shrimps that clean fishes. Bull. Mar. Sci. Gulf Caribb. 11: 237–57.

Palaemon macrodactylus Rathbun, §334
LITTLE, G. 1969. The larval development of the shrimp, *Palaemon macrodactylus* Rathbun, reared in the laboratory, and the effect of eyestalk extirpation on development. Crustaceana 17: 69–87.
NEWMAN, W. A. 1963. On the introduction of an edible oriental shrimp (Caridea, Palaemonidae) to San Francisco Bay. Crustaceana 5: 119–32.
SIEGFRIED, C. A. 1980. Seasonal abundance and distribution of *Crangon franciscorum* and *Palaemon macrodactylus* (Decapoda, Caridea) in the San Francisco Bay–Delta. Biol. Bull. 159: 177–92. □ *Palaemon macrodactylus* seems more tolerant of varied environmental conditions, occurring in the same habitats as *Crangon franciscorum* but also in additional ones. Neither shrimp is common at salinities less than 1 percent.
SITTS, R. M., and A. W. KNIGHT. 1979. Predation by the estuarine shrimps *Crangon franciscorum* Stimpson and *Palaemon macrodactylus* Rathbun. Biol. Bull. 156: 356–68. □Both species are mainly carnivorous. The most frequent prey was the mysid *Neomysis mercedis*.

Pandalus danae Stimpson, and similar, usually deep-water, shrimps, §283
BAUER, R. T. 1975. Grooming behaviour and morphology of the caridean shrimp *Pandalus danae* Stimpson (Decapoda: Natantia: Pandalidae). Zool. J. Linn. Soc. 56: 45–71.
BERKELEY, A. A. 1930. The post-embryonic development of the common pandalids of British Columbia. Contr. Can. Biol. Fish., n.s., 6(6): 79–163.
BUTLER, T. H. 1964. Growth, reproduction, and distribution of pandalid shrimps in British Columbia. J. Fish. Res. Board Can. 21: 1403–52. □Includes *Pandalus danae* and eight other species of pandalids.
RICE, R. L., K. I. McCUMBY, and H. M. FEDER. 1980. Food of *Pandalus borealis*, *Pandalus hypsinotus*, and *Pandalus goniurus* (Pandalidae, Decapoda) from lower Cook Inlet, Alaska. Proc. Natl. Shellfish. Assoc. 70: 47–54.
ROTHLISBERG, P. C. 1980. A complete larval description of *Pandalus jordani* Rathbun (Decapoda, Pandalidae) and its relation to other members of the genus *Pandalus*. Crustaceana 38: 19–48.

Spirontocaris prionota (Stimpson), §150

Infraorder PALINURA. Spiny lobsters

Panulirus interruptus (Randall), §148
BACKUS, J. 1960. Observations on the growth rate of the spiny lobster. Calif. Fish Game 46: 177–81.
LINDBERG, R. G. 1955. Growth, population dynamics, and field behavior in the spiny lobster, *Panulirus interruptus* (Randall). Univ. Calif. Publ. Zool. 59: 157–247.

Infraorder THALASSINIDEA
Ghost shrimps and Mud shrimps

POWELL, R. R. 1974. The functional morphology of the fore-guts of the thalassinid crustaceans, *Callianassa californiensis* and *Upogebia pugettensis*. Univ. Calif. Publ. Zool. 102: 1–41.
STEVENS, B. A. 1928. Callianassidae from the west coast of North America. Publ. Puget Sound Biol. Sta. 6: 315–69.
THOMPSON, L. C., and A. W. PRITCHARD. 1969. Osmoregulatory capacities of *Callianassa* and *Upogebia* (Crustacea: Thalassinidea). Biol. Bull. 136: 114–29.
THOMPSON, R. K., and A. W. PRITCHARD. 1969. Respiratory adaptations of two burrowing crustaceans, *Callianassa californiensis* and *Upogebia pugettensis* (Decapoda, Thalassinidea). Biol. Bull. 136: 274–87.

Callianassa affinis Holmes, §57

C. californiensis Dana, §292
FARLEY, R. D., and J. F. CASE. 1968. Perception of external oxygen by the burrowing shrimp, *Callianassa californiensis* Dana and *C. affinis* Dana. Biol. Bull. 134: 261–65.
HOFFMAN, C. J. 1981. Associations between the arrow goby *Clevelandia ios* (Jordan and Gilbert) and the ghost shrimp *Callianassa californiensis* Dana in natural and artificial burrows. Pacific Sci. 35: 211–16.
JOHNSON, G. E., and J. J. GONOR. 1982. The tidal exchange of *Callianassa californiensis* (Crustacea, Decapoda) larvae between the ocean and the Salmon River Estuary, Oregon. Estuarine Coastal Shelf Sci. 14: 501–16.

MacGinitie, G. E. 1934. The natural history of *Callianassa californiensis* Dana. Amer. Midl. Natur. 15: 166–77.

Torres, J. J., D. L. Gluck, and J. J. Childress. 1977. Activity and physiological significance of the pleopods in the respiration of *Callianassa californiensis* (Dana) (Crustacea: Thalassinidea). Biol. Bull. 152: 134–46. ☐The pleopods of *Callianassa* serve a respiratory function in that they generate water currents within the burrow by beating intermittently; however, their surface does not contribute directly to gas exchange.

C. gigas Dana (= *C. longimana* Stimpson), §313

Upogebia pugettensis (Dana), §313

Hart, J. F. L. 1937. Larval and adult stages of British Columbia Anomura. Can. J. Res. D15: 179–220. ☐Concerns *Upogebia pugettensis* and three species of hermit crabs (none of the latter covered here).

MacGinitie, G. E. 1930. The natural history of the mud shrimp *Upogebia pugettensis* (Dana). Ann. Mag. Natur. Hist (10)6: 36–44.

Infraorder Anomura. Hermit crabs, Porcelain crabs, and Mole crabs

Ball, E. E., Jr. 1968. Activity patterns and retinal pigment migration in *Pagurus* (Decapoda, Paguridea). Crustaceana 14: 302–6. ☐In the laboratory *Pagurus samuelis* and *Pagurus granosimanus* were active at night and inactive (often clustered with other crabs) during the day.

Bollay, M. 1964. Distribution and utilization of gastropod shells by the hermit crabs *Pagurus samuelis*, *Pagurus granosimanus*, and *Pagurus hirsutiusculus* at Pacific Grove, California. Veliger 6 (suppl.): 71–76.

Gonor, S. L., and J. J. Gonor. 1973a. Descriptions of the larvae of four North Pacific Porcellanidae (Crustacea: Anomura). U.S. Natl. Mar. Fish. Serv., Fish. Bull. 71: 189–223. ☐The four are *Petrolisthes cinctipes*, *Petrolisthes eriomerus*, *Pachycheles pubescens*, and *Pachycheles rudis*.

——. 1973b. Feeding, cleaning, and swimming behavior in larval stages of porcellanid crabs (Crustacea: Anomura). U.S. Natl. Mar. Fish. Serv., Fish. Bull. 71: 225–34. ☐The two zoeal stages are predators. At the molt to megalopa, the larva becomes a filter feeder, like the adult.

Haig, J. 1960. The Porcellanidae (Crustacea: Anomura) of the eastern Pacific. Allan Hancock Pacific Exped. 24: 1–440.

Hazlett, B. A. 1981. The behavioral ecology of hermit crabs. Ann. Rev. Ecol. Syst. 12: 1–22.

Johnson, M. W., and W. M. Lewis. 1942. Pelagic larval stages of the sand crabs *Emerita analoga* (Stimpson), *Blepharipoda occidentalis* Randall, and *Lepidopa myops* Stimpson. Biol. Bull. 83: 67–87.

McLaughlin, P. A. 1974. The hermit crabs (Crustacea Decapoda, Paguridea) of northwestern North America. Zool. Verh. (Leiden) 130: 1–396.

Orton, J. H. 1927. On the mode of feeding of the hermit crab, *Eupagurus bernhardus*, and some other Decapoda. J. Mar. Biol. Assoc. U.K. 14: 909–21.

Reese, E. S. 1962. Shell selection behaviour of hermit crabs. Anim. Behav. 10: 347–60. ☐Concerns the behavior and shell preferences of *Pagurus samuelis* and *P. hirsutiusculus*.

——. 1963. The behavioral mechanisms underlying shell selection by hermit crabs. Behaviour 21: 78–126.

Rittschof, D. 1980. Chemical attraction of hermit crabs and other attendants to simulated gastropod predation sites. J. Chem. Ecol. 6: 103–18. ☐Hermit crabs (*Clibanarius vittatus* and *Pagurus longicarpus*) are attracted by chemicals released from a wounded gastropod. As soon as the victim's shell becomes available (e.g., when the predator leaves), a flurry of shell exchanges occurs as the attracted hermits fight over the newly emptied shell.

Snyder-Conn, E. K. 1981. The adaptive significance of clustering in the hermit crab *Clibanarius digueti*. Mar. Behav. Physiol. 8: 43–53. ☐Aggregation seems to improve chances of surviving extreme cold and desiccation, and may enhance resistance to displacement by strong currents or waves.

Spight, T. M. 1977. Availability and use of shells by intertidal hermit crabs. Biol. Bull. 152: 120–33.

Taylor, P. R. 1981. Hermit crab fitness: The effect of shell condition and behavioral adaptations on environmental resistance. J. Exp. Mar. Biol. Ecol. 52: 205–18. ☐The species occurring highest on the shore, *Pagurus samuelis*, aggregated with other individuals and sought refuge in shady spots more often than the mid- to low-intertidal species, *Pagurus granosimanus* and *Pagurus hirsutiusculus*. The low-intertidal to subtidal *Pagurus hemphilli* exhibited these tendencies even less.

Vance, R. R. 1972a. Competition and mechanism of coexistence in three sympatric species of intertidal hermit crabs. Ecology 53: 1062–74. ☐Examines shell preference and the importance of shells as a limiting resource for *Pagurus beringanus*, *Pagurus granosimanus*, and *Pagurus hirsutiusculus* in the San Juan Islands.

——. 1972b. The role of shell adequacy in behavioral interactions involving hermit crabs. Ecology 53: 1075–83.

Wang, D., and D. A. Jillson. 1979. On the relevance of small gastropod shells to competing hermit crab species. Veliger 21: 488–89. ☐A brief reminder that newly metamorphosed crabs, as well as adults, must find suitable shells. Availability of small shells may limit the adult population, although little is known about this.

Blepharipoda occidentalis Randall, §186

Knight, M. D. 1968. The larval development of *Blepharipoda occidentalis* Randall and *B. spinimana* (Philippi) (Decapoda, Albuneidae). Proc. Calif. Acad. Sci. (4) 35: 337–70.

Cryptolithodes sitchensis Brandt, §132

Hart, J. F. L. 1965. Life history and larval development of *Cryptolithodes typicus* Brandt (Decapoda, Anomura) from British Columbia. Crustaceana 8: 255–76. ☐A closely related species.

Odenweller, D. B. 1972. A new range record for the umbrella crab, *Cryptolithodes sitchensis* Brandt. Calif. Fish Game 58: 240–43. ☐Extends the known range southward from Pacific Grove to Point Loma.

Emerita analoga (Stimpson), §186

BARNES, N. B., and A. M. WENNER. 1961. Seasonal variation in the sand crab *Emerita analoga* (Decapoda, Hippidae) in the Santa Barbara area of California. Limnol. Oceanogr. 13: 465–75.

BURTON, R. S. 1979. Depth regulatory behavior of the first stage zoea larvae of the sand crab *Emerita analoga* Stimpson (Decapoda: Hippidae). J. Exp. Mar. Biol. Ecol. 37: 255–70.

COX, G. W., and G. H. DUDLEY. 1968. Seasonal pattern of reproduction of the sand crab, *Emerita analoga*, in southern California. Ecology 49: 746–51.

CUBIT, J. 1969. Behavior and physical factors causing migration and aggregation of the sand crab *Emerita analoga* (Stimpson). Ecology 50: 118–23.

DILLERY, D. G., and L. V. KNAPP. 1970. Longshore movements of the sand crab, *Emerita analoga* (Decapoda, Hippidae). Crustaceana 18: 233–40.

EFFORD, I. E. 1965. Aggregation in the sand crab, *Emerita analoga* (Stimpson). J. Anim. Ecol. 34: 63–75.

———. 1966. Feeding in the sand crab, *Emerita analoga* (Stimpson) (Decapoda, Anomura). Crustaceana 10: 167–82.

———. 1967. Neoteny in sand crabs of the genus *Emerita* (Anomura, Hippidae). Crustaceana 13: 81–93.

———. 1970. Recruitment to sedentary marine populations as exemplified by the sand crab, *Emerita analoga* (Decapoda, Hippidae). Crustaceana 18: 293–308.

FUSARO, C. 1978. Growth rate of the sand crab, *Emerita analoga* (Hippidae), in two different environments. U.S. Natl. Mar. Fish. Serv., Fish. Bull. 76: 369–75. □Crabs at Santa Cruz Island, where the water was colder and food probably less abundant, grew only one-third as fast as crabs at Goleta Bay, only 42 km away.

———. 1980. Temperature and egg production by the sand crab, *Emerita analoga* (Stimpson) (Decapoda, Hippidae). Crustaceana 38: 55–60. □Egg production and development were functions of temperature in the laboratory: in warmer water, a larger proportion of females produced eggs and developmental times were shorter.

KNOX, C., and R. A. BOOLOOTIAN. 1963. Functional morphology of the external appendages of *Emerita analoga*. Bull. South. Calif. Acad. Sci. 62: 45–68.

MACGINITIE, G. E. 1938. Movements and mating habits of the sand crab, *Emerita analoga*. Amer. Midl. Natur. 19: 471–81.

PAUL, D. H. 1979. An endogenous motor program for sand crab uropods. J. Neurobiol. 10: 273–89. □One of several superb papers by D. Paul on the swimming behavior and neural organization of *Emerita*. The author's earlier works are cited in the paper.

PERRY, D. M. 1980. Factors influencing aggregation patterns in the sand crab *Emerita analoga* (Crustacea: Hippidae). Oecologia 45: 379–84.

SNODGRASS, R. E. 1952. The sand crab *Emerita talpoida* (Say) and some of its relatives. Smithsonian Misc. Collec. 117 (8): 1–34. □Presents the structural anatomy, from the viewpoint of functional adaptations, in the various sand crabs, including *Emerita analoga*.

E. rathbunae Schmitt, §186

KNIGHT, M. D. 1967. The larval development of the sand crab *Emerita rathbunae* Schmitt (Decapoda, Hippidae). Pacific Sci. 21: 58–76.

Hapalogaster cavicauda Stimpson, §132

H. mertensii Brandt, §231

MILLER, P. E., and H. G. COFFIN. 1961. A laboratory study of the developmental stages of *Hapalogaster mertensii* (Brandt) (Crustacea, Decapoda). Publ. Dept. Biol. Sci., Walla Walla College 30: 1–18.

Isocheles pilosus (Holmes) (formerly *Holopagurus*), §262

WICKSTEN, M. K. 1979. Digging by the hermit crab *Isocheles pilosus* (Holmes) (Decapoda, Anomura, Diogenidae). Crustaceana Suppl. 5: 100.

Lepidopa californica Efford (formerly *Lepidopa myops* Stimpson), §261

Oedignathus inermis (Stimpson), §132

Pachycheles rudis Stimpson, §318

KNIGHT, M. D. 1966. The larval development of *Polyonyx quadriungulatus* Glassell and *Pachycheles rudis* Stimpson (Decapoda, Porcellanidae) cultured in the laboratory. Crustaceana 10: 75–97.

MACMILLAN, F. E. 1972. The larval development of northern California Porcellanidae (Decapoda, Anomura). I. *Pachycheles pubescens* Holmes in comparison to *Pachycheles rudis* Stimpson. Biol. Bull. 142: 57–70.

Pagurus beringanus (Benedict), §214

P. dalli (Benedict), §225

P. granosimanus (Stimpson), §196

P. hemphilli (Benedict), §28

TAYLOR, P. R. 1979. An association between an amphipod, *Liljeborgia* sp., and the hermit crab, *Pagurus hemphilli* (Benedict). Mar. Behav. Physiol. 6: 185–88.

P. hirsutiusculus (Dana), §201

FITCH, B. M., and E. W. LINDGREN. 1979. Larval development of *Pagurus hirsutiusculus* (Dana) reared in the laboratory. Biol. Bull. 156: 76–92.

GHIRADELLA, H., J. CRONSHAW, and J. CASE. 1968. Fine structure of the aesthetasc hairs of *Pagurus hirsutiusculus* Dana. Protoplasma 66: 1–20.

MESCE, K. A. 1982. Calcium-bearing objects elicit shell selection behavior in a hermit crab. Science 215: 993–95.

WICKSTEN, M. K. 1977. Shells inhabited by *Pagurus hirsutiusculus* (Dana) at Coyote Point Park, San Francisco Bay, California. Veliger 19: 445–46. □Almost all of the shells utilized by hermit crabs at this location were of gastropod species introduced from the Atlantic.

P. samuelis (Stimpson), §12

COFFIN, H. G. 1960. The ovulation, embryology and developmental stages of the hermit crab *Pagurus samuelis* (Stimpson). Publ. Dept. Biol. Sci., Walla Walla College 25: 1–28. □Cites earlier work on *Pagurus samuelis* by this author.

MacMillan, F. E. 1971. The larvae of *Pagurus samuelis* (Decapoda: Anomura) reared in the laboratory. Bull. South. Calif. Acad. Sci. 70: 58–68.

Petrolisthes cinctipes Randall, §18

Glassell, S. A. 1945. Four new species of North American crabs of the genus *Petrolisthes*. J. Wash. Acad. Sci. 35: 223–29. □Describes two species occurring within our range, *Petrolisthes manimaculis* and *Petrolisthes cabrilloa*.

Hartman, H. B., and M. S. Hartman. 1977. The stimulation of filter feeding in the porcelain crab *Petrolisthes cinctipes* Randall by amino acids and sugars. Comp. Biochem. Physiol. 56A: 19–22.

Kropp, R. K. 1981. Additional porcelain crab feeding methods (Decapoda, Porcellanidae). Crustaceana 40: 307–10. □Describes *Petrolisthes cabrilloa* feeding on deposits at La Jolla.

Kurup, N. G. 1964. The intermolt cycle of an anomuran, *Petrolisthes cinctipes* Randall (Crustacea: Decapoda). Biol. Bull. 127: 97–107.

Wicksten, M. K. 1973. Feeding in the porcelain crab, *Petrolisthes cinctipes* (Randall) (Anomura: Porcellanidae). Bull. South. Calif. Acad. Sci. 72: 161–63.

P. eriomerus Stimpson, §231

Pleuroncodes planipes Stimpson, not treated

During years of unusually warm ocean temperatures, swarms of these bright-red, pelagic anomurans may be washed ashore and stranded on California beaches.

Boyd, C. M. 1967. The benthic and pelagic habits of the red crab, *Pleuroncodes planipes*. Pacific Sci. 21: 394–403.

Glynn, P. W. 1961. The first recorded mass stranding of pelagic red crabs, *Pleuroncodes planipes*, at Monterey Bay, California, since 1859, with notes on their biology. Calif. Fish Game 47: 97–101.

Longhurst, A. R., and D. L. R. Seibert. 1971. Breeding in an oceanic population of *Pleuroncodes planipes* (Crustacea, Galatheidae). Pacific Sci. 25: 426–28.

Infraorder Brachyura
True crabs: Shore crabs, Pea crabs, Spider crabs, Edible crabs, Fiddler crabs, etc.

Brown, S. C., S. R. Cassuto, and R. W. Loos. 1979. Biomechanics of chelipeds in some decapod crustaceans J. Zool. (Lond.) 188: 143–59. □Includes *Pachygrapsus crassipes* among the brachyurans studied.

Case, J. 1964. Properties of the dactyl chemoreceptors of *Cancer antennarius* Stimpson and *C. productus* Randall. Biol. Bull. 127: 428–46.

Derby, C. D., and J. Atema. 1980. Induced host odor attraction in the pea crab *Pinnotheres maculatus*. Biol. Bull. 158: 26–33.

Garth, J. S. 1958. Brachyura of the Pacific coast of America: Oxyrhyncha. Allan Hancock Pacific Exped. 21. Part 1, text; Part 2, tables and plates. 854 pp.

Garth, J. S., and W. Stephenson. 1966. Brachyura of the Pacific coast of America. Brachyrhyncha: Portunidae. Allan Hancock Monogr. Mar. Biol. 1: 1–154.

Gross, W. J. 1961. Osmotic tolerance and regulation in crabs from a hypersaline lagoon. Biol. Bull. 121: 290–301. □Both *Hemigrapsus oregonensis* and *Pachygrapsus crassipes* osmoregulate, and were thriving in a lagoon with salinities 175 percent of normal.

———. 1964. Trends in water and salt regulation among aquatic and amphibious crabs. Biol. Bull. 127: 447–66. □Cites many of the author's earlier papers on osmoregulation and salinity tolerances of intertidal crabs.

Hart, J. F. L. 1935. The larval development of British Columbia Brachyura. I. Xanthidae, Pinnotheridae (in part), and Grapsidae. Can. J. Res. D12: 411–32.

Hartnoll, R. G. 1969. Mating in the Brachyura. Crustaceana 16: 161–81.

Hines, A. H. 1982. Coexistence in a kelp forest: Size, population dynamics, and resource partitioning in a guild of spider crabs (Brachyura, Majidae). Ecol. Monogr. 52: 179–98. □Concerns the population dynamics, distribution, size, microhabitats, and foods of five species of spider crabs (*Loxorhynchus crispatus*, *Mimulus foliatus*, *Pugettia producta*, *Pugettia richii*, and *Scyra acutifrons*) in a kelp forest off Hopkins Marine Station.

Kittredge, J. S., M. Terry, and F. T. Takahashi. 1971. Sex pheromone activity of the molting hormone, crustecdysone, on male crabs (*Pachygrapsus crassipes*, *Cancer antennarius*, and *C. anthonyi*). U.S. Natl. Mar. Fish. Serv., Fish. Bull. 69: 337–43.

Knudsen, J. W. 1960a. Aspects of the ecology of the California pebble crabs (Crustacea: Xanthidae). Ecol. Monogr. 30: 165–85.

———. 1960b. Reproduction, life history, and larval ecology of the California Xanthidae, the pebble crabs. Pacific Sci. 14: 3–17.

Mir, R. D. 1961. The external morphology of the first zoeal stages of the crabs, *Cancer magister* Dana, *Cancer antennarius* Stimpson, and *Cancer anthonyi* Rathbun. Calif. Fish Game 47: 103–11.

Nations, J. D. 1975. The genus *Cancer* (Crustacea: Brachyura): Systematics, biogeography, and fossil record. Los Angeles Co. Mus. Natur. Hist., Sci. Bull. 23. 104 pp.

Pearce, J. B. 1962. Adaptation in symbiotic crabs of the family Pinnotheridae. Biologist 45: 11–15.

Rathbun, M. J. 1918. The grapsoid crabs of America. Bull. U.S. Natl. Mus. 97: 1–461.

———. 1925. The spider crabs of America. Bull. U.S. Natl. Mus. 129: 1–613.

———. 1930. The cancroid crabs of America of the families Euryalidae, Portunidae, Atelecyclidae, Cancridae, and Xanthidae. Bull. U.S. Natl. Mus. 152: 1–609.

Schmitt, W. L., J. C. McCain, and E. S. Davidson. 1973. Crustaceorum catalogus editus a H.-E. Gruner et L. B. Holthuis. 3. Decapoda I, Brachyura I, Fam. Pinnotheridae. The Hague: Junk. 160 pp.

Sulkin, S. D. 1975. The influence of light in the depth regulation of crab larvae. Biol. Bull. 148: 333–43.

Warner, G. F. 1977. The biology of crabs. New York: Van Nostrand Reinhold. 202 pp. □A good introductory, general account with a substantial list of references.

WEAR, R. G. 1970. Notes and bibliography on the larvae of xanthid crabs. Pacific Sci. 24: 84–89.

WELLS, W. W. 1940. Ecological studies on the pinnotherid crabs of Puget Sound. Univ. Wash. Publ. Oceanogr. 2: 19–50.

WICKSTEN, M. K. 1980. Decorator crabs. Sci. Amer. 242: 146–54.

WILLASON, S. W. 1981. Factors influencing the distribution and coexistence of *Pachygrapsus crassipes* and *Hemigrapsus oregonensis* (Decapoda: Grapsidae) in a California salt marsh. Mar. Biol. 64: 125–33. □ *Hemigrapsus oregonensis* and *Pachygrapsus crassipes* compete, but coexist, in a tidal marsh near Santa Barbara. Each species has certain advantages. *Hemigrapsus* excavates protective burrows, has greater tolerance of low salinity, and has greater recruitment. On the other hand (claw?), *Pachygrapsus* seems to be the more efficient forager and is not averse to eating the smaller *Hemigrapsus*.

WILLIAMSON, D. I. 1976. Larval characters and the origin of crabs (Crustacea, Decapoda, Brachyura). Thalassia Jugosl. 10: 401–14.

ZIPSER, E., and G. J. VERMEIJ. 1978. Crushing behavior of tropical and temperate crabs. J. Exp. Mar. Biol. Ecol. 31: 155–72. □ Mostly concerns tropical crabs, but *Cancer productus* and *C. oregonensis* are considered.

Cancer antennarius Stimpson, §105

CARROLL, J. C. 1982. Seasonal abundance, size composition, and growth of rock crab, *Cancer antennarius* Stimpson, off central California. J. Crustacean Biol. 2: 549–61.

C. gracilis Dana, not treated

Sometimes common on mud flats and in eelgrass beds. Similar to *Cancer magister*, but with a more rounded and somewhat smaller, smoother carapace. Intertidal, especially from Dillon Beach northward, and offshore in the south.

C. magister Dana, §152

See also the papers by Wickham under *Carcinonemertes errans*, p. 519.

BOTSFORD, L. W., and D. E. WICKHAM. 1978. Behavior of age-specific, density-dependent models and the northern California dungeness crab (*Cancer magister*) fishery. J. Fish. Res. Board Can. 35: 833–43.

BUCHANAN, D. V., and R. E. MILLEMANN. 1969. The prezoeal stage of the Dungeness crab, *Cancer magister* Dana. Biol. Bull. 137: 250–55.

BUTLER, T. H. 1967. A bibliography of the Dungeness crab, *Cancer magister* Dana. Fish. Res. Board Can., Tech. Rep. 1: 1–12.

FISHER, W. S., and D. E. WICKHAM. 1976. Mortalities and epibiotic fouling of eggs from wild populations of the Dungeness crab, *Cancer magister*. U.S. Natl. Mar. Fish. Serv., Fish. Bull. 74: 201–7.

GOTSHALL, D. W. 1977. Stomach contents of northern California Dungeness crabs, *Cancer magister*. Calif. Fish Game 63: 43–51.

———. 1978. Northern California Dungeness crab, *Cancer magister*, movements as shown by tagging. Calif. Fish Game 64: 234–54.

MACKAY, D. C. G. 1942. The Pacific edible crab, *Cancer magister*. Fish. Res. Board Can. Bull. 62: 1–32.

PEARSON, W. H., P. C. SUGARMAN, and D. L. WOODRUFF. 1979. Thresholds for detection and feeding behavior in the dungeness crab, *Cancer magister* (Dana). J. Exp. Mar. Biol. Ecol. 39: 65–78. □ The threshold concentration of clam extract was 10^{-10} g/l.

POOLE, R. L. 1966. A description of laboratory-reared zoeae of *Cancer magister* Dana, and megalopae taken under natural conditions (Decapoda: Brachyura). Crustaceana 11: 83–97.

———. 1967. Preliminary results of the age and growth study of the market crab (*Cancer magister*) in California: The age and growth of *Cancer magister* in Bodega Bay. Mar. Biol. Assoc. India, Symp. on Crustacea, Proc. Pt. II, pp. 553–67.

REED, P. H. 1969. Culture methods and effects of temperature and salinity on survival and growth of Dungeness crab (*Cancer magister*) larvae in the laboratory. J. Fish. Res. Board Can. 26: 389–97.

SNOW, C. D., and J. R. NEILSEN. 1966. Pre-mating and mating behavior of the Dungeness crab (*Cancer magister* Dana). J. Fish. Res. Board Can. 23: 1319–23.

STEVENS, B. G., D. A. ARMSTRONG, and R. CUSIMANO. 1982. Feeding habits of the Dungeness crab *Cancer magister* as determined by the index of relative importance. Mar. Biol. 72: 135–45.

WALDRON, K. D. 1958. The fishery and biology of the Dungeness crab (*Cancer magister* Dana) in Oregon waters. Fish Comm. Oregon, Contr. 24: 1–43.

WICKHAM, D. E. 1979. *Carcinonemertes errans* and the fouling and mortality of eggs of the Dungeness crab, *Cancer magister*. J. Fish. Res. Board Can. 36: 1319–24.

WILD, P. W., and R. N. TASTO, eds. 1983. Life history, environment, and mariculture studies of the Dungeness crab, *Cancer magister*, with emphasis on the central California fishery resource. Calif. Dept. Fish Game, Fish Bull. 172: 1–352.

C. oregonensis (Dana), §231

WICKSTEN, M. K. 1979. Records of *Cancer oregonensis* in California (Brachyura: Cancridae). Calif. Fish Game 65: 118–20.

C. productus Randall, §106

TRASK, T. 1970. A description of laboratory-reared larvae of *Cancer productus* Randall (Decapoda, Brachyura) and a comparison to larvae of *Cancer magister* Dana. Crustaceana 18: 133–46.

Carcinus maenas (Linnaeus) (formerly *Carcinides*), an introduced, Atlantic species, §219

DARE, P. J., and D. B. EDWARDS. 1981. Underwater television observations on the intertidal movements of shore crabs, *Carcinus maenas*, across a mudflat. J. Mar. Biol. Assoc. U.K. 61: 107–16.

Cycloxanthops novemdentatus (Lockington), §133

KNUDSEN, J. W. 1960. Life cycle studies of the Brachyura of western North America. IV. The life cycle of *Cycloxanthops novemdentatus* (Stimpson). Bull. South. Calif. Acad. Sci. 59: 1–8. □ "Stimpson" is an error for "Lockington."

Fabia subquadrata Dana, §159

ANDERSON, G. L. 1975. The effects of intertidal height and the parasitic crustacean *Fabia subquadrata* Dana on the nutrition and reproductive capacity of the California sea mussel *Mytilus californianus* Conrad. Veliger 17: 299–306.

IRVINE, J. A., and H. G. COFFIN. 1960. Laboratory culture and early stages of *Fabia subquadrata* (Dana) (Crustacea, Decapoda). Publ. Dept. Biol. Sci., Walla Walla College 28: 1–24.

PEARCE, J. B. 1966. The biology of the mussel crab, *Fabia subquadrata*, from the waters of the San Juan Archipelago, Washington. Pacific Sci. 20: 3–35.

Hemigrapsus nudus (Dana), §§27, 203

DEHNEL, P. A. 1960. Effect of temperature and salinity on the oxygen consumption of two intertidal crabs. Biol. Bull. 118: 215–49.

DEHNEL, P. A., and T. H. CAREFOOT. 1965. Ion regulation in two species of intertidal crabs. Comp. Biochem. Physiol. 15: 377–97.

DEHNEL, P. A., and D. STONE. 1964. Osmoregulatory role of the antennary gland in two species of estuarine crabs. Biol. Bull. 126: 354–72. □ *Hemigrapsus nudus* and *H. oregonensis*.

JACOBY, C. A. 1981. Behavior of the purple shore crab *Hemigrapsus nudus* Dana, 1851. J. Crust. Biol. 1: 531–44.

NAUGHTON, C. C. 1970. Setae on the chelae of shore crabs, *Hemigrapsus*: Function. Northwest Sci. 44: 253–57. □ The spongy pad of setae on the claws of *Hemigrapsus nudus* and *H. oregonensis* is apparently used to brush and moisten the crab's eyes.

PILTZ, F. M. 1969. A record of the entoniscid parasite, *Portunion conformis* Muscatine (Crustacea: Isopoda) infecting two species of *Hemigrapsus*. Bull. South. Calif. Acad. Sci. 68: 257–59.

SMITH, R. I., and P. R. RUDY. 1972. Water-exchange in the crab *Hemigrapsus nudus* measured by use of deuterium and tritium oxides as tracers. Biol. Bull. 143: 234–46.

TODD, M.-E., and P. A. DEHNEL. 1960. Effect of temperature and salinity on heat tolerance of two grapsoid crabs, *Hemigrapsus nudus* and *Hemigrapsus oregonensis*. Biol. Bull. 118: 150–72.

H. oregonensis (Dana), §285

BATIE, R. E. 1982a. Population structure of the crab *Hemigrapsus oregonensis* (Brachyura, Grapsidae) in Yaquina Bay Estuary, Oregon: 1. Reproductive cycle. Northwest Sci. 56: 20–26. □ The number of ovigerous females was high from February to May (maximum in March).

———. 1982b. Population structure of the crab *Hemigrapsus oregonensis* (Brachyura, Grapsidae) in Yaquina Bay Estuary, Oregon (USA): 2. Vertical distribution, biomass and production. Northwest Sci. 56: 241–49.

EASTON, D. M. 1972. Autotomy of walking legs in the Pacific shore crab *Hemigrapsus oregonensis*. Mar. Behav. Physiol. 1: 209–17.

SYMONS, P. E. K. 1964. Behavioral responses of the crab *Hemigrapsus oregonensis* to temperature, diurnal light variation, and food stimuli. Ecology 45: 580–91.

Heterocrypta occidentalis (Dana), §249

Lophopanopeus bellus (Stimpson), §231

Two subspecies are recognized, differing somewhat in range, morphology, and reproductive season: *Lophopanopeus bellus bellus* (Stimpson) from Resurrection Bay (Alaska) to Cayucos (San Luis Obispo Co.), and *L. bellus diegensis* Rathbun from Monterey Bay to San Diego.

KNUDSEN, J. W. 1959. Life cycle studies of the Brachyura of western North America. II. The life cycle of *Lophopanopeus bellus diegensis* (Rathbun). Bull. South. Calif. Acad. Sci. 58: 57–64.

L. frontalis (Rathbun), §132

L. leucomanus heathii Rathbun, §132

KNUDSEN, J. W. 1958. Life cycle studies of the Brachyura of western North America. I. General culture methods and the life cycle of *Lophopanopeus leucomanus leucomanus* (Lockington). Bull. South. Calif. Acad. Sci. 57: 51–59.

MENZIES, R. J. 1948. A revision of the brachyuran genus *Lophopanopeus*. Allan Hancock Found. Occ. Pap. 4: 1–45.

Loxorhynchus crispatus Stimpson, §131

WICKSTEN, M. K. 1975. Observations on decorating behavior following molting in *Loxorhynchus crispatus* Stimpson (Decapoda, Majidae). Crustaceana 29: 315–16.

———. 1977. Feeding in the decorator crab, *Loxorhynchus crispatus* (Brachyura: Majidae). Calif. Fish Game 63: 122–24.

———. 1978. Attachment of decorating materials in *Loxorhynchus crispatus* (Brachyura: Majidae). Trans. Amer. Microsc. Soc. 97: 217–20.

———. 1979. Decorating behavior in *Loxorhynchus crispatus* Stimpson and *Loxorhynchus grandis* Stimpson (Brachyura, Majidae). Crustaceana 5(suppl.): 37–46.

Malacoplax californiensis (Lockington) (formerly *Speocarcinus*), §285

Mimulus foliatus Stimpson, §132

Opisthopus transversus Rathbun, commensal with *Parastichopus*, keyhole limpets, *Cryptochiton*, etc., §76

See also Webster (1968), under *Cryptochiton*, p. 574, and Wolfson (1974), under *Conus*, p. 558.

BEONDE, A. C. 1968. *Aplysia vaccaria*, a new host for the pinnotherid crab *Opisthopus transversus*. Veliger 10: 375–78.

HOPKINS, T. S., and T. B. SCANLAND. 1964. The host relations of a pinnotherid crab, *Opisthopus transversus* Rathbun (Crustacea: Decapoda). Bull. South. Calif. Acad. Sci. 63: 175–80.

Oregonia gracilis Dana, §334

Pachygrapsus crassipes Randall, §14

BOOLOOTIAN, R. A. 1965. Aspects of reproductive biology of the striped shore crab *Pachygrapsus crassipes*. Bull. South. Calif. Acad. Sci. 64: 43–49.

BOVBJERG, R. V. 1960. Behavioral ecology of the crab, *Pachygrapsus crassipes*. Ecology 41: 668–72.

CHAPIN, D. 1968. Some observations of predation on *Acmaea* species by the crab *Pachygrapsus crassipes*. Veliger 11(suppl.): 67–68.

HIATT, R. W. 1948. The biology of the lined shore crab, *Pachygrapsus crassipes* Randall. Pacific Sci. 2: 135–213.

ROBERTS, J. L. 1957. Thermal acclimation of metabolism in the crab *Pachygrapsus crassipes* Randall. I. The influence of body size, starvation, and molting. II. Mechanisms and the influence of season and latitude. Physiol. Zool. 30: 232–42, 242–55.

SCHLOTTERBECK, R. E. 1976. The larval development of the lined shore crab, *Pachygrapsus crassipes* Randall, 1840 (Decapoda Brachyura, Grapsidae) reared in the laboratory. Crustaceana 30: 184–200.

Parapinnixa affinis Holmes, commensal in tubes of *Terebella californica*, §311

GLASSELL, S. A. 1933. Notes on *Parapinnixa affinis* Holmes and its allies. Trans. San Diego Soc. Natur. Hist. 7: 319–30.

Paraxanthias taylori (Stimpson) (formerly *Xanthias*), §133

KNUDSEN, J. W. 1959. Life cycle studies of the Brachyura of western North America. III. The life cycle of *Paraxanthias taylori* (Stimpson). Bull. South. Calif. Acad. Sci. 58: 138–45.

Pilumnus spinohirsutus (Lockington), §133

Pinnixa barnharti Rathbun, commensal in the sea cucumber *Caudina*, §264

P. eburna Wells, in tubes of *Abarenicola*, §310

P. faba (Dana), commensal in *Tresus*, §300

PEARCE, J. B. 1966. On *Pinnixa faba* and *Pinnixa littoralis* (Decapoda: Pinnotheridae) symbiotic with the clam, *Tresus capax* (Pelecypoda: Mactridae). *In* H. Barnes, ed., Some contemporary studies in marine science, pp. 565–89. London: Allen & Unwin.

ZULLO, V. A., and D. D. CHIVERS. 1969. Pleistocene symbiosis: Pinnotherid crabs and pelecypods from Cape Blanco, Oregon. Veliger 12: 72–73. □ *Pinnixa faba* was found within the valves of *Tresus capax* in a late Pleistocene deposit.

P. franciscana Rathbun, occasionally with *Urechis*; juveniles occasionally with northern terebellid worms, §232

P. littoralis Holmes, with *Tresus, Saxidomus*; juveniles with *Clinocardium*, etc., §300

P. longipes (Lockington), with *Axiothella* in southern California, §293

BOUSQUETTE, G. D. 1980. The larval development of *Pinnixa longipes* (Lockington, 1877) (Brachyura: Pinnotheridae), reared in the laboratory. Biol. Bull. 159: 592–605.

P. schmitti Rathbun, with *Echiurus*, the young with terebellids, §306.

P. tubicola Holmes, with terebellid worms, §53

SCANLAND, T. B., and T. S. HOPKINS. 1978. A supplementary description of *Pinnixa tomentosa* and comparison with the geographically adjacent *Pinnixa tubicola* (Brachyura, Pinnotheridae). Proc. Biol. Soc. Wash. 91: 636–41.

Podochela hemphilli (Lockington), §334

Portunus xantusii (Stimpson), §249

ALLY, J. R. R. 1974. A description of the laboratory-reared first and second zoeae of *Portunus xantusii*

(Stimpson) (Brachyura, Decapoda). Calif. Fish Game 60: 74–78.

Pugettia gracilis Dana, §222

P. producta (Randall) (formerly *Epialtus*), §107

KNUDSEN, J. W. 1964. Observations of the mating process of the spider crab *Pugettia producta* (Majidae, Crustacea). Bull South. Calif. Acad. Sci. 63: 38–41.

MASTRO, E. 1981. Algal preferences for decoration by the Californian kelp crab, *Pugettia producta* (Randall) (Decapoda, Majidae). Crustaceana 41: 64–70. □ *Pugettia producta* juveniles actively decorate using hooks on their rostrum (the hooks, and decorating ability, are lost when the animal molts upon reaching adulthood). In this study, 82 percent of the decorated juvenile crabs collected near Bodega Bay had decorated their rostrum with fragments of red algae. Juveniles kept isolated from food ate their algal decorations, suggesting that decorating behavior in this species may be related to feeding as well as camouflage.

ZIMMER-FAUST, R. K., and J. F. CASE. 1982. Organization of food search in the kelp crab, *Pugettia producta* (Randall). J. Exp. Mar. Biol. Ecol. 57: 237–55.

Pyromaia tuberculata (Lockington) (formerly *Inachoides*), §334

Randallia ornata (Randall), §249

Rhithropanopeus harrisii (Gould), §219

CHAMBERLAIN, N. A. 1962. Ecological studies of the larval development of *Rhithropanopeus harrisii* (Xanthidae, Brachyura). Tech. Rep. Chesapeake Bay Inst. 28: 1–47.

COSTLOW, J. D., JR. 1966. The effect of eyestalk extirpation on larval development of the mud crab, *Rhithropanopeus harrisii* (Gould). Gen. Comp. Endocrinol. 7: 255–74.

COSTLOW, J. D., JR., C. G. BOOKHOUT, and R. J. MONROE. 1966. Studies on the larval development of the crab, *Rhithropanopeus harrisii* (Gould). I. The effect of salinity and temperature on larval development. Physiol. Zool. 39: 81–100.

CRONIN, T. W., and R. B. FORWARD, JR. 1980. The effects of starvation on phototaxis and swimming of larvae of the crab *Rhithropanopeus harrisii*. Biol. Bull. 158: 283–94.

OTT, F. S., and R. B. FORWARD, JR. 1976. The effect of temperature on phototaxis and geotaxis by larvae of the crab *Rhithropanopeus harrisii* (Gould). J. Exp. Mar. Biol. Ecol. 23: 97–107.

SMITH, R. I. 1967. Osmotic regulation and adaptive reduction of water permeability in a brackish-water crab, *Rhithropanopeus harrisii* (Brachyura: Xanthidae). Biol. Bull. 133: 643–58.

WHEELER, D. E., and C. E. EPIFANIO. 1978. Behavioral response to hydrostatic pressure in larvae of two species of xanthid crabs. Mar. Biol. 46: 167–74.

Scleroplax granulata Rathbun, commensal with *Urechis, Upogebia*, and *Callianassa*, §306

Scyra acutifrons Dana, §131

Taliepus nuttallii (Randall) (formerly *Epialtus*), §107

Uca crenulata (Lockington), §284

CHRISTY, J. H. 1982. Burrow structure and use in the sand fiddler crab, *Uca pugilator* (Bosc). Anim. Behav. 30: 687–94. □Two kinds of burrows are constructed in upper-intertidal substrata in Florida: temporary burrows, used for refuge during high tide and at night; and breeding burrows, defended by courting males and the site of mating, oviposition, and egg incubation by females.

CRANE, J. 1957. Basic patterns of display in fiddler crabs (Ocypodidae, genus *Uca*). Zoologica 42: 69–82.

———. 1975. Fiddler crabs of the world—Ocypodidae: Genus *Uca*. Princeton, N.J.: Princeton University Press. 736 pp.

DEMBROWSKI, J. B. 1926. Notes on the behavior of the fiddler crab. Biol. Bull. 50: 179–201.

KATZ, L. C. 1980. Effects of burrowing by the fiddler crab, *Uca pugnax* (Smith). Estuarine Coastal Mar. Sci. 11: 233–37. □Crabs at a density of 42 per square meter can turn over 18 percent of the upper 15 cm of sediment in a year.

VALIELA, I., D. F. BABIEC, W. ATHERTON, S. SEITZINGER, and C. KREBS. 1974. Some consequences of sexual dimorphism: Feeding in male and female fiddler crabs, *Uca pugnax* (Smith). Biol. Bull. 147: 652–60.

Class CHILOPODA. Centipedes

Order GEOPHILOMORPHA

Although the overwhelming majority of centipedes are fundamentally terrestrial, some small ones are strongly halophilic, if not intertidal, and species have been described from many parts of the world. They occur in our region, but are usually overlooked. At least one species, *Nyctunguis heathii*, is common in central California; during the day, individuals hide in high-intertidal crevices, emerging at night during low tide and preying on a variety of invertebrates.

CHAMBERLIN, R. V. 1960. A new marine centipede from the California littoral. Proc. Biol. Soc. Wash. 73: 99–102.

CLOUDSLEY-THOMPSON, J. L. 1948. *Hydroschendyla submarina* (Grube) in Yorkshire: With an historical review of the marine Myriapoda. Naturalist (Lond.), Oct.–Dec. 1948: 149–52.

———. 1951. Supplementary notes on Myriapoda. Naturalist (Lond.), Jan.–Mar. 1951: 16–17. □Attempts a complete bibliography to 1951 of marine centipedes and millipedes.

———. 1968. Spiders, scorpions, centipedes and mites. Oxford: Pergamon. 278 pp.

Class INSECTA (HEXAPODA)

Only a very small percentage of insect species reside in marine environments, but because of the tremendous size of this, the largest class of animals, even a small percentage translates into a substantial number of species. Other than oceanic and coastal water striders (Gerridae: Hemiptera), marine insects are intertidal, and as a group they occupy most of the major littoral habitats. However, they are not as conspicuous as other intertidal organisms owing to their small size, low numbers, or secretive habits, and they have in general been ignored, except by taxonomists. This is unfortunate because the impact of intertidal insects in some cases may be substantial (see, for example, the accounts in Evans 1980; Richards 1982; and Robles and Cubit, 1981).

The following bibliography of intertidal insects was compiled with the very great assistance of W. G. Evans, who of course is not responsible for any errors that may be present.

CHENG, L., ed. 1976. Marine insects. New York: Elsevier/North Holland. 581 pp. □Worldwide coverage of marine and brackish-water insects.

EVANS, W. G. 1968. Some intertidal insects from western Mexico. Pan-Pac. Entomol. 44: 236–41.

———. 1980. Insecta, Chilopoda, and Arachnida: Insects and allies. *In* R. H. Morris, D. P. Abbott, and E. C. Haderlie, Intertidal invertebrates of California, pp. 641–56. Stanford, Calif.: Stanford University Press. □The discussion of marine insects in general is followed by individual accounts of biology and distributions (and color photographs) of 31 species of intertidal insects, 3 arachnids, and 1 centipede. All species discussed are from California, although many are widely distributed on the Pacific coast.

SMITH, R. I., and J. T. CARLTON, eds. 1975. Light's manual: Intertidal invertebrates of the central California coast. 3d ed. Berkeley: University of California Press. 716 pp. □Chapters on marine mites, Collembola, Hemiptera, Diptera, and Coleoptera are very useful to anyone interested in identifying insects and mites collected in the intertidal zone. Keys are provided as well as some important references, but species are not discussed individually.

USINGER, R. L., ed. 1956. Aquatic insects of California. Berkeley: University of California Press. 508 pp.

———. 1957. Marine insects. *In* J. W. Hedgpeth, ed., Treatise on marine ecology and paleoecology, Vol. 1, pp. 1177–82. New York: Geological Society of America Memoir 67. 1,296 pp.

Order COLLEMBOLA
Springtails and Bristletails

SCOTT, D. B., JR. 1956. Aquatic Collembola. *In* R. L. Usinger, ed., Aquatic insects of California, pp. 74–78. Berkeley: University of California Press.

Order THYSANURA. Silverfish

BENEDETTI, R. 1973. Notes on the biology of *Neomachilus halophila* on a California sandy beach (Thysanura: Machilidae). Pan-Pac. Entomol. 49: 246–49.

Order HEMIPTERA. True bugs

STOCK, M. W., and J. D. LATTIN. 1976. Biology of intertidal *Saldula palustris* (Douglas) on the Oregon coast. J. Kans. Entomol. Soc. 49: 313–26.

Order DERMAPTERA. Earwigs

LANGSTON, L. 1974. The maritime earwig in California (Dermaptera: Carcinophoridae). Pan-Pac. Entomol. 50: 28–34.

Order DIPTERA. Flies

Except for algal-feeding (Chironomidae, Tipulidae) and predatory (Dolichopodidae) flies, the forms associated with wrack are the ones most commonly observed in the intertidal zone. Kelp or wrack flies (Coelopidae, Canaceida, Sphaeroceridae, and Anthomyiidae) may be present in dense swarms.

EGGLISHAW, H. J. 1960. Studies on the family Coelopidae (Diptera). Trans. Roy. Entomol. Soc. Lond. 112: 109–40.

HUCKETT, H. C. 1971. The Anthomyiidae of California exclusive of the subfamily Scatophaginae (Diptera). Bull. Calif. Insect Survey 12: 1–121.

KOMPFNER, H. 1974. Larvae and pupae of some wrack dipterans on a California beach (Diptera: Coelopidae, Anthomyiidae, Sphaeroceridae). Pan-Pac. Entomol. 50: 44–52.

ROBLES, C. D., and J. CUBIT. 1981. Influence of biotic factors in an upper intertidal community: Dipteran larvae grazing on algae. Ecology 62: 1536–47.

WIRTH, W. W. 1949. A revision of the clunionine midges with descriptions of a new genus and four new species (Diptera: Tendipedidae). Univ. Calif. Publ. Entomol. 8: 151–82.

———. 1951. A revision of the dipterous family Canaceidae. Occ. Pap. Bernice P. Bishop Mus. 20: 245–75.

———. 1969. The shore flies of the genus *Canaceoides* Cresson (Diptera: Canaceidae). Proc. Calif. Acad. Sci. 36: 551–71.

Order COLEOPTERA. Beetles

This, the largest of the insect orders, also has the most marine species, with representatives of 11 families and more than 25 genera present on the Pacific coast—enough to keep biologists busy for a long time, since, with the exception of *Thinopinus* and *Thalassotrechus*, not much is known about them.

ARNETT, R. H., JR. 1968. The beetles of the United States. Washington, D.C.: Catholic University Press. 1,112 pp.

BLACKWELDER, R. E. 1932. The genus *Endeodes* LeConte (Coleoptera, Melyridae). Pan-Pac. Entomol. 8: 128–36. □ *Endeodes* is the sole genus of intertidal Melyridae.

BLAISDELL, F. E., SR. 1943. Contributions toward a knowledge of the insect fauna of Lower California. 7. Coleoptera: Tenebrionidae. Proc. Calif. Acad. Sci. 24: 171–288.

CRAIG, P. C. 1970. The behaviour and distribution of the intertidal sand beetle, *Thinopinus pictus* (Coleoptera: Staphylinidae). Ecology 51: 1012–17.

EVANS, W. G. 1976. Circadian and circatidal locomotory rhythms in the intertidal beetle *Thalassotre*

chus barbarae (Horn): Carabidae. J. Exp. Mar. Biol. Ecol. 22: 79–90.

———. 1977. Geographic variation, distribution and taxonomic status of the intertidal insect *Thalassotrechus barbarae* (Horn) (Coleoptera: Carabidae). Quaest. Entomol. 13: 83–90.

HATCH, M. 1971. The beetles of the Pacific northwest. 5. Rhipiceroidea, Sternoxi, Phytophaga, Rhynchophora and Lamellicornia. Univ. Wash. Publ. Biol. 16: 1–662.

LEECH, H. B., and H. P. CHANDLER. 1956. Aquatic Coleoptera. In R. L. Usinger, ed., Aquatic insects of California, pp. 293–371. Berkeley: University of California Press.

MOORE, I. 1956. A revision of the Pacific coast Phytosi with a review of the foreign genera (Coleoptera: Staphylinidae). Trans. San Diego Soc. Natur. Hist. 12: 103–52.

MOORE, I., and E. F. LEGNER. 1975a. A catalogue of the Staphylinidae of America north of Mexico (Coleoptera). Univ. Calif. Spec. Publ. 3015. 514 pp.

———. 1975b. Revision of the genus *Endeodes* LeConte with a tabular key to the species (Coleoptera: Melyridae). J. NY Entomol. Soc. 83: 70–81.

ORTH, R. E., I. MOORE, and T. W. FISHER. 1977. Year-round survey of Staphylinidae of a sandy beach in southern California. Wasmann J. Biol. 35: 169–95.

PERKINS, P. D. 1980. Aquatic beetles of the family Hydraenidae in the western hemisphere: Classification, biogeography and inferred phylogeny (Insecta: Coleoptera). Quaest. Entomol. 16: 3–554. □ *Ochthebius vandykei* (Limnebiidae) of Evans (1981) and others is redesignated *Neochthebius vandykei* in the family Hyraenidae.

RICHARDS, L. J. 1982. Prey selection by an intertidal beetle: Field test of an optimal diet model. Oecologia 55: 325–32. □ The staphylinid beetle *Thinopinus pictus* LeConte preys on the amphipod *Megalorchestia californiana*.

SPILMAN, T. J. 1967. The heteromerous intertidal beetles (Coleoptera: Salpingidae: Aegialitinae). Pacific Insects 9: 1–21.

———. 1972. A new genus and species of jumping shore beetle from Mexico (Coleoptera: Limnichidae). Pan-Pac. Entomol. 48: 108–15.

Subphylum CHELICERATA

Modern chelicerates are predominantly terrestrial, although the group has ancestral ties with the sea. These ties are most apparent today in members of the class Merostomata, represented by the horseshoe crabs of the Atlantic coast (not treated here), and the class Pycnogonida, the sea spiders, many of which occur on our coast. The other class, the Arachnida, consists primarily of terrestrial forms, although some of the mites (order Acarina) are truly marine and certain spiders and pseudoscorpions are regular occupants of the high shore.

Class ARACHNIDA
Spiders, Mites, Pseudoscorpions, etc.

FIRSTMAN, B. L. 1973. The relationship of the chelicerate arterial system to the evolution of the endosternite. J. Arachnol. 1: 1–54.

SAVORY, T. 1964. Arachnida. New York: Academic Press. 291 pp. ☐An introductory summary; marine species are discussed.

SNOW, K. R. 1970. The arachnids: An introduction. London: Routledge and Kegan Paul. 84 pp. ☐Brief but concise, informative, and very readable.

Order ACARINA. Mites

Mites found in marine habitats range from high in the intertidal zone to depths of over 5,000 meters (Newell 1975). A common and conspicuous intertidal mite, *Neomolgus littoralis*, is distributed in coastal areas throughout the Northern Hemisphere; individuals are relatively large (2.5–3.5 mm) and bright red in color, and they have snout-like mouthparts.

HALBERT, J. N. 1920. The Acarina of the seashore. Proc. Roy. Irish Acad. 35: 106–52. ☐Some of the nomenclature has changed, but this account is a classic treatment of marine littoral mites, some of which are European, others holarctic. Both habitats and systematics are included for most species.

NEWELL, I. M. 1975. Marine mites (Halacaridae). *In* R. I. Smith and J. T. Carlton, eds., Light's manual, pp. 425–31. Berkeley: University of California Press. ☐A concise account of this little-known group of mites; the collection and preparation of specimens are covered and a key is given to species.

Order CHELONETHIDA. Pseudoscorpions

Several of these quaint creatures, especially of the genus *Garypus*, frequent the high beaches, living among stones. The common beach species on this coast appears to be *Garypus californicus* Banks, found from Ensenada to Monterey, on San Nicolas Island, and also in the Tomales Bay area (especially on Hog Island); the body is about 3–4 mm long, making *G. californicus* fairly conspicuous for a pseudoscorpion. A monster, *Garypus giganteus* Chamberlin, occurs on Baja California beaches, and another species occurs on beaches in the Gulf of California; these animals build little dome-shaped huts of sand grains on the undersurfaces of rocks. Another pseudoscorpion, *Halobisium occidentale* Beier, occurs on *Salicornia* flats and debris-littered flats of bays and estuaries from San Francisco Bay to Alaska, possibly intergrading with *Halobisium orientale* (Redikorzev) on the Siberian coast. The curious "mating waltz" observed in some pseudoscorpions has not been reported for these halophilous forms, but would be well worth confirming. It is a ceremony that would have delighted Ed Ricketts.

CHAMBERLIN, J. C. 1921. Notes on the genus *Garypus* in North America. Can. Entomol. 53: 186–91.

———. 1924. The giant *Garypus* of the Gulf of California. Nature, Sept. 24: 171–72, 175.

———. 1931. The arachnid order Chelonethida. Stanford Univ. Publ. Biol. Sci. 7: 1–284. ☐Covers morphology and phylogeny, with a key to genera and a bibliography. Includes figures of *Garypus californicus* (Fig. 3) and *Halobisium* (Fig. 56).

LEE, V. F. 1979. The maritime pseudoscorpions of Baja California, Mexico (Arachnida: Pseudoscorpionida). Occ. Pap. Calif. Acad. Sci. 131. 38 pp. ☐The evolutionary, systematic, and ecological approach of this work provides a valuable contribution to our knowledge of marine pseudoscorpions.

WEYGOLDT, P. 1969. The biology of pseudoscorpions. Cambridge, Mass.: Harvard University Press. 145 pp. ☐A concise little book, packed with information about pseudoscorpions, including maritime species.

Class PYCNOGONIDA. Sea spiders

COLE, L. J. 1904. Pycnogonida of the west coast of North America. Harriman Alaska Exped. 10: 249–98. ☐Still an essential reference.

FRY, W. G., ed. 1978. Sea spiders (Pycnogonida). Proceedings of a meeting held in honor of Joel W. Hedgpeth on 7 October 1976 in the rooms of the Linnean Society of London. Zool. J. Linn. Soc. 63(1 and 2): 1–238. ☐Several excellent, modern papers. Includes a bibliography of the pycnogonids (pp. 197–238).

HEDGPETH, J. W. 1941. A key to the Pycnogonida of the Pacific coast of North America. Trans. San Diego Soc. Natur. Hist. 9: 253–64.

———. 1949. Report on the Pycnogonida collected by the *Albatross* in Japanese waters in 1900 and 1906. Proc. U.S. Natl. Mus. 98: 233–321.

———. 1951. Pycnogonids from Dillon Beach and vicinity, with descriptions of two new species. Wasmann J. Biol. 9: 105–17.

———. 1954. On the phylogeny of the Pycnogonida. Acta Zool. 35: 193–213.

KING, P. E. 1973. Pycnogonids. New York: St. Martins Press. 144 pp. ☐A detailed general account.

SCHMITT, W. L. 1934. Notes on certain pycnogonids, including descriptions of two new species of *Pycnogonum*. J. Wash. Acad. Sci. 24: 61–70.

ZIEGLER, A. C. 1960. Annotated list of Pycnogonida collected near Bolinas, California. Veliger 3: 19–22.

Achelia chelata (Hilton), §159

BAMBER, R. N., and M. H. DAVIS. 1982. Feeding of *Achelia echinata* Hodge (Pycnogonida) on marine algae. J. Exp. Mar. Biol. Ecol. 60: 181–87. ☐Other reports suggest that this Atlantic pycnogonid eats hydroids and bryozoans, but here it was found eating algae (e.g., *Enteromorpha*).

BENSON, P. H., and D. C. CHIVERS. 1960. A pycnogonid infestation of *Mytilus californianus*. Veliger 3: 16–18.

A. nudiuscula (Hall), §332

Ammothea hilgendorfi (Böhm) (formerly *Lecythorhynchus marginatus* Cole), §90

KRAPP, F., and R. SCONFIETTI. 1983. *Ammothea hilgendorfi* (Böhm, 1879), an adventitious pycnogonid new for the Mediterranean Sea. Mar. Ecol. 4: 123–32.

Ammothella biunguiculata (Dohrn), §143

A. tuberculata Cole, §153

Anoplodactylus erectus Cole, §333
HILTON, W. A. 1916. Life history of *Anoplodactylus erectus* Cole. J. Entomol. Zool. 8: 25–34.

Halosoma viridintestinale Cole, §316

Phoxichilidium femoratum (Rathke), §90

Pycnogonum rickettsi Schmitt, §319

P. stearnsi Ives, §319
FRY, W. G. 1965. The feeding mechanisms and preferred foods of three species of Pycnogonida. Bull Brit. Mus. (Natur. Hist.), Zool. 12: 195–223. □Includes *Pycnogonum stearnsi.*

Rhynchothorax philopsammum Hedgpeth, not treated
A curious burrowing form found in coarse sand at Tomales Bluff (Hedgpeth 1951).

Tanystylum anthomasti Hedgpeth, §173

T. californicum Hilton, §90

Phylum MOLLUSCA

The term mollusk, which is derived from Aristotle's word for the cuttlefish, is considered to mean soft-bodied. Linnaeus grossly misunderstood this phylum, and he lumped several diverse groups, including echinoderms, with his Mollusca. Our modern concept of the phylum dates from Cuvier, late in the 1700's. All these naturalists understood the Mollusca as essentially soft-bodied animals, many of which produced external shells. They are indeed animals in which the processes of evolution are manifested chiefly by the rearrangements and specializations of their soft parts; they are "visceral beings." Those students who are aware of the true nature of mollusks and who devote most of their time to studying the soft anatomy, especially in relation to function, are termed malacologists. Students for whom the principal appeal of mollusks is their shells are known as conchologists.

Since many more have responded to the aesthetic and geometrical appeal of seashells than to the zoological importance of guts and gizzards, many well-meaning souls have managed to confuse the nomenclature by inserting inadequately described species and misidentifications. Almost daily, some name has to be changed. Sometimes this is because of a legitimate reappraisal of zoological affinities that requires assignment to a different genus; but perhaps just as often it is because too many cooks have stirred the brew. This business is not, of course, a peculiarity of molluscan names; but because more people are interested in seashells than in crabs or worms, the name changes and inadequate descriptions are more apparent in this group.

Mollusca, in any event, is a large and important phylum. As important components of the intertidal biota, mollusks increasingly have attracted the interest of ecologists and students of behavior. They have also attracted the attention of experimentalists interested in convenient animals for laboratory studies in physiology and cell biology. Keeping up with all of the resulting literature is a challenge, to say the least; entire books can be overlooked, in spite of this computerized age. A comprehensive bibliography of Pacific coast mollusks is simply beyond our scope, and the references provided below are only a sample. Nevertheless, they should serve as a good introduction, and many provide useful bibliographies for further study.

ABBOTT, R. T. 1974. American seashells. 2d ed. New York: Van Nostrand Reinhold. 663 pp.

FLORKIN, M., and B. T. SCHEER, eds. 1972. Chemical zoology. Vol. 7. Mollusca. New York: Academic Press. 567 pp. □Several good chapters, but see especially "The molluscan framework," by C. R. Stasek, pp. 1–44.

HANNA, G. D. 1966. Introduced mollusks of western North America. Occ. Pap. Calif. Acad. Sci. 48. 108 pp.

KEEN, A. M. 1971. Seashells of tropical west America. 2d ed. Stanford, Calif.: Stanford University Press. 1,064 pp.

KEEN, A. M., and E. COAN. 1974. Marine molluscan genera of western North America: An illustrated key. Stanford, Calif.: Stanford University Press. 208 pp.

MACDONALD, K. B. 1969. Quantitative studies of salt marsh mollusc faunas from the North American Pacific coast. Ecol. Monogr. 39: 33–60. □Concerns gastropods and bivalves found in salt marshes from Gray Harbor (Oregon) to Black Warrior Lagoon (Baja California).

McLEAN, J. H. 1978. Marine shells of southern California. Rev. ed. Los Angeles Co. Mus. Natur. Hist., Sci. Ser. 24. 104 pp.

MORTON, J. E. 1967. Molluscs. 4th ed. London: Hutchinson University Library. 244 pp.

PURCHON, R. D. 1977. The biology of the Mollusca. 2d ed. New York: Pergamon. 560 pp.

SMITH, A. G., and M. GORDON, JR. 1948. The marine mollusks and brachiopods of Monterey Bay, California, and vicinity. Proc. Calif. Acad. Sci. (4) 26: 147–245.

WICKSTEN, M. K. 1978. Checklist of marine mollusks at Coyote Point Park, San Francisco Bay, California. Veliger 21: 127–30.

WILBUR, K. M., and C. M. YONGE, eds. 1964, 1966. Physiology of Mollusca. Vol. 1 (1964), 473 pp.; Vol. 2 (1966), 645 pp. New York: Academic Press.

YONGE, C. M. 1947. The pallial organs in the aspidobranch Gastropoda and their evolution throughout the Mollusca. Phil. Trans. Roy. Soc. Lond. B232: 443–518. □A splendid paper by a master zoologist; a classic.

YONGE, C. M., and T. E. THOMPSON. 1976. Living marine molluscs. London: Collins. 288 pp.

Class GASTROPODA. Snails, Limpets, Sea hares, Nudibranchs, etc.

By far the largest of the molluscan classes, with more than 80 percent of the species, the gastropods have established themselves in all littoral habitats. Their only rivals in terms of diversity are the crustaceans. Gastropods are united as a group by one unique feature, called torsion, which is a rotation of the visceral mass and shell 180° with respect to the head-foot axis, so that the original hind end comes to lie over the head. The precise functional advantages of torsion, which takes place in the veliger larval stage, have been the subject of much debate (see, for example, Ghiselin 1966); but it seems one must assume that the advantages, whatever they are, must be substantial. After all, can tens of thousands of snails all be wrong?

DENNY, M. 1980. Locomotion: The cost of gastropod crawling. Science 208: 1288–90. □The energetic cost of gastropod locomotion, measured here for a terrestrial slug, is high compared with other forms of locomotion, owing primarily to the cost of producing mucus.

FARMER, W. M. 1970. Swimming gastropods (Opisthobranchia and Prosobranchia). Veliger 13: 73–89.

GHISELIN, M. T. 1966. The adaptive significance of gastropod torsion. Evolution 20: 337–48.

GRENON, J.-F., and G. WALKER. 1981. The tenacity of the limpet, *Patella vulgata* L.: An experimental approach. J. Exp. Mar. Biol. Ecol. 54: 277–308.

GRIFFITH, L. M. 1967. The intertidal univalves of British Columbia. B.C. Prov. Mus. Natur. Hist. Handb. 26. 101 pp.

KITTING, C. L. 1979. The use of feeding noises to determine the algal foods being consumed by individual intertidal molluscs. Oecologia 40: 1–17.

MILLER, S. L. 1974a. Adaptive design of locomotion and foot form in prosobranch gastropods. J. Exp. Mar. Biol. Ecol. 14: 99–156.

———. 1974b. The classification, taxonomic distribution, and evolution of locomotor types among prosobranch gastropods. Proc. Malacol. Soc. Lond. 41: 233–72.

MORRIS, T. E., and C. S. HICKMAN. 1981. A method for artificially protracting gastropod radulae and a new model of radula function. Veliger 24: 85–90.

PALMER, A. R. 1980. Locomotion rates and shell form in the Gastropoda: A reevaluation. Malacologia 19: 289–96.

SOLEM, A. 1972. Malacological applications of scanning electron microscopy.—II. Radular structure and functioning. Veliger 14: 327–36.

THOMPSON, T. E. 1969. Acid secretion in Pacific Ocean gastropods. Austral. J. Zool. 17: 755–64. □Acid secretion, probably a defensive adaptation, is recorded for 25 species of Pacific Ocean gastropods (16 prosobranchs and 9 opisthobranchs).

UNDERWOOD, A. J. 1979. The ecology of intertidal gastropods. Adv. Mar. Biol. 16: 111–210. □A comprehensive review with extensive bibliography.

VERMEIJ, G. J. 1972. Intraspecific shore-level size gradients in intertidal molluscs. Ecology 53: 693–700.

———. 1973. Morphological patterns in high-intertidal gastropods: Adaptive strategies and their limitations. Mar. Biol. 20: 319–46.

———. 1975. Marine faunal dominance and molluscan shell form. Evolution 28: 656–64.

Subclass PROSOBRANCHIA

BISHOP, M. J., and S. J. BISHOP. 1973. A census of marine prosobranch gastropods at San Diego, California. Veliger 16: 143–52.

CARRIKER, M. R., and L. G. WILLIAMS. 1978. The chemical mechanism of shell dissolution by predatory boring gastropods: A review and an hypothesis. Malacologia 17: 143–56.

DUERR, F. G. 1965. Survey of digenetic trematode parasitism in some prosobranch gastropods of the Cape Arago region, Oregon. Veliger 8: 42. □Finds substantial levels of parasitism in *Littorina scutulata* and *Olivella biplicata*, but not in four other species of prosobranchs.

FEDER, H. M. 1963. Gastropod defensive responses and their effectiveness in reducing predation by starfishes. Ecology 44: 505–12.

FRETTER, V., and A. GRAHAM. 1962. British prosobranch molluscs. Their functional anatomy and ecology. London: Ray Society. 755 pp. □Still a monumental, exemplary, and indispensable work.

Order ARCHAEOGASTROPODA
Limpets, Abalones, Turbans

Superfamily PLEUROTOMARIACEA. Abalones

BOOLOOTIAN, R. A., A. FARMANFARMAIAN, and A. C. GIESE. 1962. On the reproductive cycle and breeding habits of two western species of *Haliotis*. Biol. Bull. 122: 183–93. □*Haliotis cracherodii* and H. *rufescens*.

COOPER, J., M. WIELAND, and A. HINES. 1977. Subtidal abalone populations in an area inhabited by sea otters. Veliger 20: 163–67.

COX, K. W. 1962. California abalones, Family Haliotidae. Calif. Dept. Fish Game, Fish Bull. 118: 1–133. □The standard reference for abalone fanciers.

CROFTS, D. R. 1929. *Haliotis*. Liverpool Mar. Biol. Comm. [LMBC] Memoir 29. University Press of Liverpool. 174 pp.

HANSEN, J. C. 1970. Commensal activity as a function of age in two species of California abalones (Mollusca: Gastropoda). Veliger 13: 90–94. □Describes the incidence of the sponge *Cliona celata* and the boring clam *Penitella conradi* in shells of *Haliotis cracherodii* and H. *rufescens*.

LEIGHTON, D. L. 1974. The influence of temperature on larval and juvenile growth in three species of southern California abalones. U.S. Natl. Mar. Fish. Serv., Fish. Bull. 72: 1137–45. □Concerns *Haliotis fulgens*, H. *rufescens*, and H. *corrugata*.

MINCHIN, D. 1975. The righting response in haliotids. Veliger 17: 249–50.

MORSE, D. E., H. DUNCAN, N. HOOKER, and A. MORSE. 1977. Hydrogen peroxide induces spawning in mollusks, with activation of prostaglandin endoperoxide synthetase. Science 196: 298–300.

MORSE, D. E., N. HOOKER, H. DUNCAN, and L. JENSEN. 1979. γ-aminobutyric acid, a neurotransmitter, induces planktonic abalone larvae to settle and begin metamorphosis. Science 204: 407–10.

MOTTET, M. G. 1978. A review of the fishery biology of abalones. Wash. Dept. Fish. Tech. Rep. 37. 81 pp.

OWEN, B., J. H. McLEAN, and R. J. MEYER. 1971. Hybridization in the eastern Pacific abalones (*Haliotis*). Los Angeles Co. Mus. Natur. Hist., Sci. Bull. 9. 37 pp. □Hybrids between several species, including *Haliotis rufescens*, occur naturally (approximately 2 per 1,000 abalones caught commercially in southern California), and may be produced in the laboratory.

VOLTZOW, J. 1983. Flow through and around the abalone *Haliotis kamtschatkana*. Veliger 26: 18–21. □The shell design of this, and presumably other, abalone may enable the animal to take advantage of an induced flow to move water more efficiently through its mantle cavity. In an ambient water flow, water entered the shell of a *dead* abalone at a region to the left of the left cephalic tentacle and also through the one or two anteriormost shell openings; it exited through the two or three posteriormost openings.

Haliotis cracherodii Leach, black abalone, §179

HINES, A., S. ANDERSON, and M. BRISBIN. 1980. Heat tolerance in the black abalone, *Haliotis cracherodii* Leach, 1814: Effects of temperature fluctuation and acclimation. Veliger 23: 113–18.

LEIGHTON, D., and R. A. BOOLOOTIAN. 1963. Diet and growth in the black abalone, *Haliotis cracherodii*. Ecology 44: 227–38.

WEBBER, H. H., and A. C. GIESE. 1969. Reproductive cycle and gametogenesis in the black abalone *Haliotis cracherodii* (Gastropoda: Prosobranchiata). Mar. Biol. 4: 152–59.

WRIGHT, M. B. 1975. Growth in the black abalone, *Haliotis cracherodii*. Veliger 18: 194–99.

H. fulgens Philippi, green abalone, §75

TUTSCHULTE, T., and J. H. CONNELL. 1981. Reproductive biology of three species of abalones (*Haliotis*) in southern California, U.S.A. Veliger 23: 195–206.

H. kamtschatkana Jonas, pinto abalone, §74

BREEN, P. A., and B. E. ADKINS. 1980. Spawning in a British Columbia population of northern abalone, *Haliotis kamtschatkana*. Veliger 23: 177–79.

PAUL, A. J., and J. M. PAUL. 1981. Temperature and growth of maturing *Haliotis kamtschatkana* Jonas. Veliger 23: 321–24.

PAUL, A. J., J. M. PAUL, D. W. HOOD, and R. A. NEVE. 1977. Observations on food preferences, daily ration requirements and growth of *Haliotis kamtschatkana* Jonas in captivity. Veliger 19: 303–9.

QUAYLE, D. B. 1971. Growth, morphometry and breeding in the British Columbia abalone (*Haliotis kamtschatkana* Jonas). Fish. Res. Board Can. Tech. Rep. 279. 84 pp.

H. rufescens Swainson, red abalone, §74

GIORGI, A. E., and J. D. DeMARTINI. 1977. A study of the reproductive biology of the red abalone, *Haliotis rufescens* Swainson, near Mendocino, California. Calif. Fish Game 63: 80–94.

LEIGHTON, D. L. 1961. Observations on the effect of diet on shell coloration in the red abalone, *Haliotis rufescens* Swainson. Veliger 4: 29–32.

MONTGOMERY, D. H. 1967. Responses of two haliotid gastropods (Mollusca), *Haliotis assimilis* and *Haliotis rufescens*, to the forcipulate asteroids (Echinodermata), *Pycnopodia helianthoides* and *Pisaster ochraceus*. Veliger 9: 359–68.

OLSEN, D. A. 1968. Banding patterns in *Haliotis*. II. Some behavioral considerations and the effect of diet on shell coloration for *Haliotis rufescens*, *Haliotis corrugata*, *Haliotis sorenseni*, and *Haliotis assimilis*. Veliger 11: 135–39.

YOUNG, J. S., and J. D. DeMARTINI. 1970. The reproductive cycle, gonadal histology, and gametogenesis of the red abalone *Haliotis rufescens* (Swainson). Calif. Fish Game 56: 298–309.

Superfamily FISSURELLACEA
Keyhole limpets

Diodora aspera (Rathke), rough keyhole limpet, §181

DIMOCK, R. V., JR., and J. G. DIMOCK. 1969. A possible "defense" response in a commensal polychaete. Veliger 12: 65–68.

MARGOLIN, A. S. 1964. The mantle response of *Diodora aspera*. Anim. Behav. 12: 187–94.

MURDOCK, G. R., and S. VOGEL. 1978. Hydrodynamic induction of water flow through a keyhole limpet (Gastropoda, Fissurellidae). Comp. Biochem. Physiol. 61A: 227–31. □Because of the elevated "keyhole," an ambient water current can induce flow through the mantle cavity and out the hole; similar engineering, with similar results, is seen in sponges, prairie-dog burrows, and fireplace chimneys.

Fissurella volcano Reeve, §15

Megathura crenulata (Sowerby), giant keyhole limpet, §180

ILLINGWORTH, J. F. 1902. The anatomy of *Lucapina crenulata* Gray. Zool. Jahrb. (Anat. Ontog.) 16: 449–80. □Exquisite illustrations.

Superfamily PATELLACEA. True limpets

ABBOTT, D. P., D. EPEL, J. H. PHILLIPS, I. A. ABBOTT, and R. STOHLER, eds. 1968. The biology of *Acmaea*. Veliger 11(suppl.): 1–112. □A collection of papers reporting research performed by undergraduate students in a class at Hopkins Marine Station in the spring of 1966. Several of the papers are cited individually below; for the many that are not, the reader is referred to the complete volume.

BULKLEY, P. T. 1968. Shell damage and repair in five members of the genus *Acmaea*. Veliger 11 (suppl.): 64–66.

CHAPIN, D. 1968. Some observations of predation on *Acmaea* species by the crab *Pachygrapsus crassipes*. Veliger 11(suppl.): 67–68.

CHOAT, J. H. 1977. The influence of sessile organisms on the population biology of three species of acmaeid limpets. J. Exp. Mar. Biol. Ecol. 26: 1–26. □*Collisella digitalis* and *C. paradigitalis* do not flourish in the presence of barnacles (*Chthamalus dalli* and *Balanus glandula*), whereas *Collisella scabra* does. Also, competition between *Collisella digitalis* and *Collisella paradigitalis* apparently influences the upper limit of the latter.

CHOAT, J. H., and R. BLACK. 1979. Life histories of limpets and the limpet-laminarian relationship.

J. Exp. Mar. Biol. Ecol. 41: 25–50. □Concerns the contrasting life histories of *Collisella digitalis* and *Notoacmea insessa*.

DORAN, S. R., and D. S. McKENZIE. 1972. Aerial and aquatic respiratory responses to temperature variations in *Acmaea digitalis* and *Acmaea fenestrata*. Veliger 15: 38–42.

FRANK, P. W. 1965. Growth of three species of *Acmaea*. Veliger 7: 201–2. □The three are *Collisella digitalis*, *C. pelta*, and *C. paradigitalis*.

FRITCHMAN, H. K., III. 1961, 1962. A study of the reproductive cycle in the California Acmaeidae (Gastropoda). Part I, Veliger 3 (1961): 57–63; Part II, Veliger 3 (1961): 95–101; Part III, Veliger 4 (1961): 41–47; Part IV, Veliger 4 (1962): 134–40.

HARTWICK, B. 1981. Size gradients and shell polymorphism in limpets with consideration of the role of predation. Veliger 23: 254–64. □Concerns intertidal size gradients and the color patterns of shells of *Collisella digitalis* (especially), *C. pelta*, *Notoacmea persona*, and *N. scutum* with reference to predation by the black oystercatcher *Haematopus bachmani*.

JOHNSON, S. E. 1968. Occurrence and behavior of *Hyale grandicornis*, a gammarid amphipod commensal in the genus *Acmaea*. Veliger 11(suppl.): 56–60.

LINDBERG, D. R. 1981. Acmaeidae: Gastropoda, Mollusca. Pacific Grove, Calif.: Boxwood Press. 122 pp. □An extremely useful reference work with keys, a pictorial glossary, and a section on identification techniques. Photographs and drawings of radular teeth and plates accompany the description of each species, along with notes on habitat and distribution. The volume includes an extensive bibliography by J. T. Carlton.

MARGOLIN, A. S. 1964. A running response of *Acmaea* to seastars. Ecology 45: 191–93.

MURPHY, P. G. 1978. *Collisella austrodigitalis* sp. nov.: A sibling species of limpet (Acmaeidae) discovered by electrophoresis. Biol. Bull. 155: 193–206. □Proposes, on the basis of protein electrophoresis studies, that what has been called *Collisella digitalis* in central California is actually two species: *C. digitalis*, whose approximate southern extent would be Point Conception, and a new species, *C. austrodigitalis*, whose northern limit would be Monterey Bay.

NICOTRI, M. E. 1977. Grazing effects of four marine intertidal herbivores on the microflora. Ecology 58: 1020–32. □The four are *Collisella pelta*, *C. paradigitalis*, *Notoacmea scutum*, and the mesogastropod *Littorina scutulata*.

PHILLIPS, D. W. 1975a. Distance chemoreception-triggered avoidance behavior of the limpets *Acmaea* (*Collisella*) *limatula* and *Acmaea* (*Notoacmea*) *scutum* to the predatory starfish *Pisaster ochraceus*. J. Exp. Zool. 191: 199–210.

———. 1975b. Localization and electrical activity of the distance chemoreceptors that mediate predator avoidance behaviour in *Acmaea limatula* and *Acmaea scutum* (Gastropoda, Prosobranchia). J. Exp. Biol. 63: 403–12.

———. 1976. The effect of a species-specific avoidance response to predatory starfish on the intertidal distribution of two gastropods. Oecologia 23: 83–94.

PHILLIPS, D. W., and P. CASTORI. 1982. Defensive responses to predatory seastars by two specialist limpets, *Notoacmea insessa* (Hinds) and *Collisella instabilis* (Gould), associated with marine algae. J. Exp. Mar. Biol. Ecol. 59: 23–30.

PHILLIPS, D. W., and M. L. CHIARAPPA. 1980. Defensive responses of gastropods to the predatory flatworms *Freemania litoricola* (Heath and McGregor) and *Notoplana acticola* (Boone). J. Exp. Mar. Biol. Ecol. 47: 179–89.

SHOTWELL, J. A. 1950. The vertical zonation of *Acmaea*, the limpet. Ecology 31: 647–49. □A second paper by the same author in the same year is cited.

STIMSON, J., and R. BLACK. 1975. Field experiments on population regulation in intertidal limpets of the genus *Acmaea*. Oecologia 18: 111–20.

TARR, K. J. 1977. An analysis of water-content regulation in osmoconforming limpets (Mollusca: Patellacea). J. Exp. Zool. 201: 259–68. □*Collisella digitalis*, which is more tolerant of osmotic stress than *Notoacmea scutum*, has more control of its body volume, although both species tend to be osmoconformers.

TEST, A. R. G. 1945. Ecology of California *Acmaea*. Ecology 26: 395–405.

———. 1946. Speciation in limpets of the genus *Acmaea*. Contr. Lab. Vert. Biol. Univ. Mich. 31: 1–24. □Note on nomenclature: various references to "Mrs. Grant" or "Grant (MS)" refer to her personal communications to E. F. Ricketts or to her unpublished thesis on the Pacific coast acmaeids on file in the University of California library. References to "Dr. A. R. (Grant) Test" are to the published works cited above.

WELLS, R. A. 1980. Activity pattern as a mechanism of predator avoidance in two species of acmaeid limpet. J. Exp. Mar. Biol. Ecol. 48: 151–68. □*Collisella limatula* stays out of sight during periods when predators such as *Octopus* might be foraging; *C. scabra* resides on a home scar from which it is difficult to remove and forages only when awash.

WOLCOTT, T. G. 1973. Physiological ecology and intertidal zonation in limpets (*Acmaea*): A critical look at "limiting factors." Biol. Bull. 145: 389–422.

All of the following species, except *Lottia gigantea*, were formerly placed in the genus *Acmaea*. Alas, widely different anatomical and shell characters require that we largely replace an old friend, *Acmaea*, with two new ones, *Collisella* and *Notoacmea*; the name *Acmaea* is retained only by *A. mitra* among the local forms.

Acmaea mitra Rathke, §117

Collisella asmi (Middendorff), on the snail *Tegula*, §13

ALLEMAN, L. L. 1968. Factors affecting the attraction of *Acmaea asmi* to *Tegula funebralis*. Veliger 11(suppl.): 61–63.

BREWER, B. A. 1975. Epizoic limpets on the black turban snail, *Tegula funebralis* (A. Adams, 1855). Veliger 17: 307–10.

EIKENBERRY, A. B., JR., and D. E. WICKIZER. 1964. Studies on the commensal limpet *Acmaea asmi* in relation to its host, *Tegula funebralis*. Veliger 6(suppl.): 66–70.

C. conus (Test), §25

C. digitalis (Rathke), §§6, 11, 160
See also the works by Collins (1976, 1977) and Haven (1971, 1973), listed below under *Collisella scabra*.
BREEN, P. A. 1971. Homing behavior and population regulation in the limpet *Acmaea (Collisella) digitalis*. Veliger 14: 177–83.
———. 1972. Seasonal migration and population regulation in the limpet *Acmaea (Collisella) digitalis*. Veliger 15: 133–41.
FRANK, P. W. 1965. The biodemography of an intertidal snail population. Ecology 46: 831–44.
GIESEL, J. T. 1969. Factors influencing the growth and relative growth of *Acmaea digitalis*, a limpet. Ecology 50: 1084–87.
———. 1970. On the maintenance of a shell pattern and behavior polymorphism in *Acmaea digitalis*, a limpet. Evolution 24: 98–119.
WILLOUGHBY, J. W. 1973. A field study on the clustering and movement behavior of the limpet *Acmaea digitalis*. Veliger 15: 223–30.

C. instabilis (Gould), on stipes of the algae *Laminaria* and *Pterygophora*, not treated

C. limatula (Carpenter) (*Acmaea scabra* Nuttall of Keep), file limpet, §25
EATON, C. M. 1968. The activity and food of the file limpet *Acmaea limatula*. Veliger 11(suppl.): 5–12.
MARKEL, R. P. 1974. Aspects of the physiology of temperature acclimation in the limpet *Acmaea limatula* Carpenter (1864): An integrated field and laboratory study. Physiol. Zool. 47: 99–109.
SEAPY, R. R. 1966. Reproduction and growth in the file limpet, *Acmaea limatula* Carpenter, 1864 (Mollusca: Gastropoda). Veliger 8: 300–310. □Reports on a population at Palos Verdes Peninsula in southern California.
SEGAL, E. 1956. Microgeographic variation as thermal acclimation in an intertidal mollusc. Biol. Bull. 111: 129–52.
SEGAL, E., and P. A. DEHNEL. 1962. Osmotic behavior in an intertidal limpet, *Acmaea limatula*. Biol. Bull. 122: 417–30.

C. ochracea (Dall), not treated
LINDBERG, D. R. 1979. Variations in the limpet, *Collisella ochracea*, and the northeastern Pacific distribution of *Notoacmea testudinalis* (Acmaeidae). Nautilus 93: 50–56.

C. paradigitalis (Fritchman), §6
This species, considered for a while to be a synonym of *Collisella strigatella* (Carpenter), has been reinstated; *C. strigatella* is a Panamic species.
DIXON, J. 1981. Evidence of gregarious settlement in the larvae of the marine snail *Collisella strigatella* (Carpenter). Veliger 24: 181–84.
FRITCHMAN, H. K. 1960. *Acmaea paradigitalis* sp. nov. (Acmaeidae, Gastropoda). Veliger 2: 53–57.

SEAPY, R. R., and W. J. HOPPE. 1973. Morphological and behavioral adaptations to desiccation in the intertidal limpet *Acmaea (Collisella) strigatella*. Veliger 16: 181–88.

C. pelta (Rathke) (formerly *Acmaea cassis* Eschscholtz), §§11, 195
CRAIG, P. C. 1968. The activity pattern and food habits of the limpet *Acmaea pelta*. Veliger 11(suppl.): 13–19.
JOBE, A. 1968. A study of morphological variation in the limpet *Acmaea pelta*. Veliger 11(suppl.): 69–72. □There are even more variations than the several covered here.

C. scabra (Gould), §11
COLLINS, L. S. 1976. Abundance, substrate angle, and desiccation resistance in two sympatric species of limpets. Veliger 19: 199–203.
———. 1977. Substrate angle, movement and orientation of two sympatric species of limpets, *Collisella digitalis* and *Collisella scabra*. Veliger 20: 43–48.
HAVEN, S. B. 1971. Niche differences in the intertidal limpets *Acmaea scabra* and *Acmaea digitalis* (Gastropoda) in central California. Veliger 13: 231–48.
———. 1973. Competition for food between the intertidal gastropods *Acmaea scabra* and *Acmaea digitalis*. Ecology 54: 143–51.
JESSE, W. F. 1968. Studies of homing behavior in the limpet *Acmaea scabra*. Veliger 11(suppl.): 52–55.
LINDBERG, D. R., and K. R. DWYER. 1982. The topography, formation and role of the home depression of *Collisella scabra* (Gould) (Gastropoda: Acmaeidae). Veliger 25: 229–34.
SUTHERLAND, J. P. 1970. Dynamics of high and low populations of the limpet, *Acmaea scabra* (Gould). Ecol. Monogr. 40: 169–88.
———. 1972. Energetics of high and low populations of the limpet, *Acmaea scabra* (Gould). Ecology 53: 430–37.

C. triangularis (Carpenter), §119

Lottia gigantea Sowerby, owl limpet, §7
ABBOTT, D. P. 1956. Water circulation in the mantle cavity of the owl limpet *Lottia gigantea* Gray. Nautilus 69: 79–87.
FISHER, W. K. 1904. The anatomy of *Lottia gigantea* Gray. Zool. Jahrb. (Anat.) 20: 1–66. □A beautifully illustrated and detailed account.
SIBLEY, J. B. 1981. A study of homing in the owl limpet, *Lottia gigantea* (Sowerby), by analyzing its movement in an experimental pool. Master's thesis, Biology, San Diego State University. 113 pp.
STIMSON, J. 1970. Territorial behavior of the owl limpet, *Lottia gigantea*. Ecology 51: 113–18. □But see also Wright (1982) for a contrasting view.
———. 1973. The role of the territory in the ecology of the intertidal limpet *Lottia gigantea* (Gray). Ecology 54: 1020–30.
WRIGHT, W. G. 1982. Ritualized behavior in a territorial limpet. J. Exp. Mar. Biol. Ecol. 60: 245–51.
WRIGHT, W. G., and D. R. LINDBERG. 1982. Direct observation of sex change in the patellacean limpet *Lottia gigantea*. J. Mar. Biol. Assoc. U.K. 62: 737–38.

Notoacmea depicta (Hinds), painted limpet, §274
N. fenestrata (Reeve), §117

N. insessa (Hinds), on the kelp *Egregia*, §119
BLACK, R. 1976. The effects of grazing by the limpet, *Acmaea insessa*, on the kelp, *Egregia laevigata*, in the intertidal zone. Ecology 57: 265–77.

PROCTOR, S. J. 1968. Studies on the stenotopic marine limpet *Acmaea insessa* (Mollusca: Gastropoda: Prosobranchia) and its algal host *Egregia menziesii* (Phaeophyta). Ph.D. dissertation, Biological Sciences, Stanford University. 144 pp.

N. paleacea (Gould), on surfgrass (*Phyllospadix* spp.), §119
FISHLYN, D. A., and D. W. PHILLIPS. 1980. Chemical camouflaging and behavioral defenses against a predatory seastar by three species of gastropods from the surfgrass community. Biol. Bull. 158: 34–48.

N. persona (Rathke), §11
KENNY, R. 1969. Growth characteristics of *Acmaea persona* Eschscholtz. Veliger 11: 336–39.

LINDBERG, D. R., M. G. KELLOGG, and W. E. HUGHES. 1975. Evidence of light reception through the shell of *Notoacmea persona* (Rathke, 1833) (Archaeogastropoda: Acmaeidae). Veliger 17: 383–86. □See also a follow-up note by Lindberg and Kellogg (1982, Veliger 25: 173–74) clarifying an unfortunate choice of terms regarding shell structure.

N. scutum (Rathke) (=*Acmaea testudinalis scutum* Eschscholtz), plate limpet, §25
KITTING, C. L. 1980. Herbivore-plant interactions of individual limpets maintaining a mixed diet of intertidal marine algae. Ecol. Monogr. 50: 527–50.

PHILLIPS, D. W. 1979. Ultrastructure of sensory cells on the mantle tentacles of the gastropod *Notoacmea scutum*. Tissue Cell 11: 623–32. □The mantle margin of *Notoacmea scutum* contains receptors sensitive to light, touch, and chemicals (including those eliciting the limpet's defensive responses to predatory seastars); yet only two morphologically distinct types of receptor were found.

———. 1981. Life-history features of the marine intertidal limpet *Notoacmea scutum* (Gastropoda) in central California. Mar. Biol. 64: 95–103.

ROGERS, D. A. 1968. The effects of light and tide on movements of the limpet *Acmaea scutum*. Veliger 11(suppl.): 20–24.

WEBBER, H. H., and P. A. DEHNEL. 1968. Ion balance in the prosobranch gastropod, *Acmaea scutum*. Comp. Biochem. Physiol. 25: 49–64.

Superfamily TROCHACEA
Top shells and Turbans

GRANGE, K. R. 1976. Rough water as a spawning stimulus in some trochid and turbinid gastropods. N.Z. J. Mar. Freshwater Res. 10: 203–16. □The study involves species in New Zealand, but since information is generally lacking on our species, this paper may provide a clue.

HOFFMAN, D. L. 1980. Defensive responses of marine gastropods (Prosobranchia, Trochidae) to certain predatory sea stars and the direwhelk, *Searlesia dira*. Pacific Sci. 34: 233–43. □*Calliostoma*

ligatum and three species of *Margarites* respond defensively.

LOWRY, L. F., A. J. MCELROY, and J. S. PEARSE. 1974. The distribution of six species of gastropod molluscs in a California kelp forest. Biol. Bull. 147: 386–96. □Concerns *Calliostoma annulatum*, *C. canaliculatum*, *C. ligatum*, *Tegula brunnea*, *T. montereyi*, and *T. pulligo*.

REIDMAN, M. L., A. H. HINES, and J. S. PEARSE. 1981. Spatial segregation of four species of turban snails (Gastropoda: *Tegula*) in central California. Veliger 24: 97–102. □At Hopkins Marine Station, *Tegula funebralis* is strictly intertidal, and abundant above mean lower low water. *T. brunnea* is common below 0.5 meter above MLLW, extending subtidally to about 7 m depth. *T. pulligo* and *T. montereyi* occur subtidally.

Astraea gibberosa (Dillwyn) (formerly *A. inaequalis* (Martyn)), §117

A. undosa (Wood), §117
SCHWALM, C. C. 1973. Population dynamics and energetics of *Astraea undosa*. Master's thesis, Biology, San Diego State University. 113 pp. □Intertidal and subtidal populations were studied at Point Loma in southern California. Maximum size was estimated at 11.2 cm in diameter but the average was about 5.7 cm. Growth was rapid and life span relatively short (less than 10 years) for such a large snail.

Calliostoma annulatum (Lightfoot), §182
C. canaliculatum (Lightfoot), §182
C. ligatum (formerly *C. costatum* (Martyn)), §§182, 215
HARROLD, C. 1982. Escape responses and prey availability in a kelp forest predator-prey system. Amer. Natur. 119: 132–35. □The defensive behaviors of *Tegula pulligo* and *Calliostoma ligatum* reduce the chances of being captured by the seastar *Pisaster giganteus*. The defenses of *Calliostoma*, which include a slippery, mucus-covered shell, are more effective than those of *Tegula*.

KEEN, A. M. 1975. On some west American species of *Calliostoma*. Veliger 17: 413–14.

PERRON, F. 1975. Carnivorous *Calliostoma* (Prosobranchia: Trochidae) from the northeastern Pacific. Veliger 18: 52–54.

Norrisia norrisi (Sowerby), §136
MILLER, W. E. 1975. Environmental cues and the orientation and movement of *Norrisia norrisii*. Veliger 17: 292–95.

SCHMITT, R. J., C. W. OSENBERG, and M. G. BERCOVITCH. 1983. Mechanisms and consequences of shell fouling in the kelp snail, *Norrisia norrisi* (Sowerby) (Trochidae): Indirect effects of octopus drilling. J. Exp. Mar. Biol. Ecol. 69: 267–81. □Barnacle cyprids (*Megabalanus californicus*) settled in incomplete octopus drill holes. Heavy fouling of *Norrisia* shells increased the frequency of dislodgment from kelp plants on which the snail usually resides and also decreased the snail's escape velocity away from predators.

Tegula brunnea (Philippi), §29

BELCIK, F. P. 1965. Note on a range extension and observations of spawning in *Tegula*, a gastropod. Veliger 7: 233–34. ☐ Extends known range north to Cape Arago, Oregon; notes spawning in laboratory tanks in August.

WATANABE, J. M. 1983. Anti-predator defenses of three kelp forest gastropods: Contrasting adaptations of closely-related prey species. J. Exp. Mar. Biol. Ecol. 71: 257–70. ☐ Concerns *Tegula brunnea*, *T. pulligo*, and *T. montereyi*.

T. eiseni Jordan (= *T. ligulata* (Menke)), §137

T. funebralis (A. Adams), §13

ABBOTT, D. P., L. R. BLINKS, J. H. PHILLIPS, and R. STOHLER, eds. 1964. The biology of *Tegula funebralis* (A. Adams, 1855). Veliger 6(suppl.): 1–82. ☐ Consists of 19 papers (some preliminary reports) on various aspects of *Tegula*, the results of a course at Hopkins Marine Station. Several are cited separately below; for others, the reader is referred to the complete volume.

BEST, B. 1964. Feeding activities of *Tegula funebralis* (Mollusca: Gastropoda). Veliger 6(suppl.): 42–45.

BYERS, B. A., and J. B. MITTON. 1981. Habitat choice in the intertidal snail *Tegula funebralis*. Mar. Biol. 65: 149–54.

DOERING, P. H., and D. W. PHILLIPS. 1983. Maintenance of the shore-level size gradient in the marine snail *Tegula funebralis* (A. Adams): Importance of behavioral responses to light and sea star predators. J. Exp. Mar. Biol. Ecol. 67: 159–73.

FRANK, P. W. 1965. Shell growth in a natural population of the turban snail, *Tegula funebralis*. Growth 29: 395–403.

———. 1969. Sexual dimorphism in *Tegula funebralis*. Veliger 11: 440. ☐ Snails over 1.5 cm in diameter can be sexed with 80–90 percent accuracy by examining the sole of the foot: those with lighter, cream-colored feet are males, and those with darker, orangish-brown feet are females.

———. 1975. Latitudinal variation in the life history features of the black turban snail *Tegula funebralis* (Prosobranchia: Trochidae). Mar. Biol. 31: 181–92.

GELLER, J. B. 1982a. Microstructure of shell repair materials in *Tegula funebralis* (A. Adams, 1855). Veliger 25: 155–59. ☐ A resistant, secondary shell layer is secreted in response to shell erosion by the fungus *Pharcidia balani*.

———. 1982b. Chemically mediated avoidance response of a gastropod, *Tegula funebralis* (A. Adams), to a predatory crab, *Cancer antennarius* (Stimpson). J. Exp. Mar. Biol. Ecol. 65: 19–27.

JENSEN, J. T. 1981. Distribution, activity and food habits of juvenile *Tegula funebralis* and *Littorina scutulata* (Gastropoda: Prosobranchia) as they relate to resource partitioning. Veliger 23: 333–38.

KOSIN, D. F. 1964. The light responses of *Tegula funebralis* (Mollusca: Gastropoda). Veliger 6(suppl.): 46–50.

MARKOWITZ, D. V. 1980. Predator influence on shore-level size gradients in *Tegula funebralis* (A. Adams). J. Exp. Mar. Biol. Ecol. 45: 1–13.

OVERHOLSER, J. A. 1964. Orientation and response of *Tegula funebralis* to tidal current and turbulence (Mollusca: Gastropoda). Veliger 6(suppl.): 38–41.

PAINE, R. T. 1969. The *Pisaster-Tegula* interaction: Prey patches, predator food preference, and intertidal community structure. Ecology 50: 950–61. ☐ A classic, although the author in his zeal for casting all aspects of the life history in terms of "control" by *Pisaster* predation tends, at times, to ignore alternative explanations of his data.

———. 1971. Energy flow in a natural population of the herbivorous gastropod *Tegula funebralis*. Limnol. Oceanogr. 16: 86–98.

PHILLIPS, D. W. 1978. Chemical mediation of invertebrate defensive behaviors and the ability to distinguish between foraging and inactive predators. Mar. Biol. 49: 237–43. ☐ Concerns *Tegula funebralis* (as well as the sea urchin *Strongylocentrotus purpuratus*) and cites earlier work by several other authors on the defensive responses of *Tegula*.

STOHLER, R. 1964. Studies on mollusk populations. VI. *Tegula funebralis* (A. Adams, 1855) (Mollusca: Gastropoda). Veliger 6(suppl.): 77–81.

SZAL, R. 1971. "New" sense organ of primitive gastropods. Nature 229: 490–92.

WARA, W. M., and B. B. WRIGHT. 1964. The distribution and movement of *Tegula funebralis* in the intertidal region of Monterey Bay, California (Mollusca: Gastropoda). Veliger 6(suppl.): 30–37.

WELDON, P. J., and D. L. HOFFMAN. 1975. Unique form of tool-using in two gastropod molluscs (Trochidae). Nature 256: 720–21. ☐ Specimens of *Tegula brunnea* and *T. funebralis*, if placed on their backs in aquaria provided with a gravel substratum, used small stones to assist in righting themselves.

YARNALL, J. L. 1964. The responses of *Tegula funebralis* to starfishes and predatory snails (Mollusca: Gastropoda). Veliger 6(suppl.): 56–58.

Order MESOGASTROPODA
Periwinkles, Slipper shells, Cowries, Moon snails, Horn shells, etc.

FOWLER, B. H. 1980. Reproductive biology of *Assiminea californica* (Tryon, 1865) (Mesogastropoda: Rissoacea). Veliger 23: 163–66. ☐ This small mesogastropod is a common inhabitant of coastal salt marshes from Vancouver Island to southern Baja California.

Batillaria attramentaria (Sowerby) (listed as *B. zonalis* (Bruguière) in previous editions; misidentified in earlier studies), §286

See also Driscoll (1972), MacDonald (1969), and Whitlatch and Obrebski (1980), listed below under *Cerithidea*.

WHITLATCH, R. B. 1974. Studies on the population ecology of the salt marsh gastropod *Batillaria zonalis*. Veliger 17: 47–55. ☐ Studies at two locations in Tomales Bay.

YAMADA, S. B. 1982. Growth and longevity of the mud snail *Batillaria attramentaria*. Mar. Biol. 67: 187–92. ☐ In British Columbia, *Batillaria attramentaria* attained, during its first 6 years of life, average shell lengths of 3, 9, 12, 15, 17, and 18 mm. Life span probably does not exceed 10 years.

YAMADA, S. B., and C. S. SANKURATHRI. 1977. Direct development in the intertidal gastropod *Batillaria zonalis* (Bruguière, 1792). Veliger 20: 179.

Cerithidea californica (Haldeman), §286

BRIGHT, D. B. 1958. Morphology of the common mudflat snail, *Cerithidea californica*. Bull. South. Calif. Acad. Sci. 57: 127–39.

―――. 1960. Morphology of the common mudflat snail, *Cerithidea californica* II. Bull. South. Calif. Acad. Sci. 59: 9–15.

DRISCOLL, A. L. 1972. Structure and function of the alimentary tract of *Batillaria zonalis* and *Cerithidea californica*, style-bearing mesogastropods. Veliger 14: 375–86.

MacDONALD, K. B. 1969. Quantitative studies of salt marsh faunas from the North American Pacific coast. Ecol. Mongr. 39: 33–60.

MARTIN, W. E. 1972. An annotated key to the cercariae that develop in the snail *Cerithidea californica*. Bull. South. Calif. Acad. Sci. 71: 39–43.

RACE, M. S. 1981. Field ecology and natural history of *Cerithidea californica* (Gastropoda: Prosobranchia) in San Francisco Bay. Veliger 24: 18–27.

―――. 1982. Competitive displacement and predation between introduced and native mud snails. Oecologia 54: 337–47. □Concerns the native *Cerithidea californica* and the introduced *Ilyanassa obsoleta* in San Francisco Bay.

SCOTT, D. B., and T. L. CASS. 1977. Response of *Cerithidea californica* (Haldeman) to lowered salinities and its paleoecological implications. Bull. South. Calif. Acad. Sci. 76: 60–63.

WHITLATCH, R. B., and S. OBREBSKI. 1980. Feeding selectivity and coexistence in two deposit-feeding gastropods. Mar. Biol. 58: 219–25. □Both gastropods selectively ingest diatoms, but *Cerithidea* ingests somewhat finer material.

YOSHINO, T. P. 1975. A seasonal and histologic study of larval Digenea infecting *Cerithidea californica* (Gastropoda: Prosobranchia) from Goleta Slough, Santa Barbara County, California. Veliger 18: 156–61.

Crepidula adunca Sowerby, §29

MORITZ, C. E. 1939. Organogenesis in the gastropod *Crepidula adunca* Sowerby. Univ. Calif. Publ. Zool. 43: 217–48.

PUTMAN, D. A. 1964. The dispersal of young of the commensal gastropod *Crepidula adunca* from its host, *Tegula funebralis*. Veliger 6(suppl.): 63–66. □Describes the hatching process and behavior of the young.

C. fornicata Linnaeus, §29

JOHNSON, J. K. 1972. Effect of turbidity on the rate of filtration and growth of the slipper limpet, *Crepidula fornicata* Lamarck, 1799. Veliger 14: 315–20. □Filtration rates and shell growth decreased as the level of turbidity increased, due to clogging of the filtering mechanism.

ORTON, J. H. 1912a. An account of the natural history of the slipper limpet (*Crepidula fornicata*). J. Mar. Biol. Assoc. U.K. 9: 437–43.

―――. 1912b. The mode of feeding of *Crepidula*, with an account of the current-producing mechanism in the mantle cavity, and some remarks on the mode of feeding in gastropods and lamellibranchs. J. Mar. Biol. Assoc. U.K. 9: 444–78.

PRATT, D. M. 1974. Behavioral defenses of *Crepidula fornicata* against attack by *Urosalpinx cinerea*. Mar. Biol. 27: 47–49.

C. norrisiarum Williamson, §136

MacGINITIE, N., and G. E. MacGINITIE. 1964. Habitats and breeding seasons of the shelf limpet *Crepidula norrisiarum* Williamson. Veliger 7: 34.

C. nummaria Gould, §215

C. onyx Sowerby, §296

COE, W. R. 1942. The reproductive organs of the prosobranch mollusk *Crepidula onyx* and their transformation during the change from male to female phase. J. Morphol. 70: 501–12.

―――. 1953. Influences of association, isolation, and nutrition on the sexuality of snails of the genus *Crepidula*. J. Exp. Zool. 122: 5–20. □Information on *Crepidula nivea* probably applies to *C. nummaria*; also discussed are *C. norrisiarum*, *C. fornicata*, *C. onyx*, and *C. perforans* (as *C. williamsi*).

HOAGLAND, K. E. 1977. Systematic review of fossil and recent *Crepidula* and discussion of evolution of the Calyptraeidae. Malacologia 16: 353–420.

Cypraea spadicea Swainson, §136

See also Keen (1971), under Mollusca, p. 549.

DARLING, S. D. 1965. Observations on the growth of *Cypraea spadicea*. Veliger 8: 14–15.

Epitonium tinctum (also currently as *Nitidiscala*), §64

DuSHANE, H. 1979. The family Epitoniidae (Mollusca: Gastropoda) in the northeastern Pacific. Veliger 22: 91–134. □Concerns the genera *Epitonium*, *Nitidiscala*, and *Opalia* among others.

ROBERTSON, R. 1963. Wentletraps (Epitoniidae) feeding on sea anemones and corals. Proc. Malacol. Soc. Lond. 35: 51–63.

SALO, S. 1977. Observations on feeding, chemoreception and toxins in two species of *Epitonium*. Veliger 20: 168–72.

SMITH, C. R. 1977. Chemical recognition of prey by the gastropod *Epitonium tinctum* (Carpenter, 1864). Veliger 19: 331–40.

SMITH, C. R., and A. BREYER. 1983. Comparison of northern and southern populations of *Epitonium tinctum* (Carpenter, 1864) on the California coast. Veliger 26: 37–46. □Snails from northern populations (Bodega Bay) were longer than southern snails (Carpinteria Beach), and became sexually mature at a larger size.

Lacuna porrecta Carpenter and **L. variegata** Carpenter, §274

Several other species of *Lacuna* inhabit the low zone of our coast. Among the more common are *L. marmorata* Dall, ranging from Alaska to San Diego, which occurs on rocks, algae, surfgrass (*Phyllospadix*), and eelgrass (*Zostera*); *L. unifasciata* Carpenter, Monterey to Baja California, on eelgrass and algae; and *L. vincta* (Montagu), from Puget Sound north, on kelps and other algae.

FIELD, L. H. 1974. A description and experimental analysis of Batesian mimicry between a marine gastropod and an amphipod. Pacific Sci. 28: 439–47.

FISHLYN, D. A., and D. W. PHILLIPS. 1980. Chemical camouflaging and behavioral defenses against a predatory seastar by three species of gastropods from the surfgrass *Phyllospadix* community. Biol. Bull. 158: 34–48.

WARBURTON, K. 1979. Variation in shell geometry in the genus *Lacuna* (Prosobranchia: Lacunidae). J. Natur. Hist. 13: 385–91.

Lamellarids, not treated
Members of the family Lamellariidae show remarkable concealing ability, often resembling closely the compound ascidians upon which they feed. The shell is internal or concealed by the mantle. Genera represented on the Pacific coast are *Lamellaria, Marsenina,* and *Marseniopsis.*

BEHRENS, D. W. 1980. The Lamellariidae of the northeastern Pacific. Veliger 22: 323–39.

GHISELIN, M. T. 1964. Morphological and behavioral concealing adaptations of *Lamellaria stearnsii,* a marine prosobranch gastropod. Veliger 6: 123–24.

HUNT, D. E. 1980. Predation by a rockfish, *Sebastes chrysomelas,* on *Lamellaria diegoensis* Dall, 1885. Veliger 22: 291.

LAMBERT, G. 1980. Predation by the prosobranch mollusk *Lamellaria diegoensis* on *Cystodytes lobatus,* a colonial ascidian. Veliger 22: 340–44.

Littorina keenae Rosewater (= *Littorina planaxis* Philippi), §§1, 10
Long known by most students of the shore as *Littorina planaxis,* L. *keenae* is unfortunately the correct name (see Rosewater 1978). How long it will take for the correct name to become widely used remains to be seen.

BOCK, C. E., and R. E. JOHNSON. 1967. The role of behavior in determining the intertidal zonation of *Littorina planaxis* Philippi, 1847, and *Littorina scutulata* Gould, 1849. Veliger 10: 42–54.

DAHL, A. L. 1964. Macroscopic algal foods of *Littorina planaxis* Philippi and *Littorina scutulata* Gould (Gastropoda: Prosobranchiata). Veliger 7: 139–43.

DINTER, I. 1974. Pheromonal behavior in the marine snail *Littorina littorea* Linnaeus. Veliger 17: 37–39.

FOSTER, M. S. 1964. Microscopic algal food of *Littorina planaxis* Philippi and *Littorina scutulata* Gould (Gastropoda: Prosobranchiata). Veliger 7: 149–52.

GIBSON, D. G. 1964. Mating behavior in *Littorina planaxis* Philippi (Gastropoda: Prosobranchiata). Veliger 7: 134–39.

NEALE, J. R. 1965. Rheotactic responses in the marine mollusk *Littorina planaxis* Philippi. Veliger 8: 7–10.

NORTH, W. J. 1954. Size distribution, erosive activities and gross metabolic efficiency of the marine intertidal snails, *Littorina planaxis* and *L. scutulata.* Biol. Bull. 106: 185–97.

PETERS, R. S. 1964. Function of the cephalic tentacles in *Littorina planaxis* Philippi (Gastropoda: Prosobranchiata). Veliger 7: 143–48.

RAFTERY, R. E. 1983. *Littorina* trail following: Sexual preference, loss of polarized information, and trail alterations. Veliger 25: 378–82. □Male and female *Littorina keenae* are likely to follow a trail laid by a conspecific, regardless of the latter's sex, and to follow it in the direction of travel. Information on trail polarity is gradually lost.

ROSEWATER, J. 1978. A case of double primary homonymy in eastern Pacific Littorinidae. Nautilus 92: 123–25. □Breaks the bad news that an old and familiar name, *Littorina planaxis,* must be discarded by the rules of nomenclature. *Littorina keenae* is the replacement.

SCHMITT, R. J. 1979. Mechanics and timing of egg capsule release by the littoral fringe periwinkle *Littorina planaxis* (Gastropoda: Prosobranchia). Mar. Biol. 50: 359–66.

URBAN, E. K. 1962. Remarks on the taxonomy and intertidal distribution of *Littorina* in the San Juan Archipelago, Washington. Ecology 43: 320–23.

YAMADA, S. B. 1977. Geographic range limitation of the intertidal gastropods *Littorina sitkana* and *L. planaxis.* Mar. Biol. 39: 61–65.

L. scutulata Gould and **L. plena** Gould, §10
What has been called *Littorina scutulata* for many years may actually be two species—*Littorina scutulata* and *L. plena*—nearly identical in most respects. Separation of the two, as presented in the papers below, is based on the morphology of the egg capsules and penes, on electrophoresis of body proteins, and to a limited extent on shell morphometrics.

CHOW, V. 1975. The importance of size in the intertidal distribution of *Littorina scutulata* (Gastropoda: Prosobranchia). Veliger 18: 69–78.

MASTRO, E., V. CHOW, and D. HEDGECOCK. 1982. *Littorina scutulata* and *Littorina plena:* Sibling species status of two prosobranch gastropod species confirmed by electrophoresis. Veliger 24: 239–46.

MURRAY, T. 1979. Evidence for an additional *Littorina* species and a summary of the reproductive biology of *Littorina* from California. Veliger 21: 469–74.

———. 1982. Morphological characterization of the *Littorina scutulata* species complex. Veliger 24: 233–38.

L. sitkana Philippi (sometimes erroneously written as L. *sitchana*), §195
BEHRENS, S. 1972. The role of wave impact and desiccation on the distribution of *Littorina sitkana* Philippi, 1845. Veliger 15: 129–32. □See also the 1977 paper by this author, as S. B. Yamada, listed above.

McCORMACK, S. M. D. 1982. The maintenance of shore-level size gradients in an intertidal snail (*Littorina sitkana*). Oecologia 54: 177–83. □Body size of *Littorina sitkana* increases with increasing height on the shore in British Columbia. Large snails are selectively removed from lower regions by predators (the pile perch, *Rhacochilus vacca,* being one) and possibly by wave surge. The snail's behavior was also involved in maintaining the size gradient, as transplanted snails moved toward their original ("home") level within 24 hours of release

Opalia funiculata (Carpenter) (*O. crenimarginata* (Dall) in previous editions), §64

See DuShane (1979) and Robertson (1963) listed under *Epitonium tinctum.*

Polinices draconis (Dall), §252

P. lewisii (Gould), §252
BERNARD, F. R. 1967. Studies on the biology of the naticid clam drill *Polinices lewisi* (Gould) (Gastropoda, Prosobranchiata). Fish. Res. Board Can., Tech. Rep. 42: 1–41.
MANGUM, C. P. 1979. A note on blood and water mixing in large marine gastropods. Comp. Biochem. Physiol. 63A: 389–91.
REID, R. G. B., and J. A. FRIESEN. 1980. The digestive system of the moon snail *Polinices lewisii* (Gould, 1847) with emphasis on the role of the esophageal gland. Veliger 23: 25–34.

P. reclusianus (Deshayes) (sometimes spelled *P. recluzianus*), §252

Serpulorbis squamigerus (Carpenter) (formerly *Aletes*), §47
HADFIELD, M. G. 1970. Observations on the anatomy and biology of two California vermetid gastropods. Veliger 12: 301–9. □Concerns *Serpulorbis squamigerus* and *Petaloconchus montereyensis.*
HADFIELD, M. G., and C. N. HOPPER. 1980. Ecological and evolutionary significance of pelagic spermatophores of vermetid gastropods. Mar. Biol. 57: 315–25.

Trichotropis cancellata Hinds, §215
YONGE, C. M. 1962. On the biology of the mesogastropod *Trichotropis cancellata* Hinds, a benthic indicator species. Biol. Bull. 122: 160–81.

Order NEOGASTROPODA
Whelks, Rock snails, Olives, Cones, etc.

BUTLER, A. J. 1979. Relationships between height on the shore and size distributions of *Thais* spp. (Gastropoda, Muricidae). J. Exp. Mar. Biol. Ecol. 41: 163–94.
CRANE, J. M., JR. 1969. Mimicry of the gastropod *Mitrella carinata* by the amphipod *Pleustes platypa.* Veliger 12: 200.
MURDOCK, W. W. 1969. Switching in general predators: Experiments on predator specificity and stability of prey populations. Ecol. Monogr. 39: 335–54. □Concerns the prey preferences of the predators *Nucella emarginata* and *Acanthina spirata.*
PECHENIK, J. A. 1982. Ability of some gastropod egg capsules to protect against low-salinity stress. J. Exp. Mar. Biol. Ecol. 63: 195–208. □Encapsulated embryos of *Ilyanassa obsoleta, Nucella lamellosa,* and *N. lima* tolerated water of low salinity better than embryos removed prematurely from their capsules.
PONDER, W. F. 1973. The origin and evolution of the Neogastropoda. Malacologia 12: 295–338.

Acanthina punctulata (Sowerby), §165

A. spirata (Blainville), §165
The two *Acanthina* species were long united under the name *A. spirata* (as in previous editions). Many of the studies alleged to be on *A. spirata* probably refer instead to *A. punctulata* through misidentification.
HEMINGWAY, G. T. 1978. Evidence for a paralytic venom in the intertidal snail, *Acanthina spirata* (Neogastropoda, Thaisidae). Comp. Biochem. Physiol. 60C: 79–81.
MENGE, J. L. 1974. Prey selection and foraging period of the predaceous rocky intertidal snail, *Acanthina punctulata.* Oecologia 17: 293–316. □The population studied was on Santa Cruz Island of the Santa Barbara Channel Islands.
SLEDER, J. 1981. *Acanthina punctulata* (Neogastropoda: Muricacea): Its distribution, activity, diet, and predatory behavior. Veliger 24: 172–80.

Amphissa versicolor Dall, §137
KENT, B. 1981. Behavior of the gastropod *Amphissa columbiana* (Prosobranchia: Columbellidae). Veliger 23: 275–76. □The snail shows two basic defensive responses upon contact with predatory seastars: first, there is a twisting, running escape response, and second, if this fails to free the snail, its everted proboscis is used to attack individual tube feet.

Busycotypus canaliculatus (Linnaeus) (formerly *Busycon*), §296
There is an extensive literature, not cited here, on Atlantic coast individuals of this introduced species.
STOHLER, R. 1962. *Busycotypus (B.) canaliculatus* in San Francisco Bay. Veliger 4: 211–12.

Ceratostoma foliatum (Gmelin) (known previously as *Purpura* and *Murex*), §117
KENT, B. W. 1981. Feeding and food preferences of the muricid gastropod *Ceratostoma foliatum.* Nautilus 95: 38–42.
PALMER, A. R. 1977. Function of shell sculpture in marine gastropods: Hydrodynamic destabilization in *Ceratostoma foliatum.* Science 197: 1293–95.
SPIGHT, T. M., C. BIRKELAND, and A. LYONS. 1974. Life histories of large and small murexes (Prosobranchia: Muricidae). Mar. Biol. 24: 229–42.
SPIGHT, T. M., and A. LYONS. 1974. Development and functions of the shell sculpture of the marine snail *Ceratostoma foliatum.* Mar. Biol. 24: 77–83.

C. inornatum (Recluz) (listed as *Ocenebra japonica* (Dunker) in previous editions), §219

C. nuttalli (Conrad), §200

Conus californicus Reeve, §136
KOHN, A. J. 1966. Food specialization in *Conus* in Hawaii and California. Ecology 47: 1041–43.
NYBAKKEN, J. 1970. Radular anatomy and systematics of the west American Conidae (Mollusca: Gastropoda). Amer. Mus. Novit. 2414: 1–29.
SAUNDERS, P. R., and F. WOLFSON. 1961. Food and feeding behavior in *Conus californicus* Hinds, 1844. Veliger 3: 73–76.
WOLFSON, F. H. 1974. Two symbioses of *Conus* (Mollusca: Gastropoda) with brachyuran crabs. Veliger 16: 427–29.

Fusitriton oregonensis (Redfield) (formerly *Argobuccinum*), §214
BOTTJER, D. J. 1981. Periostracum of the gastropod *Fusitriton oregonensis*: Natural inhibitor of boring and encrusting organisms. Bull. Mar. Sci. 31: 916–21.
HOWARD, F. B. 1962. Egg-laying in *Fusitriton oregonensis* (Redfield). Veliger 4: 160. □Nice photograph

of the process, and of rosette-like egg masses observed in June at about 57°N near Baranof Island, Alaska.

Ilyanassa obsoleta (Say), §§219, 286

This species, also commonly known as *Nassarius obsoletus*, was introduced from the Atlantic. It has been the subject of an exceptionally large number of studies on the east coast, only a small sample of which is listed here.

BRETZ, D. D., and R. V. DIMOCK, JR. 1983. Behaviorally important characteristics of the mucous trail of the marine gastropod *Ilyanassa obsoleta* (Say). J. Exp. Mar. Biol. Ecol. 71: 181–91.

CONNOR, M. S., and R. K. EDGAR. 1982. Selective grazing by the mud snail *Ilyanassa obsoleta*. Oecologia 53: 271–75. ☐ *Ilyanassa* is apparently selective, choosing particles with high organic content. Migratory diatoms are the most selected food item.

CRISP, M. 1969. Studies on the behavior of *Nassarius obsoletus* (Say) (Mollusca: Gastropoda). Biol. Bull. 136: 355–73.

CURTIS, L. A., and L. E. HURD. 1981a. Crystalline style cycling in *Ilyanassa obsoleta* (Say) (Mollusca: Neogastropoda): Further studies. Veliger 24: 91–96.

———. 1981b. Nutrient procurement strategy of a deposit-feeding estuarine neogastropod, *Ilyanassa obsoleta*. Estuarine Coastal Shelf Sci. 13: 277–85. ☐Concludes that, regarding its feeding habits, *Ilyanassa* is a rather nonselective biological "vacuum cleaner."

MURPHY, D. J., and E. McCAUSLAND. 1980. The relationship between freezing and desiccation tolerance in the marine snail *Ilyanassa obsoleta* (Stimpson). Estuaries 3: 318–20.

PACE, M. L., S. SHIMMEL, and W. M. DARLEY. 1979. The effect of grazing by a gastropod, *Nassarius obsoletus*, on the benthic microbial community of a salt marsh mudflat. Estuarine Coastal Mar. Sci. 9: 121–34.

PECHENIK, J. A. 1978. Adaptations to intertidal development: Studies on *Nassarius obsoletus*. Biol. Bull. 154: 282–91.

SCHELTEMA, R. S. 1964. Feeding habits and growth in the mud-snail *Nassarius obsoletus*. Chesapeake Sci. 5: 161–66.

SMITH, B. S. 1980. The estuarine mud snail, *Nassarius obsoletus*: Abnormalities in the reproductive system. J. Molluscan Stud. 46: 247–56.

STENZLER, D., and J. ATEMA. 1977. Alarm response of the marine mud snail, *Nassarius obsoletus*: Specificity and behavioral priority. J. Chem. Ecol. 3: 159–71.

TROTT, T. J., and R. V. DIMOCK, JR. 1978. Intraspecific trail following by the mud snail *Ilyanassa obsoleta*. Mar. Behav. Physiol. 5: 91–101.

Nassarius fossatus (Gould) (formerly *Nassa* or *Alectrion*), §286

DEMOND, J. 1952. The Nassaridae of the west coast of North America between Cape San Lucas, Lower California, and Cape Flattery, Washington. Pacific Sci. 6: 300–317.

GORE, R. H. 1966. Observations on the escape response in *Nassarius vibex* (Say) (Mollusca: Gastropoda). Bull. Mar. Sci. 16: 423–34. ☐Our own

nassariids behave similarly when contacted by seastars such as *Pisaster brevispinus*.

MACGINITIE, G. E. 1931. The egg-laying process of the gastropod *Alectrion fossatus* Gould. Ann. Mag. Natur. Hist. (10)8: 258–61.

McLEAN, N. 1971. On the function of the digestive gland in *Nassarius* (Gastropoda: Prosobranchia). Veliger 13: 273–74. ☐Concerns *Nassarius fossatus, N. tegula,* and *Ilyanassa obsoleta*.

N. tegula (Reeve), §270

Nucella canaliculata (Duclos) (*Thais* in previous editions; also formerly *Purpura*), §200

See the references listed under other species of *Nucella* below: Dayton (1971), Houston (1971), Lyons and Spight (1973), Paris (1960).

N. emarginata (Deshayes) (also known as *Thais*), §165

See also the references listed under other species of *Nucella*: Bertness (1977), Bertness and Schneider (1976), Houston (1971), LeBoeuf (1971), Lyons and Spight (1973), Paris (1960), Spight (1976a, 1982a).

HOUSTON, R. S. 1971. Reproductive biology of *Thais emarginata* (Deshayes, 1839) and *Thais canaliculata* (Duclos, 1832). Veliger 13: 348–57. ☐*Nucella emarginata* spawns throughout the year in California; *N. canaliculata* spawns seasonally from the end of March through the middle of May.

LeBOEUF, R. 1971. *Thais emarginata* (Deshayes): Description of the veliger and egg capsule. Veliger 14: 205–11.

N. lamellosa (Gmelin) (also as *Thais*), §200

BERTNESS, M. D. 1977. Behavioral and ecological aspects of shore-level size gradients in *Thais lamellosa* and *Thais emarginata*. Ecology 58: 86–97.

BERTNESS, M. D., and D. E. SCHNEIDER. 1976. Temperature relations of Puget Sound thaids in reference to their intertidal distribution. Veliger 19: 47–58.

CONNELL, J. H. 1970. A predator-prey system in the marine intertidal region. I. *Balanus glandula* and several predatory species of *Thais*. Ecol. Monogr. 40: 49–78.

DAYTON, P. K. 1971. Competition, disturbance, and community organization: The provision and subsequent utilization of space in a rocky intertidal community. Ecol. Monogr. 41: 351–89.

HEMINGWAY, G. T. 1976. A comment on thaisid "cannibalism." Veliger 18: 341.

LAMBERT, P., and P. A. DEHNEL. 1974. Seasonal variation in biochemical composition during the reproductive cycle of the intertidal gastropod *Thais lamellosa* Gmelin (Gastropoda: Prosobranchia). Can. J. Zool. 52: 305–18.

LYONS, A., and T. M. SPIGHT. 1973. Diversity of feeding mechanisms among embryos of Pacific northwest *Thais*. Veliger 16: 189–94.

PARIS, O. H. 1960. Some quantitative aspects of predation by muricid snails on mussels in Washington Sound. Veliger 2: 41–47.

SPIGHT, T. M. 1974. Sizes of populations of a marine snail. Ecology 55: 712–29.

———. 1976a. Ecology of hatching for marine snails. Oecologia 24: 283–94.

———. 1976b. Colors and patterns of an intertidal snail, *Thais lamellosa*. Res. Popul. Ecol. 17: 176–90.

———. 1977. Do intertidal snails spawn in the right places? Evolution 31: 682–91.

———. 1982a. Population sizes of two marine snails with a changing food supply. J. Exp. Mar. Biol. Ecol. 57: 195–217.

———. 1982b. Risk, reward, and the duration of feeding excursions by a marine snail. Veliger 24: 302–8.

STICKLE, W. B., Jr. 1973. The reproductive physiology of the intertidal prosobranch *Thais lamellosa* (Gmelin). I. Seasonal changes in the rate of oxygen consumption and body component indexes. Biol. Bull. 144: 511–24.

———. 1975. The reproductive physiology of the intertidal prosobranch *Thais lamellosa* (Gmelin). II. Seasonal changes in biochemical composition. Biol. Bull. 148: 448–60.

Olivella baetica Carpenter (formerly *Olivella pedroana* Conrad and *O. intorta* Carpenter), §253

O. biplicata (Sowerby), §253

EDWARDS, D. C. 1968. Reproduction in *Olivella biplicata*. Veliger 10: 297–304.

———. 1969a. Predators on *Olivella biplicata*, including a species-specific predator avoidance response. Veliger 11: 326–33.

———. 1969b. Zonation by size as an adaptation for intertidal life in *Olivella biplicata*. Amer. Zool. 9: 399–417.

GIFFORD, D. S., and E. W. GIFFORD. 1944. Californian *Olivella*s. Nautilus 57: 73–80. □Includes *Olivella baetica* (as *O. pedroana*), *O. biplicata*, and *O. pycna*.

HICKMAN, C. S., and J. H. LIPPS. 1983. Foraminiferivory: Selective ingestion of foraminifera and test alterations produced by the neogastropod *Olivella*. J. Foram. Res. 13: 108–14.

PHILLIPS, D. W. 1977. Activity of the gastropod mollusk *Olivella biplicata* in response to a natural light/dark cycle. Veliger 20: 137–43.

———. 1977. Avoidance and escape responses of the gastropod mollusc *Olivella biplicata* (Sowerby) to predatory asteroids. J. Exp. Mar. Biol. Ecol. 28: 77–86.

STOHLER, R. 1969. Growth study in *Olivella biplicata* (Sowerby, 1825). Veliger 11: 259–67.

Pteropurpura trialata (Sowerby) (formerly *Pterynotus*), §296

Searlesia dira (Reeve), §200

LLOYD, M. C. 1971. The biology of *Searlesia dira* (Mollusca: Gastropoda) with emphasis on feeding. Ph.D. dissertation, University of Michigan. 97 pp.

LOUDA, S. M. 1979. Distribution, movement and diet of the snail *Searlesia dira* in the intertidal community of San Juan Island, Puget Sound, Washington. Mar. Biol. 51: 119–31.

RIVEST, B. R. 1983. Development and the influence of nurse egg allotment on hatching size in *Searlesia dira* (Reeve, 1846) (Prosobranchia: Buccinidae). J. Exp. Mar. Biol. Ecol. 69: 217–41.

Urosalpinx cinerea (Say), §§219, 286

This introduced species, the Atlantic oyster drill, has been much studied in connection with its feeding habits. Limited space precludes all but a few references.

CARRIKER, M. R., and D. VAN ZANDT. 1972. Predatory behavior of a shell-boring muricid gastropod. *In* H. E. Winn and B. L. Olla, eds., Behavior of marine animals, 1, Invertebrates, pp. 157–244. New York: Plenum. □A thorough review; cites a 1955 paper by Carriker summarizing general aspects of this snail's biology.

CARRIKER, M. R., D. VAN ZANDT, and T. J. GRANT. 1978. Penetration of molluscan and nonmolluscan minerals by the boring gastropod *Urosalpinx cinerea*. Biol. Bull. 155: 511–26.

FRANZ, D. R. 1971. Population age structure, growth and longevity of the marine gastropod *Urosalpinx cinerea* Say. Biol. Bull. 140: 63–72.

JILLSON, D. A. 1981. A niche analysis of coexisting *Thais lapillus* and *Urosalpinx cinerea* populations. Veliger 23: 277–81.

RITTSCHOF, D., L. G. WILLIAMS, B. BROWN, and M. R. CARRIKER. 1983. Chemical attraction of newly hatched oyster drills. Biol. Bull. 164: 493–505.

WILLIAMS, L. G., D. RITTSCHOF, B. BROWN, and M. R. CARRIKER. 1983. Chemotaxis of oyster drills *Urosalpinx cinerea* to competing prey odors. Biol. Bull. 164: 536–48.

WOOD, L. 1968. Physiological and ecological aspects of prey selection by the marine gastropod *Urosalpinx cinerea* (Prosobranchia: Muricidae). Malacologia 6: 267–320.

Subclass OPISTHOBRANCHIA
Nudibranchs, Sea hares, Sea slugs, etc.

See especially Behrens (1980), Hurst (1967), MacFarland (1966), and Marcus (1961).

BEHRENS, D. W. 1980. Pacific coast nudibranchs: A guide to the opisthobranchs of the northeastern Pacific. Los Osos, Calif.: Sea Challengers. 112 pp. □An excellent little book, complete with 162 color plates of excellent quality.

BELCIK, F. P. 1975. Additional opisthobranch mollusks from Oregon. Veliger 17: 276–77.

BERTSCH, H. 1981. Correct authorship and date citations for several Californian opisthobranch gastropod taxa described in publications by Cooper and Cockerell. Veliger 24: 49–50.

FAULKNER, D. J., and M. T. GHISELIN. 1983. Chemical defense and evolutionary ecology of dorid nudibranchs and some other opisthobranch gastropods. Mar. Ecol. Prog. Ser. 13: 295–301.

GHISELIN, M. T. 1965. Reproductive function and the phylogeny of opisthobranch gastropods. Malacologia 3: 327–78.

GOSLINER, T. M. 1981. Origins and relationships of primitive members of the Opisthobranchia (Mollusca: Gastropoda). Biol. J. Linn. Soc. 16: 197–225.

GOSLINER, T. M., and G. C. WILLIAMS. 1970. The opisthobranch mollusks of Marin County, California (Gastropoda). Veliger 13: 175–80.

HURST, A. 1965. Studies on the structure and function of the feeding apparatus of *Philine aperta* with a comparative consideration of some other opisthobranchs. Malacologia 2: 281–347.

———. 1967. The egg masses and veligers of thirty northeast Pacific opisthobranchs. Veliger 9: 255–88.

LANCE, J. R. 1966. New distributional records of some northeastern Pacific Opisthobranchiata (Mollusca, Gastropoda) with descriptions of two new species. Veliger 9: 69–81. □Extends known range of several opisthobranchs southward into southern California and the Gulf of California.

MACFARLAND, F. M. 1966. Studies of opisthobranchiate mollusks of the Pacific coast of North America. Mem. Calif. Acad. Sci. 6: 1–546. □For years—for decades—we had awaited this great work; and finally, some 15 years after the author's death, it was published, according to the will of Mrs. MacFarland. There are many fine things in this book; but unfortunately it is not without flaws, since many of the species proposed in this work had already been named by others in the interim, and it will take specialists as many more years to unscramble the complications (see Roller 1970 for a start). However, all this, and much more, has been set forth with marvelous tact and restraint by G. Dallas Hanna in the introduction to the volume. For those who do not read prefaces, we can only say *caveat emptor*.

MARCUS, ER. 1961. Opisthobranch mollusks from California. Veliger 3(suppl.): 1–85. □Provides definitive descriptions, with guts and gizzards, of most of the common species and some of the rare ones. Many sea slugs, alas, cannot be known from their pretty exteriors alone.

MILLEN, S. V. 1983. Range extensions of opisthobranchs in the northeastern Pacific. Veliger 25: 383–86.

MORTON, J. E. 1972. The form and functioning of the pallial organs in the opisthobranch *Akera bullata* with a discussion on the nature of the gill in Notaspidea and other tectibranchs. Veliger 14: 337–49. □Good general discussion of the gill in opisthobranchs.

ROBILLIARD, G. A. 1969. A method of color preservation in opisthobranch mollusks. Veliger 11: 289–91. □Works for some colors, some species, for a while. Unlikely to replace, but may supplement, color photography and good old-fashioned field notes.

ROLLER, R. A. 1970. A list of recommended nomenclatural changes for MacFarland's "Studies of opisthobranchiate mollusks of the Pacific coast of North America." Veliger 12: 371–74.

ROLLER, R. A., and S. J. LONG. 1969. An annotated list of opisthobranchs from San Luis Obispo County, California. Veliger 11: 424–30.

SPHON, G. G. 1972. Some opisthobranchs (Mollusca: Gastropoda) from Oregon. Veliger 15: 153–57. □Records 19 species of opisthobranchs from intertidal locations in Oregon. Only one had been reported previously from Oregon, although Oregon was within the recorded range of all but one species.

SPHON, G. G., and J. R. LANCE. 1968. An annotated list of nudibranchs and their allies from Santa Barbara County, California. Proc. Calif. Acad. Sci. (4)36: 73–84.

THOMPSON, T. E. 1976. Biology of opisthobranch molluscs. Vol. 1. London: Ray Society. 206 pp.

Order CEPHALASPIDEA

GOSLINER, T. M. 1980. Systematics and phylogeny of the Aglajidae (Opisthobranchia: Mollusca). Zool. J. Linn. Soc. 68: 325–60. □Includes treatment of the genera *Aglaja*, *Melanochlamys*, and *Navanax*. Reinstates *Navanax* as a genus distinct from *Aglaja*.

Aglaja ocelligera (Bergh), §274

Bulla gouldiana Pilsbry, §287

ROBLES, L. J. 1975. The anatomy and functional morphology of the reproductive system of *Bulla gouldiana* (Gastropoda: Opisthobranchia). Veliger 17: 278–91.

Haminoea vesicula (Gould), §274

Melanochlamys diomedea (Bergh), §274

M. diomedea is also listed sometimes as a species of *Aglaja*.

Navanax inermis (Cooper) (also known as *Aglaja* or *Chelidonura*), §294

BERTSCH, H., and A. A. SMITH. 1970. Observations on opisthobranchs of the Gulf of California. Veliger 13: 171–74.

PAINE, R. T. 1963. Food recognition and predation on opisthobranchs by *Navanax inermis* (Gastropoda: Opisthobranchia). Veliger 6: 1–9.

———. 1965. Natural history, limiting factors and energetics of the opisthobranch *Navanax inermis*. Ecology 46: 603–19.

SLEEPER, H. L., V. J. PAUL, and W. FENICAL. 1980. Alarm pheromones from the marine opisthobranch *Navanax inermis*. J. Chem. Ecol. 6: 57–70. □When molested, this opisthobranch secretes a bright-yellow mixture of compounds into its mucous trail. The compounds induce any *Navanax* following the trail to cease and turn away. The function in nature is unclear.

SUSSWEIN, A. J., and M. V. L. BENNETT. 1979. Plasticity of feeding behavior in the opisthobranch mollusc *Navanax*. J. Neurobiol. 10: 521–34.

SUSSWEIN, A. J., M. S. CAPPELL, and M. V. L. BENNETT. 1982. Distance chemoreception in *Navanax inermis*. Mar. Behav. Physiol. 8: 231–41.

Rictaxis punctocaelatus (Carpenter) (formerly *Acteon*), §287

Order ANASPIDEA

BEEMAN, R. D. 1968. The order Anaspidea. Veliger 3(suppl.): 87–102.

Aplysia californica Cooper (*Tethys* in older works), §104

Aplysia has been the subject of a tremendous number of laboratory studies, many dealing with cellular aspects of neural and behavioral integration. Its popularity is to a large extent the result of a

central nervous system (composed of very large cells consistently located) that is nearly ideal for experimentation, a feature *Aplysia* shares with other opisthobranchs. For an adequate survey of the literature dealing with this animal as a model nervous system, the interested reader must consult accounts other than the present one; Kandel's excellent book (1979) is a good start. A sample of papers with a natural history bent is provided below.

AMBROSE, H. W., III, R. P. GIVENS, R. CHEN, and K. P. AMBROSE. 1979. Distastefulness as a defense mechanism in *Aplysia brasiliana* (Mollusca: Gastropoda). Mar. Behav. Physiol. 6: 57–64.

AUDESIRK, T. E. 1975. Chemoreception in *Aplysia californica*. I. Behavioral localization of distance chemoreceptors used in food-finding. Behav. Biol. 15: 45–55.

————. 1979. A field study of growth and reproduction in *Aplysia californica*. Biol. Bull. 157: 407–21. □In southern California, small specimens appeared between February and May. Breeding activity was highest during July and August, and the number of animals decreased during the summer. Few adults were found from October through December, and it was suggested that mature animals died after the summer breeding period.

AUDESIRK, T. E., and G. J. AUDESIRK. 1977. Chemoreception in *Aplysia californica*—II. Electrophysiological evidence for detection of the odor of conspecifics. Comp. Biochem. Physiol. 56A: 267–70.

DiMATTEO, T. 1982. The ink of *Aplysia dactylomela* (Rang, 1828) (Gastropoda: Opisthobranchia) and its role as a defensive mechanism. J. Exp. Mar. Biol. Ecol. 57: 169–80.

KANDEL, E. R. 1979. Behavioral biology of *Aplysia*. San Francisco: Freeman. 463 pp.

KANDEL, P., and T. R. CAPO. 1979. The packaging of ova in the egg cases of *Aplysia californica*. Veliger 22: 194–98.

KRIEGSTEIN, A. R. 1977. Stages in the post-hatching development of *Aplysia californica*. J. Exp. Zool. 199: 275–88.

KRIEGSTEIN, A. R., V. CASTELLUCCI, and E. R. KANDEL. 1974. Metamorphosis of *Aplysia californica* in laboratory culture. Proc. Natl. Acad. Sci. USA 71: 3654–58.

MACGINITIE, G. E. 1934. The egg-laying activities of the sea hare *Tethys californicus* (Cooper). Biol. Bull. 67: 300–303.

ROBERTS, M. H., and G. D. BLOCK. 1982. Dissection of circadian organization of *Aplysia* through connective lesions and electrophysiological recording. J. Exp. Zool. 219: 39–50. □*Aplysia californica* displays a circadian activity rhythm, the timing mechanism for which seems to be localized in the cerebral and buccal ganglia and the sensory structures served by these ganglia.

WINKLER, L. R., and E. Y. DAWSON. 1963. Observations and experiments on the food habits of California sea hares of the genus *Aplysia*. Pacific Sci. 17: 102–5.

Phyllaplysia taylori Dall, §274

BEEMAN, R. D. 1963. Variation and synonymy of *Phyllaplysia* in the northeastern Pacific (Mollusca: Opisthobranchia). Veliger 6: 43–47.

————. 1970a. An ecological study of *Phyllaplysia taylori* Dall, 1900 (Gastropoda: Opisthobranchia) with an emphasis on its reproduction. Vie et Milieu (A) Biol. Mar. 21: 189–211.

————. 1970b. An autoradiographic study of sperm exchange and storage in a sea hare, *Phyllaplysia taylori*, a hermaphroditic gastropod (Opisthobranchia: Anaspidea). J. Exp. Zool. 175: 125–32.

————. 1970c. The anatomy and functional morphology of the reproductive system in the opisthobranch mollusk *Phyllaplysia taylori* Dall, 1900. Veliger 13: 1–31. □This paper and the one above have untangled and illuminated the structure and function of this animal's complicated hermaphroditic reproductive system, in which a single duct may simultaneously contain incoming sperm, outgoing sperm, and a departing string of fertilized eggs.

BRIDGES, C. B. 1975. Larval development of *Phyllaplysia taylori* Dall, with a discussion of development in the Anaspidea (Opisthobranchiata: Anaspidea). Ophelia 14: 161–84. □*Phyllaplysia* is unusual for an aplysiid in that its development is direct. Development within the capsule takes about 30 days at 14.5°C.

Order SACOGLOSSA

This order contains various interesting beasts (seven or eight species on our coast) that often escape notice: they tend to be small, cryptically colored, and patchy in distribution. These animals are usually found associated with algae, upon which they feed by slitting open cells of the plant with a narrow radula and sucking out the contents. Several sacoglossans contain chloroplasts, derived from their food and maintained in a symbiotic association.

GONOR, J. J. 1961. Observations on the biology of *Hermaeina smithi*, a sacoglossan opisthobranch from the west coast of North America. Veliger 4: 85–98. □Now known as *Aplysiopsis smithi*.

GREENE, R. W. 1968. The egg masses and veligers of southern California sacoglossan opisthobranchs. Veliger 11: 100–104.

————. 1970a. Symbiosis in sacoglossan opisthobranchs: Symbiosis with algal chloroplasts. Malacologia 10: 357–68.

————. 1970b. Symbiosis in sacoglossan opisthobranchs: Translocation of photosynthetic products from chloroplast to host tissue. Malacologia 10: 369–80.

HAND, C., and J. STEINBERG. 1955. On the occurrence of the nudibranch *Alderia modesta* (Loven, 1844) on the central California coast. Nautilus 69: 22–28.

JENSEN, K. R. 1980. A review of sacoglossan diets, with comparative notes on radular and buccal anatomy. Malacol. Rev. 13: 55–78.

MILLEN, S. V. 1980. Range extensions, new distribution sites, and notes on the biology of sacoglossan opisthobranchs (Mollusca: Gastropoda) in British Columbia. Can. J. Zool. 58: 1207–9.

WILLIAMS, G. C., and T. M. GOSLINER. 1973. Range extensions for four sacoglossan opisthobranchs from the coasts of California and the Gulf of California (Mollusca: Gastropoda). Veliger 16: 112–16.

Order NOTASPIDEA

Pleurobranchaea californica MacFarland, not treated
Although subtidal, this form deserves mention because of the research that has been done, and continues to be done, on its behavior and related neural circuitry. Besides, it is a large (to 36 cm in length!) and interesting animal, sure to fascinate whoever sees it.
CHIVERS, D. D. 1967. Observations on *Pleurobranchaea californica* MacFarland, 1966 (Opisthobranchia, Notaspidea). Proc. Calif. Acad. Sci. (4)32: 515–21. □Describes egg ribbon and capsules.
DAVIS, W. J., G. J. MPITSOS, and J. M. PINNEO. 1974. The behavioral hierarchy of the mollusk *Pleurobranchaea*. I, The dominant position of the feeding behavior. J. Comp. Physiol. 90: 207–24. II, Hormonal suppression of feeding associated with egg-laying. J. Comp. Physiol. 90: 225–43.
MPITSOS, G. J., S. D. COLLINS, and A. D. MC-CLELLAN. 1978. Learning: A model system for physiological studies. Science 199: 497–506.

Tylodina fungina Gabb, §98
DuSHANE, H. 1966. Range extension for *Tylodina fungina* Gabb, 1865 (Gastropoda). Veliger 9: 86.

Order NUDIBRANCHIA

BEHRENS, D. W. 1982. *Sakuraeolis enosimensis* (Baba, 1930) (Nudibranchia: Aeolidacea) in San Francisco Bay. Veliger 24: 359–63. □Easily confused with *Hermissenda*, this nudibranch has been introduced from Japan.
BERTSCH, H. 1977. The Chromodoridinae nudibranchs from the Pacific coast of America. I. Investigative methods and supra-specific taxonomy. Veliger 20: 107–18.
———. 1978a. The Chromodoridinae nudibranchs from the Pacific coast of America. II. The genus *Chromodoris*. Veliger 20: 307–27. □Includes *Chromodoris macfarlandi*.
———. 1978b. The Chromodoridinae nudibranchs from the Pacific coast of America. III. The genera *Chromolaichma* and *Mexichromis*. Veliger 21: 70–86. □Includes *Mexichromis porterae*.
BERTSCH, H., T. GOSLINER, R. WHARTON, and G. WILLIAMS. 1972. Natural history and occurrence of opisthobranch gastropods from the open coast of San Mateo County, California. Veliger 14: 302–14.
BLOOM, S. A. 1976. Morphological correlations between dorid nudibranch predators and sponge prey. Veliger 18: 289–301.
———. 1981. Specialization and noncompetitive resource partitioning among sponge-eating dorid nudibranchs. Oecologia 49: 305–15. □Six species of dorid nudibranchs in the San Juan Islands appear to partition available habitat and sponge food

resources without competing: *Archidoris montereyensis, Archidoris odhneri, Cadlina luteomarginata, Diaulula sandiegensis, Anisodoris nobilis,* and *Discodoris heathi.*
BLOOM, S. A., and C. F. BLOOM. 1977. Radular variation in two species of sponge-rasping dorid nudibranchs. J. Molluscan Stud. 43: 296–300. □The number of teeth per row, the number of rows in the radula, and the length of teeth increase as specimens of *Archidoris montereyensis* and *Anisodoris nobilis* grow, casting some doubt on the taxonomic utility of these features.
DAY, R. M., and L. G. HARRIS. 1978. Selection and turnover of coelenterate nematocysts in some aeolid nudibranchs. Veliger 21: 104–9. □Only certain of the nematocyst types present in cnidarians taken as prey are retained in the cerata of these nudibranchs. The nematocysts apparently have a short storage life and are replaced within 5 days.
EDMUNDS, M. 1966. Protective mechanisms in the Eolidacea (Mollusca, Nudibranchia). J. Linn. Soc. Lond. 46: 27–71.
FUHRMAN, F. A., G. J. FUHRMAN, and K. DE-RIEMER. 1979. Toxicity and pharmacology of extracts from dorid nudibranchs. Biol. Bull. 156: 289–99. □Extracts of several dorids are shown to be lethal when injected into shore crabs and mice.
GODDARD, J. H. R. 1981. Range extension and notes on the food, morphology, and color pattern of the dorid nudibranch *Hallaxa chani*. Veliger 24: 155–58.
HARRIS, L. G. 1973. Nudibranch associations. *In* T. C. Cheng, ed., Current topics in comparative pathobiology, Vol. 2, pp. 213–315. New York: Academic Press.
McBETH, J. W. 1971. Studies on the food of nudibranchs. Veliger 14: 158–61. □Describes food and feeding of *Hypselodoris californiensis, Anisodoris nobilis, Hopkinsia rosacea, Triopha catalinae* (as *T. carpenteri*), *Dendrodoris fulva,* and *Flabellinopsis iodinea* in the San Diego area.
McDONALD, G. R., and J. W. NYBAKKEN. 1978. Additional notes on the food of some California nudibranchs with a summary of known food habits of California species. Veliger 21: 110–19.
———. 1980. Guide to the nudibranchs of California. Melbourne, Fla.: American Malacologists. 72 pp. □Contains keys, color photographs, and notes on biology and taxonomy.
NYBAKKEN, J. 1974. A phenology of the smaller dendronotacean, arminacean and aeolidacean nudibranchs at Asilomar State Beach over a twenty-seven-month period. Veliger 16: 370–73.
———. 1978. Abundance, diversity and temporal variability in a California intertidal nudibranch assemblage. Mar. Biol. 45: 129–46. □A three-and-a-half-year study of the abundances of 31 species of nudibranchs in the Pacific Grove area.
NYBAKKEN, J., and G. McDONALD. 1981. Feeding mechanisms of west American nudibranchs feeding on Bryozoa, Cnidaria and Ascidiacea, with special respect to the radula. Malacologia 20: 439–49.
ROBILLIARD, G. A. 1970. The systematics and

some aspects of the ecology of the genus *Dendronotus*. Veliger 12: 433–79.

––––––. 1971. A new species of *Polycera* (Opisthobranchia: Mollusca) from the northeastern Pacific, with notes on other species. Syesis 4: 235–43.

RUSSELL, H. D. 1971. Index Nudibranchia: A catalog of the literature, 1554–1965. Greenville: Delaware Museum of Natural History. 141 pp.

THOMPSON, T. E. 1971. Tritoniidae from the North American Pacific coast (Mollusca: Opisthobranchia). Veliger 13: 333–38.

WILLIAMS, L. G. 1980. Development and feeding of larvae of the nudibranch gastropods *Hermissenda crassicornis* and *Aeolidia papillosa*. Malacologia 20: 99–116.

Aegires albopunctatus MacFarland, §96

BERTSCH, H. 1980. The nudibranch *Aegires albopunctatus* (Polyceratacea: Aegiretidae) preys on *Leucilla nuttingi* (Porifera: Calcarea). Veliger 22: 222–24.

Aeolidia papillosa (Linnaeus), §109

BERTSCH, H. 1974. Descriptive study of *Aeolidia papillosa* with scanning electron micrographs of the radula. Tabulata 7: 3–6.

HALL, S. J., C. D. TODD, and A. D. GORDON. 1982. The influence of ingestive conditioning on the prey species selection in *Aeolidia papillosa* (Mollusca: Nudibranchia). J. Anim. Ecol. 51: 907–21.

WATERS, V. L. 1973. Food-preference of the nudibranch *Aeolidia papillosa*, and the effect of the defenses of the prey on predation. Veliger 15: 174–92.

Anisodoris nobilis (MacFarland), §§109, 178

KITTING, C. L. 1981. Feeding and pharmacological properties of *Anisodoris nobilis* individuals (Mollusca; Nudibranchia) selecting different noxious sponges plus detritus. Biol. Bull. 161: 126–40. □Taken as a group, the *Anisodoris* studied fed on at least five sponge species. However, different individuals tended to choose different sponge species.

Archidoris montereyensis (Cooper), §109

COOK, E. F. 1962. A study of food choices of two opisthobranchs, *Rostanga pulchra* MacFarland and *Archidoris montereyensis* (Cooper). Veliger 4: 194–96.

Armina californica (Cooper), §251

BERTSCH, H. 1968. Effect of feeding by *Armina californica* on the bioluminescence of *Renilla kollikeri*. Veliger 10: 440–41.

KASTENDIEK, J. 1976. Behavior of the sea pansy *Renilla kollikeri* Pfeffer (Coelenterata: Pennatulacea) and its influence on the distribution and biological interactions of the species. Biol. Bull. 151: 518–37. □Describes predation on the sea pansy *Renilla* by *Armina* and a behavioral escape response of the prey.

Cadlina luteomarginata MacFarland (*C. marginata* MacFarland in previous editions), §109

Chromodoris macfarlandi Cockerell (*Glossodoris* in previous editions), §109

See Bertsch (1978a) above.

Corambe pacifica MacFarland and O'Donoghue, §110

Diaulula sandiegensis (Cooper), §35

ELVIN, D. W. 1976. Feeding of a dorid nudibranch, *Diaulula sandiegensis*, on the sponge *Haliclona permollis*. Veliger 19: 194–98.

Dirona albolineata Cockerell and Eliot, §109

ROBILLIARD, G. A. 1971. Predation by the nudibranch *Dirona albolineata* on three species of prosobranchs. Pacific Sci. 25: 429–35. □*Dirona* will eat ectoprocts, hydroids, small crustaceans, sponges, and tunicates, apparently depending on what is available. One population in the San Juan Islands was feeding primarily on bryozoans. Another population, the one for which data are presented, was feeding on small gastropods (*Lacuna carinata* and two species of *Margarites*), which were crushed in the nudibranch's jaws.

Doridella steinbergae (Lance), §110

ANDERSON, G. B. 1971. A contribution to the biology of *Doridella steinbergae* and *Corambe pacifica*. Master's thesis, California State University, Hayward. 48 pp.

PERRON, F. E., and R. D. TURNER. 1977. Development, metamorphosis, and natural history of the nudibranch *Doridella obscura* Verrill (Corambidae: Opisthobranchia). J. Exp. Mar. Biol. Ecol. 27: 171–85. □An Atlantic species.

Eubranchus olivaceus (O'Donoghue) (formerly *Galvina*), §271

Hermissenda crassicornis (Eschscholtz), §§109, 295

BURGIN, U. F. 1965. The color pattern of *Hermissenda crassicornis* (Eschscholtz, 1831) (Gastropoda: Opisthobranchia: Nudibranchia). Veliger 7: 205–15.

HARRIGAN, J. F., and D. L. ALKON. 1978. Larval rearing, metamorphosis, growth and reproduction of the eolid nudibranch *Hermissenda crassicornis* (Eschscholtz, 1831) (Gastropoda: Opisthobranchia). Biol. Bull. 154: 430–39. □The animal reproduces all year and can be raised through its complete life cycle in the laboratory. The generation time (egg to egg) may be as short as 2.5 months, and the life span about 4 months.

LEDERHENDLER, I. I., E. S. BARNES, and D. L. ALKON. 1980. Complex responses to light of the nudibranch *Hermissenda crassicornis* (Gastropoda: Opisthobranchia). Behav. Neural Biol. 28: 218–30. □The eye and light responses of *Hermissenda* have received considerable attention by cell biologists because the eye complex conveniently contains only 5 photoreceptors plus 12 ganglion cells, all of which can be penetrated by electrodes. In addition to its discussion of behavior, this paper includes pertinent references to studies of a cellular nature.

LONGLEY, R. D., and A. J. LONGLEY. 1982. *Hermissenda*: Agonistic behavior or mating behavior? Veliger 24: 230–31. □Describes mating behavior, which may have been confused with agonistic behavior by earlier workers.

RUTOWSKI, R. L. 1983. Mating and egg mass production in the aeolid nudibranch *Hermissenda crassicornis* (Gastropoda: Opisthobranchia). Biol. Bull. 165: 276–85.

ZACK, S. 1974. The effects of food deprivation on

agonistic behavior in an opisthobranch mollusc, *Hermissenda crassicornis*. Behav. Biol. 12: 223–32. □Deprivation increased the proportion of aggressive encounters and the proportion of aggressive responses involving biting.

――――. 1975. A description and analysis of agonistic behavior patterns in an opisthobranch mollusc, *Hermissenda crassicornis*. Behaviour 53: 238–67.

Hopkinsia rosacea MacFarland, §109

Hypselodoris californiensis (Bergh) (also as *Glossodoris* and *Chromodoris*), §109

Melibe leonina (Gould) (formerly *Chioraera*), §275

AJESKA, R. A., and J. NYBAKKEN. 1976. Contributions to the biology of *Melibe leonina* (Gould, 1852) (Mollusca: Opisthobranchia). Veliger 19: 19–26. □Information on feeding and mating behavior, diet, predators, and commensals. Includes photographs of the animal, and cites the 1923 Agersborg morphological paper on this form.

BICKELL, L. R., and S. C. KEMPF. 1983. Larval and metamorphic morphogenesis in the nudibranch *Melibe leonina* (Mollusca: Opisthobranchia). Biol. Bull. 165: 119–38.

GUBERLET, J. E. 1928. Observations on the spawning habits of *Melibe*. Publ. Puget Sound Biol. Sta. 6: 262–70.

HURST, A. 1968. The feeding mechanism and behaviour of the opisthobranch *Melibe leonina*. Symp. Zool. Soc. Lond. 22: 151–66.

Mexichromis porterae (Cockerell) (also as *Glossodoris*, *Chromodoris*, or *Hypselodoris*), §109

See Bertsch (1978b) above.

Rostanga pulchra MacFarland, §33

See also Cook (1962), under *Archidoris montereyensis*.

ANDERSON, E. S. 1971. The association of the nudibranch *Rostanga pulchra* MacFarland, 1905, with the sponges *Ophlitaspongia pennata*, *Esperiopsis originalis*, and *Plocamia karykina*. Ph.D. dissertation, Biology, University of California, Santa Cruz. 161 pp.

CHIA, F.-S., and R. KOSS. 1978. Development and metamorphosis of the planktotrophic larvae of *Rostanga pulchra* (Mollusca: Nudibranchia). Mar. Biol. 46: 109–19.

Triopha catalinae (Cooper) (*T. carpenteri* (Stearns) in many papers), §§34, 109

FERREIRA, A. J. 1977. A review of the genus *Triopha* (Mollusca: Nudibranchia). Veliger 19: 387–402.

NYBAKKEN, J., and J. EASTMAN. 1977. Food preference, food availability and resource partitioning in *Triopha maculata* and *Triopha carpenteri* (Opisthobranchia: Nudibranchia). Veliger 19: 279–89.

T. maculata MacFarland, §109

Subclass PULMONATA

Trimusculus reticulatus (Sowerby) (formerly *Gadinia*), §147

WALSBY, J. R. 1975. Feeding and the radula in the marine pulmonate limpet *Trimusculus reticulatus*. Veliger 18: 139–45.

YONGE, C. M. 1958. Observations in life on the pulmonate limpet *Trimusculus* (*Gadinia*) *reticulatus* (Sowerby). Proc. Malacol. Soc. Lond. 33: 31–37.

――――. 1960. Further observations on *Hipponix antiquatus* with notes on north Pacific pulmonate limpets. Proc. Calif. Acad. Sci. (4)31: 111–19.

Class BIVALVIA (PELECYPODA). Clams, Cockles, Mussels, Oysters, Shipworms, etc.

In many habitats, bivalves are the dominant organisms; dense beds of mussels, clams, and oysters come readily to mind. The group, the second largest of the molluscan classes, is diverse, having undergone extensive adaptive radiation. Bivalve taxonomy itself has undergone an evolution of sorts in recent years, and the practice of dividing the bivalves into the large orders Filibranchia and Eulamellibranchia (plus the Protobranchia and Septibranchia) has been abandoned in this edition. In its place is a scheme not quite as descriptive or convenient in some ways as the old one, but one that reflects the current thinking of systematists.

BERNARD, F. R. 1983. Catalogue of the living Bivalvia of the eastern Pacific Ocean: Bering Strait to Cape Horn. Can. Spec. Publ. Fish. Aquat. Sci. 61. 102 pp.

BREUM, O. 1970. Stimulation of burrowing activity by wave action in some marine bivalves. Ophelia 8: 197–207.

FITCH, J. E. 1953. Common marine bivalves of California. Calif. Dept. Fish Game, Fish Bull. 90: 1–102.

GORDON, J., and M. R. CARRIKER. 1978. Growth lines in a bivalve mollusk: Subdaily patterns and dissolution of the shell. Science 202: 519–21.

HADERLIE, E. C. 1980. Stone boring marine bivalves from Monterey Bay, California. Veliger 22: 345–54.

――――. 1981. Growth rates of *Penitella penita* (Conrad, 1837), *Chaceia ovoidea* (Gould, 1851) (Bivalvia: Pholadidae) and other rock boring bivalves in Monterey Bay. Veliger 24: 109–14.

HADERLIE, E. C., J. C. MELLOR, C. S. MINTER, and G. C. BOOTH. 1974. The sublittoral benthic fauna and flora off Del Monte Beach, Monterey, California. Veliger 17: 185–204. □Includes information on bivalve borers and nestlers of subtidal shale.

HOLLAND, D. A., and K. K. CHEW. 1974. Reproductive cycle of the Manila clam (*Venerupis japonica*), from Hood Canal, Washington. Proc. Natl. Shellfish. Assoc. 64: 53–58. □Introduced from Japan, this is now a commercially important species in Washington.

LOOSANOFF, V. L., H. C. DAVIS, and P. E. CHANLEY. 1966. Dimensions and shapes of larvae of some marine bivalve molluscs. Malacologia 4: 351–435. □Describes the larvae of 20 species of bivalves, with an eye toward the identification of larvae in plankton tows.

MACGINITIE, G. E. 1941. On the method of feeding of four pelecypods. Biol. Bull. 80: 18–25.

Maurer, D. 1967. Mode of feeding and diet, and synthesis of studies on marine pelecypods from Tomales Bay, California. Veliger 10: 72–76. ☐Summarizes and cites previous work by author on the bivalves *Transennella tantilla*, *Tellina modesta* (as *T. buttoni*), and *Tellina nuculoides* (as *T. salmonea*).

———. 1969. Pelecypod-sediment association in Tomales Bay, California. Veliger 11: 243–49.

Mohlenberg, F., and H. U. Riisgard. 1978. Efficiency of particle retention in 13 species of suspension feeding bivalves. Ophelia 17: 239–46.

Peterson, C. H. 1977. Competitive organization of the soft-bottom macrobenthic communities of southern California lagoons. Mar. Biol. 43: 343–59. ☐Concerns *Cryptomya californica*, *Nuttallia nuttallii* (as *Sanguinolaria*), *Protothaca staminea*, *Saxidomus nuttalli*, *Tagelus californianus*, and *Tresus nuttallii*.

Quayle, D. B. 1973. The intertidal bivalves of British Columbia. B. C. Prov. Mus. Natur. Hist. Handb. 17. 104 pp.

Reid, R. G. B. 1971. Criteria for categorizing feeding types in bivalves. Veliger 13: 358–59.

Trueman, E. R. 1968. The burrowing activities of bivalves. Symp. Zool. Soc. Lond. 22: 167–86.

Yonge, C. M. 1971. On functional morphology and adaptive radiation in the bivalve superfamily Saxicavacea (*Hiatella* (= *Saxicava*), *Saxicavella*, *Panomya*, *Panope*, *Cyrtodaria*). Malacologia 11: 1–44.

Subclass PTERIOMORPHA

Order Mytiloida

Van Winkle, W., Jr. 1970. Effect of environmental factors on byssal thread formation. Mar. Biol. 7: 143–48. ☐In *Ischadium demissum* (as *Modiolus*) the number of byssal threads formed per hour decreased with increasing mussel size, and prior exposure to air for 1 to 2 days enhanced subsequent thread formation. In *Mytilus edulis* no byssal threads were produced at a salinity of 16 parts per thousand and few at 32 parts per thousand.

Ischadium demissum (Dillwyn), §285
The generic placement of this animal, listed as *Volsella* in the preceding edition, is uncertain; it is also placed at times in the genera *Geukensia*, *Modiolus*, or *Arcuatula*.

Bertness, M. D. 1980. Growth and mortality in the ribbed mussel *Geukensia demissa* (Bivalvia: Mytilidae). Veliger 23: 62–69.

Booth, C. E., and C. P. Mangum. 1978. Oxygen uptake and transport in the lamellibranch mollusc *Modiolus demissus*. Physiol. Zool. 51: 17–32.

Kuenzler, E. J. 1961. Structure and energy flow of a mussel population in a Georgia salt marsh. Limnol. Oceanogr. 6: 191–204.

Lent, C. M. 1968. Air-gaping by the ribbed mussel, *Modiolus demissus* (Dillwyn): Effects and adaptive significance. Biol. Bull. 134: 60–73.

———. 1969. Adaptations of the ribbed mussel, *Modiolus demissus* (Dillwyn), to the intertidal habitat. Amer. Zool. 9: 283–92.

Lithophaga plumula (Hanley) (*L. subula* (Reeve) of some authors), §241

See also Haderlie (1980), listed above under Bivalvia.

Hodgkin, N. M. 1962. Limestone boring by the mytilid *Lithophaga*. Veliger 4: 123–29.

Jaccarina, V., W. H. Bannister, and H. Micallef. 1968. The pallial glands and rock boring in *Lithophaga lithophaga* (Lamellibranchia, Mytilidae). J. Zool. (Lond.) 154: 397–401.

Yonge, C. M. 1955. Adaptations to rock boring in *Botula* and *Lithophaga* (Lamellibranchia, Mytilidae) with a discussion on the evolution of this habit. Quart. J. Microsc. Sci. 96: 383–410.

Modiolus capax (Conrad), fat horse mussel, §§175, 332

M. modiolus (Linnaeus), horse mussel, §§159, 175
Musculus senhousia (Benson) (formerly *Volsella*), §285

Mytilus californianus Conrad, §158
See the references under *Mytilus edulis*.

M. edulis Linnaeus, §197
An enormous amount of work has been done on the ecology and physiology of *Mytilus californianus* and *M. edulis*. Fortunately, the literature has recently been summarized by Bayne (1976), and readers are referred to that volume and its bibliography for earlier work; papers of particular interest are those by Coe, Field, Fox, Harger, Reish, Ross, Seed, and White. Work on these common and important bivalves has not diminished since 1976, however, and only a small sample of the many recent, worthy papers appears here.

Bayne, B. L., ed. 1976. Marine mussels: Their ecology and physiology. Cambridge: Cambridge University Press. 506 pp.

Bayne, B. L., and C. M. Worrall. 1980. Growth and production of mussels *Mytilus edulis* from two populations. Mar. Ecol. Prog. Ser. 3: 317–28.

Chan, G. L. 1973. Subtidal mussel beds in Baja California with a new record size for *Mytilus californianus*. Veliger 16: 239–40.

Elvin, D. W., and J. J. Gonor. 1979. The thermal regime of an intertidal *Mytilus californianus* Conrad population on the central Oregon coast. J. Exp. Mar. Biol. Ecol. 39: 265–79.

Harger, J. R. E. 1972. Competitive coexistence among intertidal invertebrates. Amer. Sci. 60: 600–607. ☐Summarizes and cites earlier work by Harger on *Mytilus californianus* and *M. edulis* published originally as a series of papers in *The Veliger*, volumes 11–14 (1968–72); also cited in Bayne (1976) above.

Jorgensen, C. B. 1981. Mortality, growth, and grazing impact of a cohort of bivalve larvae, *Mytilus edulis* L. Ophelia 20: 185–92. ☐A cohort of bivalve larvae (mainly *Mytilus edulis*) in Danish waters was followed. The initial density was 3,000 per liter. The pelagic stage lasted one month and daily mortality was 13 percent. The larvae cleared 40–50 percent of small food particles (mostly flagellates) daily from the surrounding water. Diatoms were too large to be consumed.

Kent, R. M. L. 1979. The influence of heavy infestations of *Polydora ciliata* on the flesh content of *Mytilus edulis*. J. Mar. Biol. Assoc. U.K. 59: 289–97.

Kopp, J. C. 1979. Growth and intertidal gradient in the sea mussel *Mytilus californianus* Conrad, 1837 (Mollusca: Bivalvia: Mytilidae). Veliger 22: 51–56.

Levinton, J. S., and T. H. Suchanek. 1978. Geographic variation, niche breadth and genetic differentiation at different geographic scales in the mussels *Mytilus californianus* and *M. edulis*. Mar. Biol. 49: 363–75.

Manahan, D. T., S. H. Wright, G. C. Stephens, and M. A. Rice. 1982. Transport of dissolved amino acids by the mussel *Mytilus edulis*: Demonstration of net uptake from natural seawater. Science 215: 1253–55.

Moore, D. R., and D. J. Reish. 1969. Studies on the *Mytilus edulis* community in Alamitos Bay, California. IV. Seasonal variation in gametes from different regions in the Bay. Veliger 11: 250–55.

Paine, R. T. 1974. Intertidal community structure: Experimental studies on the relationship between a dominant competitor and its principal predator. Oecologia 15: 93–120. □A classic paper demonstrating the best, and in places the worst, of "modern" marine ecology.

———. 1976. Size-limited predation: An observational and experimental approach with the *Mytilus-Pisaster* interaction. Ecology 57: 858–73.

Petraitis, P. S. 1978. Distributional patterns of juvenile *Mytilus edulis* and *Mytilus californianus*. Veliger 21: 288–92.

Price, H. A. 1982. An analysis of factors determining seasonal variation in the byssal attachment strength of *Mytilus edulis*. J. Mar. Biol. Assoc. U.K. 62: 147–55. □Mussels in Britain showed increases in byssal strength with corresponding seasonal increase in wave action; there was little or no time lag. Decreases in byssal strength followed a period of weak wave action by a 29-day period.

Reish, D. J., and J. L. Ayers, Jr. 1968. Studies on the *Mytilus edulis* community in Alamitos Bay, California. III. The effects of reduced dissolved oxygen and chlorinity concentrations on survival and byssus thread production. Veliger 10: 384–88.

Riisgard, H. U., and A. Randlov. 1981. Energy budgets, growth and filtration rates in *Mytilus edulis* at different algal concentrations. Mar. Biol. 61: 227–34.

Suchanek, T. H. 1978. The ecology of *Mytilus edulis* L. in exposed rocky intertidal communities. J. Exp. Mar. Biol. Ecol. 31: 105–20.

———. 1981. The role of disturbance in the evolution of life history strategies in the intertidal mussels *Mytilus edulis* and *Mytilus californianus*. Oecologia 50: 143–52.

Taylor, R. L. 1966. *Haplosporidium tumefacientis* sp. n., the etiologic agent of a disease of the California mussel, *Mytilus californianus* Conrad. J. Invert. Pathol. 8: 109–21.

Tsuchiya, M. 1983. Mass mortality in a population of the mussel *Mytilus edulis* L. caused by high temperature on rocky shores. J. Exp. Mar. Biol. Ecol. 66: 101–11.

Septifer bifurcatus (Conrad), §136

Order Pterioida

Anomia peruviana d'Orbigny, §217

Argopecten aequisulcatus (Carpenter), §§220, 255
Some consider this species, listed as a variety of *Pecten circularis* Sowerby in the preceding edition, a northern subspecies of the Panamic *Argopecten circularis* (Sowerby).

Ordzie, C. J., and G. C. Garofalo. 1980. Behavioral recognition of molluscan and echinoderm predators by the bay scallop, *Argopecten irradians* (Lamarck) at two temperatures. J. Exp. Mar. Biol. Ecol. 43: 29–37. □Describes responses of this Atlantic species to predatory gastropods and seastars.

Chlamys hericius (Gould) (*Pecten* in previous editions), §220

Bloom, S. A. 1975. The motile escape response of a sessile prey: A sponge-scallop mutualism. J. Exp. Mar. Biol. Ecol. 17: 311–21.

Crassostrea gigas (Thunberg), Japanese or Pacific oyster, §219
See also *Ostrea lurida*, below.

Bamford, D. R., and R. Gingles. 1974. Absorption of sugars in the gill of the Japanese oyster, *Crassostrea gigas*. Comp. Biochem. Physiol. 49A: 637–46.

Berg, C. J., Jr. 1971. A review of possible causes of mortality of oyster larvae of the genus *Crassostrea* in Tomales Bay, California. Calif. Fish Game 57: 69–75. □Adults spawn but larvae generally do not survive (however, see Span 1978, below). No single factor is likely accountable for the absence of spat, but high turbidity in the water, lack of proper food, toxic blooms of dinoflagellates, and other factors are seen as likely to contribute.

Bernard, F. R. 1973. Crystalline style formation and function in the oyster *Crassostrea gigas* (Thunberg, 1795). Ophelia 12: 159–70.

———. 1974a. Annual biodeposition and gross energy budget of mature Pacific oysters, *Crassostrea gigas*. J. Fish. Res. Board Can. 31: 185–90.

———. 1974b. Particle sorting and labial palp function in the Pacific oyster *Crassostrea gigas* (Thunberg, 1795). Biol. Bull. 146: 1–10.

Quayle, D. B. 1969. Pacific oyster culture in British Columbia. Bull. Fish. Res. Board Can. 169: 1–192.

Span, J. A. 1978. Successful reproduction of giant Pacific oysters in Humboldt Bay and Tomales Bay, California. Calif. Fish Game 64: 123–24.

C. virginica (Gmelin), §219

Galtsoff, P. S. 1964. The American oyster *Crassostrea virginica* Gmelin. U.S. Fish Wildl. Serv., Fish. Bull. 64: 1–480.

Nelson, T. C. 1924. The attachment of oyster larvae. Biol. Bull. 46: 143–51.

Newell, R. I. E., and S. J. Jordan. 1983. Preferential ingestion of organic material by the American oyster *Crassostrea virginica*. Mar. Ecol. Prog. Ser. 13: 47–53.

Ortega, S. 1981. Environmental stress, competition and dominance of *Crassostrea virginica* near Beaufort, North Carolina, USA. Mar. Biol. 62: 47–56.

Hinnites giganteus Gray (= *H. multirugosus* (Gale)), rock scallop, §125

GRAU, G. 1959. Pectinidae of the eastern Pacific. Allan Hancock Pacific Exped. 23: 1–308.

LAURÉN, D. J. 1982. Oogenesis and protandry in the purple-hinge rock scallop, *Hinnites giganteus*, in upper Puget Sound, Washington, USA. Can. J. Zool. 60: 2333–36.

MONICAL, J. B., JR. 1980. Comparative studies on growth of the purple-hinge rock scallop *Hinnites multirugosus* (Gale) in the marine waters of southern California. Proc. Natl. Shellfish. Assoc. 70: 14–21.

YONGE, C. M. 1951. Studies on Pacific coast mollusks. III. Observations on *Hinnites multirugosus* (Gale). Univ. Calif. Publ. Zool. 55: 409–20.

Leptopecten latiauratus (Conrad) (formerly *Pecten*), §276

CLARK, G. R., II. 1971. The influence of water temperature on the morphology of *Leptopecten latiauratus* (Conrad, 1837). Veliger 13: 269–72.

Ostrea edulis Linnaeus, §339

CRANFIELD, H. J. 1973. Observations on the function of the glands of the foot of the pediveliger of *Ostrea edulis* during settlement. Mar. Biol. 22: 211–23. ☐Third of three consecutive papers by Cranfield in the same volume; the series details various behavioral, morphological, and histological aspects of settlement and attachment.

———. 1974. Observations on the morphology of the mantle folds of the pediveliger of *Ostrea edulis* L. and their function during settlement. J. Mar. Biol. Assoc. U.K. 54: 1–12.

LEONARD, V. K. 1969. Seasonal gonadal changes in two bivalve mollusks in Tomales Bay, California. Veliger 11: 382–90. ☐*Ostrea edulis* spawns during the summer in Tomales Bay but fails to propagate; high turbidity seems largely responsible for larval mortality.

MORTON, B. 1971. The diurnal rhythm and tidal rhythm of feeding and digestion in *Ostrea edulis*. Biol. J. Linn. Soc. 3: 329–42.

O. lurida Carpenter, native or Olympia oyster, §219

BARRETT, E. M. 1963. The California oyster industry. Calif. Dept. Fish Game, Fish Bull. 123: 1–103.

BONNOT, P. 1937. Settling and survival of spat of the Olympia oyster, *Ostrea lurida*, on upper and lower horizontal surfaces. Calif. Fish Game 23: 224–28.

CHAPMAN, W. M., and A. H. BANNER. 1949. Contributions to the life history of the Japanese oyster drill (*Tritonalia japonica*) with notes on other enemies of the Olympia oyster (*Ostrea lurida*). Washington Dept. Fish. Biol. Rep. 49A: 167–200.

COE, W. R. 1932. Development of the gonads, and the sequence of the sexual phases, in the California oyster. Bull. Scripps Inst. Oceanogr. Tech. Ser. 3: 119–44.

———. 1934. Alternation of sexuality in oysters. Amer. Natur. 68: 236–51.

HOPKINS, A. E. 1935. Attachment of larvae of the Olympia oyster, *Ostrea lurida*, to plane surfaces. Ecology 16: 82–87.

———. 1937. Experimental observations on spawning, larval development and settling in the Olympia oyster, *Ostrea lurida*. Bull. U.S. Bur. Fish. 48: 439–503.

QUAYLE, D. B. 1969. Pacific oyster culture in British Columbia. Bull. Fish. Res. Board Can. 169: 1–192.

YONGE, C. M. 1960. Oysters. London: Collins. 209 pp.

Pododesmus cepio (Gray), §217

Listed as *Pododesmus macroschisma* (Deshayes), an Alaskan species, in previous editions; the range of *Pododesmus cepio* is British Columbia to the Gulf of California.

LEONARD, V. K. 1969. Seasonal gonadal changes in two bivalve mollusks in Tomales Bay, California. Veliger 11: 382–90. ☐Concerns *Pododesmus cepio* and *Ostrea edulis*.

YONGE, C. M. 1977. Form and evolution in the Anomiacea (Mollusca: Bivalvia)—*Pododesmus, Anomia, Patro, Enigmonia* (Anomiidae): *Placunanomia, Placuna* (Placunidae fam. nov.). Phil. Trans. Roy. Soc. Lond. B276: 453–527.

Subclass HETERODONTA

Order VENEROIDA

BERNARD, F. R. 1976. Living Chamidae of the eastern Pacific (Bivalvia: Heterodonta). Los Angeles Co. Mus. Natur. Hist. Contr. Sci. 278: 1–43.

COAN, E. 1971. The northwest American Tellinidae. Veliger 14(suppl.): 1–63. ☐Species of *Macoma* and *Tellina*.

———. 1973. The northwest American Psammobiidae. Veliger 16: 40–57. ☐Concerns *Nuttallia nuttallii* and *Tagelus californianus*.

NARCHI, W. 1971. Structure and adaptation in *Transennella tantilla* (Gould) and *Gemma gemma* (Totten) (Bivalvia: Veneridae). Bull. Mar. Sci. 21: 866–85.

PETERSON, C. H. 1982. The importance of predation and intra- and interspecific competition in the population biology of two infaunal suspension-feeding bivalves, *Protothaca staminea* and *Chione undatella*. Ecol. Monogr. 52: 437–75.

———. 1983. Interactions between two infaunal bivalves, *Chione undatella* (Sowerby) and *Protothaca staminea* (Conrad), and two potential enemies, *Crepidula onyx* Sowerby and *Cancer anthonyi* (Rathbun). J. Exp. Mar. Biol. Ecol. 68: 145–58.

POHLO, R. 1969. Confusion concerning deposit feeding in the Tellinacea. Proc. Malacol. Soc. Lond. 38: 361–64.

Chama arcana Bernard, §126

Listed in previous editions as *Chama pellucida* Sowerby, which is a South American species.

VANCE, R. R. 1978. A mutualistic interaction between a sessile marine clam and its epibionts. Ecology 59: 679–85.

YONGE, C. M. 1967. Form, habit, and evolution in the Chamidae (Bivalvia) with reference to conditions in the rudists (Hippuritacea). Phil. Trans. Roy. Soc. Lond. B252: 49–105. ☐This study is based in large part on observations of California species at Pacific Grove.

Chione undatella (Sowerby), §260

JONES, C. C. 1979. Anatomy of *Chione cancellata* and some other chionines (Bivalvia: Veneridae). Malacologia 19: 157–99. □ *Chione undatella* is included.

Clinocardium nuttallii (Conrad), basket cockle (*Cardium corbis* (Martyn) of some authors), §259

COOKE, W. J. 1975. The occurrence of an endozoic green alga in the marine mollusc, *Clinocardium nuttallii* (Conrad, 1837). Phycologia 14: 35–39.

EVANS, J. W. 1972. Tidal growth increments in the cockle *Clinocardium nuttalli*. Science 176: 416–17.

FEDER, H. M. 1967. Organisms responsive to predatory sea stars. Sarsia 29: 371–94. □ Spectacular photographs of the escape response of *Cardium echinatum* to *Asterias rubens*. Our *Clinocardium* is capable of similar gymnastics when presented with a similar stimulus.

FRASER, C. M. 1931. Notes on the ecology of the cockle, *Cardium corbis* Martyn. Trans. Roy. Soc. Can., Ser. 3, 25: 59–72.

GALLUCCI, V. F., and B. B. GALLUCCI. 1982. Reproduction and ecology of the hermaphroditic cockle *Clinocardium nuttallii* (Bivalvia: Cardiidae) in Garrison Bay. Mar. Ecol. Prog. Ser. 7: 137–45.

TAYLOR, C. C. 1960. Temperature, growth and mortality—the Pacific cockle. J. Cons. Int. Explor. Mer 26: 117–24.

Donax gouldii Dall, bean clam, §188

COAN, E. 1973. The northwest American Donacidae. Veliger 16: 130–39.

COE, W. R. 1955. Ecology of the bean clam, *Donax gouldi*, on the coast of California. Ecology 36: 512–14.

IRWIN, T. H. 1973. The intertidal behavior of the bean clam, *Donax gouldii* Dall, 1921. Veliger 15: 206–12.

POHLO, R. H. 1967. Aspects of the biology of *Donax gouldi* and a note on evolution in Tellinacea (Bivalvia). Veliger 9: 330–37.

Gemma gemma (Totten), gem clam, not treated

This small clam (shell to 5 mm long), introduced from the Atlantic, may be rare to abundant on mud bottoms in the low intertidal or subtidal from Puget Sound to San Diego. The clam, which broods its young, usually lives in estuarine habitats, and may be confused with another small clam, *Transennella tantilla*, more commonly found in sandy sediments.

GREEN, R. H., and K. D. HOBSON. 1970. Spatial and temporal structure in a temperate intertidal community with special emphasis on *Gemma gemma* (Pelecypoda: Mollusca). Ecology 51: 999–1011.

SELLMER, G. P. 1967. Functional morphology and ecological life history of the gem clam, *Gemma gemma* (Eulamellibranchia: Veneridae). Malacologia 5: 137–223.

SHULENBERGER, E. 1970. Responses of *Gemma gemma* to a catastrophic burial (Mollusca: Pelecypoda). Veliger 13: 163–70.

THOMPSON, J. K. 1982. Population structure of *Gemma gemma* (Bivalvia: Veneridae) in south San Francisco Bay, with a comparison to some northeastern United States estuarine populations. Veliger 24: 281–90.

Glans carpenteri (Lamy) (*Cardita* in previous editions; *Glans subquadrata* (Carpenter) of some authors), §136

COAN, E. 1977. Preliminary review of the northwest American Carditidae. Veliger 19: 375–86.

YONGE, C. M. 1969. Functional morphology and evolution within the Carditacea (Bivalvia). Proc. Malacol. Soc. Lond. 38: 493–527.

Macoma balthica (Linnaeus) (= *Macoma inconspicua* (Broderip and Sowerby)), not treated

Common in middle- and low-intertidal zones of muddy bays, Bering Strait to San Francisco.

McGREER, E. R. 1983. Growth and reproduction of *Macoma balthica* (L.) on a mud flat in the Fraser River estuary, British Columbia. Can. J. Zool. 61: 887–94.

NICHOLS, F. H., and J. K. THOMPSON. 1982. Seasonal growth in the bivalve *Macoma balthica* near the southern limit of its range. Estuaries 5: 110–20. □ Growth in San Francisco Bay is highly seasonal, being rapid between March and May after a January–February spawning period.

VASSALLO, M. T. 1969. The ecology of *Macoma inconspicua* (Broderip and Sowerby, 1829) in central San Francisco Bay. I. The vertical distribution of the *Macoma* community. Veliger 11: 223–34.

———. 1971. The ecology of *Macoma inconspicua* (Broderip and Sowerby, 1829) in central San Francisco Bay. II. Stratification of the *Macoma* community within the substrate. Veliger 13: 279–84.

M. nasuta (Conrad), bent-nosed clam, §303

GALLUCCI, V. F., and J. HYLLEBERG. 1976. A quantification of some aspects of growth in the bottom-feeding bivalve *Macoma nasuta*. Veliger 19: 59–67.

HYLLEBERG, J., and V. F. GALLUCCI. 1975. Selectivity in feeding by the deposit-feeding bivalve *Macoma nasuta*. Mar. Biol. 32: 167–78.

M. secta (Conrad), §260

RAE, J. G. 1978. Reproduction in two sympatric species of *Macoma* (Bivalvia). Biol. Bull. 155: 207–19. □ Concerns *Macoma nasuta* and *M. secta* near Tomales Bay.

———. 1979. The population dynamics of two sympatric species of *Macoma* (Mollusca: Bivalvia). Veliger 21: 384–99.

REID, R. G. B., and A. REID. 1969. Feeding processes of members of the genus *Macoma* (Mollusca: Bivalvia). Can. J. Zool. 47: 649–57. □ Concludes that *Macoma secta* is a deposit feeder, but that, in contrast to the findings of Hylleberg and Gallucci (1975), *M. nasuta* is a suspension feeder.

Mercenaria mercenaria (Linnaeus), quahog, not treated

Introduced from the Atlantic, *Mercenaria mercenaria* (alas for the lovely Linnaean name *Venus mercenaria*) is found in sand or sandy mud in San Francisco Bay, Humboldt Bay, and Colorado Lagoon (Los Angeles Co.); small breeding colonies seem to have become established in the latter two localities. This commercially important clam (on the east coast) has been much investigated, but its scarcity on this coast warrants but a mention and a few references.

BELDING, D. L. 1911. The life history and growth of the quahog (*Venus mercenaria*). Rep. Mass. Comm. Fish Game 1910: 18–128.

CRANE, J. M., L. G. ALLEN, and C. EISEMANN. 1975. Growth rate, distribution, and population density of the northern quahog *Mercenaria mercenaria* in Long Beach, California. Calif. Fish Game 61: 68–81.

EVERSOLE, A. G., W. K. MICHENER, and P. J. ELDRIDGE. 1980. Reproductive cycle of *Mercenaria mercenaria* in a South Carolina estuary. Proc. Natl. Shellfish. Assoc. 70: 22–30.

MACKENZIE, C. L., JR. 1977. Predation on hard clam (*Mercenaria mercenaria*) populations. Trans. Amer. Fish. Soc. 106: 530–37.

Nuttallia nuttallii (Conrad) (formerly *Sanguinolaria*), §260

POHLO, R. H. 1972. Feeding and associated morphology in *Sanguinolaria nuttallii* (Bivalvia: Tellinacea). Veliger 14: 298–301. □Concludes that *Nuttallia* is a nonselective suspension feeder.

Protothaca staminea (Conrad), rock cockle, littleneck clam (formerly *Paphia, Tapes,* and *Venerupis*), §206

FEDER, H. M., J. C. HENDEE, P. HOLMES, G. J. MUELLER, and A. J. PAUL. 1979. Examination of a reproductive cycle of *Protothaca staminea* using histology, wet weight–dry weight ratios, and condition indices. Veliger 22: 182–87.

PAUL, A. J., J. M. PAUL, and H. M. FEDER. 1976a. Recruitment and growth in the bivalve *Protothaca staminea*, at Olsen Bay, Prince William Sound, ten years after the 1964 earthquake. Veliger 18: 385–92.

———. 1976b. Growth of the littleneck clam, *Protothaca staminea*, on Porpoise Island, southeast Alaska. Veliger 19: 163–66.

PETERSON, C. H., and M. L. QUAMMEN. 1982. Siphon nipping: Its importance to small fishes and its impact on growth of the bivalve *Protothaca staminea* (Conrad). J. Exp. Mar. Biol. Ecol. 63: 249–68.

QUAYLE, D. D. 1943. Sex, gonad development and seasonal gonad changes in *Paphia staminea* Conrad. J. Fish. Res. Board Can. 6: 140–51.

SCHMIDT, R. R., and J. E. WARME. 1969. Population characteristics of *Protothaca staminea* (Conrad) from Mugu Lagoon, California. Veliger 12: 193–99.

SMITH, S. H. 1974. The growth and mortality of the littleneck clam, *Protothaca staminea*, in Tia Juana Slough, California. Master's thesis, Biology, San Diego State University. 116 pp. □Specimens approached their maximum size at a faster rate than littleneck clams in other studies (Alaska, British Columbia, and Mugu Lagoon, California), but the maximum size eventually attained was intermediate between other reported values.

Psephidia lordi (Baird), §206

Pseudopythina rugifera (Carpenter) (also as *Orobitella*), §313

Saxidomus giganteus (Deshayes), butter clam, §301

FRASER, C. M. 1929. The spawning and free-

swimming larval periods of *Saxidomus* and *Paphia*. Trans. Roy. Soc. Can. 23: 195–98.

FRASER, C. M., and G. M. SMITH. 1928. Notes on the ecology of the butter clam, *Saxidomus giganteus*. Trans. Roy. Soc. Can. 22: 271–86.

S. nuttalli Conrad, butter clam, §301

See also Fraser (1929), above.

Siliqua patula (Dixon), northern razor clam, §189

See also Pohlo (1963), below under *Solen sicarius*.

BOURNE, N., and D. B. QUAYLE. 1970. Breeding and growth of razor clams in British Columbia. Fish. Res. Board Can. Tech. Rept. 132. 41 pp.

BRESSE, W. P., and A. ROBINSON. 1981. Razor clams, *Siliqua patula* (Dixon): Gonadal development, induced spawning and larval rearing. Aquaculture 22: 27–33. □Phytoplankton at a concentration of 2–2.5 million cells per milliliter induced spawning of *Siliqua patula* and several other bivalves.

HIRSCHHORN, G. 1962. Growth and mortality rates of the razor clam (*Siliqua patula*) on Clatsop beaches, Oregon. Oregon Fish Comm., Contr. 27. 55 pp.

NICKERSON, R. B. 1975. A critical analysis of some razor clam (*Siliqua patula*, Dixon) populations in Alaska. Alaska Dept. Fish Game. 294 pp.

WEYMOUTH, F. W., H. C. McMILLIN, and H. B. HOLMES. 1931. Relative growth and mortality of the Pacific razor clam (*Siliqua patula* Dixon), and their bearing on the commercial fishery. Bull. U.S. Bur. Fish. 46: 543–67.

YONGE, C. M. 1952. Studies on Pacific coast mollusks. IV. Observations on *Siliqua patula* Dixon and on evolution within the Solenidae. Univ. Calif. Publ. Zool. 55: 421–38.

Solen rosaceus Carpenter, jackknife clam, §260

S. sicarius Gould, jackknife clam, §279

POHLO, R. H. 1963. Morphology and mode of burrowing in *Siliqua patula* and *Solen rosaceus* (Mollusca: Bivalvia). Veliger 6: 98–104.

SCHNEIDER, D. 1982. Escape response of an infaunal clam *Ensis directus* Conrad 1843, to a predatory snail, *Polinices duplicatus* Say 1822. Veliger 24: 371–72. □When stimulated by predatory *Polinices*, this Atlantic razor clam apparently leaves the sand and propels itself over the surface "by a series of vigorous lashes with the foot, combined with rapid ejections of water along the ventral side of the foot." It is not known whether our razor clams respond similarly.

TRUEMAN, E. R. 1966. The dynamics of burrowing in *Ensis* (Bivalvia). Proc. Roy. Soc. Lond. 166: 459–76.

Tagelus californianus (Conrad), jackknife clam, §304

POHLO, R. H. 1966. A note on the feeding behavior in *Tagelus californianus* (Bivalvia: Tellinacea). Veliger 8: 225.

Tapes japonica Deshayes, Japanese littleneck or Manila clam, §206

Listed as *Protothaca semidecussata* (Reeve) in previous editions; at various times placed in the genera *Paphia, Venerupis, Ruditapes,* or *Protothaca*.

WILLIAMS, J. G. 1980a. Growth and survival in newly settled spat of the Manila clam, *Tapes ja-*

ponica. U.S. Natl. Mar. Fish Serv., Fish. Bull. 77: 891–900.

————. 1980b. The influence of adults on the settlement of spat of the clam, *Tapes japonica.* J. Mar. Res. 38: 729–42.

Tellina bodegensis Hinds, and two related species, §260

STEPHENS, A. C. 1929. Notes on the rate of growth of *Tellina tenuis* da Costa in the Firth of Clyde. J. Mar. Biol. Assoc. U.K. 16: 117–29. ☐Refers to Stephens's 1928 work on natural history.

YONGE, C. M. 1949. On the structure and adaptations of the Tellinacea, deposit-feeding Eulamellibranchia. Phil. Trans. Roy. Soc. Lond. B234: 29–76.

Tivela stultorum (Mawe), Pismo clam, §190

COE, W. R. 1947. Nutrition, growth and sexuality of the Pismo clam (*Tivela stultorum*). J. Exp. Zool. 104: 1–24.

COE, W. R., and J. E. FITCH. 1950. Population studies, local growth rates and reproduction of the Pismo clam (*Tivela stultorum*). J. Mar. Res. 9: 188–92.

FITCH, J. E. 1950. The Pismo clam. Calif. Fish Game 36: 285–312.

HERRINGTON, W. C. 1930. The Pismo clam: Further studies of its life history and depletion. Calif. Dept. Fish Game, Fish Bull. 18: 1–69.

STEPHENSON, M. D. 1977. Sea otter predation on Pismo clams in Monterey Bay. Calif. Fish Game 63: 117–20.

WEYMOUTH, F. W. 1923. The life-history and growth of the Pismo clam (*Tivela stultorum* Mawe). Calif. Dept. Fish Game, Fish Bull. 7: 1–120.

Transennella tantilla (Gould), not treated

A small clam of sand flats and of viviparous, protandrous habit. Confusable with *Gemma gemma.* See also Narchi (1971), under Veneroida, p. 568, and Maurer (1967), under Bivalvia, p. 566.

HANSEN, B. 1953. Brood protection and sex ratio of *Transennella tantilla* (Gould), a Pacific bivalve. Vidensk. Medd. Dan. Naturhist. Foren. 115: 313–24.

Tresus capax (Gould) and **T. nuttallii** (Conrad) (formerly *Schizothaerus*), horse clams, §300

BOURNE, N., and D. W. SMITH. 1972. Breeding and growth of the horse clam, *Tresus capax* (Gould), in southern British Columbia. Proc. Natl. Shellfish. Assoc. 62: 38–46.

BREED-WILLEKE, G. M., and D. R. HANCOCK. 1980. Growth and reproduction of subtidal and intertidal populations of the gaper clam *Tresus capax* (Gould) from Yaquina Bay, Oregon. Proc. Natl. Shellfish. Assoc. 70: 1–13.

CLARK, P., J. NYBAKKEN, and L. LAURENT. 1975. Aspects of the life history of *Tresus nuttallii* in Elkhorn Slough. Calif. Fish Game 61: 215–27.

KATANSKY, S. C., R. W. WARNER, and R. L. POOLE. 1969. On the occurrence of larval cestodes in the Washington clam, *Saxidomus nuttalli,* and the gaper clam, *Tresus nuttalli,* from Drakes Estero, California. Calif. Fish Game. 55: 317–22.

PEARCE, J. B. 1965. On the distribution of *Tresus nuttallii* and *Tresus capax* (Pelecypoda: Mactridae) in the waters of Puget Sound and the San Juan Archipelago. Veliger 7: 166–70. ☐Concerns the distribution of these clams, and their association (*Tresus capax*) or lack of association (*T. nuttallii*) with pinnotherid crabs.

POHLO, R. H. 1964. Ontogenetic changes of form and mode of life in *Tresus nuttalli* (Bivalvia: Mactridae). Malacologia 1: 321–30.

REID, R. G. B., and J. A. FRIESEN. 1980. The digestive system of the moon snail *Polinices lewisii* (Gould, 1847) with emphasis on the role of the oesophageal gland. Veliger 23: 25–34. ☐One of the most frequent prey of *Polinices* at Vancouver Island were small clams (*Tresus nuttallii*). The clams were not bored, but were attacked and eaten through the permanent gape in the shell at the siphon base and ventral mantle edge.

STOUT, W. E. 1970. Some associates of *Tresus nuttallii* (Conrad, 1837) (Pelecypoda: Mactridae). Veliger 13: 67–70. ☐Examination of the clams' siphonal plates revealed 50 species of associates representing 10 animal phyla and several plant divisions.

SWAN, E. F., and J. H. FINUCANE. 1952. Observations on the genus *Schizothaerus.* Nautilus 66: 19–26. ☐Concerns distinctions between *Tresus capax* and *T. nuttallii.*

WENDELL, F., J. D. DEMARTINI, P. DINNEL, and J. SIECKE. 1976. The ecology of the gaper or horse clam, *Tresus capax* (Gould, 1850) (Bivalvia: Mactridae), in Humboldt Bay, California. Calif. Fish Game 62: 41–64.

Order MYOIDA

COE, W. R. 1941. Sexual phases in wood-boring mollusks. Biol. Bull. 81: 168–76. ☐Includes *Bankia setacea, Lyrodus pedicellatus,* and *Teredo navalis.*

KENNEDY, G. L. 1974. West American Cenozoic Pholadidae (Mollusca: Bivalvia). San Diego Natur. Hist. Mus. Mem. 8: 1–127.

MENZIES, R. J., J. MOHR, and C. M. WAKEMAN. 1963. The seasonal settlement of wood-borers in Los Angeles–Long Beach harbors. Wasmann J. Biol. 21: 97–120.

MORTON, B. 1978. Feeding and digestion in shipworms. Oceanogr. Mar. Biol. Ann. Rev. 16: 107–44.

TURNER, R. D. 1954–1955. The family Pholadidae in the western Atlantic and eastern Pacific. I. Pholadinae. II. Martesiinae, Jouannetiinae and Xylophaginae. Johnsonia 3: 1–160.

————. 1966. A survey and illustrated catalogue of the Teredinidae (Mollusca: Bivalvia). Cambridge, Mass.: Museum of Comparative Zoology of Harvard University. 265 pp.

Bankia setacea (Tryon), Pacific shipworm, §328

DEAN, R. C. 1978. Mechanisms of wood digestion in the shipworm *Bankia gouldi* Bartsch: Enzyme degradation of celluloses, hemicelluloses, and wood cell walls. Biol. Bull. 155: 297–316.

HADERLIE, E. C. 1982. Long-term natural resistance of some Central American hardwoods to attacks by the shipworm *Bankia setacea* (Tryon) and the gribble *Limnoria quadripunctata* Holthuis in Monterey Harbor. Veliger 25: 182–84.

HADERLIE, E. C., and J. C. MELLOR. 1973. Settlement, growth rates and depth preference of the shipworm *Bankia setacea* (Tryon) in Monterey Bay. Veliger 15: 265–86.

Barnea subtruncata (Sowerby) (formerly *Barnea pacifica* (Stearns)), §147

Cryptomya californica (Conrad), §313
YONGE, C. M. 1951. Studies on Pacific coast mollusks. I. On the structure and adaptations of *Cryptomya californica* (Conrad). Univ. Calif. Publ. Zool. 55: 395–400.

Hiatella arctica (Linnaeus) and **H. pholadis** (Linnaeus), §147
Probably the same species, referable to *H. arctica*; listed as species of *Saxicava* in previous editions. See also Yonge (1971), under Bivalvia, p. 566.
ALI, R. M. 1970. The influence of suspension density and temperature on the filtration rate of *Hiatella arctica*. Mar. Biol. 6: 291–302.
HUNTER, W. R. 1949. The structure and behavior of "*Hiatella gallicana*" (Lamarck) and "*H. arctica*" (L.), with special reference to the boring habit. Proc. Roy. Soc. Edinb. B63: 271–89.

Lyrodus pedicellatus (Quatrefages) (formerly *Teredo diegensis* Bartsch), shipworm, §338
ECKELBARGER, K. J., and D. J. REISH. 1972a. A first report of self-fertilization in the wood-boring family Teredinidae (Mollusca: Bivalvia). Bull. South. Calif. Acad. Sci. 71: 48–50.
————. 1972b. Effects of varying temperatures and salinities on settlement, growth, and reproduction of the wood-boring pelecypod, *Lyrodus pedicellatus*. Bull. South. Calif. Acad. Sci. 71: 116–27.

Mya arenaria Linnaeus, soft-shell clam, §302
See also Vassallo (1969, 1971), under *Macoma balthica*, p. 569.
BROUSSEAU, D. J. 1978a. Spawning cycle, fecundity, and recruitment in a population of soft-shell clam, *Mya arenaria*, from Cape Ann, Massachusetts. U.S. Natl. Mar. Fish. Serv., Fish. Bull. 76: 155–66.
————. 1978b. Population dynamics of the soft-shell clam *Mya arenaria*. Mar. Biol. 50: 63–71.
————. 1979. Analysis of growth rate in *Mya arenaria* using the Von Bertalanffy equation. Mar. Biol. 51: 221–27.
FEDER, H. M., and A. J. PAUL. 1974. Age, growth and size-weight relationships of the soft-shell clam, *Mya arenaria*, in Prince William Sound, Alaska. Proc. Natl. Shellfish. Assoc. 64: 45–52.
GOSHIMA, S. 1982. Population dynamics of the soft clam, *Mya arenaria* L., with special reference to its life history pattern. Publ. Amakusa Mar. Biol. Lab. 6: 119–65. □Reports on a Japanese population.
MACDONALD, B. A., and M. L. H. THOMAS. 1980. Age determination of the soft-shell clam *Mya arenaria* using shell internal growth lines. Mar. Biol. 58: 105–9.
MATTHIESSEN, G. C. 1961. Intertidal zonation in populations of *Mya arenaria*. Limnol. Oceanogr. 5: 381–88.

Panopea generosa Gould, geoduck clam, §299
GOODWIN, L. 1977. The effects of season on visual and photographic assessment of subtidal geoduck clam (*Panope generosa* Gould) populations. Veliger

20: 155–58. □In summer the animals are actively pumping and the burrow openings are obvious. During winter the animals are less active and the burrow opening may be filled with sediment, making detection difficult. A similar phenomenon has been observed (Suer 1982) for burrows of the echiuran *Urechis caupo*.

Parapholas californica (Conrad), §147

Penitella conradi Valenciennes (formerly *Pholadidea parva* (Tryon)), §74
MEREDITH, S. E. 1968. Notes on the range extension of the boring clam *Penitella conradi* Valenciennes and its occurrence in the shell of the California mussel. Veliger 10: 281–82.
SMITH, E. H. 1969. Functional morphology of *Penitella conradi* relative to shell-penetration. Amer. Zool. 9: 569–80. □Presents some evidence that *Penitella conradi* erodes by chemical means the molluscan shells into which it bores; the animal is also known, however, to bore into noncalcareous rock.

P. gabbii (Tryon), §147

P. penita (Conrad), §242
Three additional papers by Evans are cited in those listed here.
EVANS, J. W. 1968. Growth rate of the rock-boring clam *Penitella penita* (Conrad 1837) in relation to hardness of rock and other factors. Ecology 49: 619–28.
————. 1970. Sexuality in the rock-boring clam *Penitella penita*. Can. J. Zool. 48: 625–27.
EVANS, J. W., and M. H. LEMESSURIER. 1972. Functional micromorphology and circadian growth of the rock-boring clam *Penitella penita*. Can. J. Zool. 50: 1251–58.
HADERLIE, E. C. 1981. Influence of terminal end of burrow on callum shape in the rock boring clam *Penitella penita* (Conrad, 1837) (Bivalvia: Pholadidae). Veliger 24: 51–53.

Platyodon cancellatus (Conrad), §147
YONGE, C. M. 1951. Studies on Pacific coast mollusks. II. Structure and adaptations for rock boring in *Platyodon cancellatus* (Conrad). Univ. Calif. Publ. Zool. 55: 401–7.

Teredo navalis (Linnaeus), shipworm, §§328, 337
CARPENTER, E. J., and J. L. CULLINEY. 1975. Nitrogen fixation in marine shipworms. Science 187: 551–52.
GRAVE, B. H. 1928. Natural history of shipworm, *Teredo navalis*, at Woods Hole, Massachusetts. Biol. Bull. 55: 260–82.
KRISTENSEN, E. S. 1979. Observations on growth and life cycle of the shipworm *Teredo navalis* L. (Bivalvia, Mollusca) in the Isefjord, Denmark. Ophelia 18: 235–42.
MILLER, R. C. 1924. The boring mechanism of *Teredo*. Univ. Calif. Publ. Zool. 26: 41–80.

Zirfaea pilsbryi Lowe, §305

Class POLYPLACOPHORA
Chitons, or Sea cradles

The chitons are a small but distinctive group (of perhaps 500 living species worldwide) with two characteristic features: a dorsal set of eight overlapping plates, and a series of small gills along a ventral mantle groove. Although not often found in large numbers, chitons are nevertheless common, and in some locations the impact of their herbivory is readily apparent.

ANDRUS, J. K., and W. B. LEGARD. 1975. Description of the habitats of several intertidal chitons (Mollusca: Polyplacophora) found along the Monterey Peninsula of central California. Veliger 18 (suppl.): 3–8. □Notes on habitats and associates of 12 species of chiton.

BARNAWELL, E. B. 1960. The carnivorous habit among the Polyplacophora. Veliger 2: 85–88. □Although chitons are generally considered to be primarily herbivorous, three species of the genus *Mopalia* in San Francisco Bay consumed animal food (bryozoans, hydroids, barnacles) to various degrees: *Mopalia muscosa* (15 percent animal food), *Mopalia ciliata* (45 percent), and *Mopalia hindsii* (59 percent).

BAXTER, J. M., and A. M. JONES. 1981. Valve structure and growth in the chiton *Lepidochitona cinereus* (Polyplacophora: Ischnochitonidae). J. Mar. Biol. Assoc. U.K. 61: 65–78.

BERRY, S. S. 1917. Notes on west American chitons—I. Proc. Calif. Acad. Sci. 7: 229–48.

———. 1919. Notes on west American chitons—II. Proc. Calif. Acad. Sci. 9: 1–36.

———. 1922. Fossil chitons of western North America. Proc. Calif. Acad. Sci. 11: 399–526. □Includes many Recent species.

BOYLE, P. R. 1972. The aesthetes of chitons. I. Role in the light response of whole animals. Mar. Behav. Physiol. 1: 171–84.

———. 1977. The physiology and behavior of chitons (Mollusca: Polyplacophora). Mar. Biol. Ann. Rev. 15: 461–509. □A good general review with bibliography, although few references after 1970 are included.

BURGHARDT, G. E., and L. E. BURGHARDT. 1969. A collector's guide to west coast chitons. Spec. Publ. 4, San Francisco Aquarium Soc., Calif. 45 pp.

FERREIRA, A. J. 1982. The family Lepidochitonidae Iredale, 1914 (Mollusca: Polyplacophora) in the eastern Pacific. Veliger 25: 93–138.

GIESE, A. C., J. S. TUCKER, and R. A. BOOLOOTIAN. 1959. Annual reproductive cycles of the chitons, *Katharina tunicata* and *Mopalia hindsii*. Biol. Bull. 117: 81–88.

HYMAN, L. H. 1967. The invertebrates: Mollusca. Vol. VI. New York: McGraw-Hill. 792 pp. (pp. 70–142 deal with the Polyplacophora).

LEISE, E. M., and R. A. CLONEY. 1982. Chiton integument: Ultrastructure of the sensory hairs of *Mopalia muscosa* (Mollusca: Polyplacophora). Cell. Tissue Res. 223: 43–59.

LINSENMEYER, T. A. 1975. The resistance of five species of polyplacophorans to removal from natural and artificial surfaces. Veliger 18(suppl.): 83–86. □The lateral force required to dislodge a chiton was least for *Stenoplax heathiana*, then increased in order for *Mopalia lignosa*, *M. muscosa*, *Katharina tunicata*, and *Nuttallina californica*. The amount of force required was directly related to the degree of wave exposure characteristic of a species' habitat.

LOWENSTAM, H. A. 1962. Magnetite in denticle capping in Recent chitons (Polyplacophora). Bull. Geol. Soc. Amer. 73: 435–38.

MILLER, R. L. 1977. Chemotactic behavior of the sperm of chitons (Mollusca: Polyplacophora). J. Exp. Zool. 202: 203–12. □The sperm of all five chiton species tested were attracted to egg extracts; however, the attraction was not species-specific.

OMELICH, P. 1967. The behavioral role and the structure of the aesthetes of chitons. Veliger 10: 77–82.

PUTMAN, B. F. 1982. The littoral and sublittoral Polyplacophora of Diablo Cove and vicinity, San Luis Obispo County, California. Veliger 24: 364–66.

SMITH, A. G. 1966. The larval development of chitons (Amphineura). Proc. Calif. Acad. Sci. 32: 433–46.

———. 1973. Polyplacophora—a selected bibliography. Of Sea and Shore 4: 201–6, 208.

———. 1977. Rectification of west coast chiton nomenclature (Mollusca: Polyplacophora). Veliger 19: 215–58.

THORPE, S. R., JR. 1962. A preliminary report on spawning and related phenomena in California chitons. Veliger 4: 202–10. □Mostly about species of *Mopalia*.

TOMLINSON, J., D. REILLY, and R. BALLERING. 1980. Magnetic radular teeth and geomagnetic responses in chitons. Veliger 23: 167–70.

TUCKER, J. S., and A. C. GIESE. 1959. Shell repair in chitons. Biol. Bull. 116: 318–22. □Concerns *Cryptochiton stelleri*, *Katharina tunicata*, and *Mopalia hindsii*.

YONGE, C. M. 1939. On the mantle cavity and its contained organs in the Loricata (Placophora). Quart. J. Microsc. Sci. 81: 367–90.

Callistochiton crassicostatus Pilsbry, §128

FERREIRA, A. J. 1979. The genus *Callistochiton* Dall, 1879 (Mollusca: Polyplacophora) in the eastern Pacific, with the description of a new species. Veliger 21: 444–66.

Cryptochiton stelleri (Middendorff), §76

MACGINITIE, G. E., and N. MACGINITIE. 1968. Notes on *Cryptochiton stelleri* (Middendorff, 1846). Veliger 11: 59–61.

MCDERMID, K. 1981. Preliminary studies on the association between *Pleonosporium squarrosum* (Rhodophyta) and *Cryptochiton stelleri* (Polyplacophora). Veliger 23: 317–20. □This alga, and many others, has been found growing on the back of *Cryptochiton*, contrary to an earlier report suggesting that little or nothing could grow on this chiton's back.

OKUDA, S. 1947. Notes on the post-larval development of the giant chiton, *Cryptochiton stelleri* (Midd.). J. Fac. Sci. Hokkaido Univ., Zool. 9: 267–75.

PALMER, J. B., and P. W. FRANK. 1974. Estimates of growth of *Cryptochiton stelleri* (Middendorff, 1846). Veliger 16: 301–4.

PETERSEN, J. A., and K. JOHANSEN. 1973. Gas exchange in the giant sea cradle *Cryptochiton stelleri* (Middendorff). J. Exp. Mar. Biol. Ecol. 12: 27–43.

ROBBINS, K. B. 1975. Active absorption of D-glucose and D-galactose by intestinal tissue of the chiton, *Cryptochiton stelleri* (Middendorff, 1846). Veliger 18: 122–27. □Much work has been done on intestinal absorption in the gut of *Cryptochiton*. Interested readers are referred to the references cited in this paper.

TUCKER, J. S., and A. C. GIESE. 1962. Reproductive cycle of *Cryptochiton stelleri* (Middendorff). J. Exp. Zool. 150: 33–43.

WEBSTER, S. K. 1968. An investigation of the commensals of *Cryptochiton stelleri* (Middendorff, 1847) in the Monterey Peninsula area, California. Veliger 11: 121–25.

Ischnochiton regularis (Carpenter), §128

Katharina tunicata (Wood), §177

HIMMELMAN, J. 1978. The reproductive cycle of *Katharina tunicata* Wood and its controlling factors. J. Exp. Mar. Biol. Ecol. 31: 27–41. □Concerns *Katharina* on the coast of British Columbia.

HIMMELMAN, J. H., and T. H. CAREFOOT. 1975. Seasonal changes in calorific value of three Pacific coast seaweeds, and their significance to some marine invertebrate herbivores. J. Exp. Mar. Biol. Ecol. 18: 139–51. □Although *Katharina* consumed less of the brown alga *Hedophyllum sessile* in the fall than during the spring and summer, the chiton's caloric intake did not decrease substantially in the fall because the caloric value of the alga was higher at that time.

LAWRENCE, J. M., and A. C. GIESE. 1969. Changes in lipid composition of the chiton, *Katharina tunicata*, with the reproductive and nutritional state. Physiol. Zool. 42: 353–60. □One of the later papers in a series on the reproductive cycle and chemical composition of *Katharina* in central California; earlier work is cited in this paper's bibliography.

RUMRILL, S. S., and R. A. CAMERON. 1983. Effects of gamma-aminobutyric acid on the settlement of larvae of the black chiton *Katharina tunicata*. Mar. Biol. 72: 243–47.

Lepidochitona dentiens (Gould) (also as *Cyanoplax*), §237

Cyanoplax has recently been considered a synonym of *Lepidochitona* (see Ferreira 1982, above). Although the former name continues to be used by many chiton workers, the latter is apparently correct and will be used here.

GOMEZ, R. L. 1975. An association between *Nuttallina californica* and *Cyanoplax hartwegii*, two west coast polyplacophorans (chitons). Veliger 18 (suppl.): 28–29. □Describes the occurrence of *Lepidochitona* (as *Cyanoplax dentiens*) beneath the larger chiton *Nuttallina californica*. The article's title is in error: *Cyanoplax hartwegii* should read *Cyanoplax dentiens*.

KUES, B. S. 1974. A new subspecies of *Cyanoplax dentiens* (Polyplacophora) from San Diego, California. Veliger 16: 297–300.

L. hartwegii (Carpenter) (also as *Cyanoplax*), §36

CONNOR, M. S. 1975. Niche apportionment among the chitons *Cyanoplax hartwegii* and *Mopalia muscosa* and the limpets *Collisella limatula* and *Collisella pelta* under the brown alga *Pelvetia fastigiata*. Veliger 18(suppl.): 9–17.

DEBEVOISE, A. E. 1975. Predation on the chiton *Cyanoplax hartwegii* (Mollusca: Polyplacophora). Veliger 18(suppl.): 47–50. □The seastars *Pisaster ochraceus* and *Leptasterias pusilla* and the crabs *Pachygrapsus crassipes* and *Hemigrapsus nudus* feed on *Lepidochitona* under artificial conditions. The algal covering of *Pelvetia* may protect the chiton from *Pisaster* predation.

LYMAN, B. W. 1975. Activity patterns of the chiton *Cyanoplax hartwegii* (Mollusca: Polyplacophora). Veliger 18(suppl.): 63–69.

MCGILL, V. L. 1975. Response to osmotic stress in the chiton *Cyanoplax hartwegii* (Mollusca: Polyplacophora). Veliger 18(suppl.): 109–12.

ROBB, M. F. 1975. The diet of the chiton *Cyanoplax hartwegii* in three intertidal habitats. Veliger 18 (suppl.): 34–37.

L. keepiana Berry, §237

BERRY, S. S. 1948. Two misunderstood west American chitons. Leaflets in Malacol. 1: 13–15.

Lepidozona cooperi (Pilsbry), §128

L. mertensii (Middendorff), §128

FERREIRA, A. J. 1978. The genus *Lepidozona* (Mollusca: Polyplacophora) in the temperate eastern Pacific, Baja California to Alaska, with the description of a new species. Veliger 21: 19–44.

HELFMAN, E. S. 1968. A ctenostomatous ectoproct epizoic on the chiton *Ischnochiton mertensii*. Veliger 10: 290–91. □The bryozoan *Farella elongata* is attached to the ventral surface of the girdle.

Mopalia ciliata (Sowerby), §128

M. hindsii (Reeve), §128

M. lignosa (Gould), §128

FULTON, F. T. 1975. The diet of the chiton *Mopalia lignosa* (Gould, 1846) (Mollusca: Polyplacophora). Veliger 18(suppl.): 38–41.

M. muscosa (Gould), §§128, 221

BOOLOOTIAN, R. A. 1964. On growth, feeding and reproduction in the chiton *Mopalia muscosa* of Santa Monica Bay. Helgol. Wiss. Meeresunters. 11: 186–99.

FITZGERALD, W. J. 1975. Movement pattern and phototactic response of *Mopalia ciliata* and *Mopalia muscosa* in Marin County, California. Veliger 18: 37–39.

HIMMELMAN, J. H. 1980. Reproductive cycle patterns in the chiton genus *Mopalia* (Polyplacophora). Nautilus 94: 39–49. □Concerns *Mopalia ciliata*, *M. hindsii*, *M. laevior*, *M. lignosa*, and *M. muscosa* on the coast of British Columbia. A diversity of reproductive patterns was evident; for example, *M. ciliata* spawned synchronously with the spring phytoplankton bloom, whereas *M. muscosa* spawned irregularly throughout the year.

MORAN, W. M., and R. E. TULLIS. 1980. Ion and water balance of the hypo- and hyperosmotically stressed chiton *Mopalia muscosa*. Biol. Bull. 159: 364–75.

SMITH, S. Y. 1975. Temporal and spatial activity patterns of the intertidal chiton *Mopalia muscosa*. Veliger 18(suppl.): 57–62.

WATANABE, J. M., and L. R. COX. 1975. Spawning

behavior and larval development in *Mopalia lignosa* and *Mopalia muscosa* (Mollusca: Polyplacophora) in central California. Veliger 18(suppl.): 18–27.

WESTERSUND, K. R. 1975. Exogenous and endogenous control of movement in the chiton *Mopalia muscosa* (Mollusca: Polyplacophora). Veliger 18 (suppl.): 70–73.

Nuttallina californica (Reeve), §169

MOORE, M. M. 1975. Foraging of the western gull *Larus occidentalis* and its impact on the chiton *Nuttallina californica*. Veliger 18(suppl.): 51–53.

NISHI, R. 1975. The diet and feeding habits of *Nuttallina californica* (Reeve, 1847) from two contrasting habitats in central California. Veliger 18 (suppl.): 30–33. □*Nuttallina* consumes large pieces of algae, whose size averages 12 percent of the chiton's body length.

ROBBINS, B. A. 1975. Aerial and aquatic respiration in the chitons *Nuttallina californica* and *Tonicella lineata*. Veliger 18(suppl.): 98–102.

SIMONSEN, M. 1975. Response to osmotic stress in vertically separated populations of an intertidal chiton, *Nuttallina californica* (Mollusca; Polyplacophora). Veliger 18(suppl.): 113–16.

Placiphorella velata Dall, §128

McLEAN, J. H. 1962. Feeding behaviour of the chiton *Placiphorella*. Proc. Malacol. Soc. Lond. 35: 23–26.

Stenoplax conspicua (Pilsbry), §127
S. heathiana Berry, §127

Tonicella lineata (Wood), §36

BARNES, J. R. 1972. Ecology and reproductive biology of *Tonicella lineata* (Wood, 1815) (Mollusca: Polyplacophora). Ph.D. dissertation, Oregon State University. 149 pp.

BARNES, J. R., and J. J. GONOR. 1973. The larval settling response of the lined chiton *Tonicella lineata*. Mar. Biol. 20: 259–64.

DEMOPULOS, P. A. 1975. Diet, activity and feeding in *Tonicella lineata* (Wood, 1815). Veliger 18(suppl.): 42–46.

HIMMELMAN, J. H. 1975. Phytoplankton as a stimulus for spawning in three marine invertebrates. J. Exp. Mar. Biol. Ecol. 20: 199–214. □*Tonicella lineata* is one of the three whose cue for spawning may be the spring phytoplankton bloom.

———. 1979. Factors regulating the reproductive cycles of two northeast Pacific chitons, *Tonicella lineata* and *T. insignis*. Mar. Biol. 50: 215–25.

SEIFF, S. R. 1975. Predation upon subtidal *Tonicella lineata* of Mussel Point, California (Mollusca: Polyplacophora). Veliger 18(suppl.): 54–56. □Seastars that will consume other species of chiton rarely eat *Tonicella* in aquaria.

Class CEPHALOPODA. Octopods (or Octopuses), Squids, *Nautilus*

The *-pus* of *octopus* is from Greek *pous* ("foot"), and is usually rendered *-pod* or *-poda* in English. The word *octopi*, though now listed in many dictionaries, is an incorrect plural, which apparently arose by false analogy of *octopus* to Latin second-declension nouns (such as *amicus*, "friend") whose nominative plurals end in *-i*.

In any event, there is an extensive literature on these animals, and a small sample is provided, although many worthy papers have not been cited in consideration of space. In particular, virtually none of the work on cephalopod physiology has been included. Interested readers are referred to the accounts and bibliographies in Hochberg and Fields (1980) and Wells (1978).

ALTMAN, J. S., and M. NIXON. 1970. Use of the beaks and radula by *Octopus vulgaris* in feeding. J. Zool. (Lond.) 161: 25–38.

ARNOLD, J. M., and K. O. ARNOLD. 1969. Some aspects of hole-boring predation by *Octopus vulgaris*. Amer. Zool. 9: 991–96.

BERRY, S. S. 1912. A review of the cephalopods of western North America. Bull. U.S. Bur. Fish. 30: 269–336.

———. 1953. Preliminary diagnoses of six west American species of *Octopus*. Leaflets in Malacol. 1: 51–58.

COUSTEAU, J.-Y., and P. DIOLE. 1973. Octopus and squid: The soft intelligence. Garden City, N.Y.: Doubleday. 304 pp. □Many beautiful photographs; reads like a television script.

HOCHBERG, F. G. 1976. Benthic cephalopods of the eastern Pacific. Proc. Tax. Stand. Prog. 4: 3–8, 14–25.

HOCHBERG, F. G., and W. G. FIELDS. 1980. Cephalopoda: The squids and octopuses. *In* R. H. Morris, D. P. Abbott, and E. C. Haderlie, Intertidal invertebrates of California, pp. 429–44. Stanford, Calif.: Stanford University Press.

LANE, F. W. 1960. Kingdom of the octopus: The life-history of the Cephalopoda. New York: Sheridan. 300 pp. (Paperback ed. New York: Pyramid.)

MESSENGER, J. B. 1977. Evidence that *Octopus* is colour blind. J. Exp. Biol. 70: 49–55. □Although many species of *Octopus* display remarkable changes in color in response to food, mates, or changes in background, there is accumulating evidence that they are themselves colorblind. Experiments on *Octopus vulgaris* (described in this and several other papers) suggest, instead, that discriminations are more likely based on differences in "brightness," not color.

NIXON, M., and J. B. MESSENGER, eds. 1977. The biology of cephalopods. Symp. Zool. Soc. Lond. 38. New York: Academic Press. 615 pp. □The title of this symposium volume implies a more general topical coverage than is actually the case; however, there are some useful reviews.

PACKARD, A. 1972. Cephalopods and fish: The limits of convergence. Biol. Rev. 47: 241–307.

PACKARD, A., and F. G. HOCHBERG. 1977. Skin patterning in *Octopus* and other genera. Symp. Zool. Soc. Lond. 38: 191–231. □An excellent review with details on *Octopus bimaculoides*, *O. dofleini*, and *O. rubescens*.

ROBSON, G. C. 1929–1932. A monograph of the recent Cephalopoda. Part 1, 236 pp.; Part 2, 359 pp. London: British Museum (Natur. Hist.).

Voss, G. L., and R. F. Sisson. 1971. Shy monster, the octopus. Natl. Geogr. Mag. 140: 776–99.

Wells, M. J. 1962. Brain and behaviour in cephalopods. Stanford, Calif.: Stanford University Press. 171 pp.

——. 1978. Octopus: Physiology and behaviour of an advanced invertebrate. London: Chapman and Hall. 417 pp.

Wodinsky, J. 1969. Penetration of the shell and feeding on gastropods by octopus. Amer. Zool. 9: 997–1010.

Wood, F. G. 1971. An octopus trilogy. Natur. Hist. 80: 14–24, 84–87. ☐Mostly historical (with a touch of biological) sleuthing about a gigantic mass of tissue cast ashore in 1896 near St. Augustine, Florida; the mass may have been the remains of an octopus measuring 200 feet from the tip of one tentacle to the tip of another—then again, maybe not.

Loligo opalescens Berry, market squid, not treated
Specimens of this pelagic, commercially harvested cephalopod are commonly seen in fish markets and are occasionally cast up on sandy beaches.

Fields, W. G. 1965. The structure, development, food relations, reproduction, and life history of the squid *Loligo opalescens* Berry. Calif. Dept. Fish Game, Fish Bull. 131: 1–108.

Hanlon, R. T., R. F. Hixon, W. H. Hulet, and W. T. Yang. 1979. Rearing experiments on the California market squid *Loligo opalescens* Berry, 1911. Veliger 21: 428–31.

Hobson, E. S. 1965. Spawning in the Pacific coast squid, *Loligo opalescens*. Underwater Natur. 3: 20–21.

Hurley, A. C. 1974. Feeding behavior, food consumption, growth, and respiration of the squid *Loligo opalescens* raised in the laboratory. Calif. Dept. Fish Game, Fish Bull. 74: 176–82.

——. 1978. School structure of the squid *Loligo opalescens*. U.S. Natl. Mar. Fish Serv., Fish. Bull. 76: 433–42.

Kato, S., and J. E. Hardwick. 1975. The California squid fishery. F.A.O. Fish. Rep. 170, Suppl. 1: 107–27.

Recksiek, C. W., and H. W. Frey, eds. 1978. Biological, oceanographic, and acoustic aspects of the market squid, *Loligo opalescens* Berry. Calif. Dept. Fish Game, Fish Bull. 169: 1–185.

Octopus bimaculoides Pickford and McConnaughey and **O. bimaculatus** Verrill, two-spotted octopuses, §135
The work of Pickford and McConnaughey (1949) established two sympatric species of two-spotted octopus, differing principally in egg size and ecological habitat (they also differ in their assemblage of mesozoans!). *Octopus bimaculoides* occurs in the low intertidal, especially on mud flats, but also in protected microhabitats on rocky shores; its recorded range is San Simeon to Ensenada. *O. bimaculatus* generally occurs subtidally on rocky shores, although intertidal specimens are found, particularly in the southern portion of its range, from Santa Barbara to the southern tip of Baja California. The *Octopus bimaculatus* of earlier editions is actually *O. bimaculoides*.

Ambrose, R. F. 1981. Observations on the embryonic development and early post-embryonic behavior of *Octopus bimaculatus* (Mollusca: Cephalopoda). Veliger 24: 139–46.

——. 1982. Shelter utilization by the molluscan cephalopod *Octopus bimaculatus*. Mar. Ecol. Prog. Ser. 7: 67–73. ☐Suitable shelters (holes, crevices, and spaces under rocks) were always in excess during this study at Santa Catalina Island. Nevertheless, nearly half of the individuals inhabited their shelter for more than a month.

Muller, A. 1971. Characteristics of coloration on *Octopus bimaculatus*, the Sea of Cortez blue-dot octopus. Sea of Cortez, Inst. Biol. Res. Newsl. 5: 1–5.

Pickford, G. E., and B. H. McConnaughey. 1949. The *Octopus bimaculatus* problem: A study in sibling species. Bull. Bingham Oceanogr. Coll. 12: 1–66.

Pilson, M. E. Q., and P. B. Taylor. 1961. Hole drilling by octopus. Science 134: 1366–68.

Taylor, P. B., and L.-C. Chen. 1969. The predator-prey relationship between the octopus (*Octopus bimaculatus*) and the California scorpionfish (*Scorpaena guttata*). Pacific Sci. 23: 311–16. ☐California scorpionfish eat small octopuses but avoid large ones, presumably because large octopuses turn the tables and eat scorpionfish. Most observations were on *Octopus bimaculatus*, a few on *O. dofleini*.

O. dofleini (Wülker) and **O. rubescens** Berry, §135
Small specimens of both species may be found intertidally on the Pacific coast, and they are easily confused. Indeed, older authors, including earlier editions of this book, that refer to *Octopus apollyon* (or *Octopus hongkongensis*) more likely refer to *Octopus dofleini* in part and to *O. rubescens* in part; often the subject's identity remains unclear.
At present, three subspecies of *Octopus dofleini*, the giant Pacific octopus, are recognized (see Pickford 1964); the most likely to be encountered is *O. dofleini martini* Pickford, whose range is British Columbia to Monterey. The range of *O. rubescens* is Alaska to Baja California.

Brocco, S. L., and R. A. Cloney. 1980. Reflector cells in the skin of *Octopus dofleini*. Cell Tissue Res. 205: 167–86.

Fisher, W. K. 1923. Brooding-habits of a cephalopod. Ann. Mag. Natur. Hist., ser. 9, 12: 147–49.

——. 1925. On the habits of an octopus. Ann. Mag. Natur. Hist., ser. 9, 15: 411–14.

Gabe, S. H. 1975. Reproduction in the giant octopus of the North Pacific, *Octopus dofleini martini*. Veliger 18: 146–50.

Hartwick, E. B., and G. Thorarinsson. 1978. Den associates of the giant Pacific octopus, *Octopus dofleini* (Wülker). Ophelia 17: 163–66. ☐Several scavengers, including the seastar *Pycnopodia helianthoides* and the snail *Amphissa columbiana*, may be attracted to the remains of octopus feeding, which accumulate around dens.

Hartwick, E. B., G. Thorarinsson, and L. Tulloch. 1978a. Antipredator behavior in *Octopus dofleini* (Wülker). Veliger 21: 263–64.

——. 1978b. Methods of attack by *Octopus do-*

fleini (Wülker) on captured bivalve and gastropod prey. Mar. Behav. Physiol. 5: 193–200. ☐Accumulated shells found around the dens of *Octopus dofleini* suggest that this octopus attacks bivalves and gastropods by several methods—drilling, breaking, and pulling apart—apparently depending on the type of prey and the previous experience of the predator.

HARTWICK, E. B., L. TULLOCH, and S. MacDONALD. 1981. Feeding and growth of *Octopus dofleini* (Wülker). Veliger 24: 129–38.

HIGH, W. L. 1976. The giant Pacific octopus. Mar. Fish. Rev. 38: 17–22.

KYTE, M. A., and G. W. COURTNEY. 1977. A field observation of aggressive behavior between two North Pacific octopus, *Octopus dofleini martini*. Veliger 19: 427–28.

MANN, T., A. W. MARTIN, and J. B. THIERSCH. 1970. Male reproductive tract, spermatophores and spermatophoric reaction in the giant octopus of the North Pacific, *Octopus dofleini martini*. Proc. Roy. Soc. Lond. B175: 31–61.

MARLIAVE, J. B. 1981. Neustonic feeding in early larvae of *Octopus dofleini* (Wülker). Veliger 23: 350–51.

PENNINGTON, H. 1979. New fishery for Alaskans: The giant Pacific octopus. Alaska Seas and Coasts 7: 1–3, 12.

PICKFORD, G. E. 1964. *Octopus dofleini* (Wülker), the giant octopus of the North Pacific. Bull. Bingham Oceanogr. Coll. 19: 1–70.

WARREN, L. R., M. F. SHEIER, and D. A. RILEY. 1974. Colour changes of *Octopus rubescens* during attacks on unconditioned and conditioned stimuli. Anim. Behav. 22: 211–19.

WINKLER, L. R., and L. M. ASHLEY. 1954. The anatomy of the common octopus of northern Washington. Publ. Dept. Biol. Sci., Walla Walla College 10: 1–29.

Phylum **PHORONIDA**

Three phyla—Phoronida, Brachiopoda, and Ectoprocta (Bryozoa)—compose the lophophorates, a group united by the presence of a set of ciliated, food-gathering tentacles containing extensions of the coelom. A few general papers on lophophore function and the disputed phylogenetic affinities of these animals are listed below.

BULLIVANT, J. S. 1968. The method of feeding of lophophorates (Bryozoa, Phoronida, Brachiopoda). N.Z. J. Mar. Freshwater Res. 2: 135–46.

EMIG, C. C. 1977. Notes on the localization, ecology, and taxonomy of the phoronideans. Tethys 7: 357–64.

FARMER, J. D., J. W. VALENTINE, and R. COWEN. 1973. Adaptive strategies leading to the ectoproct ground-plan. Syst. Zool. 22: 233–39.

GILMOUR, T. H. J. 1978. Ciliation and function of the food-collecting and waste-rejecting organs of lophophorates. Can. J. Zool. 56: 2142–55. ☐See Strathmann (1982) for an alternative view.

NIELSEN, C. 1977. Phylogenetic considerations: The protostomian relationships. *In* R. M. Woollacott and R. L. Zimmer, eds., Biology of bryozoans, pp. 519–34. New York: Academic Press.

STRATHMANN, R. 1973. Function of lateral cilia in suspension feeding of lophophorates (Brachiopoda, Phoronida, Ectoprocta). Mar. Biol. 23: 129–36.

———. 1982. Cinefilms of particle capture by an induced local change of beat of the lateral cilia of a bryozoan. J. Exp. Mar. Biol. Ecol. 62: 225–36.

The phoronids are a small phylum, consisting of fewer than 16 described species. Eight of these are known from the west coast, of which four are listed here.

EMIG, C. C. 1974. The systematics and evolution of the phylum Phoronida. Z. Zool. Syst. Evol. 12: 128–51.

———. 1977. Embryology of Phoronida. Amer. Zool. 17: 21–37.

MARSDEN, J. R. 1959. Phoronidea from the Pacific coast of North America. Can. J. Zool. 37: 87–111. ☐Unites *Phoronopsis viridis* and *P. harmeri* as one species. A number of current workers believe the species are distinct.

ZIMMER, R. L. 1967. The morphology and function of accessory reproductive glands in the lophophores of *Phoronis vancouverensis* and *Phoronopsis harmeri*. J. Morphol. 121: 159–78.

———. 1973. Morphological and developmental affinities of the lophophorates. *In* G. P. Larwood, ed., Living and fossil Bryozoa—recent advances in research, pp. 593–99. Second Conf. Internat. Bryozool. Assoc. New York: Academic Press.

Phoronis vancouverensis Pixell, §229

MARSDEN, J. R. 1957. Regeneration in *Phoronis vancouverensis*. J. Morphol. 101: 307–23.

Phoronopsis californica Hilton, §297

P. harmeri Pixell and **P. viridis** Hilton, §297

RATTENBURY, J. C. 1953. Reproduction in *Phoronopsis viridis*: The annual cycle in the gonads, maturation and fertilization of the ovum. Biol. Bull. 104: 182–96.

———. 1954. The embryology of *Phoronopsis viridis*. J. Morphol. 95: 289–349.

RONAN, T. E., JR. 1978. Food-resources and the influence of spatial pattern on feeding in the phoronid *Phoronopsis viridis*. Biol. Bull. 154: 472–84. ☐In Bodega Harbor, intertidal populations of *Phoronopsis viridis* may form dense clusters of up to 150,000 individuals per square meter. In such clusters, insufficient room exists for all the lophophores to expand at the surface, so they are stratified.

Phylum **BRACHIOPODA**

Brachiopods superficially resemble bivalve mollusks in that the bodies of both are encased between two shells. The resemblances end there, however: the internal anatomy differs radically between the two, and the shells of brachiopods are ventral-dorsal whereas those of bivalves are right-left. Although brachiopods are now relatively few in number (fewer than 300 living species are known), they are common in the fossil record. More than 30,000 fossil species have been described, with some genera extending as far back as the Cambrian; this rich history is reflected in several of the references below.

BERNARD, F. R. 1972. The living Brachiopoda of British Columbia. Syesis 5: 73–82.

GUTMANN, W. F., K. VOGEL, and H. ZORN. 1978. Brachiopods: Biomechanical interdependences governing their origin and phylogeny. Science 199: 890–93.

HERTLEIN, L. G., and U. S. GRANT IV. 1944. The Cenozoic Brachiopoda of western North America. Publ. Math. Phys. Sci., Univ. Calif. Los Angeles 3: 1–236.

LaBARBERA, M. 1977. Brachiopod orientation to water movement: 1. Theory, laboratory behavior, and field orientations. Paleobiology 3: 270–87.

LOEB, M. J., and G. WALKER. 1977. Origin, composition, and function of secretions from pyriform organs and internal sacs of four settling cheilo-ctenostome bryozoan larvae. Mar. Biol. 42: 37–46.

McCAMMON, H. M., and W. A. REYNOLDS. 1976. Experimental evidence for direct nutrient accumulation by the lophophore of articulate brachiopods. Mar. Biol. 34: 41–51.

REYNOLDS, W. A., and H. M. McCAMMON. 1977. Aspects of the functional morphology of the lophophore in articulate brachiopods. Amer. Zool. 17: 121–32.

RUDWICK, M. J. S. 1970. Living and fossil brachiopods. London: Hutchinson University Library. 199 pp.

STEELE-PETROVIC, H. M. 1976. Brachiopod food and feeding processes. Paleontology 19: 417–36.

Glottidia albida (Hinds), tongue "clam," §291

JONES, G. F., and J. L. BARNARD. 1963. The distribution and abundance of the inarticulate brachiopod *Glottidia albida* (Hinds) on the mainland shelf of southern California. Pacific Natur. 4: 27–52.

Terebratalia transversa (Sowerby), lamp shell, §227

PAINE, R. T. 1969. Growth and size distribution of the brachiopod *Terebratalia transversa* Sowerby. Pacific Sci. 23: 337–43.

Phylum ECTOPROCTA (BRYOZOA)

Several reviews of bryozoan biology have appeared in the last 10 years, and two of particular note are those of Ryland (1976) and Woollacott and Zimmer (1977). The most comprehensive systematic treatment of Pacific coast species remains that of Osburn (1950–53).

ANNOSCIA, E., ed. 1968. Proceedings of the First International Conference of Bryozoa. Atti Soc. Ital. Sci. Natur. Mus. Civ. Stor. Nat. Milano 108: 1–377.

NIELSEN, C. 1981. On morphology and reproduction of '*Hippodiplosia*' insculpta and *Fenestrulina malusii* (Bryozoa, Cheilostomata). Ophelia 20: 91–125.

OSBURN, R. C. 1950–53. Bryozoa of the Pacific coast of America. I, Cheilostomata—Anasca. Allan Hancock Pac. Exped. 14 (1950): 1–269. II, Cheilostomata—Ascophora. Allan Hancock Pac. Exped. 14

(1952): 271–611. III, Cyclostomata, Ctenostomata, Entoprocta, and addenda. Allan Hancock Pac. Exped. 14 (1953): 613–841.

PINTER, P. 1969. Bryozoan-algal associations in southern California waters. Bull. South. Calif. Acad. Sci. 68: 199–218. □Records 17 species of bryozoans encrusting 62 species of algae. The bryozoans display preferences, and certain of the algae are seen as likely to be producing anti-fouling substances that discourage larval settlement.

REED, C. G., and R. A. CLONEY. 1982a. The larval morphology of the marine bryozoan *Bowerbankia gracilis* (Ctenostomata: Vesicularioidea). Zoomorphology 100: 23–54.

———. 1982b. The settlement and metamorphosis of the marine bryozoan *Bowerbankia gracilis* (Ctenostomata: Vesicularioidea). Zoomorphology 101: 103–32. □A detailed light and electron microscopic study.

ROSS, J. R. P. 1970. Keys of the Recent cyclostome Ectoprocta of marine waters of northwest Washington State. Northwest Sci. 44: 154–69.

RYLAND, J. S. 1974 (publ. 1976). Behaviour, settlement and metamorphosis of bryozoan larvae: A review. Thalassia Jugosl. 10: 239–62.

———. 1976. Physiology and ecology of marine bryozoans. Adv. Mar. Biol. 14: 285–443.

SOULE, J. D., and D. F. SOULE. 1969. Systematics and biogeography of burrowing bryozoans. Amer. Zool. 9: 791–802.

WINSTON, J. E. 1978. Polypide morphology and feeding behavior in marine ectoprocts. Bull. Mar. Sci. 28: 1–31. □Includes details on several Pacific coast species, as well as being a good general account.

WOOLLACOTT, R. M., and W. J. NORTH. 1971. Bryozoans of California and northern Mexico kelp beds. *In* W. J. North, ed., The biology of giant kelp beds (*Macrocystis*) in California, pp. 455–79. Nova Hedwigia 32: 1–600.

WOOLLACOTT, R. M., and R. L. ZIMMER. 1977. Biology of bryozoans. New York: Academic Press. 566 pp. □A superb collection of reviews on various aspects of bryozoan biology.

———. 1978. Metamorphosis of cellularoid bryozoans. *In* F.-S. Chia and M. E. Rice, eds., Settlement and metamorphosis of marine invertebrate larvae, pp. 49–63. New York: Elsevier.

Alcyonidium polyoum (Hassall) (formerly *A. mytili* Dalyell), §316

THORPE, J. P., J. S. RYLAND, and J. A. BEARDMORE. 1978. Genetic variation and biochemical systematics in the marine bryozoan *Alcyonidium mytili*. Mar. Biol. 49: 343–50. □Electrophoretic evidence gathered on populations in British waters suggest that *Alcyonidium polyoum* and *A. mytili* are separate species.

Bowerbankia gracilis (O'Donoghue), §316

Bugula californica Robertson, §113

B. neritina (Linnaeus), §324

MAWATARI, S. 1951. Natural history of a common fouling bryozoan *Bugula neritina* (Linnaeus). Misc. Rep. Res. Inst. Nat. Resources, Tokyo 19–21: 47–54.

REED, C. G., and R. M. WOOLLACOTT. 1982.

Mechanisms of rapid morphogenetic movements in the metamorphosis of the bryozoan *Bugula neritina* (Cheilostomata, Cellularioidea). 1. Attachment to the substratum. J. Morphol. 172: 335–48.

WOOLLACOTT, R. M., and R. L. ZIMMER. 1971. Attachment and metamorphosis of the cheiloctenostome bryozoan *Bugula neritina* (Linné). J. Morphol. 134: 351–82.

———. 1975. A simplified placenta-like system for the transport of extraembryonic nutrients during embryogenesis of *Bugula neritina* (Bryozoa). J. Morphol. 147: 355–77.

B. pacifica Robertson, §114
Cryptosula pallasiana (Moll), §324
Dendrobeania lichenoides (Robertson), §229
D. murrayana (Johnston), §229
Diaperoecia californica (d'Orbigny) (formerly *Idmonea*), §270
Eurystomella bilabiata (Hincks), §140
Flustrellidra corniculata (Smith) (formerly *Flustrella cervicornis* (Robertson)), §111
Hippodiplosia insculpta (Hincks), §140

Membranipora membranacea (Linnaeus), **M. tuberculata** (Bosc), and **M. villosa** Hincks, §110
DEBURGH, M. E., and P. V. FANKBONER. 1978. A nutritional association between the bull kelp *Nereocystis luetkeana* and its epizooic bryozoan *Membranipora membranacea*. Oikos 31: 69–72.

YOSHIOKA, P. M. 1982a. Role of planktonic and benthic factors in the population dynamics of the bryozoan *Membranipora membranacea*. Ecology 63: 457–68.

———. 1982b. Predator-induced polymorphism in the bryozoan *Membranipora membranacea* (L.) J. Exp. Mar. Biol. Ecol. 61: 233–42. □Predation by the nudibranchs *Corambe pacifica* and *Doridella steinbergae* induce the formation of spines. As a result, three described species of *Membranipora* are considered to be only phenotypic variants of *Membranipora membranacea*.

Phidolopora pacifica (Robertson), §140
Tricellaria occidentalis (Trask), and related species, §112
Tubulipora flabellaris (Fabricius), §229
Zoobotryon verticillatum (Delle Chiaje) (formerly *Z. pellucidum* Ehrenberg), §229
BULLIVANT, J. S. 1968a. Attachment and growth of the stoloniferous ctenostome bryozoan *Zoobotryon verticillatum*. Bull. South. Calif. Acad. Sci. 67: 199–202.

———. 1968b. The rate of feeding of the bryozoan *Zoobotryon verticillatum*. N.Z. J. Mar. Freshwater Res. 2: 111–34.

Phylum ENTOPROCTA

Although many older and a few recent systematic reports on bryozoans include these animals, entoprocts and ectoprocts are very different in structure and organization. Species are included in Part III (1953) of Osburn's monograph (listed under Ectoprocta).

MARISCAL, R. N. 1975. Entoprocta. In A. C. Giese and J. S. Pearse, eds., Reproduction of marine invertebrates. 2. Entoprocts and lesser coelomates, pp. 1–41. New York: Academic Press.

NIELSEN, C. 1971. Entoproct life-cycles and the entoproct/ectoproct relationship. Ophelia 9: 209–341.

———. 1977. The relationships of Entoprocta, Ectoprocta, and Phoronida. Amer. Zool. 17: 149–50. □Nielsen believes that the Entoprocta and Ectoprocta are closely related and should be united as classes of one phylum. Although this view is held by a minority of zoologists, minority views sometimes turn out to be correct.

SOULE, D. F., and J. D. SOULE. 1965. Two new species of *Loxosomella*, Entoprocta, epizoic on Crustacea. Allan Hancock Found. Publ. Occ. Pap. 29: 1–19.

Barentsia benedeni (Foettinger) and **B. gracilis** (M. Sars), §316
MARISCAL, R. N. 1965. The adult and larval morphology and life history of the entoproct *Barentsia gracilis* (M. Sars, 1835). J. Morphol. 116: 311–38. □The species studied was actually *Barentsia benedeni*, which is easily confused with *B. gracilis*.

B. ramosa (Robertson), §115
Pedicellina cernua (Pallas), §115

Phylum ECHINODERMATA

The echinoderms are almost exclusively marine, except for a few species that tolerate brackish water. Aside from their hard internal skeletal elements, which in many species project outward to form spines, the most distinctive echinoderm feature is a unique system of water-filled tubes (the water-vascular system) that terminates in many blind-ending hydraulic structures called tube feet, or podia; these podia serve a variety of functions including locomotion, food gathering, and respiratory gas exchange. According to an abundant fossil record, echinoderms reached their pinnacle of diversity in the early Paleozoic, when at least 20 classes existed. Today, only five classes of echinoderms have living representatives, and one of these, the Crinoidea (sea lilies), is not treated here because representatives occur only in deep water along our coast. The other four classes are conspicuous inhabitants of the shore.

BINYON, J. 1972. Physiology of echinoderms. New York: Pergamon. 264 pp.

BOOLOOTIAN, R. A. 1966. Physiology of Echinodermata. New York: Interscience. 822 pp.

HOLLAND, N. D., and L. Z. HOLLAND. 1969. A bibliography of echinoderm biology, continuing Hyman's 1955 bibliography through 1965. Publ. Staz. Zool. Napoli 37: 441–53.

HYMAN, L. H. 1955. The invertebrates: Echinodermata. Vol. 4. New York: McGraw-Hill. 763 pp.

MILLOTT, N., ed. 1967. Echinoderm biology. Symp. Zool. Soc. London 20. London: Academic Press. 240 pp. □A symposium volume of contributed chapters dealing with several specific aspects of echinoderm biology; however, several basic aspects are left untouched.

NICHOLS, D. 1969. Echinoderms. 4th ed. London: Hutchinson University Library. 192 pp.

STRATHMANN, R. R. 1971. The feeding behavior of planktotrophic echinoderm larvae: Mechanisms, regulation, and rates of suspension-feeding. J. Exp. Mar. Biol. Ecol. 6: 109–60.

———. 1978. Length of pelagic period in echinoderms with feeding larvae from the northeast Pacific. J. Exp. Mar. Biol. Ecol. 34: 23–27.

Class ASTEROIDEA. Seastars, or Starfish

BIRKELAND, C. 1974. Interactions between a sea pen and seven of its predators. Ecol. Monogr. 44: 211–32. ☐Four of the predators on the sea pen are seastars (*Dermasterias imbricata*, *Crossaster papposus*, *Hippasteria spinosa*, and *Mediaster aequalis*); they, in turn, are prey for another asteroid, *Solaster dawsoni*.

FERGUSON, J. C. 1971. Uptake and release of free amino acids by starfishes. Biol. Bull. 141: 122–29.

FISHER, W. K. 1911–30. Asteroidea of the north Pacific and adjacent waters. 1, Phanerozonia and Spinulosa. Bull. U.S. Natl. Mus. 76(1) (1911): 1–419. 2, Forcipulata (part). Bull. U.S. Natl. Mus. 76(2) (1928): 1–245. 3, Forcipulata (concluded). Bull. U.S. Natl. Mus. 76(3) (1930): 1–356. ☐Still the primary source for the morphology, distribution, and habits of many of our Pacific coast seastars.

HARROLD, C., and J. S. PEARSE. 1980. Allocation of pyloric caecum reserves in fed and starved sea stars, *Pisaster giganteus* (Stimpson): Somatic maintenance comes before reproduction. J. Exp. Mar. Biol. Ecol. 48: 169–83. ☐One recent example of a series of studies on the relationship between the pyloric caeca, growth, and reproduction.

HOPKINS, T. S., and G. F. CROZIER. 1966. Observations on the asteroid echinoderm fauna occurring in the shallow water of southern California (intertidal to 60 meters). Bull. South. Calif. Acad. Sci. 65: 129–45.

LAMBERT, P. 1981. The sea stars of British Columbia. B.C. Prov. Mus. Natur. Hist. Handb. 39. 152 pp.

LANDENBERGER, D. E. 1969. The effects of exposure to air on Pacific starfish and its relationship to distribution. Physiol. Zool. 42: 220–30. ☐Concerns the relative effects of exposure on *Pisaster ochraceus*, *P. giganteus*, and *P. brevispinus*.

MAUZEY, K. P., C. BIRKELAND, and P. K. DAYTON. 1968. Feeding behavior of asteroids and escape responses of their prey in the Puget Sound region. Ecology 49: 603–19.

POLLS, I., and J. GONOR. 1975. Behavioral aspects of righting in two asteroids from the Pacific coast of North America. Biol. Bull. 148: 68–84.

SLOAN, N. A. 1980. Aspects of the feeding biology of asteroids. Oceanogr. Mar. Biol. Ann. Rev. 18: 57–124. ☐A comprehensive review.

WOBBER, D. R. 1975. Agonism in asteroids. Biol. Bull. 148: 483–96.

Astrometis sertulifera (Xantus), §72

ROBILLIARD, G. A. 1971. Feeding behaviour and prey capture in an asteroid, *Stylasterias forreri*. Sye-

sis 4: 191–95. ☐Describes, in a related species, capture of large, active prey by means of pedicellariae.

TURK, T. R. 1978. The ecological role of the predatory seastar *Astrometis sertulifera* in a San Diego rocky intertidal community. Ph.D. dissertation, Ecology, University of California, Riverside, and San Diego State University. 127 pp. ☐Approximately 80 percent of the field diet of the seastar consisted of the gastropod *Tegula eiseni* and the chiton *Stenoplax conspicua*.

Astropecten armatus Gray, §250

CHRISTENSEN, A. M. 1970. Feeding biology of the sea-star *Astropecten irregularis* Pennant. Ophelia 8: 1–134. ☐Comprehensive study of the feeding biology of this Atlantic species.

See also Edwards (1969a), under *Olivella biplicata*, p. 560, and Kastendiek (1976), under *Renilla kollikeri*, pp. 513–14.

Dermasterias imbricata (Grube), §211

ROSENTHAL, R. J., and J. R. CHESS. 1972. A predator-prey relationship between the leather star, *Dermasterias imbricata*, and the purple urchin, *Strongylocentrotus purpuratus*. U.S. Natl. Mar. Fish. Serv., Fish. Bull. 70: 205–16.

WAGNER, R. H., D. W. PHILLIPS, J. D. STANDING, and C. HAND. 1979. Commensalism or mutualism: Attraction of a sea star toward its symbiotic polychaete. J. Exp. Mar. Biol. Ecol. 39: 205–10.

Evasterias troschelii (Stimpson), §207

CHRISTENSEN, A. M. 1957. The feeding behavior of the sea-star *Evasterias troschelii* Stimpson. Limnol. Oceanogr. 2: 180–97.

Henricia leviuscula (Stimpson), §§69, 211

Leptasterias hexactis (Stimpson) (includes *L. aequalis* (Stimpson)), §§68, 210

CHIA, F.-S. 1966a. Brooding behavior of a six-rayed starfish, *Leptasterias hexactis*. Biol. Bull. 130: 304–15.

———. 1966b. Systematics of the six-rayed sea star, *Leptasterias*, in the vicinity of San Juan Island, Washington. Syst. Zool. 15: 300–306.

———. 1969. Histology of the pyloric caeca and its changes during brooding and starvation in a starfish, *Leptasterias hexactis*. Biol. Bull. 136: 185–92.

MENGE, B. A. 1972. Competition for food between two intertidal starfish species and its effect on body size and feeding. Ecology 53: 635–44. ☐Concerns *Leptasterias* and *Pisaster ochraceus*.

———. 1972. Foraging strategy of a starfish in relation to actual prey availability and environmental predictability. Ecol. Monogr. 42: 25–50.

———. 1974. The effect of wave action and competition on brooding and reproductive effort in the seastar, *Leptasterias hexactis*. Ecology 55: 84–93.

MENGE, J. L., and B. A. MENGE. 1974. Role of resource allocation, aggression and spatial heterogeneity in coexistence of two competing intertidal starfish. Ecol. Monogr. 44: 189–209.

SHIRLEY, T. C., and W. B. STICKLE. 1982. Responses of *Leptasterias hexactis* (Echinodermata: Asteroidea) to low salinity. I. Survival, activity, feeding, growth and absorption efficiency. Mar. Biol. 69: 147–54. ☐Although most echinoderms are thought to be stenohaline, restricted to nearly full-strength seawater, populations of *Leptasterias hexactis* in southern Alaska are exposed seasonally to low salinities. In the laboratory, *L. hexactis* survived and grew at salinities of 20 parts per thousand for at least 3 weeks.

L. pusilla (Fisher), §30

Linckia columbiae Gray, §71

MONKS, S. P. 1904. Variability and autotomy of *Phataria*. Proc. Acad. Natur. Sci. Phila. 51: 596–600.

Orthasterias koehleri (de Loriol), §208

Patiria miniata (Brandt) (formerly *Asterina*), §26

ANDERSON, J. M. 1959. Studies on the cardiac stomach of a starfish, *Patiria miniata* (Brandt). Biol. Bull. 117: 185–201.

Pisaster brevispinus (Stimpson), §209

HADERLIE, E. C. 1980. Sea star predation on rock-boring bivalves. Veliger 22: 400.

SMITH, L. S. 1961. Clam-digging behavior in the starfish, *Pisaster brevispinus* (Stimpson, 1857). Behaviour 18: 148–53.

VAN VELDHUIZEN, H. D. 1978. Feeding biology of subtidal *Pisaster brevispinus* on soft substrate in Bodega Harbor, California. Ph.D. dissertation, Zoology, University of California, Davis. 197 pp.

VAN VELDHUIZEN, H. D., and D. W. PHILLIPS. 1978. Prey capture by *Pisaster brevispinus* (Asteroidea: Echinodermata) on soft substrate. Mar. Biol. 48: 89–97.

P. giganteus (Stimpson), §157

LANDENBERGER, D. E. 1966. Learning in the Pacific starfish *Pisaster giganteus*. Anim. Behav. 14: 414–18.

————. 1968. Studies on selective feeding in the Pacific starfish *Pisaster* in southern California. Ecology 49: 1062–75. □Concerns *Pisaster giganteus* and *P. ochraceus*.

ROSENTHAL, R. J. 1971. Trophic interaction between the sea star *Pisaster giganteus* and the gastropod *Kelletia kelletii*. U.S. Natl. Mar. Fish. Serv., Fish. Bull. 69: 669–79. □Describes the diet of *Pisaster giganteus* in the subtidal off San Diego.

P. giganteus capitatus (Stimpson), the southern California subspecies, §142

P. ochraceus f. confertus (Stimpson), of quiet water, §199

P. ochraceus f. ochraceus (Brandt), chiefly on shores subject to wave impact, §157

FEDER, H. M. 1955. On the methods used by the starfish *Pisaster ochraceus* in opening three types of bivalve molluscs. Ecology 36: 764–67.

————. 1959. The food of the starfish, *Pisaster ochraceus*, along the California coast. Ecology 40: 721–24.

————. 1970. Growth and predation by the ochre sea star, *Pisaster ochraceus* (Brandt), in Monterey Bay, California. Ophelia 8: 161–85.

FRASER, A., J. GOMEZ, E. B. HARTWICK, and M. J. SMITH. 1981. Observations on the reproduction and development of *Pisaster ochraceus* (Brandt). Can. J. Zool. 59: 1700–1707. □In British Columbia.

MAUZEY, K. P. 1966. Feeding behavior and repro-

ductive cycles in *Pisaster ochraceus*. Biol. Bull. 131: 127–44. □In San Juan Islands.

PAINE, R. T. 1969. The *Pisaster-Tegula* interaction: Prey patches, predator food preference, and intertidal community structure. Ecology 50: 950–61. □Where mussels are absent and *Tegula funebralis* is abundant, *Pisaster ochraceus* eats lots of snails.

————. 1974. Intertidal community structure: Experimental studies on the relationship between a dominant competitor and its principal predator. Oecologia 15: 93–120. □The "dominant competitor" is *Mytilus californianus*; the predator is *Pisaster ochraceus*.

PEARSE, J. S., and D. J. EERNISSE. 1982. Photoperiodic regulation of gametogenesis and gonadal growth in the sea star *Pisaster ochraceus*. Mar. Biol. 67: 121–25.

P. ochraceus segnis Fisher, of southern California, §157

Pycnopodia helianthoides (Brandt), §66

DAYTON, P. K. 1973. Two cases of resource partitioning in an intertidal community: Making the right prediction for the wrong reason. Amer. Natur. 107: 662–70. □Concerns, among other things, the relationship between *Pycnopodia*, its sea urchin prey, and the inadvertent beneficiary of this relationship, the sea anemone *Anthopleura xanthogrammica*. A similar relationship between *Pisaster ochraceus*, *Mytilus californianus*, and *A. xanthogrammica* is also described.

PAUL, A. J., and H. M. FEDER. 1975. The food of the sea star *Pycnopodia helianthoides* (Brandt) in Prince William Sound, Alaska. Ophelia 14: 15–22.

Solaster dawsoni Verrill, §67

VAN VELDHUIZEN, H. D., and V. J. OAKES. 1981. Behavioral responses of seven species of asteroids to the asteroid predator, *Solaster dawsoni*. Oecologia 48: 214–20.

S. stimpsoni Verrill, §67

Class OPHIUROIDEA
Brittle stars, Serpent stars

AUSTIN, W. C. 1966. Feeding mechanisms, digestive tracts and circulatory systems in the ophiuroids *Ophiothrix spiculata* Le Conte, 1851, and *Ophiura luetkeni* (Lyman, 1860). Ph.D. dissertation, Biological Sciences, Stanford University, Stanford, Calif. 289 pp.

BOOLOOTIAN, R. A., and D. LEIGHTON. 1966. A key to the species of Ophiuroidea (brittle stars) of the Santa Monica Bay and adjacent areas. Los Angeles Co. Mus. Natur. Hist. Contr. Sci. 93. 20 pp.

CLARK, H. L. 1911. North Pacific ophiurans in the collection of the United States National Museum. Bull. U.S. Natl. Mus. 75. 302 pp.

HENDLER, G. 1975. Adaptational significance of the patterns of ophiuroid development. Amer. Zool. 15: 691–715.

KYTE, M. A. 1969. A synopsis and key to the recent Ophiuroidea of Washington State and southern British Columbia. J. Fish. Res. Board Can. 26: 1727–41.

WILKIE, I. C. 1978. Functional morphology of the autotomy plane of the brittlestar *Ophiocomina nigra* (Abildgaard) (Ophiuroidea, Echinodermata). Zoomorphologie 91: 289–305. □A light-microscope study revealing a series of features interpreted as adaptations for the process of autotomy.

WOODLEY, J. D. 1975. The behaviour of some amphiurid brittle-stars. J. Exp. Mar. Biol. Ecol. 18: 29–46. □Describes behaviors associated with burrowing, respiration, feeding, and spawning in several species of North Atlantic amphiurid brittle stars. Pacific coast amphiurids (e.g., *Amphiodia* and *Amphipholis*) presumably behave similarly.

Amphiodia occidentalis (Lyman), §44
SUTTON, J. E. 1976 (publ. 1978). Partial revision of the genus *Amphiodia* (Ophiuroidea: Amphiuridae) from the west coast of North America. Thalassia Jugosl. 12: 341–48.

A. urtica (Lyman) (formerly *A. barbarae* (Lyman)), §263

Amphipholis pugetana (Lyman) (formerly *Axiognathus*), §19

A. squamata (Delle Chiaje) (formerly *Axiognathus*), §19
BREHM, P., and J. G. MORIN. 1977. Localization and characterization of luminescent cells in *Ophiopsila californica* and *Amphipholis squamata* (Echinodermata: Ophiuroidea). Biol. Bull. 152: 12–25.

MARTIN, R. B. 1968. Aspects of the ecology and behaviour of *Axiognathus squamata* (Echinodermata, Ophiuroidea). Tane 14: 65–81.

PENTREATH, R. J. 1970. Feeding mechanisms and the functional morphology of podia and spines in some New Zealand ophiuroids (Echinodermata). J. Zool. (Lond.) 161: 395–429. □Includes *Amphipholis squamata* (as *Axiognathus*).

Ophioderma panamense Lutken, §121

Ophionereis annulata (Le Conte), §121
MUSCAT, A. M. 1975. Reproduction and growth in the ophiuroid, *Ophionereis annulata*. Master's thesis, Biology, San Diego State University. 83 pp. □At False Point, La Jolla, *Ophionereis annulata* is one of the most common ophiuroids, occurring at a density of about 70 per square meter. It is apparently a slow grower with a life span of 20 to 25 years. Spawning, which may occur in females with disk diameters of only 4 mm, peaked from mid-September to late October.

Ophiopholis aculeata (Linnaeus) var. **kennerlyi** (Lyman), §122
LaBARBERA, M. 1978. Particle capture by a Pacific brittle star: Experimental test of the aerosol suspension feeding model. Science 201: 1147–49. □*Ophiopholis aculeata* is a suspension feeder, capable of removing particles from seawater by a mechanism other than conventional sieving. Particles adhere directly to the mucus-covered tube feet.

LITVINOVA, N. M. 1981. Behavior of *Ophiopholis aculeata* (Ophiuroidea) in the time of reproduction. (In Russian, with English summary.) Zool. Zh. 60: 942–45. □These brittle stars take cover in empty mussel shells during the spawning season and repulse other ophiuroids; they do not, however, cease feeding.

Ophioplocus esmarki Lyman, §45
Ophiopteris papillosa (Lyman), §124
Ophiothrix spiculata Le Conte, §123
WARNER, G. F., and J. D. WOODLEY. 1975. Suspension-feeding in the brittle-star *Ophiothrix fragilis*. J. Mar. Biol. Assoc. U.K. 55: 199–210.

Class ECHINOIDEA. Sea urchins, Sand dollars, and Heart urchins

DURHAM, J. W., and R. V. MELVILLE. 1957. A classification of echinoids. J. Paleontol. 31: 242–72.

GRANT, U. S., IV, and L. G. HERTLEIN, 1938. The west American Cenozoic Echinoidea. Univ. Calif. Los Angeles Publ. Math. Phys. Sci. 2: 1–225.

LAWRENCE, J. M. 1975. On the relationships between marine plants and sea urchins. Oceanogr. Mar. Biol. Ann. Rev. 13: 213–86.

McCAULEY, J. E., and A. G. CAREY, JR. 1967. Echinoidea of Oregon. J. Fish. Res. Board Can. 24: 1385–1401.

MORTENSEN, T. 1928–1951. Monograph of the Echinoidea. 5 volumes, 17 parts. Copenhagen: Reitzel.

PEARSE, J. S., and V. B. PEARSE. 1975. Growth zones in the echinoid skeleton. Amer. Zool. 15: 731–53.

Dendraster excentricus (Eschscholtz), §246
BIRKELAND, C., and F.-S. CHIA. 1971. Recruitment risk, growth, age and predation in two populations of sand dollars, *Dendraster excentricus* (Eschscholtz). J. Exp. Mar. Biol. Ecol. 6: 265–78. □Populations in Puget Sound were studied.

CHIA, F.-S. 1969. Some observations on the locomotion and feeding of the sand dollar *Dendraster excentricus* (Eschscholtz). J. Exp. Mar. Biol. Ecol. 3: 162–70.

———. 1973. Sand dollar: A weight belt for the juvenile. Science 181: 73–74.

HIGHSMITH, R. C. 1982. Induced settlement and metamorphosis of sand dollar (*Dendraster excentricus*) larvae in predator-free sites: Adult sand dollar beds. Ecology 63: 329–37. □Larvae of *Dendraster* settle preferentially in sand occupied and somehow "scented" by adults. Such settlement behavior is beneficial to the juvenile sand dollars because adults, through their sediment reworking activities, exclude one of the predators on juvenile *Dendraster*, the tube-building tanaid *Leptochelia dubia*.

MERRILL, R. J., and E. S. HOBSON. 1970. Field observations of *Dendraster excentricus*, a sand dollar of western North America. Amer. Midl. Natur. 83: 595–624. □Populations in southern California.

NIESEN, T. M. 1977. Reproductive cycles of two populations of the Pacific sand dollar *Dendraster excentricus*. Mar. Biol. 42: 365–73.

TIMKO, P. L. 1976. Sand dollars as suspension feeders: A new description of feeding in *Dendraster excentricus*. Biol. Bull. 151: 247–59.

D. laevis H. L. Clark, sand dollar, §246
D. vizcainoensis Grant and Hertlein, sand dollar, §246
Echinarachnius parma (Lamarck), §246

Lovenia cordiformis A. Agassiz, §247

Lytechinus anamesus H. L. Clark (includes *L. pictus* (Verrill) of previous editions), §§218, 269
Although there may remain some small doubt, *Lytechinus anamesus* and *L. pictus* are very likely ecological variants of the same species and are so considered here.
Lytechinus has become a popular subject for embryological and physiological studies.

Strongylocentrotus droebachiensis (O. F. Müller), §212

S. franciscanus (A. Agassiz), §73

S. purpuratus (Stimpson), §174
The three species of *Strongylocentrotus* have been the subject of physiological, embryological, and ecological study for years. Of the hundreds of papers relating to them, we have space for only a pitifully small sample from the recent literature.

BERNARD, F. R. 1977. Fishery and reproductive cycle of the red sea urchin, *Strongylocentrotus franciscanus*, in British Columbia. J. Fish. Res. Board Can. 34: 604–10.

CAMERON, R. A., and S. C. SCHROETER. 1980. Sea urchin (*Strongylocentrotus* spp.) recruitment: Effect of substrate selection on juvenile distribution. Mar. Ecol. Prog. Ser. 2: 243–48.

COCHRAN, R. C., and F. ENGELMANN. 1975. Environmental regulation of the annual reproductive season of *Strongylocentrotus purpuratus* (Stimpson). Biol. Bull. 148: 393–401.

EBERT, T. A. 1967. Growth and repair of spines in the sea urchin *Strongylocentrotus purpuratus* (Stimpson). Biol. Bull. 133: 141–49.

———. 1968. Growth rates of the sea urchin *Strongylocentrotus purpuratus* related to food availability and spine abrasion. Ecology 49: 1075–91.

———. 1977. An experimental analysis of sea urchin dynamics and community interactions on a rock jetty. J. Exp. Mar. Biol. Ecol. 27: 1–22.

FARMANFARMAIAN, A., and A. C. GIESE. 1963. Thermal tolerance and acclimation in the western purple sea urchin, *Strongylocentrotus purpuratus*. Physiol. Zool. 36: 237–43.

GONOR, J. J. 1972. Gonad growth in the sea urchin, *Strongylocentrotus purpuratus* (Stimpson) (Echinodermata: Echinoidea) and the assumptions of gonad index methods. J. Exp. Mar. Biol. Ecol. 10: 89–103.

HIMMELMAN, J. H. 1978. Reproductive cycle of the green sea urchin, *Strongylocentrotus droebachiensis*. Can. J. Zool. 56: 1828–36.

JENSEN, M. 1974. The Strongylocentrotidae (Echinoidea), a morphologic and systematic study. Sarsia 57: 113–48.

LARSEN, B. R., R. L. VADAS, and M. KESER. 1980. Feeding and nutritional ecology of the sea urchin *Strongylocentrotus drobachiensis* in Maine, USA. Mar. Biol. 59: 49–62.

MATTISON, J. E., J. D. TRENT, A. L. SHANKS, T. B. AKIN, and J. S. PEARSE. 1977. Movement and feeding activity of red sea urchins (*Strongylocentrotus franciscanus*) adjacent to a kelp forest. Mar. Biol. 39: 25–30.

MOITOZA, D. J., and D. W. PHILLIPS. 1979. Prey defense, predator preference, and nonrandom diet: The interactions between *Pycnopodia helianthoides*

and two species of sea urchins. Mar. Biol. 53: 299–304.

TEGNER, M. J., and P. K. DAYTON. 1981. Population structure, recruitment and mortality of two sea urchins (*Strongylocentrotus franciscanus* and *S. purpuratus*) in a kelp forest. Mar. Ecol. Prog. Ser. 5: 255–68.

VADAS, R. L. 1977. Preferential feeding: An optimization strategy in sea urchins. Ecol. Monogr. 47: 337–71.

Class HOLOTHUROIDEA. Sea cucumbers

BAKUS, G. J. 1974. Toxicity in holothurians: A geographical pattern. Biotropica 6: 229–36. □Holothurians of tropical waters tend to be more toxic (to fish) than those of temperate waters, and predation by fish is also more intense in the tropics; the author suggests that the two observations are linked.

CLARK, H. L. 1924. Some holothurians from British Columbia. Can. Field Natur. 38: 54–57.

HANSEN, M. D. 1978. Food and feeding behavior of sediment feeders as exemplified by sipunculids and holothurians. Helgol. Wiss. Meeresunters. 31: 191–221.

Caudina arenicola (Stimpson) (formerly *Molpadia*), §264

Chiridota albatrossi Edwards, §245

Cucumaria curata Cowles and C. **pseudocurata** Deichmann, §176

BRUMBAUGH, J. H. 1965. The anatomy, diet, and tentacular feeding mechanism of the dendrochirote holothurian *Cucumaria curata* Cowles, 1907. Ph.D. dissertation, Biological Sciences, Stanford University. 127 pp.

———. 1980. Holothuroidea: The sea cucumbers. *In* R. H. Morris, D. P. Abbott, and E. C. Haderlie, Intertidal invertebrates of California, pp. 136–45. Stanford, Calif.: Stanford University Press.

RUTHERFORD, J. C. 1973. Reproduction, growth, and mortality of the holothurian *Cucumaria pseudocurata*. Mar. Biol. 22: 167–76.

———. 1977a. Variation in egg numbers between populations and between years in the holothurian *Cucumaria curata*. Mar. Biol. 43: 175–80.

———. 1977b. Geographical variation in morphological and electrophoretic characters in the holothurian *Cucumaria curata*. Mar. Biol. 43: 165–74. □Concludes that *Cucumaria pseudocurata* should be a synonym of *C. curata*. Not everyone agrees; the situation is complicated and confusing.

SMITH, E. H. 1962. Studies of *Cucumaria curata* Cowles, 1907. Pacific Natur. 3: 233–46.

C. lubrica Clark, §176

ENGSTROM, N. A. 1974. Population dynamics and prey-predation relations of a dendrochirote holothurian, *Cucumaria lubrica*, and sea stars in the genus *Solaster*. Ph.D. dissertation, Zoology, University of Washington, Seattle. 172 pp.

C. miniata Brandt, §235

C. piperata (Stimpson), §235

C. vegae Theel, §235

Eupentacta pseudoquinquesemita Deichmann, §235

E. quinquesemita (Selenka) (= *Cucumaria chronhjelmi* Theel), §235

Leptosynapta albicans (Selenka) (= *L. inhaerens* (O. F. Müller)), §55

GLYNN, P. W. 1965. Active movements and other aspects of the biology of *Astichopus* and *Leptosynapta* (Holothuroidea). Biol. Bull. 129: 106–27. □ *Leptosynapta albicans* is capable of swimming. The path of animals observed swimming near the surface in Monterey Bay was sinusoidal.

SMITH, G. N. 1971. Regeneration in the sea cucumber *Leptosynapta*. I. The process of regeneration. II. The regenerative capacity. J. Exp. Zool. 177: 319–42.

Lissothuria nutriens (Clark), §141

PAWSON, D. L. 1967. The psolid holothurian genus *Lissothuria*. Proc. U.S. Natl. Mus. 122: 1–17.

Pachythyone rubra (Clark) (formerly *Thyone*), §141

Parastichopus californicus (Stimpson) (formerly *Stichopus*), §77

MARGOLIN, A. S. 1976. Swimming of the sea cucumber *Parastichopus californicus* (Stimpson) in response to sea stars. Ophelia 15: 105–14.

SWAN, E. F. 1961. Seasonal evisceration in the sea cucumber, *Parastichopus californicus* (Stimpson). Science 133: 1078–79.

P. parvimensis (Clark) (formerly *Stichopus*), §77

DIMOCK, R. V., JR. 1977. Effects of evisceration on oxygen consumption by *Stichopus parvimensis* Clark (Echinodermata: Holothuroidea). J. Exp. Mar. Biol. Ecol. 28: 125–32. □ Near Santa Barbara, evisceration of *Parastichopus parvimensis* is seasonal, occurring during October and November. No significant effect on oxygen consumption was noticed.

YINGST, J. Y. 1982. Factors influencing rates of sediment ingestion by *Parastichopus parvimensis* (Clark), an epibenthic deposit-feeding holothurian. Estuarine Coastal Shelf Sci. 14: 119–34.

Psolus chitonoides Clark, §226

FANKBONER, P. V. 1978. Suspension-feeding mechanisms of the armoured sea cucumber *Psolus chitonoides* Clark. J. Exp. Mar. Biol. Ecol. 31: 11–25.

YOUNG, C. M., and F.-S. CHIA. 1982. Factors controlling spatial distribution of the sea cucumber *Psolus chitonoides*: Settling and post-settling behavior. Mar. Biol. 69: 195–205.

Phylum HEMICHORDATA (ENTEROPNEUSTA)
Acorn or Tongue worms, etc.

Some two dozen or more species of enteropneusts apparently exist on the Pacific coast. Most remain undescribed, although a monograph (based on the material of Ritter and several subsequent workers) that should remedy this situation has been promised for the near future. In the meantime, we should note that several forms may be locally common in addition to those specifically listed here. Additional species of *Saccoglossus* have been recorded from sand or mud flats at Popof Island (Alaska), Newport Bay (California), and Punta

Banda (Baja California), and from among subtidal kelp holdfasts at several locations in California. Specimens of *Schizocardium* have been found in mud at low tide in Mugu Lagoon and in Newport Bay. Two different species of *Glossobalanus* have been recorded from intertidal pools at La Jolla and in the vicinity of Nanaimo (British Columbia). Another (?) species of *Glossobalanus* has been found in eelgrass beds, where the worms form sinuous tunnels (not permanent burrows) that weave in and around the roots.

BARRINGTON, E. J. W. 1965. The biology of Hemichordata and Protochordata. San Francisco: Freeman. 176 pp. □ Concerns balanoglossids, ascidians, and amphioxus.

BULLOCK, T. H. 1945. The anatomical organization of the nervous system of *Enteropneusta*. Quart. J. Microsc. Sci. 86: 55–111.

BURDON-JONES, C. 1951. Observations on the spawning behaviour of *Saccoglossus horsti* Brambell and Goodhart, and of other Enteropneusta. J. Mar. Biol. Assoc. U.K. 29: 625–38.

BURDON-JONES, C., and A. M. PATIL. 1960. A revision of the genus *Saccoglossus* (Enteropneusta) in British waters. Proc. Zool. Soc. Lond. 134: 635–45.

RITTER, W. E. 1900. Papers from the Harriman Alaska Expedition. II. *Harrimania maculosa*, a new genus and species of Enteropneusta from Alaska, with special regard to the character of its notochord. Proc. Wash. Acad. Sci. 2: 111–32. □ Records a very common intertidal form, which should be noted here, since it seems to be a feature of underrock collecting at Kodiak, Prince William Sound, Orca, and Valdez. This thick, dark-brown acorn worm, up to 15 cm long, does not burrow as do most enteropneusts, but lies under stones after the fashion of holothurians. Other species are mentioned, but these remain manuscript species to this day.

———. 1902. The movements of the Enteropneusta and the mechanism by which they are accomplished. Biol. Bull. 3: 255–61. □ Two manuscript species are mentioned: *Saccoglossus pusillus*, described in Horst (1930), below; and *Balanoglossus occidentalis*, stated to be abundant in Puget Sound.

RITTER, W. E., and B. M. DAVIS. 1904. Studies on the ecology, morphology, and speciology of the young of some Enteropneusta of western North America. Univ. Calif. Publ. Zool. 1: 171–210.

Balanoglossus occidentalis (a manuscript species), §267

Saccoglossus pusillus (Horst), §267

DAVIS, B. M. 1908. The early life-history of *Dolichoglossus pusillus* Ritter. Univ. Calif. Publ. Zool. 4: 187–226.

HILTON, W. A. 1918. *Dolichoglossus pusillus* Ritter. J. Entomol. Zool. (Pomona) 10. 76 pp.

HORST, C. J. VAN DER. 1930. Observations on some Enteropneusta. Papers from Dr. Th. Mortensen's Pacific expedition, 1914–16, II. Vidensk. Medd. Dan. Naturhist. Foren. 87: 135–200. □ Ritter's original description is on p. 154.

Phylum CHORDATA

Subphylum UROCHORDATA (TUNICATA)
Sea squirts, Compound ascidians, Tunicates

The extensive literature on Pacific coast tunicates, with emphasis on those occurring in California, has been recently reviewed and catalogued by Abbott and Newberry (1980), bringing Van Name's classic monograph (1945) 35 years forward. Additional useful reviews of a more general nature include those of Cloney (1978), Goodbody (1974), and Millar (1971). For details, readers should consult these works and their bibliographies.

ABBOTT, D. P., and A. T. NEWBERRY. 1980. Urochordata: The tunicates. *In* R. H. Morris, D. P. Abbott, and E. C. Haderlie, Intertidal invertebrates of California, pp. 177–226. Stanford, Calif.: Stanford University Press.

ABBOTT, D. P., and W. B. TRASON. 1968. Two new colonial ascidians from the west coast of North America. Bull. South. Calif. Acad. Sci. 67: 143–54.

BARRINGTON, E. J. W., and R. P. S. JEFFERIES, eds. 1975. Protochordates. Symp. Zool. Soc. Lond. 36. New York: Academic Press. 361 pp. □A symposium volume containing a few papers on amphioxus along with ones on hemichordates, ascidians, and pogonophorans.

CLONEY, R. A. 1978. Ascidian metamorphosis: Review and analysis. *In* F.-S. Chia and M. E. Rice, eds., Settlement and metamorphosis of marine invertebrate larvae, pp. 255–82. New York: Elsevier.

FAY, R. C., and J. V. JOHNSON. 1971. Observations on the distribution and ecology of the littoral ascidians of the mainland coast of southern California. Bull. South. Calif. Acad. Sci. 70: 114–24.

GOODBODY, I. 1974. The physiology of ascidians. Adv. Mar. Biol. 12: 1–149.

HUNTSMAN, A. G. 1912. Holosomatous ascidians from the coast of western Canada. Contr. Can. Biol. 1906–10: 103–85.

LAMBERT, G. 1979. Early post-metamorphic growth, budding and spicule formation in the compound ascidian *Cystodytes lobatus*. Biol. Bull. 157: 464–77. □ This ascidian is abundant in some low-intertidal stretches of open coast; the range is British Columbia to Baja California.

MILLAR, R. H. 1966. Evolution in ascidians. *In* H. Barnes, ed., Some contemporary studies in marine science, pp. 519–34. London: Allen & Unwin.

———. 1971. The biology of ascidians. Adv. Mar. Biol. 9: 1–100. □An excellent review of the feeding, breeding, life cycles, and ecology of ascidians.

REVERBERI, G. 1971. Ascidians. *In* G. Reverberi, ed., Experimental embryology of marine and fresh-water invertebrates, pp. 507–50. Amsterdam: North-Holland.

RITTER, W. E., and R. A. FORSYTH. 1917. Ascidians of the littoral zone of southern California. Univ. Calif. Publ. Zool. 16: 439–512.

TOKIOKA, T. 1963. Contributions to Japanese ascidian fauna. XX. The outline of Japanese fauna as compared with that of the Pacific coasts of North America. Publ. Seto Mar. Biol. Lab. 11: 131–56.

TRASON, W. B. 1963. The life cycle and affinities of the colonial ascidian *Pycnoclavella stanleyi*. Univ. Calif. Publ. Zool. 65: 283–326.

VAN NAME, W. G. 1945. The North and South American ascidians. Bull. Amer. Mus. Natur. Hist. 84: 1–476.

WEST, A. B., and C. C. LAMBERT. 1976. Control of spawning in the tunicate *Styela plicata* by variations in a natural light regime. J. Exp. Zool. 195: 263–70. □ *Styela plicata* and many other tunicates normally spawn shortly after dawn; in the laboratory, they can be induced to spawn by placing them in the light after a period of dark adaptation.

YAMAGUCHI, M. 1970. Spawning periodicity and settling time in ascidians, *Ciona intestinalis* and *Styela plicata*. Rec. Oceanogr. Works Japan 10: 147–55.

YOUNG, C. M., and L. F. BRAITHWAITE. 1980. Larval behavior and post-settling morphology in the ascidian, *Chelyosoma productum* Stimpson. J. Exp. Mar. Biol. Ecol. 42: 157–69.

Aplidium californicum (Ritter and Forsyth) (formerly *Amaroucium*), §99

NAKAUCHI, M. 1979. Development and budding in the oozoids of polyclinid ascidians. 3. *Aplidium solidum*. Annot. Zool. Japon. 52: 40–49. □Fairly common in relatively calm water, this ascidian inhabits the low intertidal zone from British Columbia to San Diego.

Archidistoma psammion (Ritter and Forsyth) and
A. diaphanes (Ritter and Forsyth) (formerly *Eudistoma*), §99

LEVINE, E. P. 1962. Studies on the structure, reproduction, development, and accumulation of metals in the colonial ascidian *Eudistoma ritteri* Van Name, 1945. J. Morphol. 111: 105–37.

Ascidia ceratodes (Huntsman), §322
Botrylloides diegensis Ritter and Forsyth, §335
Ciona intestinalis (Linnaeus), §335

DYBERN, B. I. 1963. Biotope choice in *Ciona intestinalis* (L.). Influence of light. Zool. Bidrag Upps. 35: 589–601.

Clavelina huntsmani Van Name, §101
Cnemidocarpa finmarkiensis (Kiaer), §216
Corella sp., §322

In California the predominant species is *Corella willmeriana*; in Puget Sound, there is a mix of *C. willmeriana* and *C. inflata*.

LAMBERT, G. 1968. The general ecology and growth of a solitary ascidian, *Corella willmeriana*. Biol. Bull. 135: 296–307.

LAMBERT, G., C. C. LAMBERT, and D. P. ABBOTT. 1981. *Corella* species in the American Pacific Northwest: Distinction of *C. inflata* Huntsman, 1912 from *C. willmeriana* Herdman, 1898 (Ascidiacea, Phlebobranchia). Can. J. Zool. 59: 1493–1504.

Distaplia occidentalis Bancroft, §99
Euherdmania claviformis (Ritter), §102

TRASON, W. B. 1957. Larval structure and development of the oozoid in the ascidian *Euherdmania claviformis*. J. Morphol. 100: 509–45.

Metandrocarpa taylori Huntsman, §103

ABBOTT, D. P. 1953. Asexual reproduction in the colonial ascidian *Metandrocarpa taylori* Huntsman. Univ. Calif. Publ. Zool. 61: 1–78.

————. 1955. Larval structure and activity in the ascidian *Metandrocarpa taylori*. J. Morphol. 97: 569–94.

HAVEN, N. D. 1971. Temporal patterns of sexual and asexual reproduction in the colonial ascidian *Metandrocarpa taylori* Huntsman. Biol. Bull. 140: 400–415.

NEWBERRY, A. T. 1965a. The structure of the circulatory apparatus of the test and its role in budding in the polystyelid ascidian *Metandrocarpa taylori* Huntsman. Mem. Acad. Roy. Belg., Sci. (2)16: 1–59.

————. 1965b. Vascular structures associated with budding in the polystyelid ascidian *Metandrocarpa taylori*. Ann. Soc. Roy. Zool. Belg. 95: 57–74.

WATANABE, H., and A. T. NEWBERRY. 1976. Budding by oozooids in the polystyelid ascidian *Metandrocarpa taylori*. J. Morphol. 148: 161–76.

Molgula manhattensis DeKay, §335

SAFFO, M. B. 1982. Distribution of the endosymbiont *Nephromyces* Giard within the ascidian family *Molgulidae*. Biol. Bull. 162: 95–104.

SAFFO, M. B., and W. L. DAVIS. 1982. Modes of infection of the ascidian *Molgula manhattensis* by its endosymbiont *Nephromyces* Giard. Biol. Bull. 162: 105–12.

Perophora annectans Ritter, §103
Polyclinum planum (Ritter and Forsyth), §103
Pyura haustor (Stimpson), §216
P. haustor johnsoni (Ritter), §335
Styela clava Herdman (formerly *S. barnharti* Ritter and Forsyth), §322
S. gibbsii (Stimpson), §216
S. montereyensis (Dall), §322

ABBOTT, D. P., and J. V. JOHNSON. 1972. The ascidians *Styela barnharti, S. plicata, S. clava,* and *S. montereyensis* in Californian waters. Bull. South. Calif. Acad. Sci. 71: 95–105. □Compares these four species of *Styela* and clears away the taxonomic confusion that has accompanied "*S. barnharti.*"

YOUNG, C. M., and L. F. BRAITHWAITE. 1980. Orientation and current-induced flow in the stalked ascidian *Styela montereyensis*. Biol. Bull. 159: 428–40.

Subphylum CEPHALOCHORDATA
Lancelets

HUBBS, C. L. 1922. A list of the lancelets of the world with diagnoses of five new species of *Branchiostoma*. Occ. Pap. Mus. Zool., Univ. Mich., No. 105. 16 pp.

Branchiostoma californiense Andrews, §266

Subphylum VERTEBRATA (CRANIATA)

Class CHONDRICHTHYES
Sharks and Rays

Class OSTEICHTHYES. Bony fishes

Consideration of the fishes is for the most part outside the scope of this account. Certainly, many species visit the intertidal region during high tide; judging from their gut contents, these undoubtedly have some impact, in most cases unquantified, on the intertidal invertebrates that do provide our focus. However, because these transient species are unlikely to be seen by human visitors to the shore, they are not being included here (for accounts of common fishes of coastal waters, see Miller and Lea 1972 and Eschmeyer, Herald, and Hammann 1983). A few fishes, however, will be encountered commonly under rocks or in tide pools, and some of the more conspicuous ones are listed below.

Of the following references, most deal with the bony fishes (Osteichthyes); a few deal with the sharks and rays (Chondrichthyes). A third class of fishes, the lampreys and hagfishes (Agnatha), plays little or no role in the intertidal and is not considered.

BARTON, M. 1982a. Comparative distribution and habitat preferences of two species of stichaeoid fishes in Yaquina Bay, Oregon. J. Exp. Mar. Biol. Ecol. 59: 77–87. □ *Anoplarchus purpurescens* was restricted to rocky substrata at the mouth of the estuary closest to oceanic conditions. *Pholis ornata* was found both in rocky areas and above mud flats and occurred in the upper bay closest to freshwater conditions.

————. 1982b. Intertidal vertical distribution and diets of five species of central California stichaeoid fishes. Calif. Fish Game 68: 174–82.

BAYER, R. D. 1981. Shallow-water intertidal ichthyofauna of the Yaquina estuary, Oregon. Northwest Sci. 55: 182–93.

BOLIN, R. L. 1944. A review of the marine cottid fishes of California. Stanford Ichthyol. Bull. 3: 1–135.

BURGESS, T. J. 1978. The comparative ecology of two sympatric polychromatic populations of *Xererpes fucorum* Jordan and Gilbert (Pisces: Pholididae) from the rocky intertidal zone of central California. J. Exp. Mar. Biol. Ecol. 35: 43–58. □This fish, probably the most common gunnel present intertidally in central California, occurs as red or green individuals. During low tide *Xererpes* is found in exposed (but damp) algal masses that match its body color.

DeMARTINI, E. E., and B. G. PATTEN. 1979. Egg guarding and reproductive biology of the red Irish lord, *Hemilepidotus hemilepidotus* (Tilesius). Syesis 12: 41–55. □Benthic, adhesive egg masses of this littoral cottid were spawned in the intertidal zone (Washington) and protected from predatory fishes by both sexes. Information is also presented on other sculpin species.

ESCHMEYER, W. N., E. S. HERALD, and H. HAMMANN. 1983. A field guide to Pacific coast fishes of North America. Boston: Houghton Mifflin. 336 pp. □One of the pocket "Peterson field guides" sponsored by the National Audubon Society. Many illustrations, some in color.

FEDER, H. M., C. H. TURNER, and C. LIMBAUGH. 1974. Observations on fishes associated with kelp beds in southern California. Calif. Dept. Fish Game, Fish Bull. 160: 1–138.

GIBSON, R. N. 1982. Recent studies on the biology of intertidal fishes. Oceanogr. Mar. Biol. Ann. Rev. 20: 363–414.

GREEN, J. M. 1971. Local distribution of *Oligocottus maculosus* Girard and other tidepool cottids of the west coast of Vancouver Island, British Columbia. Can. J. Zool. 49: 1111–28.

———. 1973. Evidence for homing in the mosshead sculpin (*Clinocottus globiceps*). J. Fish. Res. Board Can. 30: 129–30.

GROSSMAN, G. D. 1979. Demographic characteristics of an intertidal bay goby (*Lepidogobius lepidus*). Environ. Biol. Fishes 4: 207–18. □A common intertidal (and subtidal) form of estuaries and bays along our coast. The population studied was in Morro Bay.

———. 1980. Food, fights, and burrows: The adaptive significance of intraspecific aggression in the bay goby (Pisces: Gobiidae). Oecologia 45: 261–66.

GROSSMAN, G. D., R. COFFIN, and P. B. MOYLE. 1980. Feeding ecology of the bay goby (Pisces: Gobiidae). Effects of behavioral, ontogenetic, and temporal variation on diet. J. Exp. Mar. Biol. Ecol. 44: 47–59. □Major prey items of *Lepidogobius lepidus*, an estuarine gobiid, were polychaetes, copepods, and amphipods; a variety of less-important prey was also taken. Larger fish had more diverse diets.

HALDORSON, L., and M. MOSER. 1979. Geographic patterns of prey utilization in two species of surfperch (Embiotocidae). Copeia 1979: 567–72. □The pile surfperch (*Damalichthys vacca*) eats hard-shelled forms such as mollusks, crabs, and barnacles. The striped surfperch (*Embiotoca lateralis*) eats amphipods and ectoprocts. Both species of surfperch range from Alaska to Baja California.

HART, J. L. 1973. Pacific fishes of Canada. Fish. Res. Board Can. Bull. 180. 740 pp.

HORN, M. H., S. N. MURRAY, and T. W. EDWARDS. 1982. Dietary selectivity in the field and food preferences in the laboratory for two herbivorous fishes (*Cebidichthys violaceus* and *Xiphister mucosus*) from a temperate intertidal zone. Mar. Biol. 67: 237–46.

HORN, M. H., and K. C. RIEGLE. 1981. Evaporative water loss and intertidal vertical distribution in relation to body size and morphology of stichaeoid fishes from California. J. Exp. Mar. Biol. Ecol. 50: 273–88. □The five fish species studied included *Anoplarchus purpurescens*, *Xiphister mucosus*, and *X. atropurpureus*.

HUBBS, C. 1952. A contribution to the classification of the blennioid fishes of the family Clinidae, with a partial revision of the eastern Pacific forms. Stanford Ichthyol. Bull. 4: 41–165.

JONES, A. C. 1962. The biology of the euryhaline fish *Leptocottus armatus armatus* Girard (Cottidae). Univ. Calif. Publ. Zool. 67: 321–68. □Based on studies at Walker Creek at Tomales Bay.

MARLIAVE, J. B. 1977. Substratum preferences of settling larvae of marine fishes reared in the laboratory. J. Exp. Mar. Biol. Ecol. 27: 47–60. □Includes *Gobiesox maeandricus* and *Xiphister atropurpureus* among the six species considered.

———. 1981. High intertidal spawning under rockweed, *Fucus distichus*, by the sharpnose sculpin, *Clinocottus acuticeps*. Can. J. Zool. 59: 1122–25. □Near Vancouver, Canada, spawning of this species occurs at tide levels exposed to air about half the time. Most eggs are deposited so that clumps of algae cover them at low tide. Removal of algal cover resulted in increased mortality.

MILLER, D. J., and R. N. LEA. 1972. Guide to the coastal marine fishes of California. Calif. Dept. Fish Game, Fish Bull. 157: 1–235. □Lists 76 species of fish that occur intertidally in California. Includes keys, drawings, and descriptions.

MITCHELL, D. F. 1953. An analysis of stomach contents of California tide pool fishes. Amer. Midl. Natur. 49: 862–71.

MORING, J. R. 1976. Estimates of population size for tidepool sculpins, *Oligocottus maculosus*, and other intertidal fishes, Trinidad Bay, Humboldt County, California. Calif. Fish Game 62: 65–72.

MORRIS, R. W. 1951. Early development of the cottid fish, *Clinocottus recalvus* (Greeley). Calif. Fish Game 37: 281–300. □One of the common tidepool sculpins of the central California coast.

PHILLIPS, J. B. 1957. A review of the rockfishes of California (family Scorpaenidae). Calif. Dept. Fish Game, Fish Bull. 104: 1–158.

QUAST, J. C. 1968. Observations on the food of the kelp-bed fishes. *In* W. J. North and C. L. Hubbs, eds., Utilization of kelp-bed resources in southern California, pp. 109–42. Calif. Dept. Fish Game, Fish Bull. 139: 1–264.

RICHKUS, W. A. 1978. A quantitative study of intertidepool movement of the wooly sculpin *Clinocottus analis*. Mar. Biol. 49: 277–84.

———. 1981. Laboratory studies of intraspecific behavioral interactions and factors influencing tidepool selection of the wooly sculpin, *Clinocottus analis*. Calif. Fish Game 67: 187–95.

RUSSO, R. A. 1975. Observations on the food habits of leopard sharks (*Triakis semifasciata*) and brown smoothhounds (*Mustelus henlei*). Calif. Fish Game 61: 95–103. □Intertidal invertebrates are found in the guts of both shark species in San Francisco Bay.

TALENT, L. G. 1976. Food habits of the leopard shark, *Triakis semifasciata*, in Elkhorn Slough, Monterey Bay, California. Calif. Fish Game 62: 286–98. □This shark, common around jetties and bays in central and southern California, consumes crabs, clams, fish, fish eggs, and *Urechis caupo*. In Elkhorn Slough, *Urechis* was the most important prey of sharks over 90 cm in length; for sharks below 90 cm, the crab *Hemigrapsus oregonensis* was most important.

WILLIAMS, G. C. 1957. Homing behavior of California rocky shore fishes. Univ. Calif. Publ. Zool. 59: 249–84.

YOSHIYAMA, R. M. 1980. Food habits of three species of rocky intertidal sculpins (Cottidae) in central California. Copeia 1980: 515–25. □*Oligocottus snyderi, Clinocottus analis,* and *Artedius lateralis.*

———. 1981. Distribution and abundance patterns of rocky intertidal fishes in central California. Environ. Biol. Fish. 6: 315–32. □Concerns 24 species of fish found in the rocky intertidal, including nearly all of those covered in our account.

Anoplarchus purpurescens Gill, high cockscomb, §240

YOSHIYAMA, R. M., and J. D. S. DARLING. 1982. Grazing by the intertidal fish *Anoplarchus purpurescens* upon a distasteful polychaete worm. Environ. Biol. Fish. 7: 39–45.

Clevelandia ios (Jordan and Gilbert), arrow goby of mud flats, §306

Gibbonsia elegans (Cooper), **G. metzi** Hubbs, and **G. montereyensis** Hubbs, §51

Gillichthys mirabilis Cooper, longjaw mudsucker, §266

BARLOW, G. W. 1963. Species structure of the gobiid fish *Gillichthys mirabilis* from coastal sloughs of the eastern Pacific. Pacific Sci. 17: 47–72.

DE VLAMING, V. L. 1971. The effects of food deprivation and salinity changes on reproductive function in the estuarine gobiid fish, *Gillichthys mirabilis.* Biol. Bull. 141: 458–71.

———. 1972. Reproductive cycling in the estuarine gobiid fish, *Gillichthys mirabilis.* Copeia 1972: 278–91.

TODD, E. S., and A. W. EBELING. 1966. Aerial respiration in the longjaw mudsucker *Gillichthys mirabilis* (Teleostei: Gobiidae). Biol. Bull. 130: 265–88.

Gobiesox maeandricus (Girard) (formerly *Sicyogaster*), northern clingfish, §51

Hypsypops rubicundus (Girard), garibaldi, §149

Leuresthes tenuis (Ayres), grunion, §192

Oligocottus maculosus Girard, tidepool sculpin, tidepool Johnny, §149

CRAIK, G. J. S. 1981. The effects of age and length on homing performance in the intertidal cottid, *Oligocottus maculosus* Girard. Can. J. Zool. 59: 598–604. □Apparently, juvenile *Oligocottus maculosus* move extensively between tidepools in the lower intertidal region before adopting the homing habit and a home tidepool (or group of pools).

GREEN, J. M. 1971a. Field and laboratory activity patterns of the tidepool cottid *Oligocottus maculosus* Girard. Can. J. Zool. 49: 255–64.

———. 1971b. High tide movements and homing behavior of the tidepool sculpin *Oligocottus maculosus.* J. Fish. Res. Board Can. 28: 383–89. □This sculpin often leaves its pool at high tide, returning home before low tide. Excursions were more common from pools in protected habitats and during summer months. Fish experimentally displaced to a distant pool found their way back to the home pool.

NAKAMURA, R. 1976a. Experimental assessment of factors influencing microhabitat selection by the two tidepool fishes *Oligocottus maculosus* and *O. snyderi.* Mar. Biol. 37: 97–104.

———. 1976b. Temperature and the vertical distribution of two tidepool fishes (*Oligocottus maculosus, O. snyderi*). Copeia 1976: 143–52. □*Oligocottus snyderi* is less tolerant of wide temperature extremes, and its upper distributional limit may be set by tide-pool temperatures. *O. maculosus,* which is concentrated in upper tide pools, is more tolerant of extreme temperatures.

Porichthys notatus Girard, plainfin midshipman, grunter, §240

ARORA, H. L. 1948. Observations on the habits and early life history of the batrachoid fish, *Porichthys notatus* Girard. Copeia 1948: 89–93.

Typhlogobius californiensis Steindachner, blind goby, §58

MACGINITIE, G. E. 1939. The natural history of the blind goby, *Typhlogobius californiensis* Steindachner. Amer. Midl. Natur. 21: 489–505.

Xiphister atropurpureus (Kittlitz) (formerly *Epigeichthys*), black prickleback, §51

MARLIAVE, J. B. 1975. Seasonal shifts in the spawning site of a northeast Pacific intertidal fish. J. Fish. Res. Board Can. 32: 1687–91. □*Xiphister atropurpureus* spawns during the winter on rocky beaches protected from wave action; on more exposed stretches of coast, spawning occurs in the spring, when wave action diminishes.

MARLIAVE, J. B., and E. E. DEMARTINI. 1977. Parental behavior of intertidal fishes of the stichaeid genus *Xiphister.* Can. J. Zool. 55: 60–63. □Egg masses of *Xiphister atropurpureus* and *X. mucosus* are tended by a single male. Sometimes a male is found tending more than one egg mass, each in a different stage of development, suggesting repeated spawnings by males.

WOURMS, J. P., and D. EVANS. 1974. The annual reproductive cycle of the black prickleback, *Xiphister atropurpureus,* a Pacific coast blennioid fish. Can. J. Zool. 52: 795–802. □Spawning was recorded during a 6-week period from late April to May, a time when exposure to environmental hazards of wave shock, reduced salinity, and temperature extremes was generally minimal.

X. mucosus (Girard), rock prickleback, §51

Class AVES. Birds

Although birds, like most fish, are outside the scope of this account, they will often be seen by visitors to the shore, at least from a distance, and their impact on prey populations can be considerable. On rocky shores the black oystercatcher (*Haematopus bachmani*) feeds on limpets and mussels; at some sites each bird on a foraging excursion consumes one limpet about every three minutes. And on sandy and muddy beaches, a variety of shorebirds, including sanderlings, willets, curlews, plovers, and gulls, takes a toll on populations of crustaceans, polychaetes, bivalves, and gastropods. The references below include several general identification guides and distribution studies, followed by papers on the diets, habits, and impact of shorebirds, mostly from the Pacific coast. For many additional reports on related Atlantic species, interested readers are directed to the references in Quammen (1982).

General references on distribution:

ANGELL, T., and K. C. BALCOMB III. 1982. Marine birds and mammals of Puget Sound. Seattle: University of Washington Press. 145 pp.

COGSWELL, H. L. 1977. Water birds of California. Berkeley: University of California Press. 399 pp.

GABRIELSON, I. N., and S. G. JEWETT. 1970. Birds of the Pacific Northwest: With special reference to Oregon. New York: Dover. 650 pp.

GRINNELL, J., and A. H. MILLER. 1944. The distribution of the birds of California. Pac. Coast Avifauna 27. 608 pp. ☐With excellent habitat notes.

MUNRO, J. A., and I. McT. COWAN. 1947. A review of the bird fauna of British Columbia. B. C. Prov. Mus. Natur. Hist. Spec. Publ. No. 2: 1–285.

PETERSON, R. T. 1961. A field guide to western birds. 2d ed. Boston: Houghton Mifflin. 366 pp.

UDVARDY, M. D. F. 1977. The Audubon Society field guide to North American birds: Western region. New York: Knopf. 855 pp.

Selected references on the ecological relations of birds of inshore waters and the intertidal zone:

BURGER, J., M. A. HOWE, D. C. HAHN, and J. CHASE. 1977. Effects of tide cycles on habitat selection and habitat partitioning by migrating shorebirds. Auk 94: 743–58. ☐Concerns several species on the New Jersey shore. Feeding peaked during the first 2 hours after low tide.

CONNORS, P. G., J. P. MEYERS, C. S. W. CONNORS, and F. A. PITELKA. 1981. Interhabitat movements by sanderlings in relation to foraging profitability and the tidal cycle. Auk 98: 49–64.

EVANS, P. R. 1978. Adaptations shown by foraging shorebirds to cyclical variations in the activity and availability of their intertidal invertebrate prey. *In* E. Naylor and R. G. Hartnoll, eds., Cyclic phenomena in marine plants and animals, pp. 357–66. Proc. Thirteenth European Marine Biology Symposium.

FRANK, P. W. 1982. Effects of winter feeding on limpets by black oystercatchers, *Haematopus bachmani*. Ecology 63: 1352–62. ☐During the fall and winter, oystercatchers at an Oregon site fed mostly on limpets at an average rate of a third of a limpet per bird per minute. Most of the limpets accessible to oystercatchers were eaten by the end of the winter.

GRANT, J. 1981. A bioenergetic model of shorebird predation on infaunal amphipods. Oikos 37: 53–62.

HARTWICK, E. B. 1974. Breeding ecology of the black oystercatcher (*Haematopus bachmani* Audubon). Syesis 7: 83–92.

———. 1976. Foraging strategy of the black oystercatcher (*Haematopus bachmani* Audubon). Can. J. Zool. 54: 142–55.

———. 1978. Some observations of foraging by

black oystercatchers (*Haematopus bachmani* Audubon). Syesis 11: 55–60.

———. 1981. Size gradients and shell polymorphism in limpets with consideration of the role of predation. Veliger 23: 254–64.

KENYON, K. W. 1949. Observations on behavior and populations of oystercatchers in Lower California. Condor 51: 193–99.

LAWRENCE, G. E. 1950. The diving and feeding activity of the western grebe on the breeding grounds. Condor 52: 3–16. ☐Limpets are taken as prey on wintering waters.

LEGG, K. 1954. Nesting and feeding of the black oystercatcher near Monterey, California. Condor 56: 359–60.

LINDBERG, D. R., and E. W. CHU. 1983. Western gull predation on owl limpets: Different methods at different localities. Veliger 25: 347–48.

MILLER, A. H. 1943. Census of a colony of Caspian terns. Condor 45: 220–25.

MOFFITT, J. 1941. Notes on the food of the California clapper rail. Condor 43: 270–73.

MOORE, M. M. 1975. Foraging of the western gull *Larus occidentalis* and its impact on the chiton *Nuttallina californica*. Veliger 18(suppl.): 51–53.

MORRELL, S. H., H. R. HUBER, I. J. LEWIS, and D. G. AINLEY. 1979. Feeding ecology of black oystercatchers on south Farallon Island, California. Stud. Avian Biol. 2: 185–86.

MUDGE, G. P., and P. N. FERNS. 1982. The feeding ecology of five species of gulls (Aves: Larini) in the inner Bristol Channel. J. Zool. (Lond.) 197: 497–510. ☐Food taken by gulls in English littoral areas included polychaetes, cephalopod and gastropod mollusks, shore crabs, and shrimps.

MYERS, J. P., S. L. WILLIAMS, and F. A. PITELKA. 1980. An experimental analysis of prey availability for sanderlings (Aves: Scolopacidae) feeding on sandy beach crustaceans. Can. J. Zool. 58: 1564–74. ☐Laboratory experiments at Bodega Bay indicate that larger prey (the isopods *Excirolana* spp. and the sand crab *Emerita analoga*) were more vulnerable than smaller prey and that prey within 10 mm of the surface were most vulnerable to predation by sanderlings.

QUAMMEN, M. L. 1982. Influence of subtle substrate differences on feeding by shorebirds on intertidal mudflats. Mar. Biol. 71: 339–43.

RECHER, H. F. 1966. Some aspects of the ecology of migrant shorebirds. Ecology 47: 393–407.

REEDER, W. G. 1951. Stomach analysis of a group of shorebirds. Condor 53: 43–45. ☐Records prey taken at Sunset Beach and Point Mugu, California.

STENZEL, L. E., H. R. HUBER, and G. W. PAGE. 1976. Feeding behavior and diet of the long-billed curlew and willet. Wilson Bull. 88: 314–32. ☐At Bolinas Lagoon, California, long-billed curlews (*Numenius americanus*) ate primarily the crab *Hemigrapsus oregonensis* and the mud shrimps *Callianassa californiensis* and *Upogebia pugettensis*. Willets (*Cataptrophorus semipalmatus*) ate a great variety of small and large gastropods, bivalves, polychaetes, and crustaceans.

Class MAMMALIA

Man is not the only mammal that raids the intertidal for food; the raccoon (*Procyon lotor*, various subspecies) visits the shore at low tide from Puget Sound to Baja California, feeding principally on crabs and leaving its tracks along the sand. In Baja California, where the barren backcountry offers little nourishment, the coyote is a consistent intertidal feeder. Various mice, ground squirrels, and possibly the mink also visit the shore. But by and large, the reports of terrestrial mammals foraging on the shore are mostly incidental to other works and little is known quantitatively about the impact of these visits. To the limpet or crab just eaten, however, our numbers mean little; the predators are quite real.

As for the strictly marine mammals, only the sea otter, *Enhydra lutris nereis* (a member of the weasel family, Mustelidae), has an impact on shore invertebrates worthy of note. Once almost banished from California shores by overharvesting for its valuable fur, sea otters still exist in scattered herds along parts of the coast from Monterey to the Channel Islands; the animal is now fully protected by law, and populations are growing. Sea otters feed principally on subtidal abalone, clams, sea urchins, and crabs, but sometimes they forage in the intertidal during high tide. In either case, their appetite is great, and their impact on prey populations can be considerable, facts not lost on the commercial- and sport-fishing interests with which the sea otter sometimes competes. A bountiful harvest of literature is available on this species.

A few other marine mammals might be mentioned for general interest, since they may be seen by shorebound observers, especially those equipped with binoculars. Among these are the harbor seal (*Phoca vitulina*), the Steller sea lion (*Eumetopias stelleri*), the smaller California sea lion (*Zalophus californianus*), and the sea elephant (*Mirounga angustirostris*); the California gray whale (*Eschrichtius robustus*) may also be seen from shore during its annual migration.

ANDREWS, R. C. 1914. Monograph of the Pacific Cetacea, I. The California gray whale (*Rachianectes glaucus* Cope). Mem. Amer. Mus. Natur. Hist., n.s., 1: 227–87. □The California gray whale, which calves in bays and lagoons, was once thought to be nearly extinct, but it is gradually returning; the present population is several thousand. Calving is now restricted to the lagoons of Baja California, but during the winter months these whales migrate southward along the California coast, and may often be seen from shore.

BARTHOLOMEW, G. A., JR. 1952. Reproductive and social behavior of the northern elephant seal. Univ. Calif. Publ. Zool. 47: 369–472.

BOLIN, R. L. 1938. Reappearance of the southern sea otter along the California coast. J. Mammal. 19: 301–6.

BONNOT, P. 1951. The sea lions, seals, and sea otter of the California coast. Calif. Fish Game 37: 371–89.

BOOLOOTIAN, R. A. 1961. The distribution of the California sea otter. Calif. Fish Game 47: 287–92.

BREEN, P. A., T. A. CARSON, J. B. FOSTER, and E. A. STEWART. 1982. Changes in subtidal community structure associated with British Columbia sea otter transplants. Mar. Ecol. Prog. Ser. 7: 13–20.

DAUGHERTY, A. E. 1965. Marine mammals of California. Sacramento: California Department of Fish and Game. 87 pp. □A convenient pocket-sized handbook; well illustrated.

EBERT, E. E. 1968. A food habits study of the southern sea otter, *Enhydra lutris nereis*. Calif. Fish Game 54: 33–42.

ESTES, J. A., R. J. JAMESON, and E. B. RHODE. 1982. Activity and prey selection in the sea otter: Influence of population status on community structure. Amer. Natur. 120: 242–58. □Concerns feeding at Amchitka Island in the Aleutians and at Cape Blanco, Oregon. Otters, 93 in all, were transplanted from Amchitka to sites near Port Orford and Cape Arago in 1970 and 1971.

HINES, A. H., and J. S. PEARSE. 1982. Abalones, shells, and sea otters: Dynamics of prey populations in central California. Ecology 63: 1547–60.

JONES, R. E. 1981. Food habits of smaller marine mammals from northern California. Proc. Calif. Acad. Sci. 42: 409–33.

MATHISEN, O. A., R. T. BAADE, and R. J. LOPP. 1962. Breeding habits, growth and stomach contents of the Steller sea lion in Alaska. J. Mammal. 43: 469–77.

MORRIS, R., D. V. ELLIS, and B. P. EMERSON. 1981. The British Columbia transplant of sea otters *Enhydra lutris*. Biol. Conserv. 20: 291–95.

NORRIS, K. S., ed. 1966. Whales, dolphins, and porpoises. Berkeley: University of California Press. 789 pp. □The reports of a symposium; practically everyone active in studying cetaceans was in some way involved in this four-pound leviathan.

NORRIS, K. S., and J. R. PRESCOTT. 1961. Observations on Pacific cetaceans of Californian and Mexican waters. Univ. Calif. Publ. Zool. 63: 291–402.

OGDEN, A. 1941. The California sea otter trade, 1784–1848. Berkeley: University of California Press. 251 pp. □Primarily concerned with the economics of the trade, but including life-history information, an excellent photograph of a sea-otter herd (opposite p. 146), and a color frontispiece.

ORR, R. T., and T. C. POULTER. 1965. The pinniped population of Año Nuevo Island, California. Proc. Calif. Acad. Sci., ser. 4, 32: 377–404. □This is the northern outpost of the elephant seal.

———. 1967. Some observations on reproduction, growth and social behavior in the Steller sea lion. Proc. Calif. Acad. Sci., ser. 4, 35: 193–226. □With good views and a map of Año Nuevo Island, and with a fine photograph of the junior author being accepted as a mother by a deserted pup.

OSTFELD, R. S. 1982. Foraging strategies and prey switching in the California sea otter. Oecologia 53: 170–78.

PACKARD, J. M., and C. A. RIBIC. 1982. Classification of the behavior of sea otters (*Enhydra lutris*). Can. J. Zool. 60: 1362–73.

PETERSON, R. S., and G. A. BARTHOLOMEW, JR. 1967. The California sea lion. Special Publication No. 1. Stillwater, Okla.: American Society of Mammalogists. 79 pp.

SCAMMON, C. M. 1874. The marine mammals of the northwestern coast of North America, described and illustrated: Together with an account of the American whale-fishery. San Francisco: J. H. Carman. 319 pp. ☐This classic work, long an expensive collector's item, has been reproduced by Dover (1968) with a biographical sketch of Captain Scammon by Victor Scheffer.

SCHEFFER, V. B. 1958. Seals, sea lions, and walruses: A review of the Pinnipedia. Stanford, Calif.: Stanford University Press. 179 pp. ☐The tooth marks illustrated on Plate 32 are not those of a killer whale, but of a shark.

SCOFIELD, W. L. 1941. The sea otters of California did not reappear. Calif. Fish Game 27: 35–38. ☐(They were there all the time.)

General Bibliography

When bibliographies support text citations, they are, of course, a key to the source of information. Few references given here, however, have been directly quoted in the text; many of them have been included simply because they seem to belong here. In other words, this section is intended to be a guide to further reading, rather than simply a list of "references cited."

This bibliography has been subdivided into four areas: (1) essential general references related to seashore life; (2) other useful texts and general references; (3) selected papers on marine ecology and related matters; (4) books and papers on geology, paleontology, and zoogeography. Of course, additional references, those dealing primarily with specific animals or groups, are listed in the Annotated Systematic Index.

Most of the works listed were published in or after 1968. The Fourth Edition contains an extensive annotated bibliography of works published before then, and, except for a few references of special importance, most of these previously cited works have not been reprinted here, in an attempt to conserve space.

Even with emphasis placed on works published since 1968, this bibliography is not to be considered a complete list of every recent account of marine invertebrates on the Pacific coast. Such a bibliography would fill an entire book. I have purposely omitted many items, and I am sure I have unintentionally missed others, but I hope that enough is included to enable the interested student to work into the literature and ferret them out. In any event, I thank all who have sent me their papers, and I only hope I have managed to include all those deserving mention.

Essential General References Related to Seashore Life

ABBOTT, I. A., and G. J. HOLLENBERG. 1976. Marine algae of California. Stanford, Calif.: Stanford University Press. 827 pp. □Directly or indirectly, algae provide food, shelter, and attachment sites for many marine invertebrates. Yet most invertebrate zoologists have little botanical training. This book, comprehensive and nicely illustrated, is the essential reference for identifying algae on the Pacific coast.

KOZLOFF, E. N. 1974. Keys to the marine invertebrates of Puget Sound, the San Juan

Archipelago, and adjacent regions. Seattle: University of Washington Press. 226 pp.

LEWIS, J. R. 1964. The ecology of rocky shores. London: English Universities Press. 323 pp. ☐An exhaustively thorough treatment, zone by zone, of the shores of the British Isles and of the factors influencing the distributions of organisms there. One would expect (hope?) that organisms and communities on our Pacific shores are influenced by at least some of the same factors.

MacGINITIE, G. E., and N. MacGINITIE. 1968. Natural history of marine animals. 2d ed. New York: McGraw-Hill. 523 pp. ☐Although perhaps becoming somewhat dated, this classic contains a great many interesting observations, recorded in a very readable manner.

MORRIS, R. H., D. P. ABBOTT, and E. C. HADERLIE. 1980. Intertidal invertebrates of California. Stanford, Calif.: Stanford University Press. 690 pp. + 200 pp. color. ☐This long-awaited compendium, containing 900 color photographs and summary accounts of most of the conspicuous invertebrates of the California intertidal zone, is an essential scientific reference (despite its "coffee-table book" appearance). The bibliography is extensive, and important references are indicated for each species. Since there are no keys, the book does not replace identification manuals such as *Light's* (see below), but rather it complements them.

NEWELL, R. C. 1979. Biology of intertidal animals. 3d ed. (expanded). Faversham, Kent, U.K.: Marine Ecological Surveys Ltd. 781 pp. ☐Probably the most detailed and comprehensive text on the ecological physiology and biology of seashore life.

SMITH, R. I., and J. T. CARLTON, eds. 1975. Light's manual: Intertidal invertebrates of the central California coast. 3d ed. Berkeley: University of California Press. 716 pp. ☐The standard work for those who wish to identify their creatures. The keys and useful illustrations cover a larger number of California invertebrates than the volume by Morris, Abbott, and Haderlie (which covers each species in more detail), including many of the smaller and less-common forms. Users should beware, however, that the keys are not strictly applicable very far north or south of central California.

STEPHENSON, T. A., and A. STEPHENSON. 1972. Life between tidemarks on rocky shores. San Francisco: Freeman. 425 pp. ☐Presents a unified, primarily descriptive, account of worldwide studies on the biota of rocky shores.

YONGE, C. M. 1949. The sea shore. London: Collins, New Naturalist Series. 311 pp. ☐Perhaps still the finest book of this genre yet published; a must for the bookshelf of any who go to the shore.

Other Useful Texts and General References

ALLEE, W. C. 1931. Animal aggregations. Chicago: University of Chicago Press. 431 pp.

ASTRO, R. 1973. John Steinbeck and Edward F. Ricketts: The shaping of a novelist. Minneapolis: University of Minnesota Press. 259 pp.

BARNES, R. D. 1980. Invertebrate zoology. 4th ed. Philadelphia: Saunders. 1,089 pp.

BARRINGTON, E. J. W. 1979. Invertebrate structure and function. 2d ed. New York: Wiley. 765 pp.

BASCOM, W. 1964. Waves and beaches: The dynamics of the ocean surface. Garden City, N.Y.: Anchor Books, Doubleday. 267 pp.

BRUSCA, G. J., and R. C. BRUSCA. 1978. A naturalist's seashore guide. Eureka, Calif.: Mad River Press. 205 pp.

BRUSCA, R. C. 1980. Common intertidal invertebrates of the Gulf of California. 2d ed. Tucson: University of Arizona Press. 513 pp.

BUCHSBAUM, R. I. 1976 (reissue). Animals without backbones. Chicago: University of Chicago Press. 392 pp. □Highly readable. A complete revision is in progress.

CAREFOOT, T. 1977. Pacific seashores: A guide to intertidal ecology. Seattle: University of Washington Press. 208 pp.

CHIA, F.-S., and M. E. RICE, eds. 1978. Settlement and metamorphosis of marine invertebrate larvae. New York: Elsevier. 290 pp.

CLARK, R. B. 1964. Dynamics in metazoan evolution. The origin of the coelom and segments. Oxford: Clarendon. 313 pp.

COSTELLO, D. P., and C. HENLEY. 1971. Methods for obtaining and handling marine eggs and embryos. 2d ed. Woods Hole, Mass.: Marine Biological Laboratory. 247 pp.

FLORKIN, M., and B. T. SCHEER, eds. 1967– . Chemical zoology. Several volumes. New York: Academic Press.

FRETTER, V., and A. GRAHAM. 1976. A functional anatomy of invertebrates. New York: Academic Press. 589 pp.

FREY, H. W., ed. 1971. California living marine resources and their utilization. Sacramento: California Department of Fish and Game. 140 pp.

GARDINER, M. S. 1972. The biology of invertebrates. New York: McGraw-Hill. 954 pp.

GATES, D. E., and H. W. FREY. 1974. Designated common names of certain marine organisms in California. Calif. Dept. Fish Game, Fish Bull. 161: 55–90.

GIESE, A. C., and J. S. PEARSE, eds. 1974– . Reproduction of marine invertebrates. Several volumes. New York: Academic Press.

GOTSHALL, D. W., and L. L. LAURENT. 1979. Pacific coast subtidal marine invertebrates: A fishwatcher's guide. Los Osos, Calif.: Sea Challengers. 107 pp.

HAUSER, H., and B. EVANS. 1978. The living world of the reef. New York: Walker. 96 pp.

HEDGPETH, J. W., ed. 1957. Treatise on marine ecology and paleoecology. Vol. 1. Ecology. Geol. Soc. Amer. Mem. 67. 1,296 pp.

———. 1962. Introduction to seashore life of the San Francisco Bay region and the coast of northern California. Berkeley: University of California Press. 136 pp.

———. 1978. The outer shores. Part 1. Ed Ricketts and John Steinbeck explore the Pacific coast (128 pp.). Part 2. Breaking through (182 pp.). Eureka, Calif.: Mad River Press.

HINTON, S. 1969. Seashore life of southern California. Berkeley: University of California Press. 181 pp.

HYMAN, L. H. 1940–1967. The invertebrates. Vols. 1–6. New York: McGraw-Hill.

JOHNSON, M. E., and H. J. SNOOK. 1927. Seashore animals of the Pacific coast. New York: Macmillan. (Reprinted 1967, New York: Dover.) 659 pp.

KOZLOFF, E. N. 1983. Seashore life of the northern Pacific coast. Seattle: University of Washington Press. 370 pp.

LEVINTON, J. S. 1982. Marine ecology. Englewood Cliffs, N.J.: Prentice-Hall. 526 pp.

MEGLITSCH, P. A. 1972. Invertebrate zoology. 2d ed. New York: Oxford University Press. 834 pp.

NICOL, J. A. C. 1960. The biology of marine animals. New York: Wiley-Interscience. 707 pp.

PICKARD, G. L. 1964. Descriptive physical oceanography: An introduction. New York: Macmillan. 199 pp.

RUSSELL, F. S., and C. M. YONGE. 1975. The seas. 4th ed. New York: F. Warne and Co. 283 pp.

RUSSELL-HUNTER, W. D. 1979. A life of invertebrates. New York: Macmillan. 650 pp.

SMITH, D. L. 1977. A guide to marine coastal plankton and marine invertebrate larvae. Dubuque, Iowa: Kendall/Hunt. 161 pp.

SMITH, L. 1962. Common seashore life of the Pacific northwest. Healdsburg, Calif.: Naturegraph. 66 pp.

SOUTHWARD, A. J. 1965. Life on the seashore. London: Heinemann. 153 pp.

STEINBECK, J., and E. F. RICKETTS. 1941. Sea of Cortez: A leisurely journal of travel and research. New York: Viking. 598 pp.

THORSON, G. 1971. Life in the sea. New York: McGraw-Hill. 256 pp.

VOGEL, S. 1981. Life in moving fluids: The physical biology of flow. Boston: W. Grant Press. 352 pp.

Selected Papers on Marine Ecology and Related Matters

ALLEE, W. C. 1923. Studies in marine ecology, I and II. Biol. Bull. 44: 157–253.

ANDERSON, G. L. 1975. The effects of intertidal height and the parasitic crustacean *Fabia subquadrata* Dana on the nutrition and reproductive capacity of the California sea mussel *Mytilus californianus* Conrad. Veliger 17: 299–306.

BATZLI, G. O. 1969. Distribution of biomass in rocky intertidal communities on the Pacific coast of the United States. J. Anim. Ecol. 38: 531–46.

BEAUCHAMP, K. A., and M. M. GOWING. 1982. A quantitative assessment of human trampling effects on a rocky intertidal community. Mar. Environ. Res. 7: 279–93. □ A general pattern of higher density and diversity occurred at the less-trampled sites at Santa Cruz.

BERNSTEIN, B. B., B. E. WILLIAMS, and K. H. MANN. 1981. The role of behavioral responses to predators in modifying urchins' (*Strongylocentrotus droebachiensis*) destructive grazing and seasonal foraging patterns. Mar. Biol. 63: 39–49.

BERTNESS, M. D. 1977. Behavioral and ecological aspects of shore-level size gradients in *Thais lamellosa* and *Thais emarginata*. Ecology 58: 86–97.

———. 1981a. Interference, exploitation, and sexual components of competition in a tropical hermit crab assemblage. J. Exp. Mar. Biol. Ecol. 49: 189–202.

———. 1981b. Predation, physical stress, and the organization of a tropical rocky intertidal hermit crab community. Ecology 62: 411–25.

BERTNESS, M. D., S. D. GARRITY, and S. C. LEVINGS. 1981. Predation pressure and gastropod foraging: A tropical-temperate comparison. Evolution 35: 995–1007.

BIRCH, D. W. 1981. Dominance in marine ecosystems. Amer. Natur. 118: 262–74.

BIRCH, L. C. 1957. The meanings of competition. Amer. Natur. 91: 5–18.

BLACK, R. 1976. The effects of grazing by the limpet, *Acmaea insessa*, on the kelp, *Egregia laevigata*, in the intertidal zone. Ecology 57: 265–77.

Bowman, R. S., and J. R. Lewis. 1977. Annual fluctuations in the recruitment of *Patella vulgata* L. J. Mar. Biol. Assoc. U.K. 57: 793–815.

Branch, G. M. 1981. The biology of limpets: Physical factors, energy flow, and ecological interactions. Oceanogr. Mar. Biol. Ann. Rev. 19: 235–380.

Brattstrom, H. 1980. Rocky-shore zonation in the Santa Marta area, Colombia. Sarsia 65: 163–226.

Brenchley, G. A. 1981. Disturbance and community structure: An experimental study of bioturbation in marine soft-bottom environments. J. Mar. Res. 39: 767–90.

———. 1982. Mechanisms of spatial competition in marine soft-bottom communities. J. Exp. Mar. Biol. Ecol. 60: 17–33.

Carefoot, T. H. 1967. Growth and nutrition of *Aplysia punctata* feeding on a variety of marine algae. J. Mar. Biol. Assoc. U.K. 47: 565–89.

———. 1981. A tide simulator and examples of its use. Can. J. Zool. 59: 1459–63.

Carlton, J. T. 1979. History, biogeography, and ecology of the introduced marine and estuarine invertebrates of the Pacific coast of North America. Ph.D. dissertation, Ecology, University of California, Davis. 904 pp.

Carriker, M. R., and D. van Zandt. 1972. Predatory behavior of a shell-boring muricid gastropod. *In* H. E. Winn and B. L. Olla, eds., Behavior of marine animals, Vol. 1: Invertebrates, pp. 157–244. New York: Plenum.

Castenholz, R. W. 1961. The effect of grazing on marine littoral diatom populations. Ecology 42: 783–94.

Chalmer, P. N. 1982. Settlement patterns of species in a marine fouling community and some mechanisms of succession. J. Exp. Mar. Biol. Ecol. 58: 73–85.

Choat, J. H. 1977. The influence of sessile organisms on the population biology of three species of acmaeid limpets. J. Exp. Mar. Biol. Ecol. 26: 1–26.

Coe, W. R. 1956. Fluctuations in populations of littoral marine invertebrates. J. Mar. Res. 15: 212–32.

Colman, J. 1933. The nature of the intertidal zonation of plants and animals. J. Mar. Biol. Assoc. U.K., n.s. 18: 435–76.

Connell, J. H. 1961a. Effects of competition, predation by *Thais lapillus*, and other factors on natural populations of the barnacle *Balanus balanoides*. Ecol. Monogr. 31: 61–104.

———. 1961b. The influence of interspecific competition and other factors on the distribution of the barnacle *Chthamalus stellatus*. Ecology 42: 710–23.

———. 1970. A predator-prey system in the marine intertidal region. I. *Balanus glandula* and several predatory species of *Thais*. Ecol. Monogr. 40: 49–78.

———. 1972. Community interactions on marine rocky intertidal shores. Ann. Rev. Ecol. Syst. 3: 169–92.

———. 1974. Field experiments in marine ecology. *In* R. N. Mariscal, ed., Experimental marine biology, pp. 21–54. New York: Academic Press.

Connell, J. H., and R. O. Slatyer. 1977. Mechanisms of succession in natural communities and their role in community stability and organization. Amer. Natur. 111: 1119–44.

Creese, R. G. 1980. An analysis of distribution and abundance of populations of the high-shore limpet, *Notoacmea petterdi* (Tenison-Woods). Oecologia 45: 252–60.

Croll, R. P., and R. Chase. 1977. A long-term memory for food odors in the land snail, *Achatina fulica*. Behav. Biol. 19: 261–68.

CUBIT, J. D. 1969. Behavior and physical factors causing migration and aggregation of the sand crab *Emerita analoga* (Stimpson). Ecology 50: 118–23.

———. 1974. Interactions of seasonally changing physical factors and grazing affecting high intertidal communities on a rocky shore. Ph.D. dissertation, Ecology, University of Oregon, Eugene. 133 pp.

CURREY, J. D., and R. N. HUGHES. 1982. Strength of the dogwhelk *Nucella lapillus* and the winkle *Littorina littorea* from different habitats. J. Anim. Ecol. 51: 47–56.

DAY, R. W., and R. W. OSMAN. 1981. Predation by *Patiria miniata* (Asteroidea) on bryozoans: Prey diversity may depend on the mechanism of succession. Oecologia 51: 300–309.

DAYTON, P. K. 1971. Competition, disturbance, and community organization: The provision and subsequent utilization of space in a rocky intertidal community. Ecol. Monogr. 41: 351–89.

———. 1973. Two cases of resource partitioning in an intertidal community: Making the right prediction for the wrong reason. Amer. Natur. 107: 662–70.

———. 1975. Experimental evaluation of ecological dominance in a rocky intertidal algal community. Ecol. Monogr. 45: 137–59.

DAYTON, P. K., and J. S. OLIVER. 1980. An evaluation of experimental analyses of population and community patterns in benthic marine environments. *In* K. R. Tenore and B. C. Coull, eds., Marine benthic dynamics, pp. 93–120. Columbia: University of South Carolina Press.

DEAN, T. A. 1981. Structural aspects of sessile invertebrates as organizing forces in an estuarine fouling community. J. Exp. Mar. Biol. Ecol. 53: 163–80.

DEAN, T. A., and L. E. HURD. 1980. Development in an estuarine fouling community: The influence of early colonists on later arrivals. Oecologia 46: 295–301.

DOERING, P. H., and D. W. PHILLIPS. 1983. Maintenance of the shore-level size gradient in the marine snail *Tegula funebralis* (A. Adams): Importance of behavioral responses to light and sea star predators. J. Exp. Mar. Biol. Ecol. 67: 159–73.

DOTY, M. S. 1946. Critical tide factors that are correlated with the vertical distribution of marine algae and other organisms along the Pacific coast. Ecology 27: 315–28.

———. 1957. Rocky intertidal surfaces. *In* J. W. Hedgpeth, ed., Treatise on marine ecology and paleoecology, Vol. 1: Ecology, pp. 535–85. Geol. Soc. Amer. Mem. 67.

DRING, M. J., and F. A. BROWN. 1982. Photosynthesis of intertidal brown algae during and after periods of emersion: A renewed search for physiological causes of zonation. Mar. Ecol. Prog. Ser. 8: 301–8.

DRUEHL, L. D. 1967. Vertical distribution of some benthic marine algae in a British Columbia inlet, as related to some environmental factors. J. Fish. Res. Board Can. 24: 33–46.

DRUEHL, L. D., and J. M. GREEN. 1982. Vertical distribution of intertidal seaweeds as related to patterns of submersion and emersion. Mar. Ecol. Prog. Ser. 9: 163–70.

DUGGINS, D. O. 1981. Interspecific facilitation in a guild of benthic marine herbivores. Oecologia 48: 157–63.

DUNGAN, M. L., T. E. MILLER, and D. A. THOMSON. 1982. Catastrophic decline of a top carnivore in the Gulf of California rocky intertidal zone. Science 216: 989–91.

EMSON, R. H., and R. J. FALLER-FRITSCH. 1976. An experimental investigation into the effect of crevice availability on abundance and size-structure in a population of *Littorina rudis* (Maton): Gastropoda: Prosobranchia. J. Exp. Mar. Biol. Ecol. 23: 285–97.

EVANS, R. G. 1947. The intertidal ecology of selected localities in the Plymouth neighborhood. J. Mar. Biol. Assoc. U.K. 27: 173–218.

FAGER, E. W. 1968. A sand-bottom epifaunal community of invertebrates in shallow water. Limnol. Oceanogr. 13: 448–64.

FENCHEL, T. 1975a. Factors determining the distribution patterns of mud snails (Hydrobiidae). Oecologia 20: 1–17.

———. 1975b. Character displacement and coexistence in mud snails (Hydrobiidae). Oecologia 20: 19–32.

FILICE, F. P. 1958. Invertebrates from the estuarine portion of San Francisco Bay and some factors influencing their distribution. Wasmann J. Biol. 16: 159–211.

FISHLYN, D. A., and D. W. PHILLIPS. 1980. Chemical camouflaging and behavioral defenses against a predatory seastar by three species of gastropods from the surfgrass *Phyllospadix* community. Biol. Bull. 158: 34–48.

FOSTER, B. A. 1969. Tolerance of high temperatures by some intertidal barnacles. Mar. Biol. 4: 326–32.

———. 1971. Desiccation as a factor in the intertidal zonation of barnacles. Mar. Biol. 8: 12–29.

FOTHERINGHAM, N. 1974. Trophic complexity in a littoral boulderfield. Limnol. Oceanogr. 19: 84–91.

FOX, L. R., and P. A. MORROW. 1981. Specialization: Species property or local phenomenon? Science 211: 887–93.

FRANK, P. W. 1982. Effects of winter feeding on limpets by black oystercatchers, *Haematopus bachmani*. Ecology 63: 1352–62.

GAINES, S. D., and J. LUBCHENCO. 1982. A unified approach to marine plant-herbivore interactions. II. Biogeography. Ann. Rev. Ecol. Syst. 13: 111–38.

GARRITY, S. D., and S. C. LEVINGS. 1981. A predator-prey interaction between two physically and biologically constrained tropical rocky shore gastropods: Direct, indirect and community effects. Ecol. Monogr. 51: 267–86.

GISLÈN, T. 1930. Epibioses of the Gullmar Fjord, II. Kristin. Zool. Stat. 1877–1927, 4: 1–380.

GLYNN, P. W. 1965. Community composition, structure, and interrelationships in the marine intertidal *Endocladia muricata–Balanus glandula* association in Monterey Bay, California. Beaufortia 12: 1–198.

GRANT, J. 1981a. A bioenergetic model of shorebird predation on infaunal amphipods. Oikos 37: 53–62.

———. 1981b. Sediment transport and disturbance on an intertidal sandflat: Infaunal distribution and recolonization. Mar. Ecol. Prog. Ser. 6: 249–55.

GREENE, C. H., and A. SCHOENER. 1982. Succession on marine hard substrata: A fixed lottery. Oecologia 55: 289–97.

GREENE, C. H., A. SCHOENER, and E. CORETS. 1983. Succession on marine hard substrata: The adaptive significance of solitary and colonial strategies in temperate fouling communities. Mar. Ecol. Prog. Ser. 13: 121–29.

GROSBERG, R. K. 1982. Intertidal zonation of barnacles: The influence of planktonic zonation of larvae on vertical distribution of adults. Ecology 63: 894–99.

GUNNILL, F. C. 1980a. Demography of the intertidal brown alga *Pelvetia fastigiata* in southern California, USA. Mar. Biol. 59: 169–79.

———. 1980b. Recruitment and standing stocks in populations of one green alga and five brown algae in the intertidal zone near La Jolla, California during 1973–1977. Mar. Ecol. Prog. Ser. 3: 231–43.

———. 1982a. Macroalgae as habitat patch islands for *Scutellidium lamellipes* (Copepoda: Harpacticoida) and *Ampithoe tea* (Amphipoda: Gammaridae). Mar. Biol. 69: 103–16.

———. 1982b. Effects of plant size and distribution on the numbers of invertebrate species and individuals inhabiting the brown alga *Pelvetia fastigiata*. Mar. Biol. 69: 263–80.

———. 1983. Seasonal variations in the invertebrate faunas of *Pelvetia fastigiata* (Fucaceae): Effects of plant size and distribution. Mar. Biol. 73: 115–30.

HARGER, J. R. E. 1968. The role of behavioral traits in influencing the distribution of two species of sea mussel, *Mytilus edulis* and *Mytilus californianus*. Veliger 11: 45–49.

———. 1972. Competitive coexistence among intertidal invertebrates. Amer. Sci. 60: 600–607.

HARROLD, C. 1982. Escape responses and prey availability in a kelp forest predator-prey system. Amer. Natur. 119: 132–35.

HARTWICK, B. 1981. Size gradients and shell polymorphism in limpets with consideration of the role of predation. Veliger 23: 254–64.

HAUENSCHILD, C. 1960. Lunar periodicity. Cold Spring Harbor Symp. Quant. Biol. 25: 491–97.

HAVEN, S. B. 1971. Effects of land-level changes on intertidal invertebrates, with discussion of postearthquake ecological succession. *In* The great Alaska earthquake of 1964: Biology, pp. 82–126. Washington, D.C.: National Academy of Sciences.

———. 1973. Competition for food between the intertidal gastropods, *Acmaea scabra* and *Acmaea digitalis*. Ecology 54: 143–51.

HAY, M. E. 1981. The functional morphology of turf-forming seaweeds: Persistence in stressful marine habitats. Ecology 62: 739–50.

HECK, K. L., JR., and T. A. THOMAN. 1981. Experiments on predator-prey interactions in vegetated aquatic habitats. J. Exp. Mar. Biol. Ecol. 53: 125–34.

HEDGPETH, J. W. 1976. The living edge. Geoscience and Man 14: 17–51.

HEWATT, W. G. 1935. Ecological succession in the *Mytilus californianus* habitat as observed in Monterey Bay, California. Ecology 16: 244–51.

———. 1937. Ecological studies on selected marine intertidal communities of Monterey Bay, California. Amer. Midl. Natur. 18: 161–206.

———. 1946. Marine ecological studies on Santa Cruz Island, California. Ecol. Monogr. 16: 185–210.

HILL, C. L., and C. A. KOFOID, eds. 1927. Marine borers and their relation to marine construction on the Pacific coast. Final rep. San Francisco: San Francisco Bay Marine Piling Committee. 357 pp.

HIMMELMAN, J. H., and T. H. CAREFOOT. 1975. Seasonal changes in calorific value of three Pacific coast seaweeds, and their significance to some marine invertebrate herbivores. J. Exp. Mar. Biol. Ecol. 18: 139–51.

HINES, A., S. ANDERSON, and M. BRISBIN. 1980. Heat tolerance in the black abalone, *Haliotis cracherodii* Leach, 1814: Effects of temperature fluctuation and acclimation. Veliger 23: 113–18.

HODGKIN, E. P. 1960. Patterns of life on rocky shores. J. Roy. Soc. West. Austral. 43: 35–43.

HORN, M. H., S. N. MURRAY, and R. R. SEAPY. 1983. Seasonal structure of a central California rocky intertidal community in relation to environmental variations. Bull. South. Calif. Acad. Sci. 82: 79–94.

HRUBY, T. 1975. Seasonal changes in two algal populations from the coastal waters of Washington State. J. Ecology 63: 881–89. □Concerns many species in two protected sites.

———. 1976. Observations of algal zonation resulting from competition. Estuarine Coastal Mar. Sci. 4: 231–33.

JACKSON, J. B. C. 1977. Competition on marine hard substrata: The adaptive significance of solitary and colonial strategies. Amer. Natur. 111: 743–67.

JOHNSON, P. T. 1968. Population crashes in the bean clam, *Donax gouldi*, and their significance to the study of mass mortalities in other marine invertebrates. J. Invert. Pathol. 12: 349–58.

JOHNSON, R. G. 1967a. Salinity of interstitial water in a sandy beach. Limnol. Oceanogr. 12: 1–7.

———. 1967b. The vertical distribution of the infauna of a sand flat. Ecology 48: 571–78.

JOHNSON, S. E. 1975. Microclimate and energy flow in the marine rocky intertidal. *In* D. W. Gates and R. B. Schmerl, eds., Perspectives of biophysical ecology, pp. 559–87. New York: Springer-Verlag.

JOHNSON, W. S., A. GIGON, S. L. GULMON, and H. A. MOONEY. 1974. Comparative photosynthetic capacities of intertidal algae under exposed and submerged conditions. Ecology 55: 450–53.

JONES, W. E., and A. DEMETROPOULOS. 1968. Exposure to wave action: Measurements of an important ecological parameter on rocky shores on Anglesey. J. Exp. Mar. Biol. Ecol. 2: 46–63.

JORGENSEN, C. B. 1966. Biology of suspension feeding. London: Pergamon Press.

KEEN, S. L., and W. E. NEILL. 1980. Spatial relationships and some structuring processes in benthic intertidal animal communities. J. Exp. Mar. Biol. Ecol. 45: 139–56.

KENSLER, C. B. 1967. Desiccation resistance of intertidal crevice species as a factor in their zonation. J. Anim. Ecol. 36: 391–406.

KINNE, O., ed. 1980. Diseases of marine animals. 1, General aspects, Protozoa to Gastropoda. Chichester, Eng.: Wiley. 466 pp.

KITCHING, J. A., and F. J. EBLING. 1967. Ecological studies at Lough Ine. Adv. Ecol. Res. 4: 197–291.

KNIGHT-JONES, E. W., and J. MOYSE. 1961. Intraspecific competition in sedentary marine animals. *In* Mechanisms in biological competition, pp. 72–95. Symposia Society for Exp. Biology 15. Cambridge: Cambridge University Press.

KOHN, A. J., and P. J. LEVITEN. 1976. Effect of habitat complexity on population density and species richness in tropical intertidal predatory gastropod assemblages. Oecologia 25: 199–210.

LANDENBERGER, D. E. 1966. Learning in the Pacific starfish, *Pisaster giganteus*. Anim. Behav. 14: 414–18.

————. 1969. The effects of exposure to air on Pacific starfish and its relationship to distribution. Physiol. Zool. 42: 220–30.

LEIGHTON, D. L. 1966. Studies of food preference in algivorous invertebrates of southern California kelp beds. Pacific Sci. 20: 104–13.

LIE, U. 1968. A quantitative study of benthic infauna in Puget Sound, Washington, U.S.A., in 1963–1964. With a section on polychaetes by Karl Banse, Katharine D. Hobson, and Frederick H. Nichols. Fiskeridir. Skr. Ser. Havunders. 14: 229–556.

LITTLE, C., and L. P. SMITH. 1980. Vertical zonation on rocky shores in the Severn estuary (UK). Estuarine Coastal Mar. Sci. 11: 651–70.

LITTLER, M. M., D. R. MARTZ, and D. S. LITTLER. 1983. Effect of recurrent sand deposition on rocky intertidal organisms: Importance of substrate heterogeneity in a fluctuating environment. Mar. Ecol. Prog. Ser. 11: 129–39.

LOI, T.-N. 1981. Environmental stresses and intertidal assemblages on hard substrates in the Port of Long Beach, California, USA. Mar. Biol. 63: 197–211.

LUBCHENCO, J. 1978. Plant species diversity in a marine intertidal community: Importance of herbivore food preference and algal competitive abilities. Amer. Natur. 112: 23–39.

LUBCHENCO, J., and J. CUBIT. 1980. Heteromorphic life histories of certain marine algae as adaptations to variations in herbivory. Ecology 61: 676–87.

LUBCHENCO, J., and S. D. GAINES. 1981. A unified approach to marine plant-herbivore interactions. I. Populations and communities. Ann. Rev. Ecol. Syst. 12: 405–37.

LUBCHENCO, J., and B. A. MENGE. 1978. Community development and persistence in a low rocky intertidal zone. Ecol. Monogr. 48: 67–94.

LUCKENS, P. A. 1970. Breeding, settlement and survival of barnacles at artificially modified shore levels at Leigh, New Zealand. N.Z. J. Mar. Freshwater Res. 4: 497–514.

LYNCH, M. 1977. Fitness and optimal body size in zooplankton populations. Ecology 58: 763–74.

MacGINITIE, G. E. 1935. Ecological aspects of a California marine estuary. Amer. Midl. Natur. 16: 629–765.

————. 1938. Notes on the natural history of some marine animals. Amer. Midl. Natur. 19: 207–19.

————. 1939. Littoral marine communities. Amer. Midl. Natur. 21: 28–55.

MACKIE, A. M., and P. T. GRANT. 1974. Interspecies and intraspecies chemoreception by marine invertebrates. *In* P. T. Grant and A. M. Mackie, eds., Chemoreception in marine organisms, pp. 105–41. New York: Academic Press.

MARKEL, R. P. 1974. Aspects of the physiology of temperature acclimation in the limpet *Acmaea limatula* Carpenter (1864): An integrated field and laboratory study. Physiol. Zool. 47: 99–109.

McLEAN, J. H. 1962. Sublittoral ecology of kelp beds of the open coast area near Carmel, California. Biol. Bull. 122: 95–114.

McQUAID, C. D. 1981. The establishment and maintenance of vertical size gradients in populations of *Littorina africana knysnaensis* (Philippi) on an exposed rocky shore. J. Exp. Mar. Biol. Ecol. 54: 77–89.

MEADOWS, P. S., and J. I. CAMPBELL. 1972. Habitat selection by aquatic invertebrates. Adv. Mar. Biol. 10: 271–382.

MENGE, B. A. 1976. Organization of the New England rocky intertidal community:

Role of predation, competition, and environmental heterogeneity. Ecol. Monogr. 46: 355–93.

———. 1978. Predation intensity in a rocky intertidal community: I, Relation between predator foraging activity and environmental harshness. Oecologia 34: 1–16. II, Effect of an algal canopy, wave action and desiccation on predator foraging rates. Oecologia 34: 17–35.

MENGE, B. A., and J. LUBCHENCO. 1981. Community organization in temperate and tropical rocky intertidal habitats: Prey refuges in relation to consumer pressure gradients. Ecol. Monogr. 51: 429–50.

MENGE, J. L. 1974. Prey selection and foraging period of the predaceous rocky intertidal snail, *Acanthina punctulata*. Oecologia 17: 293–316.

MENGE, J. L., and B. A. MENGE. 1974. Role of resource allocation, aggression and spatial heterogeneity in coexistence of two competing intertidal starfish. Ecol. Monogr. 44: 189–209.

MILEIKOVSKY, S. A. 1971. Types of larval development in marine bottom invertebrates, their distribution and ecological significance: A re-evaluation. Mar. Biol. 10: 193–213.

MOITOZA, D. J., and D. W. PHILLIPS. 1979. Prey defense, predator preference, and nonrandom diet: The interactions between *Pycnopodia helianthoides* and two species of sea urchins. Mar. Biol. 53: 299–304.

MOKYEVSKY, O. B. 1960. Geographical zonation of marine littoral types. Limnol. Oceanog. 5: 389–96.

MORIN, P. J. 1980. Natural physical disturbance and predation: Their importance in structuring a marine sessile community. Austral. J. Ecol. 5: 193–200.

———. 1981. Predatory salamanders reverse the outcome of competition among three species of anuran tadpoles. Science 212: 1284–86.

MURPHY, D. J. 1979. The relationship between the lethal freezing temperatures and the amounts of ice formed in the foot muscle of marine snails (Mollusca: Gastropoda). Cryobiology 16: 292–300.

MURPHY, D. J., and E. MCCAUSLAND. 1980. The relationship between freezing and desiccation tolerance in the marine snail *Ilyanassa obsoleta* (Stimpson). Estuaries 3: 318–20.

NEWELL, R. C. 1976. Adaptations to intertidal life. *In* R. C. Newell, ed., Adaptation to environment: Essays on the physiology of marine animals, pp. 1–82. London: Butterworths.

NICHOLS, F. H. 1977. Infaunal biomass and production on a mudflat, San Francisco Bay, California. *In* B. C. Coull, ed., Ecology of marine benthos, pp. 339–57. Belle W. Baruch Library in Marine Science 6. Columbia: University of South Carolina Press.

OSMAN, R. W. 1977. The establishment and development of a marine epifaunal community. Ecol. Monogr. 47: 37–63.

PAINE, R. T. 1969. The *Pisaster-Tegula* interaction: Prey patches, predator food preference, and intertidal community structure. Ecology 50: 950–61.

———. 1974. Intertidal community structure. Experimental studies on the relationship between a dominant competitor and its principal predator. Oecologia 15: 93–120.

———. 1976. Size-limited predation: An observational and experimental approach with the *Mytilus-Pisaster* interaction. Ecology 57: 858–73.

———. 1977. Controlled manipulations in the marine intertidal zone, and the contributions to ecological theory. *In* C. E. Goulden, ed., Changing scenes in natural sciences, 1776–1976, pp. 245–70. Spec. Publ. 12, Academy of Natural Sciences, Philadelphia.

———. 1979. Disaster, catastrophe, and local persistence of the sea palm *Postelsia palmaeformis*. Science 205: 685–87.

PAINE, R. T., and S. A. LEVIN. 1981. Intertidal landscapes: Disturbance and the dynamics of pattern. Ecol. Monogr. 51: 145–78.

PETERS, R. H. 1976. Tautology in evolution and ecology. Amer. Natur. 110: 1–12.

PETERSON, C. H. 1977a. Competitive organization of the soft-bottom macrobenthic communities of southern California lagoons. Mar. Biol. 43: 343–59.

———. 1977b. Species diversity and perturbations: Predictions of a non-interactive model. Oikos 29: 239–44.

———. 1979. Predation, competitive exclusion, and diversity in the soft-sediment benthic communities of estuaries and lagoons. *In* R. J. Livingston, ed., Ecological processes in coastal and marine systems, Marine Science 10, pp. 233–64. New York: Plenum.

PETERSON, C. H., and S. V. ANDRE. 1980. An experimental analysis of interspecific competition among marine filter feeders in a soft-sediment environment. Ecology 61: 129–39.

PHILLIPS, D. W. 1976. The effect of a species-specific avoidance response to predatory starfish on the intertidal distribution of two gastropods. Oecologia 23: 83–94.

———. 1978. Chemical mediation of invertebrate defensive behaviors and the ability to distinguish between foraging and inactive predators. Mar. Biol. 49: 237–43.

PURSCHKE, G. 1981. Tolerance to freezing and supercooling of interstitial Turbellaria and Polychaeta from a sandy tidal beach of the island of Sylt (North Sea). Mar. Biol. 63: 257–67.

PYKE, G. H., H. R. PULLIAM, and E. L. CHARNOV. 1977. Optimal foraging: A selective review of theory and tests. Quart. Rev. Biol. 52: 137–54.

QUAMMEN, M. L. 1981. Use of exclosures in studies of predation by shorebirds on intertidal mudflats. Auk 98: 812–17.

QUINN, J. F. 1982. Competitive hierarchies in marine benthic communities. Oecologia 54: 129–35.

RAY, D. L., ed. 1956. Marine boring and fouling organisms. Seattle: University of Washington Press. 543 pp.

RHOADES, D. C. 1974. Organism-sediment relations on the muddy sea floor. Oceanogr. Mar. Biol. Ann. Rev. 12: 263–300.

RIEDL, R. 1980. Marine ecology—a century of changes. Mar. Ecol. 1: 3–46.

RIGG, G. B., and R. C. MILLER. 1949. Intertidal plant and animal zonation in the vicinity of Neah Bay, Washington. Proc. Calif. Acad. Sci. 26: 323–51.

ROBLES, C. 1982. Disturbance and predation in an assemblage of herbivorous Diptera and algae on rocky shores. Oecologia 54: 23–31.

ROBLES, C. D., and J. CUBIT. 1981. Influence of biotic factors in an upper intertidal community: Dipteran larvae grazing on algae. Ecology 62: 1536–47.

ROHDE, K. 1982. Ecology of marine parasites. St. Lucia: University of Queensland Press. 245 pp.

RONAN, T. E. 1975. Structural and paleo-ecological aspects of a modern marine soft-sediment community: An experimental field study. Ph.D. dissertation, University of California, Davis. 220 pp.

ROSENTHAL, R. J., W. D. CLARKE, and P. K. DAYTON. 1974. Ecology and natural history of a stand of giant kelp, *Macrocystis pyrifera*, off Del Mar, California. Calif. Dept. Fish Game, Fish Bull. 72: 670–84.

SCHALL, J. J., and E. R. PIANKA. 1980. Evolution of escape behavior diversity. Amer. Natur. 115: 551–66.

SCHEUER, P. J. 1977. Chemical communication of marine invertebrates. BioScience 27: 664–68.

SCHMITT, R. J. 1981. Contrasting anti-predator defenses of sympatric marine gastropods (Family Trochidae). J. Exp. Mar. Biol. Ecol. 54: 251–63.

———. 1982. Consequences of dissimilar defenses against predation in a subtidal marine community. Ecology 63: 1588–1601.

SCHNEIDER, D. 1978. Equalisation of prey numbers by migratory shorebirds. Nature 271: 353–54.

SCHOENER, T. W. 1982. The controversy over interspecific competition. Amer. Sci. 70: 586–95.

SEAPY, R. R., and C. L. KITTING. 1978. Spatial structure of an intertidal molluscan assemblage on a sheltered sandy beach. Mar. Biol. 46: 137–45.

SEAPY, R. R., and M. M. LITTLER. 1978. The distribution, abundance, community structure, and primary productivity of macroorganisms from two central California rocky intertidal habitats. Pacific Sci. 32: 293–314.

———. 1982. Population and species diversity fluctuations in a rocky intertidal community relative to severe aerial exposure and sediment burial. Mar. Biol. 71: 87–96.

SHELFORD, V. E., et al. 1935. Some marine biotic communities of the Pacific coast of North America. Ecol. Monogr. 5: 251–354. □Cites papers on barnacles in Shelford et al., 1930, Publ. Puget Sound Biol. Stat. 7.

SHIMEK, R. L. 1981. *Neptunea pribiloffensis* (Dall, 1919) and *Tealia crassicornis* (Muller, 1776): On a snail's use of babysitters. Veliger 24: 62–66.

SINDERMANN, C. J., and A. ROSENFIELD. 1967. Principal diseases of commercially important marine bivalve Mollusca and Crustacea. U.S. Fish Wild. Serv., Fish. Bull. 66: 335–85.

SLOCUM, C. J. 1980. Differential susceptibility to grazers in two phases of an intertidal alga: Advantages of heteromorphic generations. J. Exp. Mar. Biol. Ecol. 46: 99–110.

SMEDES, G. W., and L. E. HURD. 1981. An empirical test of community stability: Resistance of a fouling community to a biological patch-forming disturbance. Ecology 62: 1561–72.

SMITH, S. V., and E. C. HADERLIE. 1969. Growth and longevity of some calcareous fouling organisms, Monterey Bay, California. Pacific Sci. 23: 447–51.

SOMMER, H.-H. 1981. Perception of constant hydrostatic pressure: A physiological basis for the vertical stratification of marine habitats. Experientia 37: 141–43.

SOUSA, W. P. 1979a. Disturbance in marine intertidal boulder fields: The nonequilibrium maintenance of species diversity. Ecology 60: 1225–39.

———. 1979b. Experimental investigations of disturbance and ecological succession in a rocky intertidal algal community. Ecol. Monogr. 49: 227–54.

———. 1980. The responses of a community to disturbance: The importance of successional age and species' life histories. Oecologia 45: 72–81.

SPARKS, A. K. 1972. Invertebrate pathology: Noncommunicable diseases. New York: Academic Press. 387 pp.

SPIGHT, T. M. 1974. Sizes of populations of a marine snail. Ecology 55: 712–29.

———. 1975. On a snail's chances of becoming a year old. Oikos 26: 9–14.

———. 1981. How three rocky shore snails coexist on a limited food resource. Res. Pop. Ecol. 23: 245–61.

———. 1982a. Population sizes of two marine snails with a changing food supply. J. Exp. Mar. Biol. Ecol. 57: 195–217.

———. 1982b. Risk, reward, and the duration of feeding excursions by a marine snail. Veliger 24: 302–8. □See also additional papers by this author listed under *Nucella lamellosa*, pp. 559–60.

STEBBING, A. R. D. 1972. Preferential settlement of a bryozoan and serpulid larvae on the younger parts of *Laminaria* fronds. J. Mar. Biol. Assoc. U.K. 52: 765–72.

STENECK, R. S., and L. WATLING. 1982. Feeding capabilities and limitation of herbivorous molluscs: A functional group approach. Mar. Biol. 68: 299–319.

STEPHENSON, T. A., and A. STEPHENSON. 1961. Life between tidemarks in North America. IVa, Vancouver Island, I. J. Ecology 49: 1–29. IVb, Vancouver Island, II. J. Ecology 49: 227–43.

STEWART, J. G., and B. MYERS. 1980. Assemblages of algae and invertebrates in southern California (USA) *Phyllospadix* dominated intertidal habitats. Aquat. Botany 9: 73–94.

STRATHMANN, R. R., E. S. BRANSCOMB, and K. VEDDER. 1981. Fatal errors in set as a cost of dispersal and the influence of intertidal flora on set of barnacles. Oecologia 48: 13–18.

STRATHMANN, R. R., and M. F. STRATHMANN. 1982. The relationship between adult size and brooding in marine invertebrates. Amer. Natur. 119: 91–101.

SUCHANEK, T. H. 1978. The ecology of *Mytilus edulis* L. in exposed rocky intertidal communities. J. Exp. Mar. Biol. Ecol. 31: 105–20.

———. 1981. The role of disturbance in the evolution of life history strategies in the intertidal mussels *Mytilus edulis* and *Mytilus californianus*. Oecologia 50: 143–52.

SUTHERLAND, J. P. 1970. Dynamics of high and low populations of the limpet, *Acmaea scabra* (Gould). Ecol. Monogr. 40: 169–88.

———. 1974. Multiple stable points in natural communities. Amer. Natur. 108: 859–73.

———. 1981. The fouling community at Beaufort, North Carolina: A study in stability. Amer. Natur. 118: 499–519.

SWEDMARK, B. 1964. The interstitial fauna of marine sand. Biol. Rev. 39: 1–42.

SWINBANKS, D. D. 1982. Intertidal exposure zones: A way to subdivide the shore. J. Exp. Mar. Biol. Ecol. 62: 69–86.

TALLMARK, B., and G. NORRGREN. 1976. The influence of parasitic trematodes on the ecology of *Nassarius reticulatus* in Gullmar Fjord. Zoon 4: 149–56.

TAYLOR, P. R. 1982. Environmental resistance and the ecology of coexisting hermit crabs: Thermal tolerance. J. Exp. Mar. Biol. Ecol. 57: 229–36.

TAYLOR, P. R., and M. M. LITTLER. 1982. The roles of compensatory mortality, physical disturbance, and substrate retention in the development and organization of a sand-influenced, rocky-intertidal community. Ecology 63: 135–46.

THISTLE, D. 1981. Natural physical disturbances and communities of marine soft bottoms. Mar. Ecol. Prog. Ser. 6: 223–28.

THORSON, G. 1950. Reproductive and larval ecology of marine bottom invertebrates. Biol. Rev. 25: 1–45.

TURNER, C. H., E. E. EBERT, and R. R. GIVEN. 1969. Man-made reef ecology. Calif. Dept. Fish Game, Fish Bull. 146: 1–221.

TURNER, T. 1983a. Complexity of early and middle successional stages in a rocky intertidal surfgrass community. Oecologia 60: 56–65.

———. 1983b. Facilitation as a successional mechanism in a rocky intertidal community. Amer. Natur. 121: 729–38.

UNDERWOOD, A. J. 1978a. A refutation of critical tidal levels as determinants of the structure of intertidal communities on British shores. J. Exp. Mar. Biol. Ecol. 33: 261–76.

———. 1978b. An experimental evaluation of competition between three species of intertidal prosobranch gastropods. Oecologia 33: 185–202.

———. 1979. The ecology of intertidal gastropods. Adv. Mar. Biol. 16: 111–210.

———. 1980. The effects of grazing by gastropods and physical factors on the upper limits of distribution of intertidal macroalgae. Oecologia 46: 201–13.

VADAS, R. L. 1977. Preferential feeding: An optimization strategy in sea urchins. Ecol. Monogr. 47: 337–71.

VANBLARICOM, G. R. 1982. Experimental analyses of structural regulation in a marine sand community exposed to oceanic swell. Ecol. Monogr. 52: 283–305.

VANCE, R. R. 1972. Competition and mechanism of coexistence in three sympatric species of intertidal hermit crabs. Ecology 53: 1062–74.

VAN LOENHOUD, P. J., and J. C. P. M. VAN DE SANDE. 1977. Rocky shore zonation in Aruba and Curacao (Netherlands Antilles), with the introduction of a new general scheme of zonation: I. and II. Proc. K. Ned. Akad. Wet. Ser. C Biol. Med. Sci. 80: 437–74.

VAN WINKLE, W., JR. 1970. Effect of environmental factors on byssal thread formation. Mar. Biol. 7: 143–48.

VIRNSTEIN, R. W. 1977. The importance of predation by crabs and fishes on benthic infauna in Chesapeake Bay. Ecology 58: 1199–1217.

WELLS, R. A. 1980. Activity pattern as a mechanism of predator avoidance in two species of acmaeid limpet. J. Exp. Mar. Biol. Ecol. 48: 151–68.

WIESER, W. 1959. The effect of grain size on the distribution of small invertebrates inhabiting the beaches of Puget Sound. Limnol. Oceanogr. 4: 181–94.

WILLIAMS, I. C., and C. ELLIS. 1975. Movements of the common periwinkle, *Littorina littorea* (L.), on the Yorkshire coast in winter and the influence of infection with larval digenea. J. Exp. Mar. Biol. Ecol. 17: 47–58.

WOOD, L. 1968. Physiological and ecological aspects of prey selection by the marine gastropod *Urosalpinx cinerea* (Prosobranchia: Muricidae). Malacologia 6: 267–320.

WOODIN, S. A. 1974. Polychaete abundance patterns in a marine soft-sediment environment: The importance of biological interactions. Ecol. Monogr. 44: 171–87.

———. 1976. Adult-larval interactions in dense infaunal assemblages: patterns of abundance. J. Mar. Res. 34: 25–41.

———. 1978. Refuges, disturbance, and community structure: A marine soft-bottom example. Ecology 59: 274–84.

———. 1981. Disturbance and community structure in a shallow water sand flat. Ecology 62: 1052–66.

YOUNG, C. M., and F.-S. CHIA. 1981. Laboratory evidence for delay of larval settlement in response to a dominant competitor. Int. J. Invert. Reprod. 3: 221–26.

Books and Papers on Geology, Paleontology, and Zoogeography

ADEGOKE, O. S. 1967. Earliest Tertiary West American species of *Platyodon* and *Penitella*. Proc. Calif. Acad. Sci., Ser. 4, 35: 1–22.

AGER, D. V. 1963. Principles of paleoecology. An introduction to the study of how and where animals and plants lived in the past. New York: McGraw-Hill. 371 pp.

ANDERSON, C. A., et al. 1950. 1940 *E. W. Scripps* cruise to the Gulf of California. Geol. Soc. Amer. Mem. 43.

ANDERSON, F. M. 1938. Lower Cretaceous deposits in California and Oregon. Geol. Soc. Amer. Spec. Pap. 19. 339 pp.

BAILEY, E. H., ed. 1966. Geology of northern California. Calif. Div. Mines and Geol. Bull. 190. 507 pp.

BRIGGS, J. C. 1974. Marine zoogeography. New York: McGraw-Hill. 475 pp.

BRUSCA, R. C., and B. R. WALLERSTEIN. 1979. Zoogeographic patterns of idoteid isopods in the northeast Pacific, with a review of shallow water zoogeography of the area. Bull. Biol. Soc. Wash. 3: 67–105.

DAETWYLER, C. C. 1966. Marine geology of Tomales Bay, central California. Pac. Mar. Sta. Res. Rep. 6. 169 pp.

EHLEN, J. 1967. Geology of state parks near Cape Arago, Coos County, Oregon. Ore Bin 29: 61–82.

FAIRBRIDGE, R. W. 1961. Eustatic changes in sea level. *In* Physics and chemistry of the earth, Vol. 4, pp. 99–185. New York: Pergamon.

GLEN, W. 1959. Pliocene and Lower Pleistocene fauna of the western part of the San Francisco Peninsula. Univ. Calif. Publ. Geol. Sci. 36: 147–98.

HALL, C. A., JR. 1964. Shallow-water marine climates and molluscan provinces. Ecology 45: 226–34.

IMBRIE, J., and N. NEWELL. 1964. Approaches to paleoecology. New York: Wiley and Sons. 432 pp.

JENKINS, O. P. 1943. Geologic formations and economic development of the oil and gas fields of California. Calif. Div. Mines Bull. 118. 773 pp.

———, ed. 1951. Geologic guidebook of the San Francisco Bay counties: History, landscape, geology, fossils, minerals, industry, and routes to travel. Calif. Div. Mines Bull. 154: 392 pp.

KEEN, A. M., and H. BENTSON. 1944. Checklist of California Tertiary marine Mollusca. Geol. Soc. Amer. Spec. Pap. 56. 280 pp.

KUENEN, P. H., 1950. Marine geology. New York: Wiley and Sons. 568 pp.

KULM, L. D., and J. V. BYRNE. 1966. Sedimentary response to hydrography in an Oregon estuary. Mar. Geol. 4: 85–118.

KUMMEL, B., and D. RAUP, eds. 1965. Handbook of paleontological techniques. San Francisco: Freeman. 852 pp.

LADD, H. S., ed. 1957. Treatise on marine ecology and paleoecology, II: Paleoecology. Geol. Soc. Amer. Mem. 67. 1,077 pp.

MacCLINTOCK, C. 1967. Shell structure of patelloid and bellerophontid gastropods. Peabody Mus. Natur. Hist. Yale Univ., Bull. 22. 140 pp.

McKEE, B. 1972. Cascadia, the geologic evolution of the Pacific northwest. New York: McGraw-Hill. 394 pp.

MENARD, H. W. 1964. Marine geology of the Pacific. New York: McGraw-Hill. 271 pp.

MOORE, R. C., ed. 1953– . Treatise on invertebrate paleontology. Lawrence, Kans.: Geological Society of America and University of Kansas. ☐ Many volumes. Systematic treatment of all invertebrates with skeletal parts; much ecology.

NATIONS, J. D. 1975. The genus *Cancer* (Crustacea: Brachyura): Systematics, biogeography, and fossil record. Los Angeles Co. Mus. Natur. Hist., Sci. Bull. 23. 104 pp.

NEWELL, I. M. 1948. Marine molluscan provinces of western North America: A critique and a new analysis. Proc. Amer. Philo. Soc. 92: 155–66.

NEWMAN, W. A. 1979. The Californian transition zone: Significance of short-range endemics. *In* J. Gray and A. Boucot, eds., Historical biogeography, plate tectonics, and the changing environment, pp. 399–416. 37th Ann. Biol. Colloq. Corvallis: Oregon State University Press.

NORTH, W. B., and J. V. BYRNE. 1965. Coastal landslides of northern Oregon. Ore Bin 27: 217–41.

OBERLING, J. J. 1964. Observations on some structural features of the pelecypod shell. Mitt. Naturforsch. Ges. Bern, n.f. 20: 1–60.

ORR, W. N., and E. L. ORR. 1981. Handbook of Oregon plant and animal fossils. Eugene, Ore.: privately printed. 285 pp.

SCHENCK, H. G., and A. M. KEEN. 1950. California fossils for the field geologist. Stanford, Calif.: Stanford University Press. 88 pp.

SEAPY, R. R., and M. M. LITTLER. 1980. Biogeography of rocky intertidal macroinvertebrates of the southern California islands. *In* D. M. Power, ed., The California islands: Proceedings of a multidisciplinary symposium, pp. 307–23. Santa Barbara, Calif.: Santa Barbara Museum of Natural History.

SHEPARD, F. P. 1948. Submarine geology. New York: Harper. 348 pp.

———. 1967. The earth beneath the sea. Baltimore, Md.: Johns Hopkins University Press. 242 pp.

SHEPARD, F. P., and K. O. EMERY. 1941. Submarine topography off the California coast: Canyons and tectonic interpretation. Geol. Soc. Amer. Spec. Pap. 31. 171 pp.

STANTON, R. J., JR., and J. R. DODD. 1976. The application of trophic structure of fossil communities in paleoenvironmental reconstruction. Lethaia 9: 327–42.

TAYLOR, J. D., W. J. KENNEDY, and A. HALL. 1969. The shell structure and mineralogy of the Bivalvia: Introduction: Nuculacea—Trigonacea. Bull. Brit. Mus. (Natur. Hist.), Zool. Suppl. 3: 1–125.

———. 1973. The shell structure and mineralogy of the Bivalvia. II. Lucinacea—Clavagellacea: Conclusions. Bull. Brit. Mus. (Natur. Hist.), Zool. 22: 253–94.

VALENTINE, J. W. 1961. Paleoecologic molluscan geography of the Californian Pleistocene. Univ. Calif. Publ. Geol. Sci. 34: 309–442.

———. 1966. Numerical analysis of marine molluscan ranges on the extratropical northeastern Pacific shelf. Limnol. Oceanogr. 11: 198–211.

———. 1973. Evolutionary paleoecology of the marine biosphere. Englewood Cliffs, N.J.: Prentice-Hall. 511 pp.

VERMEIJ, G. J., D. E. SCHINDEL, and E. ZIPSER. 1981. Predation through geological time: Evidence from gastropod shell repair. Science 214: 1024–26.

WEAVER, C. E. 1943. Paleontology of the marine Tertiary formations of Oregon and Washington. Univ. Wash. Publ. Geol. 5. 788 pp.

General Index

All animals mentioned in the text and Systematic Index are indexed below. Both common names and scientific names are listed. To aid those with a limited knowledge of zoological classification, each entry for the common name of an animal or group of animals ("crabs," e.g.) includes, in parentheses, the lowest-level taxonomic heading in the Systematic Index that embraces all the animals referred to by that common name; readers should consult that section in the Systematic Index whenever cross-references are given below to "individual species by name." Authors cited in the text are indexed, as are all authors whose works are listed in the Systematic Index and the General Bibliography; the reader will thus be able to locate all the works of any author and to locate full data for works cited by author and date in the text. For localities mentioned a great many times, only references to the Systematic Index are listed.

In the General Index we have used, e.g., "57f" to mean separate references on pages 57 and 58 and "57ff" similarly for pages 57, 58, and 59. For a discussion spanning pages 57 and 58, we use "57–58." *Passim* is used for a cluster of references in close but not consecutive sequence.

abalone (Pleurotomariacea), 101–3, 291,
  550–51. *See also species of Haliotis*
Abarenicola, 545
  pacifica, 389f, 521, 522
  vagabunda (= *A. claparedii*), 389f
  vagabunda oceanica, 389n, 522
  vagabunda vagabunda, 389n, 522
Abbott, D. P., 143, 501, 509, 551, 553, 555,
  585f, 594
Abbott, I. A., 438, 508, 551, 593
Abbott, R. T., 549
Abietinaria, 68, 117, 303, 509
  amphora, 113, 512
  anguina, 113, 238, 512
  greenei, 113, 512
  turgida, 238, 512
Acanthina, 34
  punctulata, 230, 558
  spirata, 230, 558
Acanthocyclops vernalis, 528
Acanthomysis costata, 85, 532
  sculpta, 85, 532

Acanthoptilum gracile, 365f, 513
  scalpelifolium, 513
Acarina, 547f
Ache, B. W., 538
Achelia chelata, 220, 221–22, 548
  nudiuscula, 417, 548
Acholoe pulchra, see Arctonoe pulchra
Acmaea, 27, 552. *See also species of Collisella,
  Notoacmaea*
  cassis, see Collisella pelta
  mitra, 27, 56, 150–51, 240, 272, 552
  persona of older accounts, 28. *See also
    Collisella digitalis*
  scabra Nuttall of keep, *see Collisella
    limatula*
  testudinalis scutum, see Notoacmea scutum
Acoela, 84, 517
acorn barnacles (Thoracica), 24, 222, 234,
  272. *See also* barnacles (Thoracica)
acorn worms (Hemichordata), 584. *See also
    individual species by name*
acrorhagi, 54, 515

Dungan, M. L., 598–99
Dungeness crabs (Brachyura), *see Cancer magister*
Dunn, D. F., 515
Durham, J. Wyatt, 68, 582
DuShane, H., 556, 563
Dutch Harbor, Alaska, 119, 312
Duxbury Reef, Calif., 89, 221, 315, 403
Dwyer, K. R., 553
Dybern, B. I., 585

earwigs (Dermaptera), 547
eastern mud snails (Neogastropoda), *see Ilyanassa obsoleta*
eastern oysters (Pterioida), *see Crassostrea virginica*
eastern soft-shelled clams (Myoida), *see Mya arenaria*
Eastman, J., 565
Easton, D. M., 544
Eaton, C. M., 553
Ebeling, A. W., 588
Ebert, E. E., 590, 606
Ebert, T. A., 583
Ebling, F. J., 601
ecdysis, *see* molting
Echinarachnius parma, 318, 582
Echinodermata, 579–84
Echinoidea, 582–83
Echiura, 527. *See also individual species by name*
Echiuris echiuris alaskensis, 316, 385n, 387, 527, 545
Eckelbarger, K. J., 525, 572
Ectoprocta, 577, 578–79. *See also* bryozoans
Ectyodoryx parasitica, see Myxilla incrustans
Ecuador, 155, 172, 323
Edgar, R. K., 559
Edmonds, S. J., 527
Edmunds, M., 563
Edmunds, S. J., 526
Edwards, C., 352, 512
Edwards, D. B., 543
Edwards, D. C., 560
Edwards, R. R. C., 538
Edwards, T. W., 587
Edwardsiella, 369, 388
californica, 365f, 515
sipunculoides, 316, 515
eelgrass and associates, 198, 279f, 298, 305, 334, 341–53, 360, 363, 372, 395, 420, 584. *See also Zostera*
Eernisse, D. J., 581
Efford, I. E., 253, 541
Egglishaw, H. J., 547
egg masses: of chitons, 161f; of fishes, 79, 313; of flatworms, 23, 84; of octopuses, 177; of opisthobranchs, 63ff, 137,

343–47 *passim*, 363, 372; of polychaetes, 390; of snails, 20, 34, 90, 149f, 230ff, 273, 276f, 326–27, 360–61, 362, 373, 558–59
egg production: of barnacles, 26, 222; of chitons, 162; of clams, 259; of crabs, 43, 58, 136, 202, 402, 541; of hydrozoans, 351; of isopods, 227, 251, 413; of opisthobranchs, 131–32, 137, 372; of polychaetes, 390; of seastars, 94; of snails, 20, 149, 230f, 276, 327, 361; of sponges, 61–62
Egregia menziesii (alga), 36, 51, 134, 153f, 158, 237, 433f, 441, 463, 492, 525, 554
Ehlen, J., 608
Eikenberry, A. B., Jr., 553
Eisemann, C., 570
Elasmopus, 48
mutatus, 233n, 537
rapax, 233, 537
serricatus, 233n, 537
elbow crabs (Brachyura), *see Heterocrypta occidentalis*
Eldridge, P. J., 570
elephant seals (Mammalia; *Mirounga angustirostris*), 205, 590
El Estero de Punta Banda, Baja Calif., 321, 324, 336–42 *passim*, 368
Elkhorn Slough, Calif., 75, 82, 131, 270, 278, 287, 290, 296–305 *passim*, 312, 338–45 *passim*, 349, 351, 358, 362–81 *passim*, 386–95 *passim*, 418–24 *passim*, 536
Ellis, C., 460n, 607
Ellis, D. V., 520, 524, 590
Ellis, J., 120
Elminius modestus, 270
Elvin, D. W., 61, 510, 564, 566
elytra, 78
Embiotica lateralis, 587
embryology and development: of amphipods, 118; of barnacles, 24–26, 59; of chitons, 162; of clams, 257; of corals, 68; of crabs, 201, 253–54, 278, 541; of echiurans, 386; of hydrozoans, 109–10, 351, 364, 418; of holothurians, 242; of octopuses, 177–79; of oysters, 292f, 294–95; of scyphozoans, 303; of sea anemones, 68, 514, 516; of sea hares, 132; of sea spiders, 118; of seastars, 56, 61, 95; of sea urchins, 100; of shrimps, 539; of snails, 59, 75–76, 230f, 273, 276; of sipunculids, 80; of sponges, 61–62; of tunicates, 129. *See also* brooding; egg masses; larvae
Emerita, 220, 249f
analoga, 252–54, 336, 537, 541, 589
rathbunae, 254, 541
emersion: and feeding, 456; and intertidal

gravel beaches, 205, 273–76 *passim*, 281, 308, 351, 387, 391
gray whale (Mammalia; *Eschrichtius robustus*), 590
Green, J., 528
Green, J. M., 587f, 598
Green, K., 509
Green, R. H., 569
green abalone (Pleurotomariacea), *see Haliotis fulgens*
green algae, 17, 32, 89, 279, 438. *See also Enteromorpha; Ulva*
green anemones (Actiniaria), *see species of Anthopleura*
Greene, C. H., 599
Greene, R. W., 562
gregarines ("Protozoa"), 509
Grenon, J.-F., 27n, 550
gribbles (Isopoda), *see species of Limnoria*
Griffith, L. M., 550
Grigg, R. W., 514
Grinnell, J., 589
Gromia oviformis, 123–24, 507
Grosberg, R. K., 600
Gross, W. J., 537, 542
Grossman, G. D., 587
growth rates, 1, 54, 61, 220, 274, 281–82, 333, 461, 516, 570. *See also* age and growth; embryology and development
Grubeopolynoe tuta, 307f, 524
grunion (Osteichthyes), *see Leuresthes tenuis*
grunter (Osteichthyes), *see Porichthys notatus*
Guberlet, J. E., 347, 565
Guberlet, M. L., 508
Gulf of Alaska, 300, 418
Gulf of California, 42, 75, 98, 163, 291, 305, 341, 345, 347f, 532, 548, 561, 568
Gulf of Mexico, 324
gulls (Aves), 217, 235, 259–60
Gulmon, S. L., 601
gumboot chiton (Polyplacophora), *see Cryptochiton stelleri*
Gunnill, F. C., 528, 600
Gupta, A. P., 527
Gutmann, W. F., 578

Haderlie, E. C., 143, 217, 517, 525, 565, 571–72, 581, 594, 605
Hadfield, M. G., 558
Haematopus bachmani, 588f. *See also* oyster-catchers
Hahn, D. C., 589
Haig, J., 537, 540
Halbert, J. N., 548
Haldorson, L., 587
Halecium spp., 108, 512
Half Moon Bay, Calif., 5, 260
Halichondria panicea, 124, 298, 510

Haliclona, 61n, 91, 124, 275, 510
  oculata, 119
  permollis, 61–62, 66, 120, 510
  rufescens, 275, 510
Haliclystus auricula, 154–55, 343–44, 513
  stejnegeri, 344n, 513
Haliotis corrugata, 550
  cracherodii, 8, 103, 243–45, 550f
  fulgens, 103, 550f
  kamtschatkana, 103, 245, 551
  rufescens, 87, 100–103, 550f
Haliplanella luciae, 299, 515
Halisarca sacra, 312, 510
Hall, A., 609
Hall, C. A., Jr., 608
Hall, S. J., 564
Halobisium occidentale, 548
  orientale, 548
Halosaccion glandiforme (alga), 433f, 441, 492, 529
Halosoma viridintestinale, 400, 401, 549
Halosydna brevisetosa, 77–78, 82, 166, 229, 307, 404, 521, 524
  californica, 78
  johnsoni, 77–78, 82, 524
  lordi, *see Arctonoe vittata*
  pulchra, *see Arctonoe pulchra*
Haminoea vesicula, 345, 372, 561
Hammann, H., 587
Hamner, W. M., 513, 535
Hancock, D. R., 571
Hand, C., 512, 514, 517, 522, 562, 580
Hanlon, R. T., 576
Hanna, G. D., 549, 561
Hannan, C. A., 521
Hansen, B., 571
Hansen, J. C., 550
Hansen, M. D., 526, 583
Hapalogaster cavicauda, 170f, 306, 541
  mertensii, 170, 306, 541
Haploscoloplos elongatus, 524
harbor seal (Mammalia; *Phoca vitulina*), 590
hardshell cockles (Veneroida), *see Chione undatella*
Hardwick, J. E., 576
Harenactis attenuata, 330f, 365, 513, 515
Harger, J. R. E., 494, 566, 600
Harmothoe, *see Hesperonoe*
harpacticoid copepods (Copepoda), 31, 57
Harrigan, J. F., 564
Harris, L. G., 514, 563
Harris, R. P., 529
Harrison, F. W., 509
Harrold, C., 554, 580, 600
Hart, J. F. L., 278, 306, 359, 532, 537–38, 540, 542
Hart, J. L., 587

Library of Congress Cataloging in Publication Data

Ricketts, Edward Flanders, 1896–1948.
    Between Pacific tides.

    Bibliography: p.
    Includes Index.
        1. Marine invertebrates—Pacific Coast (U.S.)
    2. Intertidal ecology—Pacific coast (U.S.)
    3. Seashore biology—Pacific Coast (U.S.)   4. Animals—
    Habitations.   I. Calvin, Jack.   II. Hedgpeth, Joel
    Walker, 1911–   III. Phillips, David W.
    IV. Title.
    QL138.R5   1985   591.926   83-40620
    ISBN 0-8047-1229-8
    ISBN 0-8047-1244-1 (Student ed.)